KEY TOPICS IN BRAIN RESEARCH

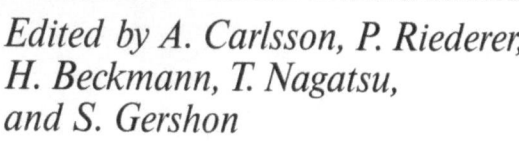

Edited by A. Carlsson, P. Riederer,
H. Beckmann, T. Nagatsu,
and S. Gershon

K. Maurer, P. Riederer, and H. Beckmann (eds.)

Alzheimer's Disease. Epidemiology, Neuropathology, Neurochemistry, and Clinics

Springer-Verlag Wien GmbH

Prof. Dr. Konrad Maurer
Prof. Dr. Peter Riederer
Prof. Dr. Helmut Beckmann

Department of Psychiatry, University of Würzburg, Würzburg,
Federal Republic of Germany

With 118 Figures (9 in color)

ISSN 0934–1420

ISBN 978-3-211-82197-8   ISBN 978-3-7091-3396-5 (eBook)
DOI 10.1007/978-3-7091-3396-5

Preface

Dementia of Alzheimer type (DAT), multiinfarct dementia (MID) and dementia occurring in the course of Parkinson's disease (PD + D) now make up one of the largest categories of chronic diseases in the elderly. In addition to the burden those illnesses impose on the affected individuals and their families they consume large socio-economic resources. In the light of all the above mentioned features, it seemed to us that a Symposium on behalf of the 125th Anniversary of Birth of Aloys Alzheimer was particularly well-suited to help to advance research on Alzheimer's disease and other dementias. This International Symposium combined with a Satellite Symposium about "Clinical Aspects of Alzheimer Dementias" took place in June 1989 in Würzburg and has been organized by the Psychiatric Departments of the Universities of Würzburg and Munich (H. Beckmann, K. Maurer, P. Riederer, H. Hippius and H. Lauter) and the Department of Pathochemistry and General Neurochemistry of the University of Heidelberg (S. Hoyer). In the chapters which follow, thorough reviews of recognized authorities in the field of dementia are given in the four main fields of epidemiology, neuropathology, neurochemistry and clinics.

This Symposium and the edition of this book would not have been possible without the generous support of E. Merck, Darmstadt. In particular we gratefully acknowledge the efforts undertaken by Mr. Hernandez-Meyer to organizing this symposium and to publish this book.

Furthermore we are grateful to the secretarial help by Mrs. Moeslein, Miss Philipp and Miss Gröbner.

Last but not least, we thank Springer-Verlag Wien New York for the excellent production of this book and the excellent cooperation over the last months.

August 1990   K. MAURER, P. RIEDERER, and H. BECKMANN

Contents

Immuncytochemistry

Neurochemistry

Clinics

Listed in Current Contents

Aloys Alzheimer 14. 06. 1864–19. 12. 1915

H. Hippius

Department of Psychiatry, University of Munich, Munich, Federal Republic of Germany

What is the reason, to celebrate the inauguration of the first Alzheimer bust here in the Department of Psychiatry of the University, in Wuerzburg, where Alzheimer has never worked in the field of psychiatry? Why not in Frankfurt, where Aloys Alzheimer observed the first patient, suffering from an illness, which – by a suggestion of Emil Kraepelin – was later to be called Alzheimer's disease? Why not in Munich, where Alzheimer had his most active and productive period as a scientist, where he published his famous studies on Alzheimer's disease and where – up to now – is the location of the house near by the charming lake Wessling built by himself for his family? Why not in Breslau, where Alzheimer in 1913 took over his first position as Head of Department and full professor (ordentlicher Professor) of Psychiatry?

The reason is that 125 years ago, on the 14th of June, 1864, Aloys Alzheimer was born nearby Wuerzburg in Marktbreit. He and his family are deeply rooted in this area around Wuerzburg, the Lower Franconia.

He grew up in Lower Franconia, went to primary school in Marktbreit, then to the Royal Humanistic Secondary School in Aschaffenburg until his final school examinations in 1883. After a couple of terms at the universities of Berlin and Tuebingen, he completed his medical studies at the university of Wuerzburg; in Wuerzburg he took his final medical exams at the beginning of 1888. He had been already promoted to doctor of medicine with a histological thesis at the University of Wuerzburg a year earlier in 1887. Alzheimer remained in Wuerzburg for a further term following his medical exams and worked at the Anatomic Institute. This step played a significant role in his life and work as a neuropathologist. However, Alzheimer's interest in psychiatry can also be traced back to this first year after his final examinations: As he wrote himself, he became the medical "travelling-companion of a mentally ill lady" (Alzheimer, 1895) for a period of five months.

Immediately after he definitely left his home-town, where he had spent most of time until then. After Wuerzburg, he went to Frankfurt, Munich and finally to Breslau, cities which became the great stations of his life.

In the foreground these cities mark Alzheimer's path in life in a purely geographical sense, but behind the names of these cities there is more than meets the eye – the encounter and work with important psychiatrists and researchers. In Frankfurt Alzheimer worked for 15 years with the clinician Emil Sioli. Here, he also met Franz Nissl. And in Munich Alzheimer became one of the closest co-workers of Emil Kraepelin.

Wuerzburg until 1888, Frankfurt until 1903, a few months in Heidelberg, Munich until 1913 and the triumvirate of the great teachers Emil Sioli, Franz Nissl and Emil Kraepelin – that was the fundament, on which Alzheimer was standing, when he was invited to take over the professorial chair for psychiatry in Breslau. After taking over his office in Breslau in 1913 Alzheimer wanted to complete his work as a clinical psychiatrist and researcher. He planned a study on anatomical alterations in the brains of patients suffering from endogenous psychoses. However, he was not able to bring this plan to realization. He became ill soon after moving to Breslau; he died at the age of only 51 years on the 19th of December, 1915.

The most important step in Alzheimer's professional life was in 1888 to leave the Wuerzburg Institute of Anatomy and to become a clinician, a psychiatrist. And his whole life he stayed a clinician – in spite of becoming prominent and famous as a neuropathologist! With 24 years of age he left the university and took on a job as assistant doctor at the Municipal Mental Hospital in Frankfurt am Main, the head of which was Emil Sioli at that time. Sioli had introduced the principle of "non-restraint" to this hospital. Therefore the hospital in Frankfurt soon earned itself the reputation of a progressive therapeutic institution for mentally ill patients. In a review of Sioli's 25-year profession as psychiatrist, Alzheimer later praised his achievements emphatically (Alzheimer, 1913). Alzheimer remained faithful to the Frankfurt hospital and his first clinical teacher, Sioli, for more than half of these 25 years (December 1888–October 1903).

The stay in Frankfurt was also decisive for Alzheimer's development as neuropathologist. When he came to Frankfurt, the pathologist and anatomist Karl Weigert had been there since 1884 as director of the Senckenberg Institute; Ludwig Edinger, who later became professor for neurology in Frankfurt, was also working there at that time. However, Alzheimer received his most important impulses during this time in the hospital from another young co-worker of Sioli, who was only four years older than Alzheimer: Shortly after Alzheimer had begun to work in Frankfurt, in 1889 Sioli fetched a co-worker of Bernhard von Gudden and Hubert Grashey from Munich as senior assistant. That was Franz Nissl (1860–1919).

Nissl described vividly how intensive the exchange of ideas between the two co-assistants and later on very close friends Nissl and Alzheimer was during their time in Frankfurt: "I was always grateful that fate led me to meet colleagues in Frankfurt, who were enthusiastic about science, and that Sioli did everything in his means to support our mutual scientific endeavours. I showed Alzheimer my technique and convinced him with my preparations and experimental results of the accuracy of my – until now by no means generally accepted – views . . . The aim of our endeavours was always sharply outlined: The greater part of the pathological process in the mentally ill should be found" (Nissl, 1916).

In April 1894 Alzheimer married the widow Caecilia Geisenheimer, nee Wallerstein, in Frankfurt. Franz Nissl was the witness. When Franz Nissl went to Kraepelin in Heidelberg in 1895, Alzheimer became his successor as senior assistant at the Frankfurt hospital. It was a heavy blow for Alzheimer that his wife, by whom he had three children, died in 1901.

During his 15-year employment at the Frankfurt hospital Alzheimer obviously did not intend to start a scientific career, although no less than Emil Kraepelin offered him a job in Heidelberg. Instead of accepting this offer, Alzheimer applied for the directorship of a mental hospital. Kraepelin said the following in his "Memoirs": "I had heard by chance some time previously that this excellent researcher (Alzheimer) was about to apply for the job of director of an asylum. A mutual friend intervened on my behalf and urgently requested him not to make such a step, but to start an academic career. Unfortunately, this advice was not successful at first and it was not until Alzheimer's attempts to become director of an asylum failed, that he came to me and I persuaded him to join our group" (Kraepelin, 1987). Before Alzheimer's promotion in Heidelberg took place, Kraepelin was appointed to the professorial chair in Munich and Alzheimer came to Munich with him.

In Munich Alzheimer stayed co-worker of Emil Kraepelin until 1912. Alzheimer took advantage of the fact that the new hospital building in Munich was not complete and would not be opened until November 1904. As soon as the building was finished, he would have many extensive clinical duties. As first assistant doctor he participated decisively in the construction and organisation of the new clinic at the Nussbaumstrasse. When Kraepelin's senior doctor, the clinician Robert Gaupp, left Munich for Tuebingen in 1906, Alzheimer became his successor. Alzheimer was Kraepelin's senior assistant and deputy until 1909.

In 1904 Alzheimer was promoted to lecturer at the Munich Medical Faculty with a thesis on "Histologische Studien zur Differentialdiagnose der progressiven Paralyse" (histological studies on the differential diagnosis of general paresis) (Alzheimer, 1904).

Kraepelin wrote the following comments in his "Memoirs" about this time: "From Heidelberg I had brought . . . Gaupp and Alzheimer to Munich, who took over the management of the splendid rooms for anatomical work. . . . Alzheimer started work in the clinic without being paid, as I had no position for him and he wanted to spend his time as he wished. In order to integrate him into the clinic, I created the class of scientific assistants. This class consisted of researchers, who were free to use the scientific facilities and equipment. Apart from Alzheimer, Ruedin and Plaut were also connected to the clinic in this manner for a long time, later also Isserlin. Without the self-sacrificing assistance of these gentlemen it would have been absolutely impossible to get the scientific work going at all" (Kraepelin, 1987).

Alzheimer's histopathological laboratory at the Munich hospital soon became a centre of research. Co-workers and scholars during this period were U. Cerletti, H. G. Creutzfeldt, A. Jakob, F. Lothmar, and G. Perusini.

Apart from the clinical and scientific work, as well as the time-consuming organisational and administrative tasks in the building phase of the hospital, from the summer term 1905 until the winter term 1912/13 Alzheimer had to give a large number of lectures. In the university prospectus of the Medical Faculty one finds his name as a docent for many topics: "Introduction in Clinical Psychiatry", "Brain and Soul", "Practice in the Exploration of the Mentally Ill", "Forensic-psychiatric Practice", "Clinical Demonstrations for Advanced Scholars", "Normal and Pathological Anatomy of the Cerebral Cortex".

A review of the many-sided, scientific work of Alzheimer proves impressively his basic attitude: In all his studies he always kept the question of the clinical relevance of his research in mind.

In an orbituary notice about Alzheimer, Spielmeyer points out that two main stages in his work differ from one another (Spielmeyer, 1916): Until his habilitation thesis about histological studies on the differential diagnosis of general paresis (1904), which nowadays is often and justifiably considered as classic, Alzheimer worked on clearly organically induced psychiatric clinical pictures (mainly with progressive paralysis and the so-called "arteriosclerotic mental disorders"). In his second productive period (1904–1915) questions as to the occurrence and relevance of anatomical alterations in the brain in endogenous psychoses are given more attention.

However, Alzheimer's name has become famous through his studies on the premature occurrence of demential processes. In November 1901, during his time at the Frankfurt hospital, Alzheimer treated a female, 51 year old patient ("Frau A. D."). This patient had become conspicuous because of a restriction in her power of comprehension and memory, aphasic symptoms, orientation disorders, incalculable behaviour, and paranoid thought contents. After the death of this patient, Alzheimer reported in 1906 at the 37. Tagung der Suedwestdeutschen Irrenaerzte (37th Conference of the South-West German Psychiatrists) on this particular case with the title "Ueber einen eigenartigen, schweren Erkrankungsprozess der Hirnrinde" (on a peculiar, serious disease process of the cerebral cortex) (Alzheimer, 1906, 107). That was the first description of a particular form of the "presenile dementia", which was later – following Kraepelin's suggestion – named "Alzheimer's disease".

As a clinical scientist, Alzheimer by no means only concentrated on histopathological problems. For example, already in 1896 he worked on forensic-psychiatric topics. In this connection Alzheimer defended the concept of the "inherited degenerative mental disorder" (Alzheimer, 1896). Further important studies dealt with the differential diagnosis of imbecility (e.g., the tuberous sclerosis and the amaurotic idiocy).

The concept of the "degeneration" earns more and more interest and scientific attention nowadays; Alzheimer was not only interested in its connection with forensic problems. The concept of degeneration was believed to have a principal significance for the etiology of all mental diseases.

Other problems, which interested Alzheimer at the beginning of the century, are nowadays also of pressing importance. In this context, Alzheimer analysed the problem of "Indikation für eine kuenstliche Schwangerschaftsunterbrechung bei Geisteskranken" (Alzheimer, 1907) (indication for an abortion in the mentally ill). He worked on various problems connected with epilepsy and was particularly interested in an exact definition of epilepsy.

Alzheimer had always seen the danger that basic research can lose its contact to the clinical side. Therefore, he emphasized continually that he real aim of psychiatric research was to obtain knowledge, which would be of advantage for clinical work. Alzheimer's personal development reflects his basic attitude: Alzheimer considered himself to be a clinician. Initially, he turned down Kraepelin's offer to start an academic career and would have preferred to take over the directorship of a mental asylum. It was not until his application for this position was not accepted that he decided to go to Kraepelin in Heidelberg! Later, he accepted the position as director of the Psychiatric Department of the Breslau University instead of continuing with his scientific work in the neuroanatomic laboratory of the Munich Psychiatric Hospital, although he had ideal conditions in Munich for his scientific work. In this respect, he differed form his 4-years older friend and colleague Franz Nissl: In 1918 Nissl gave up his clinical professorial chair in Heidelberg and returned to Munich to take over the histopathological department of the Deutsche Forschungsanstalt für Psychiatrie in the rooms of Kraepelin's hospital.

When Alzheimer left Munich in 1913, Kraepelin had "a lively feeling that the best that could be done by him for our science would disappear. The new, worrying and time-consuming work, which was awaiting him, seemed to satisfy him all the same" (Kraepelin, 1987).

During his time in Munich, Alzheimer often experienced almost bitter quarrels about the demarcation of disease units in the psychiatric field. Hoche contrasted the concept of the unspecifity of psychopathological syndromes and symptom complexes with Kraepelin's postulated differing disease units. In these arguments Alzheimer usually agreed with Kraepelin's point of view. He was of the opinion that the still unclear boundaries between various disease units can be increasingly better distinguished by research methods of a scientifically orientated psychiatry. He could refer to the example of the progressive paralysis, which he had worked on thoroughly himself. Alzheimer also criticized Wernicke's ideas on a localisation concept, which to a certain extent contradicted with Kraepelin's ideas. According to the localization theory of Wernicke, a disorder in the brain as far as the manifestation and form of a psychiatric disease is concerned is of greater importance than its etiology; Alzheimer considered this to be pure speculation and complained that this concept neglected etiological research.

Alzheimer discerned principally between the "organic" and the "endogenous" psychoses. He counted progressive paralysis and Kraepelin's dementia praecox (schizophrenia) to the organic psychoses. He considered the cause of the "endogenous" psychoses to be "mental degeneration" in the sense of the degeneration theory. He included manic-depressive illness in the endogenous psychoses as well as the hysterical disorders and the psychogenic psychoses. In Alzheimer's opinion there are greater differential diagnostic difficulties in the endogenous psychoses than in the organic ones. All the same, he thought that the "divided stems of the degeneration" must be differentiated; he imputed the existence of Kraepelin's distinct disease units.

During his last creative period Alzheimer worked intensely on the basic problems of psychiatry. Many of the problems he dealt with in his programmatic essay (Alzheimer, 1910) on "Die diagnostischen Schwierigkeiten in der Psychiatrie" (the diagnostic difficulties in psychiatry) (1910) remain of pressing importance and interest sill nowadays. Alzheimer had long intended to summarize his research in a textbook with a presentation of the histopathology of the psychoses. However, because of his early death this plan did not exceed a rough draft.

Some of Alzheimer's results in research are superseded. But two points will last for ever: Firstly his fundamental description of clinical picture and neuropathological findings in dementia of Alzheimer-type and secondly his prominent example that research in psychiatry must have its roots in clinical psychiatry. In spite of becoming more and more well-known as a neuropathologist he remained a clinician.

Therefore it is consequent that the first portrait sculpture of Alzheimer from today on will have its place in a psychiatric hospital, in the Department of Psychiatry of the Wuerzburg University.

References

Alzheimer A (1895) Curriculum Vitae. Dated 6. 10. 1895, unpublished

Alzheimer A (1896) Ein „geborener Verbrecher". Arch Psychiat 28:327–353

Alzheimer A (1903) Curriculum Vitae. Hand-written, unpublished

Alzheimer A (1904) Histologische Studien zur Differentialdiagnose der progressiven Paralyse. Habilitationsschrift, Ludwig-Maximilians-Universität München, Jena

Alzheimer A (1906) Ueber einen eigenartigen, schweren Erkrankungsprozess der Hirnrinde. Neurol Centralbl 25:1134

Alzheimer A (1907) Ueber eine eigenartige Erkrankung der Hirnrinde. Allg Z Psychiat Psych Gerichtl Med 645:146–148

Alzheimer A (1907) Indikation für eine kuenstliche Schwangerschaftsunterbrechung bei Geisteskranken. Muenchner Med Wochenschr 33:1617–1632

Alzheimer A (1910) Die diagnostischen Schwierigkeiten in der Psychiatrie. Z Gesamte Neurol Psychiat 1:1–19

Alzheimer A (1913) 25 Jahre Psychiatrie. Arch Psychiat Nervenkrankh 52:853–866

Hoff P, Hippius H (1989) Alois Alzheimer. Nervenarzt 60:332–337

Kraepelin E (1987) Memoirs. Hippius H, Peters G, Ploog D (eds) Springer, Berlin Heidelberg New York Tokyo

Nissl F (1916) Alois Alzheimer. Dtsch Med Wochenschr 14:1116–1120

Spielmeyer W (1916) Alzheimers Lebenswerk. Z Gesamte Neurol Psychiat 33:1–41

Correspondence: Dr. H. Hippius, Psychiatrische Klinik der Ludwig-Maximilians-Universität München, Nussbaumstrasse 7, D-8000 München 2, Federal Republic of Germany.

Introduction

Some philosophical aspects of Alzheimer's discovery: an American perspective

R. J. Wurtman

Department of Brain and Cognitive Sciences, Massachusetts Institute of Technology, Cambridge, U.S.A.

We are here to celebrate the 125th anniversary of the birth of Alois Alzheimer and to convey to him, posthumously, Society's gratitude for his discovery of the disease that bears his name. At present that disease remains poorly understood and even less well treated. However, the speed with which new insights into its characteristic clinical and pathological findings now arise gives us optimism that afflicted patients will, someday soon, be better served than they can be at present.

Dr. Alzheimer's contribution was considerably more than just medical: it also bore on a central concern of Western philosophers since the time of Plato and Aristotle, namely the "Mind-Body-Problem." The enunciation of Alzheimer's disease – in which a progressive *clinical* disturbance involving the ability to think could be correlated with *neuropathological* changes – provided one of the best proofs then available that the distinctly human attributes of consciousness and cognition have their origins in specific components of the brain. In the classic debate between metaphysical dualists and monists, those who held that the mind is best explored via the neurosciences had won a major victory; one consequence of this victory was that academic psychiatry in much of the world became strongly *neuro*psychiatric.

Coincidentally, Alzheimer's period of greatest productivity overlapped with that of another very different but equally influential thinker who had initially been trained as a neurologist – Sigmund Freud. Freud focussed on subjective experience and its communication by verbal expression as the principal means for understanding and treating diseases of the mind; he also proposed a formulation of the mind's structure and developmental history that, he well recognized, bore no relationship to the brain. For Freud, the brain would someday be paramount in psychiatry, but for the time being this organ was, unfortunately, irrelevant. In America, even more psychiatrists became Freudian dualists than Alzheimer monists, and the strength of this commitment has diminished only in the past few decades, largely because of the discovery of effective psychotropic drugs whose clinical actions could be correlated with their pharmacologic effects on specific brain neurotransmitters.

More broadly, much of twentieth century philosophy can be viewed as a struggle between the followers of Alzheimer and those of Freud, between those who find truth in objective *vs.* subjective phenomena; most individuals now probably utilize a synthesis of the two approaches when they attempt to understand themselves and other people. I am reminded of an MIT colleague who creates artificial intelligence algorithms by day and abstract expressionist paintings by night.

Medicine's response to Alzheimer's contribution raises questions in another domain of philosophy – that of the *philosophy of history:* when and how and why does a new intellectual construct become normative, thereafter influencing the professional behavior of the majority of workers in the field it attempts to explain? And in the case of Alzheimer's disease, why did it take seven decades for his explanation of senility to become normative, and what finally caused its widespread acceptance? Also, why has this acceptance, like that of Freud's views, come about more rapidly in the United States than in Europe? Clearly the general increase in longevity and the consequent increase in the number of aged people at high risk for developing senile dementia has magnified Society's interest in age-related cognitive disorders. But this increase in interest needn't have changed the way that senility came to be explained, i.e., not as an inevitable consequence of aging but as the reflection of a specific brain disease. Was the critical factor, at least in the United States, the establishment of funding agencies – like the National Institute on Aging – able to reinforce high-quality research efforts; or the rise of sophisticated consumer groups (like the Alzheimer's Disease and Related Diseases Association) committed to promulgating this new view of senility as a potentially-treatable disease; or perhaps the discovery of a specific cholinergic deficit, which brought Alzheimer's disease into the mainstream of contemporary neuroscience and vastly increased the pool of leading neuroscientists – especially in America – interested in working on it? Probably all of these had some effect.

Dr. Alzheimer's ideas ultimately did catch on, around the world, and they did so, we believe, because they were correct: the neuropathologic changes probably do cause the behavioral symptoms (indeed, the idea caught on so well that the question now seems to be whether *any* dementia syndromes really exist in the aged other than Alzheimer's disease). But the hypotheses that equate cerebral amyloid and/or defective cholinergic neurotransmission to senility have yet to pass the critical test of engendering a truly effective therapy, or even a reliable diagnostic and prognostic tool. That is the unarticulated agenda of scientific meetings like the present one. Dr. Alzheimer will always deserve our utmost respect and gratitude for having gotten things started.

Correspondence: Dr. R. J. Wurtman, Department of Brain and Cognitive Sciences, Massachusetts Institute of Technology, Room E 25-604, Cambridge, MA 02139, U.S.A.

The aging brain and its disorders

A. Carlsson

Department of Pharmacology, University of Göteborg, Sweden

"Last scene of all,
That ends this strange eventful history,
is second childishness, and mere oblivion,
Sans teeth, sans eyes, sans everything."
(Shakespeare, As you like it, II, vii)

Summary

In the search for the meachanisms underlying the aging process and for approaches to alleviate the resulting functional losses, the neurotransmitter strategy appears to offer great promise. This strategy is based on the hypothesis that the biological properties and vulnerabilities of nerve cells depend largely on the types of neurotransmitters they produce or are exposed to by innervation from other nerve cells. It permits an insight into the complex interactions and imbalances between neuronal systems that lead to age related functional losses. The aging process may not only cause transmitter deficiencies. Owing to failure of regulatory mechanisms neurotransmitters may prove harmful to the cells producing them as well as to cells exposed to them by innervation. The former mechanism is illustrated by the toxic potential of catecholamine autoxidation, the latter by excitotoxins.

Introduction

A severe loss of mental functions with age has been recognized for a very long time; it was aptly described by Shakespeare, as shown by the above quotation. One can wonder how old the demented people described by Shakespeare were, taking into account the dramatic "squaring off" of the survival curve so often referred to. Presumably they were a lot younger than the majority of demented people today. This is encouraging, because it gives rise to the hope that the aging process is not entirely determined by our genes. In fact, recent epidemiologic studies, for example by Svanborg et al. (1986), indicate that elderly people remain in good health up to a higher age than before, and this includes mental health. This again underlines the importance of environmental factors for the aging process.

There are many research strategies to choose between today, when trying to better understand what is going on in the aging brain. Among them the neurotransmitter strategy has perhaps proven especially fruitful so far. It can be used both for the identification of the substrates immediately responsible for functional deficits and of the underlying, long-term processes, and is thus useful for formulating strategies both for treatment and prevention. The successful research on Parkinson's disease can be appropriately referred to as a model in this context.

To illustrate what this approach can lead to in the case of normal aging, it seems suitable to start out from a robust indicator of what can be considered a normal, age-related functional stigma. I have chosen here to discuss the liability to respond to a stressful impact by getting into a confusional or delirious state. After discussing this in the short perspective, I will discuss the long-term processes underlying this reduced vitality of the aging brain. I think it is necessary to try to understand these normal phenomena before discussing the pathology of aging. This viewpoint is supported by the difficulty to draw a distinct demarcation line between normal and pathological aging. Whereas, for example, the increased liability to confusion is to be regarded a normal sign of aging, pathological dementia conditions are generally characterized by a further accentuation of this stigma.

Confusion liability as an age-related stigma: search for the underlying mechanism

There seems to be consensus that with increasing age, starting perhaps around the age of 60, it takes less and less of a stressful impact to bring an individual into a state of confusion or delirium. Once the stressful impact has been removed and adequate treatment has been given, there seems likewise to be general agreement that the individual may return to his normal baseline. One additional reason why this particular sign of reduced vitality of the brain appears especially fruitful to discuss, is that neuroleptics obviously provide a very efficient treatment of confusional and delirious states. The neuroleptics are predominantly antidopaminergic agents. In addition, some of them have alpha-adrenergic blocking properties (see Carlsson, 1978). Although this does not seem to be essential for their therapeutic efficacy, it may contribute. In favour of this assumption is the likely involvement of locus ceruleus and the central noradrenergic system in stress reactions (see Elam, 1985).

From the efficacy of the neuroleptics it seems reasonable to conclude that there exists in the brains of delirious people an imbalance involving an overweight of the dopaminergic system somewhere in the brain. Are we dealing with a hyperdopaminergic state in absolute terms? The answer is probably no. The level of dopamine goes down with age (see Fig. 1). In fact, dopaminergic neurons appear to belong to the most age-sensitive neurons of the human brain (see Carlsson, 1981). The possible mechanisms underlying this phenomenon will be discussed in a later section. The level of dopamine goes down with age in all brain regions examined post mortem, and so does tyrosine hydroxylase and the cell count of dopaminergic neurons. Recent PET data confirm the marked age dependence of the dopaminergic system (Tedroff et al., 1988).

The loss of dopaminergic neurons seems to be compensated by an increase in the physiological activity of the remaining neurons, as indicated by an increased HVA/DA ratio (Fig. 2, see Carlsson, 1988a). But there does not seem to be an overshoot. Admittedly no data are available to show how the dopaminergic neurons behave in the acute delirious state. This issue will hopefully be a future task for the imaging people

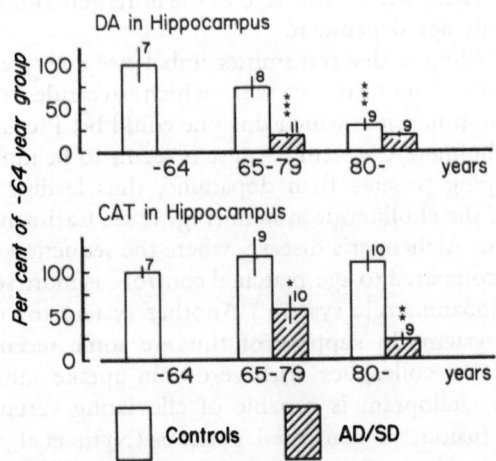

Fig. 1. Dopamine (DA) and choline acetyltransferase (CAT) in the hippocampus of three age groups of demented patients and controls, examined post mortem. Shown are the means and s.e.m. (n) * p < 0.05; ** p < 0.01; *** p < 0.001 vs. age matched controls or, in case of controls, vs. 64-year-old group (Data from a multicenter study by Gottfries C-G, Carlsson A, Eckernäs S-A, Svennerholm L; see Carlsson, 1988a)

A. Carlsson

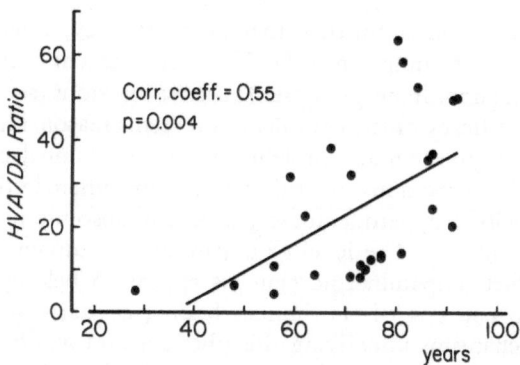

Fig. 2. The age dependence of the ratio of homovanillic acid (HVA) to dopamine (DA) in the hippocampus of controls, examined post mortem (Data from a multicenter study by Gottfries C-G, Carlsson A, Eckernäs S-A, Svennerholm L; see Carlsson, 1988a)

to resolve. However, in the following the assumption is made that the already weakened dopaminergic system is unable to bring about a severely hyperdopaminergic state.

In fact, we can probably not dismiss the possibility that a moderate, age-related involution of the dopaminergic system may be favorable by reducing the risk of transmitter imbalances, leading to confusion or psychosis. This may well be true also of the noradrenergic system, which is also markedly age dependent.

If we are dealing with a transmitter imbalance due to a deterioration of a dopamine-antagonistic system, which overrides the dopamine deficit, the question arises which this one could be. Presumably we can rule out the cholinergic system because it seems to be more resistant to the normal aging process than dopamine, thus leading to a relative overweight of the cholinergic system (Fig. 1, see Carlsson, 1981). (This is in contrast to Alzheimer's disease, where the reduction of the cholinergic system, compared to age-matched controls, is more severely affected than the dopaminergic system.) Another system to consider is the serotonergic system. In support of this are some recent findings of Gottfries and his colleagues with serotonin-uptake inhibitory. They observed that citalopram is capable of alleviating certain symptoms, including confusion, in demented patients (Nyth et al., 1987, 1988). However, other systems will also require careful consideration. This will be apparent from an examination of the various pathways interacting with the dopaminergic system.

The striatum: an important, cortex-controlled inhibitory structure?

The profound action of neuroleptics on cortical functions, as indicated by their antidelirious and antipsychotic effects, is not necessarily, or perhaps only partly, due to a primary effect of these agents on the cerebral cortex. In fact, both postmortem and more recently PET data indicate that the level of dopamine and the density of dopamine D-2 receptors, which are the most relevant subtype in connection with most of the neuroleptics in current use, are extremely low in the human cerebral cortex (for references, see Carlsson, 1988b). Thus we must seriously consider the possibility that not only the extrapyramidal but also the mental actions of neuroleptic drugs depend largely, if not entirely, on binding to receptor sites in the striatum. "Striatum" is used here in a wide sense and will thus comprise both the dorsal and the ventral, "limbic" part of this structure (see Nauta, 1989).

There is an increasing awareness of the important role of the striatum as an inhibitory structure, acting on a variety of both motor and mental functions. The main targets for the dorsal and ventral striatal complexes, which besides the striatum include the dorsal and ventral pallidum, which may be looked upon as relay stations, appear to be the thalamus and the mesencephalic reticular formation (see Fig. 3). We have proposed that the inhibitory function of the striatum is at least partly brought about by restricting the flow of sensory information relayed through the thalamus on its way to the cortex and by counteracting the

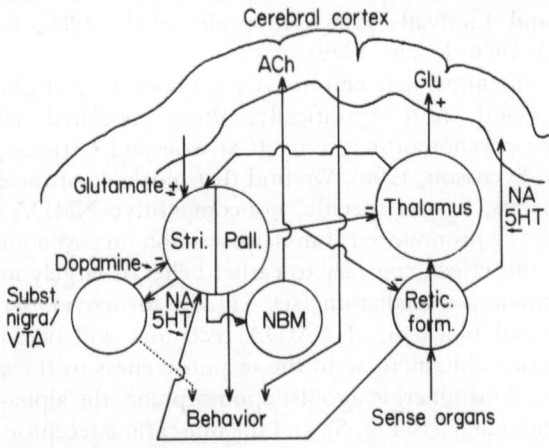

NBM = Nucleus basalis Meynert

Fig. 3. Schematic drawing serving to illustrate the hypothesis that the cerebral cortex is capable of controlling its sensory input/arousal via a dopamine-modulated feedback loop involving the striatal complexes and the thalamus/mesencephalic reticular formation

arousal induced by the mesencephalic reticular formation as well as by certain thalamic structures with similar function.

The striatum is controlled by several inputs. One is the mesostriatal dopamine system, which appears to be inhibitory on the striatum and will thus open up the flow of sensory information to the cortex via the thalamus, and increase the level of arousal. The noradrenergic and serotonergic systems, which originate in the mesencephalic reticular formation or in its close vicinity, may be part of this arousal-controlling system.

Another, apparently very powerful input to the striatum is the corticostriatal glutamatergic pathway. (Whenever "glutamatergic" is used in this context, it will comprise other endogenous excitatory amino acids and related molecules, such as aspartic acid and quinolinic acid.) This pathway is excitatory and will thus enhance the inhibitory function of the striatum and counteract the dopaminergic influence. It will be apparent then, that we are dealing here with a negative feedback loop, arising in the cerebral cortex and returning to the cortex, and, in fact largely to the same cortical area, via the striatum and the thalamus/mesencephalic reticular formation. By means of this feedback loop the cortex should be able to control the amount of incoming sensory information as well the level of arousal. Moreover, the control is likely to be selective, that is, less relevant inputs are filtered off in favor of more relevant and novel information (for recent references on the neuranatomy of the striatum and its afferent and efferent connections, see Heimer et al., 1985; Selemon and Goldman-Rakic, 1985; Goldman-Rakic and Selemon, 1986; Björklund and Lindvall, 1986; Alexander et al., 1986; Penney and Young, 1983, 1986; Nauta, 1989).

Some recent animal data emphasize the importance of glutamatergic mechanisms, and more specifically, those mediated via NMDA receptors, for psychomotor activity (Carlsson and Carlsson, 1989a, b; Carlsson and Svensson, 1990). We find that blockade of these receptors in mice and rats, by the specific, noncompetitive NMDA antagonist MK-801, exerts a pronounced stimulatory action on psychomotor activity, and that this effect, contrary to earlier belief, is largely independent of catecholaminergic mediation (see Fig. 4). Moreover, we find that already a partial blockade of NMDA receptors will be sufficient to induce a considerable increase in the responsiveness to the stimulating actions of the dopaminergic agonist apomorphine, the alpha-2-adrenergic agonist clonidine (see Fig. 5), and the muscarinic receptor antagonist atropine. We feel that the corticostriatal glutamatergic pathway is a good candidate for the antagonist to dopamine that we were looking for.

In other words, we propose that the age-related increase in the liability to confusional and delirious states could be due to an insufficient

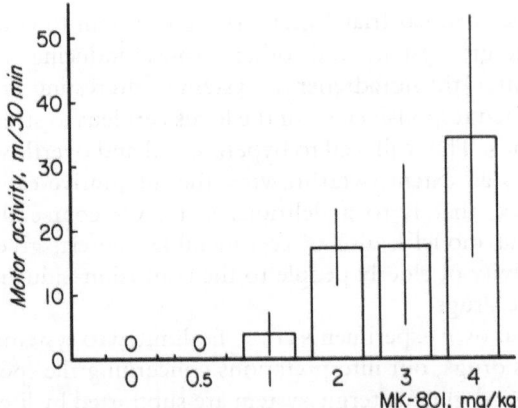

Fig. 4. Effects of various doses of the non-competitive NMDA-receptor antagonist MK-801 on motor activity in monoamine-depleted mice. Reserpine (10 mg/kg i.p.) was administered 18 h and alpha-methyltyrosine (250 mg/kg i.p.) 30 min prior to the i.p. MK-treatment. Forward locomotion was registered for 30 min, beginning 60 min after MK-801 administration. Shown are the means and SEM, N = 4. There was a significant correlation between dose and number of meters covered in 30 min (r = 0.6, p < 0.01) (Data from Carlsson M, Carlsson A, 1989 b)

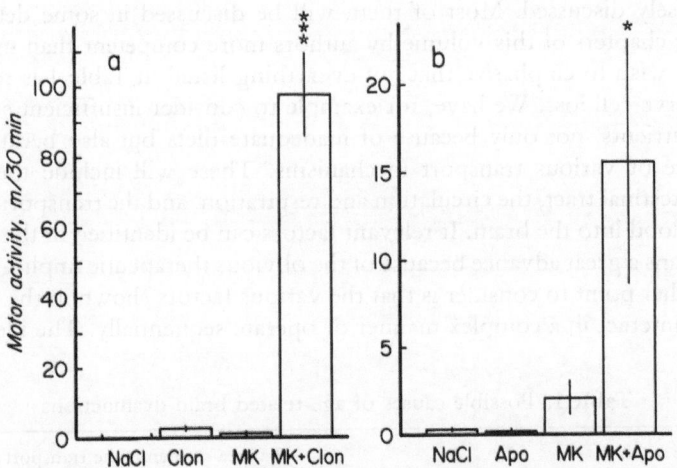

Fig. 5. a Effects of MK-801 (1 mg/kg i.p.) and clonidine (2 mg/kg i.p.), given separately or in combination, on forward locomotion in monoamine-depleted mice. **b** An analogous experiment with apomorphine (0.1 mg/kg i.p.) instead of clonidine, *** p < 0.001; * p < 0.02 vs. MK-801 (Mann-Whitney U-test) (Data from Carlsson M, Carlsson A, 1989 b)

capacity of the cortiocostriatal glutamatergic system to counterbalance the dopaminergic system and other arousal-inducing mechanisms. Among the latter, the noradrenergic system is interesting in view of the well documented responsiveness of the locus ceruleus to stressful stimuli of various kinds. This will lead to hyperarousal and overflow of sensory information to an extent overthrowing the integrative capacity of the cerebral cortex, that is to a delirious state. Of course the effect of atropine in this model is also of considerable interest, given the well-known sensitivity of elderly people to the confusion-inducing action of antimuscarinic drugs.

Whereas our own experiments are so far limited to systemic injections of the various drugs, our interpretations concerning the specific role of the corticostriatal glutamatergic system are supported by literature data, showing that stimulation and blockade of NMDA receptors in the striatum by locally applied specific agonists and antagonists will inhibit and stimulate psychomotor activity, respectively (Schmidt and Bury, 1988; Raffa et al., 1989).

Long-term aspects of the aging process

I now wish to switch gears and to discuss, very briefly, some more fundamental, long-term processes, underlying the age-related decline of brain function. Table 1 gives a list of some of the mechanisms now being intensely discussed. Most of them will be discussed in some detail in other chapters of this volume by authors more competent than myself. I just wish to emphasize that not everything listed in Table 1 is related to nerve-cell loss. We have, for example to consider insufficient supply of nutrients, not only because of inadequate diets but also because of failure of various transport mechanisms. These will include the gastrointestinal tract, the circulation and respiration, and the transport from the blood into the brain. If relevant factors can be identified in this area, it means a great advance because of the obvious therapeutic implications. Another point to consider is that the various factors shown in the Table may interact in a complex manner or operate sequentially. The identifi-

Table 1. Possible causes of age-related brain dysfunctions

– Free radicals	– Failure of membrane transport
– Endogenous toxins	– Malnutrition
– Exogenous toxins	– Circulatory dysfunction
– Deficiency of antioxidants	– Dysfunction of immune system
– Deficiency of trophic factors	

cation of such interactions and sequences depends largely on pharmacological interventions, using specific inhibitors of various kinds, for example, NMDA-receptor antagonists, calcium blockers, and agents causing degeneration of dopaminergic neurons or inhibition of dopamine
synthesis. These agents have been found capable of protecting brain
tissue from damage induced e.g. by ischemia or hypoglycemia (for
references, see Carlsson, 1989).

One further comment seems to be necessary here in view of the first
part of this paper, where I proposed that a deficient glutamatergic impact
on the striatum might be a factor favoring the development of confusional states. Here we rather see an increase in the same endogenous
compounds, that is the excitotoxins. This may look like a contradiction.
However, if we introduce the concept of a dysregulation of glutamatergic systems, it will leave room for both hyper- and hypofunctional states,
occurring on different occasions. In any event it seems clear that the
glutamatergic and related systems do indeed deserve the considerable
attention being devoted to them at present. Among other things, the
development of specific glutamatergic agonists and antagonists, suitable
for both animal and human studies, are anxiously awaited.

A new CNS model for studying age-related oxygen toxicity

The last part of this chapter will be devoted to the area of oxygen
toxicity, which is of course also attracting a lot of interest at present. I
wish to focus on one special aspect which has interested us for some
time. Even if we are dealing here with a rather special case of oxygen
toxicity, it may have some general implications and may, in fact, serve
as a model. The study of free oxygen radicals and how they are controlled by the complex antioxidant system present in tissues and body
fluids, is somewhat elusive, implying that there is indeed a need for
additional experimental models.

Figure 6 shows the autoxidation of dopamine. This is a dangerous
reaction for two reasons. Firstly superoxide anions, that is free radicals,
are formed, and secondly the autoxidation of dopamine leads to the

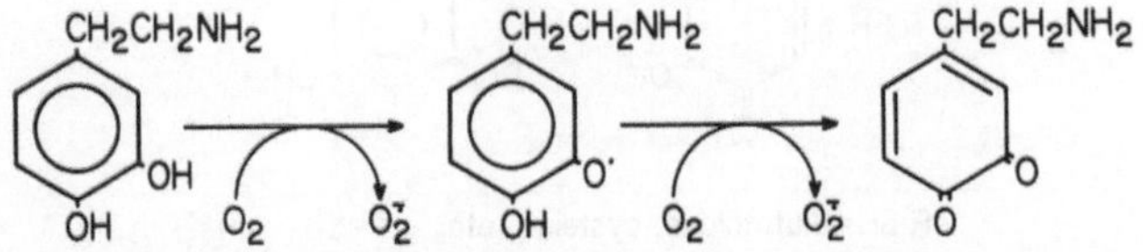

Fig. 6. Scheme illustrating the autoxidation of dopamine under formation of quinoids and
the superoxide anion (For references, see Moldéus et al., 1983)

formation of semiquinones and quinones which are also highly reactive and toxic species (see e.g. Moldéus et al., 1983). Further processing of the quinones is generally believed to result in neuromelanin, the normal occurrence of which is thus considered to be evidence in favor of the actual existence of this autoxidation process, albeit at a low rate. It has been proposed that the autoxidation process may be responsible for the age-related loss of dopaminergic neurons as well as for Parkinson's disease. It may be speculated that the autoxidation process can be speeded up by a failure of the protective antioxidant system. Alternatively there could be a failure of the scavenger system detoxifying the quinones.

To get a handle on the autoxidation mechanism we have made use of the well known fact that quinones react very promptly with thiol groups, and thus for example with glutathione and cysteine. Figure 7 demonstrates this reaction. We have been able to identify the product 5-S-cysteinyl-dopamine and some related metabolites, using HPLC (Fornstedt et al., 1986). Animal data show that 5-S-cysteinyl-dopamine accumulates when dopamine is released intraneuronally by reserpine (Fornstedt and Carlsson, 1989), but not when it is released into the extraneuronal space by amphetamine. The ratio 5-S-cysteinyl-dopamine/dopamine is higher in the human brain than in other species investigated. It is also higher in the substantia nigra than in the terminal regions, suggesting at least a partial correlation to the occurrence of neuromelanin. In human brain analyzed postmortem we have discovered that the ratio of 5-S-cysteinyl-dopamine to dopamine is higher in the brains of individuals with a significant loss of dopaminergic neurons as indicated by depigmentation of the substantia nigra (Fig. 8, Fornstedt et al., 1989). To examine this phenomenon more closely we calculated the ratio 5-S-cysteinyl-dopamine/DOPAC. The rationale for doing so is that both the formation of 5-S-cysteinyl-dopamine and DOPAC occur largely, if not exclusively, intraneuronally in the cytoplasm of dopaminergic neurons. Thus a change in this ratio would indicate a switch from one metabolic process

R·SH = Glutathione, cysteine, etc.

Fig. 7. Adduct formation from the quinonoid autoxidation product of dopamine and thiol-containing molecules, leading e.g. to formation of 5-S-cysteinyl-dopamine

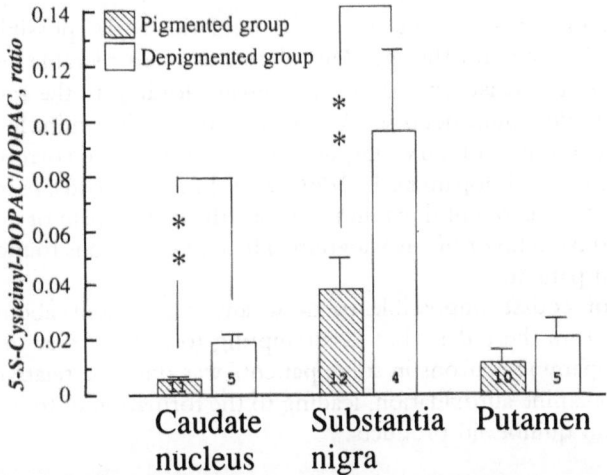

Fig. 8. Ratio of 5-S-cysteinyl-dopamine to dopamine in different brain regions analyzed post mortem. Comparison is made between a group of individuals with well pigmented and a group with poorly pigmented substantia nigra; $p < 0.01$ (Data from Fornstedt et al., 1989 b)

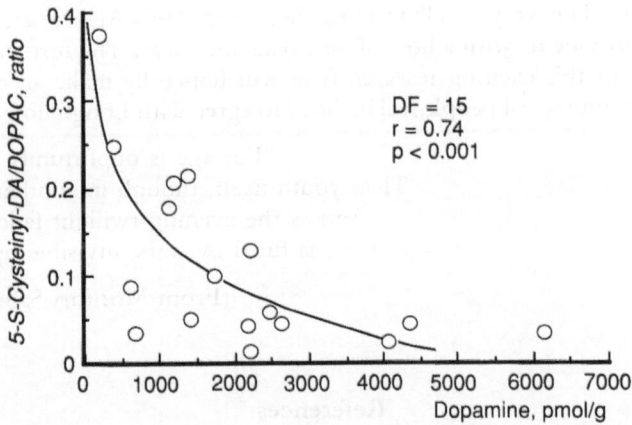

Fig. 9. Ratio of 5-S-cysteinyl-dopamine to 3,4-dihydroxyphenylacetic acid (DOPAC) versus the level of dopamine (DA) in the substantia nigra, analyzed post mortem (Data from Fornstedt et al., 1989 b)

to another. As shown in Fig. 9, there appears, in fact to be such a switch. As shown in this Fig. there was a considerable individual variation in the dopamine level in the substantia nigra of these individuals, who may be considered to have been essentially normal for the age (72–91 years), with respect to the brain, except for one individual who had Parkinson's disease, and a few patients who had mild dementia, possibly to be considered normal for the age. This illustrates a considerable individual variation in the sensitivity to the mechanisms leading to the age dependent loss of dopamine neurons. As can be seen, low dopamine levels, and thus a low number of surviving dopamine neurons, are correlated to a high 5-S-cysteinyl-dopamine/DOPAC ratio. In other words, in individuals with a severe loss of dopamine neurons the metabolism of dopamine appears to be in favor of autoxidation. The highest ratio is found for the Parkinson patient.

It is of course impossible to draw any conclusions about causal relations from these data, but it is tempting to suggest that the severe loss of dopamine neurons in some patients was due to a relatively high rate of dopamine autoxidation, leading to the formation of toxic oxygen species and quinonoid products.

Conclusion

There is a wealth of promising clues in the area of aging and brain disorders. This very timely meeting, honoring Alois Alzheimer, will no doubt provide us with a host of new data and ideas. The further development of this exciting research field will hopefully make an ever increasing number of people feel inclined to agree with Longfellow (1875):

> "For age is opportunity no less
> Than youth itself, though in another dress,
> And as the evening twilight fades away
> The sky is filled by stars, invisible by day."

(From Morituri Salutamus)

References

Alexander GE, DeLong MR, Strick PL (1986) Parallel organization of functionally segregated circuits linking basal ganglia and cortex. Ann Rev Neurosci 9:357–381

Björklund A, Lindvall O (1986) Catecholaminergic brain stem regulatory systems. In: Field J (ed) Handbook of physiol – The nervous system IV. Am Physiol Soc Washington DC, pp 155–235

Carlsson A (1978) Antipsychotic drugs, neurotransmitters and schizophrenia. Am J Psychiatry 135(2):164–173

Carlsson A (1981) Aging and brain neurotransmitters. In: Platt D (ed) Funktionsstörungen des Gehirns im Alter. Schattauer, Stuttgart New York, S 67–82

Carlsson A (1988a) Brain neurotransmitters in aging and dementia: recent findings. In: Barchas JD, Bunney WE (eds) Perspectives in psychopharmacology: a collection of papers in honor of Earl Usdin. Alan R Liss, New York, pp 209–223

Carlsson A (1988b) The current status of the dopamine hypothesis of schizophrenia. Neuropsychopharmacology 1:179–186

Carlsson A (1989) Neurotransmission and the aging brain: new vistas. In: Carlsson A, Kanowski S, Allain H, Spiegel R (eds) Cerebral insufficiency. Trends in research and treatment. Parthenon Publishing Group Ltd, Casterton Hall Carnforth, pp 1–15

Carlsson M, Carlsson A (1989a) The NMDA antagonist MK-801 causes marked locomotor stimulation in monoamine-depleted mice. J Neural Transm 75:221–226

Carlsson M, Carlsson A (1989b) Dramatic synergism between MK-801 and clonidine with respect to locomotor stimulatory effect in monoamine-depleted mice. J Neural Transm 77:65–71

Carlsson M, Svensson A (1990) Interfering with glutamatergic neurotransmission by means of MK-801 administration discloses the locomotor stimulatory potential of other transmitter systems in rats and mice. Pharmacol Biochem Behav 36 (in press)

Elam M (1985) On the physiological regulation of brain norepinephrine neurons in rat locus ceruleus. Thesis, University of Göteborg, Sweden

Fornstedt B, Carlsson A (1989) A marked rise in 5-S-cysteinyl-dopamine levels in guinea pig striatum following reserpine treatment. J Neural Transm 77:155–161

Fornstedt B, Rosengren E, Carlsson A (1986) Occurrence and distribution of 5-S-cysteinyl derivatives of dopamine, dopa and dopac in the brains of eight mammalian species. Neuropharmacology 25:451–454

Fornstedt B, Brun A, Rosengren E, Carlsson A (1989) The apparent autoxidation rate of catechols in dopamine-rich regions of human brains increases with the degree of depigmentation of substantia nigra. J Neural Transm (PD-Sect) 1:279–295

Goldman-Rakic PS, Selemon LD (1986) Topography of corticostriatal projections in nonhuman primates and implications for functional parcellation of the neostriatum. In: Jones EG, Peters A (eds) Cerebral cortex, vol 5. Plenum Publishing Corporation, New York, pp 447–466

Heimer L, Alheid GF, Zaborszky L (1985) Basal ganglia. In: Paxinos G (ed) The rat nervous system, vol 1. Forebrain and midbrain. Academic Press, New York, pp 37–86

Moldéus P, Nordenskjöld M, Bolesfoldi G, Eiche A, Haglund U, Lambert B (1983) Genetic toxicity of dopamine. Mutat Res 124:9–24

Nauta WJF (1989) Reciprocal links of the corpus striatum with the cerebral cortex and limbic system: a common substrate for movement and thought. In: Mueller J (ed) Neurology and psychiatry: a meeting of minds. Karger, Basel München Paris, pp 43–63

Nyth A-L, Balldin J, Elgen K, Gottfries C-G (1987) Behandlung med citalopram vid demens. Normalisering av DST (with summary in English). Nord Psykiat Tidskr 41:423–430

Nyth A-L, Gottfries C-G, Elgen K, Engedahl K, Harenko A, Karlsson I, Koskinen T, Larsson L, Nygaard H, Samuelsson SM, Yli-Kertula A (1988) The effect of citalopram in dementia disorders. A Scandinavian multicenter study. Poster, CINP Congress Munich, August 15–19, 1988

Penney JB Jr, Young AB (1983) Speculations on the functional anatomy of basal ganglia disorders. Ann Rev Neurosci 6:73–94

Penney JB Jr, Young AB (1986) Striatal inhomogeneities and basal ganglia function. Movement Disordes 1:3–15

Raffa RB, Ortegón ME, Robisch DM, Martin GE (1989) In vivo demonstration of the enhancement of MK-801 by L-glutamate. Life Sci 44:1593–1599

Schmidt WJ, Bury D (1988) Behavioural effects of N-methyl-D-aspartate in the anterodorsal striatum of the rat. Life Sci 43:545–549

Selemon LD, Goldman-Rakic PS (1985) Longitudinal topography and interdigitation of corticostriatal projections in the Rhesus monkey. J Neurosci 5:776–794

Svanborg A, Berg S, Mellström D, Nilsson L, Persson G (1986) Possibilities of preserving physical and mental fitness and autonomy in old age. In: Häfner H, Moschel G, Sartorius N (eds) Mental health in the elderly. Springer, Berlin Heidelberg New York, pp 195–202

Tedroff J, Aquilonius S-M, Hartvig P, Lundqvist H, Gee AG, Uhlin J, Långström B (1988) Monoamine re-uptake sites in the human brain evaluated in vivo by means of 11C-nomifensine and positron emission tomography: the effects of age and Parkinson's disease. Acta Med Scand 77:192–201

Correspondence: Prof. A. Carlsson, Department of Pharmacology, University of Göteborg, P.O. Box 33031, S-40033 Göteborg, Sweden.

Epidemiology

Epidemiology of Alzheimer's disease

H. Häfner

Central Institute of Mental Health, Mannheim, Federal Republic of Germany

Summary

A survey is given of population studies on the prevalence and incidence of late-life dementia in general, showing an overall prevalence rate of 5–6% for moderate and severe dementia at the age of 65 and over, and of Alzheimer's and vascular dementia in particular. In the European countries about 60% of all cases of late-life dementia seem to belong to Alzheimer's, about 20% to multi-infarct dementia. In Japan and the USSR the share of vascular dementia is obviously greater. Both diseases show an exponential increase of prevalence rates with age in all populations investigated, starting from 2–3% in the age group 65–70 and surmounting 30% beyond 90 years of age, MID and AD doubling about every 5 years. The incidence rates for late-life dementia also show an exponential increase from about 0.3% at the age of 60–69, tripling every ten years of life and reaching around 3–4% at the age of 80 and over. Thus the exponential increase in prevalence is primarily due to the exponential age-dependent increase in the true morbid risk. The sex distribution for both types of dementia is still under discussion. The epidemiological study of genetic and environmental risk factors indicates that up to now only Down's syndrome and a family history of AD are major risk factors for attracting AD in late life.

Introduction

On 3rd November 1906, on the occasion of the 37th meeting of south-west German psychiatrists in Tübingen, Aloys Alzheimer described the case of a 51-year-old woman with a severe loss of memory, disorientation, disturbances of language and paranoid ideas, who died after four years of the disease in a state of severest dementia. With his description he gave an account of the disease which is still valid today: the clinical aspect of a progressive loss of cognitive, mnestic and language functions and the neuropathological aspect of a degeneration of

ganglia and an agglomeration of neurofibrillary tangles and plaques in
the cerebral cortex and in the hippocampus. The establishment of an
exact diagnosis of dementia of Alzheimer's type still requires positive
findings in both clinical and neuropathological fields. The latter can
hardly be obtained from population studies. Peripheral markers which
would also in population studies allow to screen for possible cases or to
make a correct diagnosis of Alzheimer's disease, such as the evidence of
β amyloid through monoclonal antibiodies in body fluids or through
biopsy of nervous olfactorius fibres from the nasal mucosa, are not yet
at our disposal.

Prevalence of dementia in later life

The epidemiology of Alzheimer's disease is currently based on two
different approaches: first, the assessment of all cases of late-life dementia and the subsequent estimation of the share of Alzheimer's disease on
the basis of neuropathological reference data. These are still founded on
the findings of Tomlinson et al. (1970, 1976): primary dementia 53%,
vascular dementia 17%, mixed types of primary and vascular dementia
16% and the rest representing secondary dementia of different etiology.
In a prospective clinical and neuropathological study Mölsä et al. (1985)
have essentially confirmed these findings. The method of estimating the
proportions has some disadvantages. It does not allow individual case
identification and, in addition, the reference data come from only few
countries, whereas the relative prevalence of Alzheimer and vascular
dementia seems to vary across countries.

The second approach constitutes the identification of cases of
Alzheimer's disease on the basis of clearly defined clinical criteria such
as the DSM III R criteria, since the improved methods of diagnostic
discrimination through brain-imaging techniques (CCT, MRI and PET)
can hardly be applied in population studies. The most frequently used
instruments comprising operationalized clinical criteria for diagnosing
and subclassifying dementia by Alzheimer versus vascular and secondary
dementia and identifying probable or possible cases of Alzheimer's disease are the NINCDS criteria (McKhann et al., 1984), the Ischaemic
score of Hachinsky et al. (1975) and CAMDEX (Roth et al., 1988).

The overall efficiency of these instruments, insofar as they have as yet
been validated in clinical and neuropathological findings, has been considerably improved in comparison with the clinical judgement, but is
still limited with values of about 80% (Sulkava et al., 1983; Tierney
et al., 1988).

Cases of dementia in later life usually begin with slight cognitive
deficits, which cannot be clearly distinguished from psycho-organic

Table 1. Prevalence of dementia in the elderly population: results from field studies

Authors	Country	Age group	Sample size	Prevalence in %	
				Severe or moderate	Mild
Essen-Möller et al. (1956)	Sweden	60+	443	5.0	10.8
Nielsen (1962)	Denmark	65+	978*	5.9	15.4
Primrose (1962)	Great Britain	65+	222*	4.5	–
Kay et al. (1964)	Great Britain	65+	505*	5.6	5.7
Kaneko (1975)	Japan	65+	531	7.2	52.7
Hasegawa (1974)	Japan	65+	4716	3.0	1.5
Broe et al. (1976)	Scotland	65+	808	3.8	4.3
Sternberg and Gawrilowa (1978)	USSR	60+	1020	3.6	21.0
Cooper and Sosna (1983)	FRG	65+	519*	6.0	5.7
Campbell et al. (1983)	New Zealand	65+	541*	7.7	–
Weissman et al. (1985)	USA	65+	2588	3.4	12.7
Folstein et al. (1985)	USA	65+	923	6.1	–
Sulkava et al. (1985)	Finland	65+	1866	6.7	–
Copeland et al. (1987)	Great Britain	65+	1070	5.2	–
Weyerer and Dilling (1984)	FRG	65+	295	3.5	5.0

* Institutionalized elderly included.
Source: Cooper and Bickel (1989) by permission. Supplemented by the author

syndromes of different etiology and from normal memory problems in old age. Reliable epidemiological data therefore refer to moderate and severe dementias, which are defined by significant cognitive deficits affecting social competence and individual autonomy and which can thus be rather precisely diagnosed.

Table 1 presents the results of 15 virtually comparable population studies. In the ten countries referred to the prevalence rates for severe or moderately severe dementia vary between 3.0% and 7.7% of the population aged 65 and over with a mean value of approximately 5%. The high variability of the rates for mild dementia or psychoorganic syndromes from 1.5% to 52.7% reflects the difficulties of case identification mentioned and the different diagnostic procedures used.

Exponential increase with age of prevalence figures for late-life dementia

Eight studies from different countries are shown in Fig. 1. The prevalence for moderate and severe cases of dementia rises from initial values

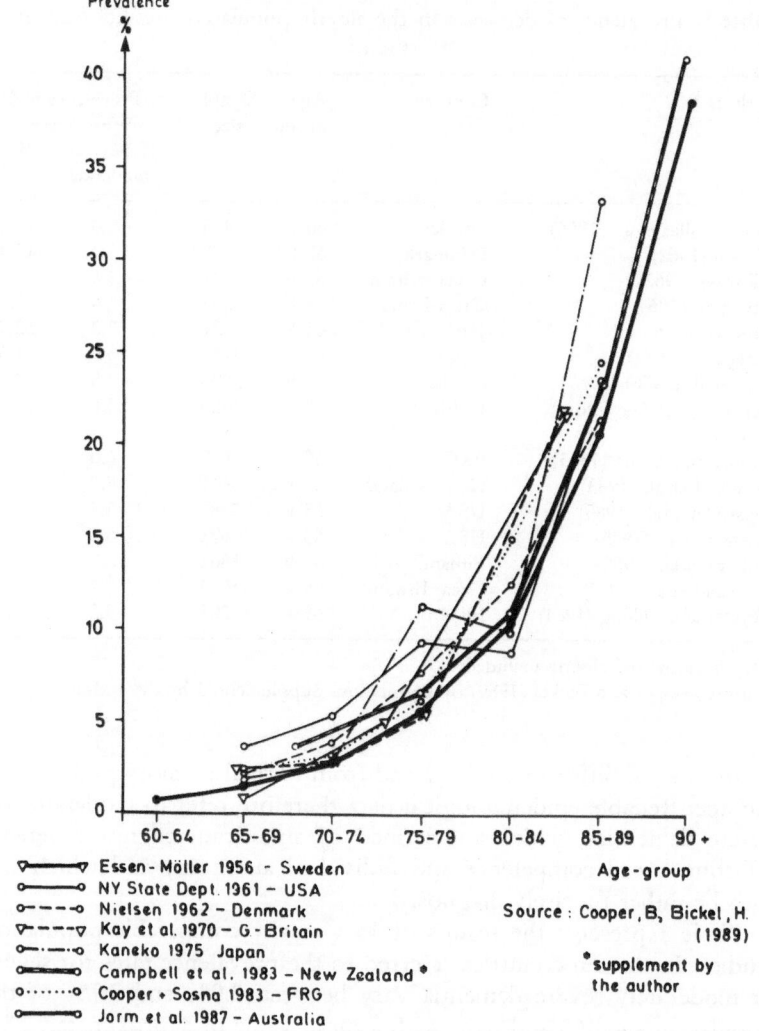

Fig. 1. Age-group related prevalence rates of moderate and severe dementia from 8 population studies

of 2–3% of the population aged 65–70 to over 20% of those aged 80 to 89 and to over 30% for those aged 90 and over.

By means of a statistical model of estimation Jorm et al. (1987) analysed 22 studies on the epidemiology of late-life dementia published between 1945 and 1985. Table 2 shows the estimated prevalence rates by 5 year age intervals from 60–64 through 90–95. According to the model

Table 2. Estimated prevalence rates for dementia by 5 year age intervals for baseline population

Age-group	Median age	Estimated prevalence (%)
60 – 64	62.5	0.7
65 – 69	67.5	1.4
70 – 74	72.5	2.8
75 – 79	77.5	5.6
80 – 84	82.0	10.5
85 – 89	87.0	20.8
90 – 95	91.5	38.6

Source: Jorm et al. (1987) by permission

of an exponential rise the prevalence rates double every 5.1 years. Whether this exponential rise continues theoretically up to the maximum of human lifespan, asymptotically approximating a prevalence of 100% – thus supporting the assumption of a physiological process of ageing – or of only about 50% – in line with the assumption of a dominant Mendelian transmission of Alzheimer's disease, which assumptions are, however, unlikely to apply to the relatively small proportion of vascular dementia – cannot be answered as yet. The presumed flattening of the increase beyond the 90th year of age has so far been based on observations of very few cases and is therefore neither proven nor at least likely.

Differentiation of prevalence data on senile dementia of Alzheimer's type and multi-infarct dementia

Jorm et al. (1987) also analysed seven studies containing age-specific data for both multi-infarct dementia and Alzheimer's disease with respect to the exponential age increase. With the exception of one US study, they were found to give a satisfactory fit to the model for both types of disease. The prevalence rates of multi-infarct dementia doubled every 5.3 years, those of Alzheimer's disease somewhat faster, namely every 4.5 years.

While the age increase of prevalence rates for late-life dementia shows a parallel course for all populations studied, the proportions of Alzheimer's dementia and multi-infarct dementia seem to vary significantly from one region to the other. Tomlinsons's reference values of approximately 60% of Alzheimer's disease and 20% of multi-infarct dementia are apparently applicable to most European studies. However, one US and two Finnish studies analysed by Jorm et al. (1987) show almost no excess of Alzheimer's disease, whereas four Japanese and two Russian studies, i.e. coming from countries with a high mortality of stroke, show a significant excess of multi-infarct dementia over Alzheimer's disease, although with different odds ratios and partly high shares of mixed dementia.

The gender difference in prevalence rates cannot be finally interpreted. In connection with the higher life expectancy of females, the crude rates for moderate and severe late-life dementia in all studies were higher for females than for males. In the meta-analysis of Jorm et al. (1987) the small number of studies showing age-adjusted prevalence rates for both sexes do not allow to infer a significant gender difference in the overall rates for late-life dementia. However, the prevalence rate for Alzheimer's disease was significantly higher for females than for males with a ratio of 1.3 to 0.77 of the values expected. There is, however, a marked variation of the sex ratios for both types of dementia, as shown in Table 3, which, up to now, does not allow to draw firm conclusions.

Jorm et al. (1988) developed a quantitative model providing detailed estimates of the prevalence rates for dementia on the basis of population projections up to the year 2025. For the Federal Republic of Germany this results in an increase of 40.73% by 2025 (see Table 4). If this estimate were to come true, we would have to reckon with a number of 850 000 dementia sufferers in 2025. About 590 000 of these would suffer from Alzheimer's disease partly combined with vascular dementia (Table 5). Dependent upon demographic trends the figures for the other countries vary considerably. To give but a few examples, by the year 2025 the prevalence of dementia will have increased in Great Britain by 34%, in the United States by 99%, in the USSR by 117%, in Australia by 162% and in Japan by 215% (see Table 6).

Table 3. Prevalence of AD and MID by sex (rates are simple averages of reported rates in percent)

Region	n of studies	AD		MID	
		Males	Females	Males	Females
Japan	4	0.8	1.5	3.3	2.7
Russia	1	1.1	3.8	6.1	2.1
Scandinavia	3	1.6	2.2	1.0	1.4
Britain	3	3.0	4.9	4.4	2.7
USA	2	0.7	2.1	3.5	0.8

Source: Jorm et al. (1987)

Table 4. Projected increases in the number of dementia cases over the base year 1980 for the Federal Republic of Germany, moderate and severe cases 60 years and over

1985	5.75%	2010	26.16%
1990	7.92%	2015	32.63%
1995	7.64%	2020	38.42%
2000	11.25%	2025	40.73%
2005	19.75%		

Total population increase 1980–2025 –13.12% (without migration movement).
Source: Jorm et al. (1988), modified

Table 5. Estimated and projected numbers of persons aged 65 and over suffering from dementia and Alzheimer's disease in the Federal Republic of Germany

	1980	2025
Late-life dementia (all forms)	680 000	850 000
Alzheimer's disease (only)	420 000	590 000

Table 6. Increase in late-life dementia by 2025 over base year 1980 for different countries

Great Britain	33.50%
Federal Republic of Germany	40.73%
USA	99.36%
USSR	116.90%
Australia	161.90%
Japan	215.40%

Source: Jorm et al. (1987), modified

Incidence of late-life dementia

In addition to the difficulties of obtaining the prevalence data, the study of incidence rates faces two problems: the small rate of demented which requires the study of a large quantity of persons and the difficulty of precisely defining the onset of a process of dementia. A simple but inaccurate method of calculating annual incidence rates is the division of annual prevalence rates by the mean duration of the disease. With a prevalence rate of approximately 6% of those aged 65 and over and a mean time of survival of about 4 years (based on Cooper and Bickel, 1989), the resulting incidence rate for this age group is around 1.5%, which is within the range of the results produced by respective field studies.

For directly calculating incidence rates in population studies the longest possible period of observation should be chosen. The denominator comprises the number of person years, i.e. the sum total of the years that all the persons of the sample survived without dementia. The numerator is the number of cases of dementia diagnosed during the total period of observation (Table 7).

The rates for those aged 65 and over vary between 10.7 and 16.3 per thousand, as shown in Table 8. The analysis by age decades shows, similar to that for prevalence rates, that there is an exponential increase with advancing age (Fig. 2). For the age decade 60–69 the incidence rate is around 0.3%, for 70–79 it lies between 1.2 and 2.3% and for 80 and over it lies between 3.4 and 4.0%. This means that starting from the 60th year of age the incidence rate triples about every ten years. The exponential increase in prevalence is therefore not only due to the accumulation

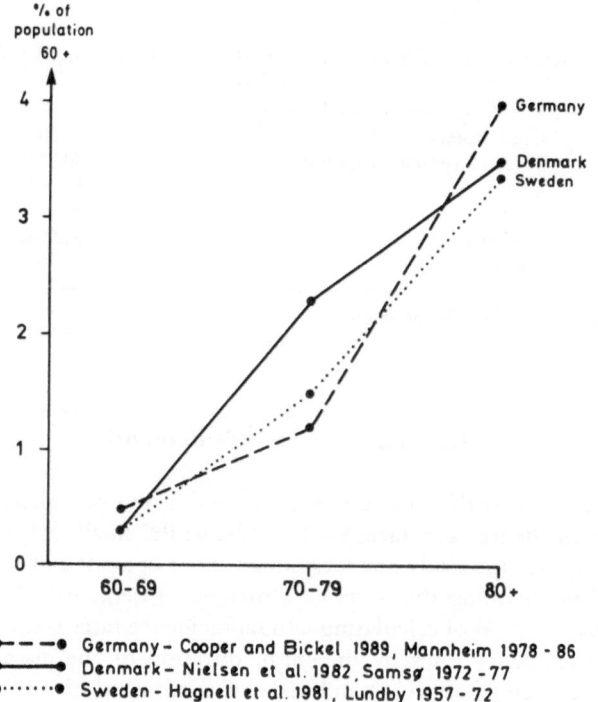

Fig. 2. Incidence rates of moderate and severe dementia by age (both sexes) from 3 population studies

Table 7. Incidence of late-life dementia

Rough estimation:

$$\text{Annual incidence rate} = \frac{\text{point prevalence rate (\%)}}{\text{mean duration of illness (years)}} = \frac{6}{4} = 1.5\%$$

Calculation:

$$\frac{\text{Number of dementia cases within population investigated}}{\text{Person years} = \text{years spent alive without dementia by all persons investigated during period of investigation}}$$

of chronic cases, but primarily to the exponential increase in the risk of developing late-life dementia.

Population surveys providing incidence rates for different types of dementia have been very scarce so far. The population study carried out by Cooper and Bickel (1989) at the Central Institute of Mental Health in Mannheim produced the following annual incidence rates 65 and over for both sexes on the basis of CAMDEX criteria: AD 9.4/1000, MID and mixed types 6.3/1000, secondary dementia 2.1/1000. These rates correspond well to the reference data of Tomlinson et al. (1976) and Mölsä et al. (1985).

Duration of the disease

The survival time of demented people is clearly reduced in comparison with the population of the same age (Bickel, 1987). Bickel gave a crude estimate of the mean duration of the disease on the basis of data from three field studies (Bergmann et al., 1971; Nielsen et al., 1982; Cooper and Bickel, 1989) by dividing the point prevalence values by the annual incidence values. With 3.7, 4.6 and 3.9 years the results were within a narrow range. Studies of deceased where the onset of the disease was defined by first signs of cognitive decline and assessed with the help of relatives (Barclay et al., 1985; Diesfeldt et al., 1986) obtained estimates of 7 to 8 years for the 65 to 80-year-old demented people. The mean duration of the disease can therefore be assessed to be approximately 7 to 8 years from the first occurrence of cognitive decline and about 4 years from the first standardized diagnosis of "dementia". In this respect there seems to be only little or no difference between vascular dementia and Alzheimer's disease.

Risk factors

Genetic determinants

An important risk factor for developing Alzheimer's disease is triso-mia 21, Down's syndrome. In virtually all Down patients who survive the age of 40, characteristic neuropathologic changes and amyloid A4 deposits can be found in the brain (Heston, 1982; Wisniewski et al., 1985). This is accounted for by the existence of an additional Alzheimer gene or a gene expressing the PreA4 amyloid.

When the strength of the genetic determination for senile dementia of Alzheimer's type is to be assessed, considerable difficulties arise in practice, because only a limited number of individuals carrying the gene and their family members live until the age of high risk. Therefore we do not as yet dispose of large-scale twin studies. Children of affected parents are too young to attract the disease (Jorm, 1987; Cooper, 1989). Negative findings of family and pedigree studies are consequently of little importance. The recent localization of a polymorphism on chromo-some 21 in patients with familial SDAT (St. George-Hyslop et al., 1987) has given new stimulation to genetic research.

Heston et al. (1981) and Whalley et al. (1982) found an increase in SDAT rates among the siblings of Alzheimer patients (Fig. 3). Three case-control studies (Heyman et al., 1984; Amaducci et al., 1986; Shalat et al., 1986) obtained slightly increased rates of Alzheimer's disease among the families of SDAT patients in comparison with control groups from the general population or from hospital patients. The evidence of these studies is only limited, since the samples examined are not very large and the statements of relatives about the dementia of long deceased are not very reliable (Amaducci et al., 1988). Heston et al. (1981) found a strong genetic determination in early-onset patients. Most of the other studies quoted also assume a stronger genetic load in cases of earlier age of onset (e.g. Amaducci before the age of 70). In late-onset cases Chandra et al. (1987) found no increase in SDAT among family members. How-ever, with respect to the sources of error mentioned, the findings are not yet sufficient to give a clear picture of the strength of genetic determina-tion in relation to age of onset, although very young age of onset is likely to be associated with a stronger genetic load.

In view of the unimodal, exponential distribution across age of the risk of first onset, the early Kraepelinian assumption of two diseases, the Mendelian inherited presenile dementia and a non-inherited senile de-mentia, must be considered unlikely.

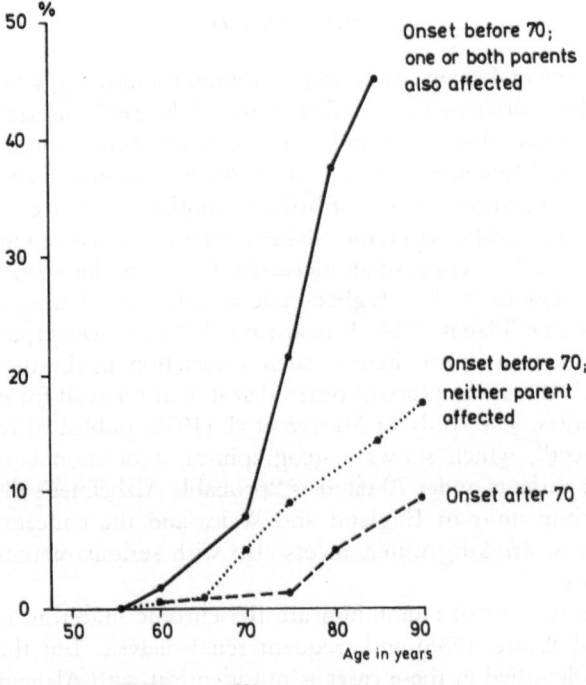

Fig. 3. Cumulative risk of DAT among siblings of DAT patients

The importance of environmental factors

Head trauma

From the dementia pugilistica of former professional boxers it has long been known that head trauma is likely to be a risk factor for Alzheimer-type changes in the brain (Corsellis et al., 1973; Rudelli et al., 1982). The findings of retrospective case control studies on this question are inconsistent. Some of them obtained an increased rate of severe head trauma before the onset of dementia (Heyman et al., 1984; Mortimer et al., 1985), whereas Amaducci et al. (1986), Chandra et al. (1987), Bharrucha et al. (1983) and Sulkava et al. (1985) found no or only a slight, insignificant connection.

Toxic substances

Environmental toxic substances like aluminium or organic solvents have also been discussed as risk factors for Alzheimer's disease; aluminium salts because they are found in low concentration especially in the centres of senile plaques (Candy et al., 1986) and because in animals they cause the generation of neurofibrillary tangles (Klatzo et al., 1965), which, however, differ from the Alzheimer tangles observed in humans. Norwegian studies report of an increased death rate due to Alzheimer's disease in regions with a higher concentration of aluminium in the drinking water (Flaten, 1988; Edwardson, 1988). However, among the doctors certifying death there is such a variation in the diagnosis of Alzheimer's disease as cause of death that it is of no avail for epidemiological studies. The study of Martyn et al. (1989) published recently in "The Lancet", which shows a geographical association between the number of patients under 70 rated as "probable Alzheimer's disease" by CCT-scanning units in England and Wales and the concentration of aluminium in drinking water, meets also with serious methodological reservations.

Further sources of aluminium are the chronic ingestion as antacid (Shore and Wyatt, 1983) and frequent renal dialysis. But the type of dementia identified in these cases is not identical with Alzheimer's disease (Heyman et al., 1984). Thus there is no evidence as yet for an increase in the risk of Alzheimer's disease due to increased absorption of aluminium. The assumption that long phenacetin abuse may be a risk factor for Alzheimer's disease as recently alleged in Australia in connection with a phenacetin epidemic in the 1930s (Murray et al., 1971) still needs to be verified. Chronic alcoholism, which may cause encephalopathia Wernicke or atrophy of the brain clearly distinguishable from Alzheimer's disease, does not seem to be an important risk factor, either.

In the field of etiology of Alzheimer's disease immune and virus theories have also been discussed. Case control studies, however, did not confirm the association with a family history of lymphoma, lymphosarcoma, with Hodgkin's disease or other immune system disorders found in a comparative study by Heston et al. (1981). Neither was there conclusive evidence for an association with prior viral diseases or possible sources of viral infections, as Amaducci et al. (1986) stated in their analysis of possible risk factors for Alzheimer's disease.

Social risk factors

In the course of their prospective study covering an average period of 7.8 years, Bickel and Cooper (1989) applied the "proportional hazard regression model" of Cox and found that neither severe physical impairment nor low social status and living alone had a significant influence on the risk of developing dementia. But if cognitive deficits are diagnosed during the first examination, the probability that dementia will occur rises by nearly 300%. This result is presumably trivial: slight cognitive decline is often an indicator of beginning dementia and therefore a powerful predictor of the occurrence of moderate or severe dementia later in life.

Thus there seem to be no clear associations with social variables over medium periods of time, insofar as confounding variables like age and physical illness are controlled (Bickel and Cooper, 1989). But replication studies with sufficiently large samples and studies on long-term risks are still needed.

Conclusions

To sum up the state of epidemiological knowledge: there is at present no conclusive evidence for the immune hypothesis or for an involvement of infectious agents and toxic exposures in the risk of developing senile dementia of Alzheimer's type. There is some indication that severe head trauma, in particular the dementia pugilistica of professional boxers, causes changes of Alzheimer's type in the brain; its quantitative contribution to the morbid risk has yet to be investigated.

Confirmed, though not sufficiently clarified, is the genetic hypothesis. The somewhat higher age-adjusted incidence of Alzheimer's disease with females should be mentioned with reservation.

The most important risk factor is old age. The exponential increase in incidence rates, which does not become obvious until the 7th decade of life, and the ensuing accumulation of prevalence rates of Alzheimer's dementia clearly indicate that this disease will continue to pose considerable human as well as economic problems, at least so long as we have not succeeded in finding an adequate preventive or therapeutic approach.

The lack of knowledge in the field of epidemiology of Alzheimer's disease, especially of peripheral markers and true risk factors for the disease, is still vast. This is partly due to the fact that research into late-life dementia has only recently started intensively. The problem itself had become evident much earlier with the ageing of populations. Apparently, scientists find it hard to quickly react to newly emerging problems. But now that the interest of researchers has been awakened and the necessary instruments are available, let us hope that governments, too, will react rapidly enough to facilitate the progress urgently needed in the research of Alzheimer's disease.

References

Alzheimer A (1907 a) Über eine eigenartige Erkrankung der Hirnrinde. All Z Psychiatr 64:146–148

Alzheimer A (1907 b) Über eine eigenartige Erkrankung der Hirnrinde. Zbl Ges Neurol Psych 18:177–179

Amaducci LA, Fratiglioni L, et al (1986) Risk factors for clinically diagnosed Alzheimer's disease: a case-control study of an Italian population. Neurology 36:922–931

Barclay LL, Zemcov A, Blass JP, Sansone J (1985) Survival in Alzheimer's disease and vascular dementias. Neurology 35:834–840

Bergmann K, Kay DWK, Foster EM, et al (1971) A follow-up study of randomly selected community residents to assess the effects of chronic brain syndrome and cerebrovascular disease. Psychiatry II. Proceedings of the 5th World Congress of Psychiatry, Mexico. Excerpta Medica, Amsterdam, pp 856–865

Bharucha NE, Schoenberg BS, Kokmen E (1983) Dementia of Alzheimer's type (DAT): a case-control study of association with medical conditions and surgical procedures. Neurology 33 [Suppl 2]:85

Bickel H (1987) Psychiatric illness and mortality among the elderly. Findings of an epidemiological study. In: Cooper B (ed) Psychiatric epidemiology: progress and prospects. Croom Helm, London, pp 192–211

Bickel H, Cooper B (1989) Incidence of dementing illness among persons aged over 65 years in an urban population. In: Cooper B, Helgason T (eds) Epidemiology and the prevention of disorders. London, Routledge

Broe GA, Akhtar AJ, Andrews GR, Caird FI, Gilmore AJJ, McLennan WJ (1976) Neurological disorders in the elderly at home. J Neurol Neurosurg Psychiatry 39:362–366

Campbell AJ, McCosh LM, Reinkew J, Allan BC (1983) Dementia in old age and the need for services. Age Ageing 12:11–16

Candy JM, Oakley AE, Klinowski J, Carpenter TA, Perry RH, Atack JR, Perry EK, Blessed G, Fairbairn A, Edwardson JA (1986) Aluminosilicates and senile plaque formation in Alzheimer's disease. Lancet i:354–357

Chandra V, Philipose V, Bell PA, Collings F, Lazaroff A, Schoenberg BS (1987) Case-control study of late-onset "probable Alzheimer's disease". Neurology 37:1295–1300

Cooper B (1984) Home and away: the disposition of mentally ill old people in an urban population. Soc Psychiatry 19:187–196

Cooper B (1989) The epidemiological contribution to research on late-life dementia. In: Williams P, Wilkinson G, Rawnsley K (eds) Scientific approaches in epidemiological and social psychiatry. Tavistock, London

Cooper B, Bickel H (1989) Prävalenz und Inzidenz von Demenzerkrankungen in der Altenbevölkerung. Ergebnisse einer populationsbezogenen Längsschnittstudie in Mannheim. Nervenarzt 60:472–482

Cooper B, Sosna U (1983) Psychische Erkrankung in der Altenbevölkerung. Nervenarzt 54:239–249

Copeland JM, Dewey ME, Wood N, Seale R, Davidson IA, McWilliam C (1987) Range of mental illness among the elderly in the community. Prevalence in Liverpool using the GMS-Agecat Package. Br J Psychiatry 150:815–823

Corsellis JAN, Bruton DJ, Freeman-Browne D (1973) The aftermath of boxing. Psychol Med 3:270–303

Diesfeldt HFA, van Houte LR, Moerkens RM (1986) Duration of survival in senile dementia. Acta Psychiatr Scand 73:366–371

Edwardson JA, Oakley AE, Taylor GA, Ward MK, Chalker P, Bishop H, Thompson J, McArthur FK, Candy JM (1988) Senile plaques, A4 amyloid deposition and trace element accumulation in brains of patients on chronic renal dialysis. Lancet

Essen-Möller E, Larsson H, Uddenberg C-E, White G (1956) Individual traits and morbidity in a Swedish rural population. Acta Psychiatr Neurol Scand 100 [Suppl]

Flaten TP (1988) Geographical associations between aluminium in drinking water and registered death rates with dementia (including Alzheimer's disease) in Norway. In: Proceedings from the Second International Symposium on Geochemistry and Health (London 1987). Geochemistry and Health Monograph Series, Science Reviews, London

Folstein MF, Anthony JC, Parked I, et al (1985) The meaning of cognitive impairment in the elderly. J Am Geriatr Soc 33:228–235

Hachinski VC, Iliff LD, Cihak E, du Boulay GH, McAllister VL, Marshall J, Russel RW, Symon L (1975) Cerebral blood flow in dementia. Arch Neurol 32:632–637

Hasegawa K (1974) Aspects of community mental health care of the elderly in Japan. Int J Ment Hlth 8:36–49

Heston LL, Mastri AR, Anderson VE, White J (1981) Dementia of the Alzheimer's type: clinical, genetics, natural history, and associated conditions. Arch Gen Psychiatry 38:1085–1090

Heyman A, Wilkinson WE, Stafford JA, Helms MJ, Sigmon AH, Weiberg T (1984) Alzheimer's disease: a study of epidemiological aspects. Ann Neurol 15:334–341

Jorm AF (1987) Understanding senile dementia. Croom Helm, London Sydney

Jorm AF, Henderson AS, Jacomb PA (1989) Regional differences in mortality from dementia in Australia: an analysis of death certificate data. Acta Psychiatr Scand art. 1138

Jorm AF, Korten AE, Henderson AS (1987) The prevalence of dementia: a quantitative integration of the literature. Acta Psychiatr Scand 76:465–479

Jorm AF, Korten AE, Jacomb PA (1988) Projected increases in the number of dementia cases for 29 developed countries: application of a new method for making projections. Acta Psychiatr Scand 78:493–500

Kaneko Z (1975) Care in Japan. In: Howells JG (ed) Modern perspectives in the psychiatry of old age. Brunner Mazell, New York, pp 519–539

Kay DWK, Beamish P, Roth M (1964) Old-age mental disorders in Newcastle upon Tyne. I. A study of prevalence. Br J Psychiatry 110:146–158

Klatzo I, Wisniewski H, Steicher E (1965) Experimental production of neurofibrillary degeneration. J Neuropathol Exp Neurol 24:187–199

Martyn CN, Osmond C, Edwardson JA, Barker DJP, Harris EC, Lacey RF (1989) Geographical relation between Alzheimer's disease and aluminium in drinking water. Lancet i:59–62

Martyn CN, Pippard EC (1988) Usefulness of mortality data in determining the geography and time trends of dementia. J Epidemiol Community Health 42:134–137

McKhann G, Drachman D, Folstein M, Katzman R, Price D, Stadlan EM (1984) Clinical diagnosis of Alzheimer's disease: report of the NINCDS-ADRDA work group under the auspices of Department of Health and Human Services Task Force on Alzheimer's disease. Neurology 34:939–944

Mölsä PK, Paljärvi L, Rinne JO, Rinne UK, Säkö E (1985) Validity of clinical diagnosis in dementia: a prospective clinicopathological study. J Neurol Neurosurg Psychiatry 48:1085–1090

Mortimer JA, French LR, Hutton JT, Schuman LM (1985) Head injury as a risk factor for Alzheimer's disease. Neurology 35:264–267

Murray RM, Greene JC, Adams JH (1971) Analgesic abuse and dementia. Lancet i:242–245

Nielsen J (1962) Geronto-psychiatric period-prevalence investigation in a geographically delimited population. Acta Psychiatr Scand 38:307–330

Nielsen JA, Biorn-Henriksen T, Bork BR (1982) Incidence and disease expectancy for senile and arteriosclerotic dementia in a geographically delimited Danish rural population. In: Magnussen J, Nielsen J, Buch J (eds) Epidemiology and prevention of mental illness in old age. Hellerup, Denmark. EGV: 52–53

Primrose EJR (1962) Psychological illness: a community study. Tavistock, London

Roth M, Tym E, Mountjoy CQ (1986) CAMDEX: a standardized instrument for the diagnosis of mental disorder in the elderly, with special reference to the early detection of dementia. Br J Psychiatry 149:698–709

Rudelli L, Strom JO, Welch PT, Ambler MW (1982) Posttraumatic premature Alzheimer's disease: neuropathologic findings and pathogenetic considerations. Arch Neurol 39:570–575

Shalat SL, Selzer B, Pidcock C, Baker EL (1986) A case-control study of medical and familial history and Alzheimer's disease. Am J Epidemiol 124:540

Shore D, Wyatt RJ (1983) Aluminium and Alzheimer's disease. J Nerv Ment Dis 171:553–558

Sternberg E, Gawrilowa S (1978) Über klinisch-epidemiologische Untersuchungen in der sowjetischen Alterspsychiatrie. Nervenarzt 49:347–353

St. George-Hyslop PH, Tanzi RE, Polinsky RJ, Haines JL, Nee L, et al (1987) The genetic defect causing familial Alzheimer's disease maps on chromosome 21. Science 235:885–890

Sulkava R, Wikström J, Aromaa A, Raitasalo R, Lehtinen V, Latheza K, Palo J (1985) Prevalence of severe dementia in Finland. Neurology 35:1025–1029

Tierney MC, Fisher RH, Jewis AJ, Zorzitto ML, Snow WG, Reid DW, Nieuwstraten P (1988) The NINCDS-ADRDA work group criteria for the clinical diagnosis of probable Alzheimer's disease: a clinicopathologic study of 57 cases. Neurology 38:359–364

Tomlinson BE, Irving D, Blessed G (1970) Observations on the brains of demented old people. J Neurol Sci 11:205–242

Tomlinson BE, Henderson G (1976) Some quantitative findings in normal and demented old people. In: Terry RD, Gershon S (eds) Neurobiology of aging. Raven Press, New York, pp 183–204

Weyerer S (1983) Mental disorders among the elderly. True prevalence and use of medical services. Arch Gerontol Geriatr 2:11–22

Whalley LJ, Carothers AD, Colleyer S, De Mey R, Frackiewicz A (1982) A study of familial factors in Alzheimer's disease. Br J Psychiatry 140:249–256

Correspondence: Prof. Dr. Dr. H. Häfner, Central Institute of Mental Health, P.O. Box 12 21 20, D-6800 Mannheim 1, Federal Republic of Germany.

Descriptive and analytic epidemiology of Alzheimer's disease

L. Amaducci[1] and A. Lippi[2]

[1] Department of Neurology and Psychiatry, University of Florence and
[2] S.M.I.D. Centre (Italian Multicentre Study on Dementia), Florence, Italy

Summary

Alzheimer's disease represents the most frequent cause of dementia in old age. Available prevalence ratios (cases per 100 population aged 65 and over) range between 0.6 in China (Shen et al., 1987), and 5.8 in Scotland (Broe et al., 1976). This variability can be influenced by different definitions of the disease under study and different case ascertainment procedures adopted. All the studies consistently showed an exponential increase of the prevalence ratios with age, and, in many studies, the age-specific prevalence ratios were consistently higher in women than in men. In analogy with the prevalence pattern, the incidence rates for Alzheimer's disease raise exponentially with age and are higher in females. The rate can be estimated around 0.01 per 1000 in the age 40–60-year group and it raises to 10 per 1000 in subjects aged 75 years and over (Mas et al., 1987).

As far as risk factors for Alzheimer's disease are concerned, our knowledges principally base on findings derived from case-control studies. A family history of dementia and a previous head trauma were found to be associated with Alzheimer's disease in case-control studies. Indications of the role of genetic factors in the etiology of Alzheimer's disease also derived from genetic studies, which demonstrated a genetic defect localized on chromosome 21 in some families with an high incidence of Alzheimer's disease. Other factors found associated with Alzheimer's disease in case-control studies are smoking, and exposure to aluminum. The comprehension of these and other hypothesized risk factors need further investigations.

Introduction

The world population aged 60 years and over was approximately 371 million in 1980. By 1985, the number had reached 415 million. It is expected to reach 1.1 billion by the year 2025 (United Nations, 1986).

This phenomenon due to reductions in fertility, in infant mortality, and in deaths from infectious diseases is becoming of great importance in developed and also in developing countries. In 1980, 6% of the world's total population, aged 60 years and over, lived in developing countries. In 1985 this percentage has reached 6.3%. By the year 2025, people aged 60 and over, living in the developing regions are expected to represent 11.9% of the world total population (United Nations, 1986).

The increasing number of elderly people will determine an increased number of subjects at risk for age-associated disorders of the nervous system including dementia. Dementia is a clinical syndrome character-ized by memory loss associated with impairment in abstract thinking and judgment, disturbances in higher cortical functions, and personality change (American Psychiatric Association, 1987). This clinical syndrome may have several causes. Dementing disorders were classified as primi-tive and secondary according to the etiology (Wells, 1979). More recent-ly the National Institutes of Health (U.S.A.) (1988) in the report of the Consensus Conference on the differential diagnosis of dementing dis-eases, held in July 1987, proposed the distinction between diseases that appear to be primary in the brain and those which are outside the brain and affected it secondarily; a fruitfull distinction for clinical purposes was also made between progressive or fixed dementing pathological states, and arrestable or reversible causes of dementias. Arrestable or reversible causes of dementia include intoxications, infections, metabolic disorders, nutritional disorders, vascular, space-occupying lesions, nor-male pressure hydrocephalus and affective disorders.

These arrestable or reversible dementias represent 10–15 per 100 of all causes of dementia and every effort must be done in recognize them, because they are potentially reversible if an appropriate therapy is ap-plied (Mardsen and Harrison, 1972).

However, the most frequent causes of dementia, as documented by clinical studies of dementia (Wells, 1979), post-mortem examinations (Tomlinson et al., 1970), and population-based surveys (Schoenberg et al., 1987; Lippi et al., 1989) are Alzheimer's disease (AD) and vascular dementia, followed by secondary dementias. Jorm (1987) observed that the relative prevalence of AD and vascular dementia changes across the countries: vascular dementia being more common in Japanese and Rus-sian studies, AD in western European countries, while no significant differences were found in Finnish and American studies. In Italy, AD seems only slightly more frequent than vascular dementia (Lippi et al., 1989).

Descriptive epidemiology of Alzheimer's disease

In descriptive epidemiology of Alzheimer's disease important problems are the choice of the population study and the type of sample, the methodology for cases ascertainment, and the accuracy of clinical diagnosis.

The study population may be obtained from the general population by randomization or by complete enumeration of the subjects. Otherwise, in community surveys all the data from hospitals, nursing homes, and the health authority will be collected. The latter approach was principally used in north-European surveys (Akesson, 1969; Molsa et al., 1982). The risk of this methodology is to underestimate the real frequency of the disease.

To cases ascertainment in population studies, two general strategies have been proposed: the one- and the two-phase approach. The first is less expensive, but the latter consents more accuracy in the diagnosis. In the two-phase approach a brief cognitive test with high sensitivity (the screening test) is administered to all people to be investigated. Subjects scoring under the cut-off level at the screening test, are extensively studied by a clinical point of view to confirm the presense of mental impairment, to diagnose dementia and classify dementias by types. However, the definite diagnosis of dementing disorders, especially AD, need of the pathological examination of the brain.

Several sets of clinical diagnostic criteria have been proposed. Recent population-based surveys utilized the NINCDS-ADRDA criteria (McKhann et al., 1984). The CAMDEX by Roth et al. (1986) provided an usefull set of diagnostic criteria for differential diagnosis of mental disturbances in the elderly, successfully used in some recent British surveys (Brayne and Calloway, 1989).

Prevalence of Alzheimer's disease

Available prevalence ratios (cases per 100 population aged 65 and over) of Alzheimer's disease range between 0.6 in the study of Shen et al. (1987) in China, and 5.8 in the study of Broe et al. (1976) in Scotland. This variability in the prevalence ratios can be influenced by different definitions of the disease under study and different case ascertainment procedures adopted. Some investigations (Akesson, 1969; Molsa et al., 1982; Sulkawa et al., 1985; Schoenberg et al., 1985) considered only severe dementia, while in other studies also mild dementia was included (Kay et al., 1964; Broe et al., 1976). Akesson (1969), who found very low prevalence ratios, adopted very restrictive diagnostic criteria: constant disorientation to time and place was required to make the diagnosis of dementia. Case-ascertainment procedure also differ in various studies. Shen et al. (1987) in Beijing, in a small population size, found a low prevalence ratio, while a more recent Chinese prevalence survey, carried out in a more representative population sample, yielded a prevalence ratio of 1.38 per 100 population aged 55 years and over (Zhang et al., 1990). Molsa et al. (1982) in Finland and Akesson (1969) in Sweden, which found lower prevalence ratios than other North-European surveys, collected data only from health and social services. Sulkawa et al. (1985) in a random sample representative of the all Finnish population, utilizing the door-to-door approach, found higher prevalence ratios than previous community based surveys.

All the studies, carried out until now, consistently showed an exponential increase of the ratios with age. In the study of Molsa et al. (1982) in Finland the prevalence ratio per 100 population increased from 0.025 in the age-class 45–54 to 6.295 in the age-class 85 and over. In an Italian prevalence survey the prevalence ratio was 6.2 per 100 for subjects aged 60 and over and rose to 8.4 per 100 in those aged 65 and over (Rocca et al., 1990). In China Shen et al. (1987) did not find dementia cases in the age group 65–74, while all cases were observed in the age-group 75 years and over.

In many studies the age-specific prevalence ratios were consistently higher in women than in men, especially in the older age-groups (Broe et al., 1976; Molsa et al., 1982; Sulkava et al., 1985). Molsa et al. (1982) found age-specific prevalence ratio to be higher for men in the age group 45 to 54 and higher for females in the other age groups. Kaneko (1975) reported in Japan prevalence ratios consistently higher for males, while a more recent Japanese survey showed a pattern consistent with those reported in European and American studies (Karasawa et al., 1982). Finally, Schoenberg et al. (1985) in a population-based survey carried out in Copiah County, Mississipi, found consistent higher prevalence ratios of AD for females and blacks.

Incidence of Alzheimer's disease

In analogy with the prevalence pattern, the incidence rate for AD raises exponentially with age. As summarized by Mas et al. (1987) the incidence rate can be estimated to be about 0.01 per 1000 in the age group 40–60 years. It raises to 1 per 1000 in the age group 65–74 years and reaches 10 per 1000 in subjects aged 75 years and over.

Akesson (1969) in Sweden in a community based survey found incidence rates rapidly increasing with age from 0.29 per 1000 in the age group 60–69 years to 4.58 per 1000 in the ages over 80. The rates were consistently higher in females in all age groups. Molsa et al. (1982) in Finland, in a community based survey, found that the age-specific annual incidence rates for AD were 0.06 cases per 1000 in the age-class 45–54 and raised to 11.44 per 1000 in the age-group 85 and over. The incidence rates were higher in females, except for the 45–54 age group. Nilsson (1984) in Sweden in a sample of 70 to 79 years old people showed a rapid increase of the incidence rates with age, from 3.6 per 1000 in subjects aged 70–74 to 13.3 in those aged 75–79, and higher rates in men. Treves et al. (1986) in Israel, using the Israeli National Neurological Register as source of cases, found lower incidence rates, but increasing with age: from 0.01 cases per 1000 in the age class 45–49, to 0.87 per 1000 in subjects aged 60 years and over. Females showed higher incidence rates after the age of 59. Moreover, in this study the age-specific indicence rates were higher in the European-American born citizens, than in those Afro-Asian-born.

No significant variations seem to have been in AD incidence rates across the years. Kokmen (1988) in Rochester, Minnesota, had determined the age- and sex-specific annual incidence rates for dementing disorders and AD starting between 1960 and 1974, showing that the incidence of dementia overall or dementia due to AD had not particularly increased in this 15 years period: Schoenberg and collegues in the period 1960–1964 in Rochester, found an annual incidence rate of 123.2 per 100000 in the population over 30 years. Kokmen in the period 1970–1974 using the same methodology found an incidence rate of 147.2 per 100000 in the same age-group. Both the studies showed higher incidence rates in women than in men.

Analytic epidemiology of Alzheimer's disease

Our knowledges on risk factors for AD are principally based on findings derived from case-control studies. In these retrospective studies one group of subjects with the disease (cases) are compared with one or more groups of subjects without the disease (controls) in respect to the exposure, in their previous life, to one or more hypothesized risk factors. Because demented people suffer from memory impairment, informations were usually recorded from a surrogate respondent and the same procedure must be used for controls for comparability reasons. Rocca et al. (1986) assessed the reliability of surrogate respondents to provide data for the specific items of a case-control study of AD conducted by Amaducci et al. (1986) in Italy. Per 100 agreement was greater than 80 for the majority of questions. Lower agreement was observed for alcohol consumption and the number of cigarettes per day, and was very poor for information about antacid drug use.

The case-control studies carried-out until now pointed out as risk factors for AD a positive family history of dementia and a previous head trauma in the history of the subject.

In the case-control study by Amaducci et al. (1986) the presence of dementia in any first-degree relative yielded an odds ratio equal to 5 in the comparison with hospital controls and over 2 in the comparison with population controls, while the presence of dementia in any siblings yielded odds ratios equal to 11 in the comparison with hospital control and over 5 in the comparison with population controls. Similar findings were found in other case-control (Heyman et al., 1984; Graves et al., 1987; Shalat et al., 1987) and comparative studies (Heston et al., 1981; Whalley et al., 1982; Breitner et al., 1984). Mortimer (1989), using the Mantel-Haenszel statistical method (Schlesselman, 1982), carried out a meta-analysis of case control studies of AD. He found a summary odds ratio (95% C.I.) for the risk factor "family history of dementia" equal to 3.96 (2.84, 5.52). Indication of the role of genetic factors in the etiology of AD derived also by pedigree studies, twin studies, and genetic studies. Rocca and Amaducci (1988) reviewed from an epidemiological point of view all the contributions to the problem of familial aggregation of AD. The authors suggested that at least 3 types should be considered regarding occurrence: autosomal dominant, familial and sporadic. The autosomal dominant type is linked to a genetic defect located on chromosome 21, at least in the families studied (St. George-Hyslop et al., 1987). The familial type individuates cases of AD with familial aggregation, but not with a clear inheritance. These cases could be have a polygenic origin. Finally sporadic cases could be due to environmental risk factors or to a combination of genetic and environmental risk factors.

Among hypothesized environmental risk factors for AD, a previous head trauma in the history of the subject was found significantly associated with the disease in case-control studies (Heyman et al., 1984; French et al., 1985; Mortimer et al., 1985; Graves et al., 1987). In some studies (Amaducci et al., 1986; Chandra et al., 1987 a, b; Shalat et al., 1987) head trauma was more frequent in cases in respect to the controls, but the difference not reached the statistical significance. On the other hand, several studies were not able to find any association between head trauma and AD. However, Mortimer (1989) observed that the association with head trauma is highly significant ($p < 0.001$) when considered across studies, despite the fact that the majority of studies reported no-significant associations for this risk factor.

Other factors found associated with AD in case-control studies are smoking (Shalat et al., 1987), and exposure to aluminum (Graves et al., 1987). These associations were negative in all the other case-control studies (Bharucha et al., 1983; Heyman et al., 1984; French et al., 1985; Amaducci et al., 1986; Chandra et al., 1987b).

References

Amaducci LA, Fratiglioni L, Rocca WA, Fieschi C, Livrea P, Pedone D, Bracco L, Lippi A, Gandolfo C, Bino G, Prencipe M, Bonatti ML, Girotti F, Carella F, Tavolato B, Ferla S, Lenzi GL, Carolei A, Gambi A, Grigoletto F, Schoenberg BS (1986) Risk factors for clinically diagnosed Alzheimer's disease: a case-control study of an Italian population. Neurology 36:922–931

American Psychiatric Association (1987) Diagnostic and statistical manual of mental disorders, 3rd edn (revised). American Psychiatric Association, Washington, p 103

Akesson HO (1969) A population study of senile and arterioslerotic psychoses. Hum Hered 19:546–566

Bharucha NE, Schoenberg BS, Kokmen E (1983) Dementia of Alzheimer's type (DAT): a case-control study of association with medical conditions and surgical procedures. Neurology 33 [Suppl 2]:85

Brayne C, Calloway P (1989) An epidemiological study of dementia in a rural population of elderly women. Br J Psychiatry 155:214–220

Breitner JCS, Folstein MF (1984) Familial Alzheimer dementia: a prevalent disorder with specific clinical features. Psychol Med 14:63–80

Broe GA, Akhtar AJ, Andrews GR, Caird FI, Gilmore AJJ, McLennan WJ (1976) Neurological disorders in the elderly at home. J Neurol Neurosurg Psychiatry 39:362–366

Chandra V, Kokmen E, Schoenberg BS (1987a) Head trauma with loss of consciousness as a risk factor for Alzheimer's disease using prospectively collected data. Neurology 37 [Suppl 1]:152

Chandra V, Philipose V, Bell PA, Lazaroff A, Schoenberg BS (1987b) Case-control study of late onset 'probable Alzheimer's disease'. Neurology 37:1295–1300

French LR, Schuman LM, Mortimer JA, Hutton JT, Boatman RA, Christians B (1985) A case-control study of dementia of the Alzheimer type. Am J Epidemiol 121:414–421

Graves AB, White E, Koepsell T, Reifler B (1987) A case-control study of Alzheimer's disease. Am J Epidemiol 126:754

Heyman A, Wilkinson WE, Stafford JA, Helms MJ, Sigmon AH, Weinberg T (1984) Alzheimer's disease: a study of epidemiological aspects. Ann Neurol 15:335–341

Heston LL, Mastri AR, Anderson VE, White J (1981) Dementia of the Alzheimer type: clinical genetics, natural history, and associated conditions. Arch Gen Psychiatry 38:1085–1090

Jorm AF, Korten AE, Henderson AS (1987) The prevalence of dementia: a quantitative integration of the literature. Acta Psychiatr Scand 76:465–479

Kaneko Z (1975) Care in Japan. In: Howells JG (ed) Modern perspectives in the psychiatry of old age. Brunner/Mazel, New York, pp 519–530

Karasawa A, Kawashima K, Kasahara H (1982) Epidemiological study of the senile in Tokio metropolitan area. In: Ohashi X (ed) Proceedings of the World Psychiatric Association Regional Symposium, Tokio 1982. The Japanese Society of Psychiatry and Neurology, pp 285–289

Kay DWK, Beamish P, Roth M (1964) Old age mental disorders in Newcastle upon Tyne. Part I. A study of prevalence. Br J Psychiatry 110:146–158

Kokmen E, Chandra V, Schoenberg BS (1988) Trends in incidence of dementing illness in Rochester, Minnesota, in three quinquennial periods, 1960–1974. Neurology 38:975–980

Mardsen CD, Harrison MJG (1972) Outcome of investigation of patients with presenile dementia. Br Med J 2:249

Mas JL, Alperovitch A, Derouesne C (1987) Epidemiologie de la demence de type Alzheimer. Rev Neurol 3:161–171

McKhann G, Drachman D, Folstein M, Katzman R, Price D, Stadlan EM (1984) Clinical diagnosis of Alzheimer's disease: report of the NINCDS-ADRDA Work Group under the auspices of Department of Health and Human Service Task Force on Alzheimer's disease. Neurology 34:939–944

Molsa PK, Marttila RJ, Rinne UK (1982) Epidemiology of dementia in a Finnish population. Acta Neurol Scand 65:541–552

Mortimer JA, French LR, Hutton JT, Schuman LM (1985) Head injury as a risk factor for Alzheimer's disease. Neurology 35:264–267

Mortimer JA (1989) Epidemiology of dementia: cross-cultural comparisons. In: Wurtman RJ, Corkin S, Growdon JH, Ritter-Walker E (eds) Alzheimer's disease: advances in basic research and therapies. Proceedings of the Fifth Meeting of the International Study Group on the Pharmacology of Memory Disorders Associated with Aging, Zürich 1989. Center for Brain Sciences and Metabolism, Charitable Trust, Cambridge MA, pp 51–74

National Institutes of Health Consensus Development Conference Statement (1988) Differential diagnosis of dementing diseases. Alz Dis Assoc Disorders 2:4–28

Nilsson LV (1984) Incidence of severe dementia in an urban sample followed from 70 to 79 years of age. Acta Psychiatr Scand 70:478–486

Rocca WA, Amaducci L (1988) The familial aggregation of Alzheimer's disease: an epidemiological review. Psychiatr Dev 1:23–36

Rocca WA, Fratiglioni L, Bracco L, Pedone D, Groppi C, Schoenberg B (1986) The use of surrogate respondents to obtain questionnaire data in case-control studies of neurological diseases. J Chron Dis 39:907–912

Rocca WA, Bonaiuto S, Lippi A, Luciani P, Turtu F, Cavarzeran F, Amaducci L (1990) Prevalence of clinically diagnosed Alzheimer's disease and other dementing disorders: a door-to-door survey in Appignano, Macerata Province, Italy. Neurology 40:626–631

Roth M, Tym E, Mountjoy CQ, Huppert FA, Hendrie H, Verma S, Goddard R (1986) CAMDEX, a standardized instrument for the diagnosis of mental disorder in the elderly with special reference to the early detection of dementia. Br J Psychiatry 149:698–709

Schlesselman JJ (1982) Case-control studies: design, conduct, analysis. Oxford University Press, New York

Schoenberg BS, Anderson DW, Haerer AF (1985) Severe dementia. Prevalence and clinical features in a biracial US population. Arch Neurol 42:740–743

Schoenberg BS, Kokmen E, Okazaki H (1987) Alzheimer's disease and other dementing illnesses in a defined United States population: incidence rates and clinical features. Ann Neurol 22:724–729

Shalat SL, Seltzer B, Pidcock C, Baker EL Jr (1987) Risk factors for Alzheimer's disease: a case-control study. Neurology 37:1630–1633

Shen YC, Li G, Chen CH, Zhao YW (1987) An epidemiological survey on age-related dementia in an urban area of Beijing. Presented at the WHO International Workshop on Epidemiology of Mental and Neurological Disorders of the Elderly. Beijing, November 16–20, 1987

St George Hyslop PH, Tanzi RE, Polinsky RJ, Haines JL, Nee L, Watkins PC, Myers RH, Feldman RG, Pollen D, Drachman D, Growdon J, Bruni A, Foncin JF, Salmon D, Frommelt P, Amaducci L, Sorbi S, Piacentini S, Stewart GD, Hobbs WJ, Conneally PM, Gusella JF (1987) The genetic defect causing familial Alzheimer's disease maps on chromosome 21. Science 235:885–890

Sulkava R, Wikstrom J, Aromaa A, Raitasalo R, Lehtinen V, Lahtela K, Palo J (1985) Prevalence of severe dementia in Finland. Neurology 35:1025–1029

Tomlinson BE, Blessed G, Roth M (1970) Observations on the brains of demented old people. J Neurol Sci 11:205

Treves T, Korczyn A, Zilber N, Kahana E, Leibowitz Y, Alter M, Schoenberg BS (1986) Presenile dementia in Israel. Arch Neurol 43:26–29

United Nations (1986) Report of the Interregional Seminar to promote the implementation of the International Plan of Action on Aging. United Nations Publication Sales No. E. 86 IV. 5

Wells CE (1979) Dementia. Davis, Philadelphia

Whalley LJ, Carothers AD, Colleyr S, De Mey R, Frackiewicz A (1982) A study of familial factors in Alzheimer's disease. Br J Psychiatry 140:249–256

Zhang M, Katzman R, Salmon D, Jin H, Cai G, Wang Z, Qu G, Grant I, Yu E, Levy P, Klauber MR, Liu WT (1990) The prevalence of dementia and Alzheimer's disease (AD) in Shangai, China: impact of age, gender, and education. Ann Neurol (in press)

Correspondence: Dr. L. Amaducci, Department of Neurological and Psychiatric Sciences, University of Florence, Viale Morgagni 85, I-50134 Florence, Italy.

A proposed classification of familial Alzheimer's disease based on analysis of 32 multigeneration pedigrees

T. D. Bird, J. P. Hughes, S. M. Sumi, D. Nochlin, G. D. Schellenberg, T. H. Lampe, and E. J. Nemens

University of Washington Medical School and VA Medical Center, Seattle, WA, U.S.A.

Summary

We propose a classification of familial Alzheimer's disease (FAD) based on our clinical and pathological analysis of 32 kindreds with multigeneration dementia. The classification can briefly be stated as follows I. Early onset FAD with mean age of onset before 50 years; II. Late onset FAD with mean age of onset after 60 years; III. Intermediate/variable onset FAD; IV. Familial dementia with atypical Alzheimer changes; V. FAD without multigeneration involvement.

Introduction

Alzheimer's disease (AD) can be briefly defined as slowly progressive dementia associated with the neuropathological changes of cortical neuronal loss, A4 amyloid plaques and neurofibrillary tangles (NFT). Within this broad definition there is considerable evidence for both clinical and pathological heterogeneity. Examples of such heterogeneity include patients with "sporadic" disease, familial disease, early age of onset, late age of onset, early and prominent aphasia, myoclonus, rigidity, primarily amyloid plaques without NFT, multiple amyloid plaques in the cerebellum and severe congophilic angiopathy. The pathogenetic significance of these differences remains to be determined. However, the rapidly accumulating literature on familial Alzheimer's disease (FAD) lacks a consistent manner in which various investigators can contrast and compare the various types of kindreds under investigation. The purpose of the present study is to propose a classification of FAD that will serve as a guideline for investigators evaluating and reporting the results of studies concerning a wide variety of FAD pedigrees.

Methods

We have systematically evaluated 32 kindreds with either classical or atypical FAD. Additional description of 24 of these families can be found in Bird et al. (1988, 1989). The major criteria for inclusion in our study were that each family have at least two generations of individuals affected with dementia and at least one neuropathological specimen from each family. Therefore, by definition, each family had a minimum of two demented individuals. When including only those family members for whom we could estimate an age of onset there were 200 such demented persons in the 32 families. The number of affected individuals in each family with an estimated age of onset ranged from 2 to 17 with a mean of 6.25 such demented persons per family. We reviewed 62 neuropathological specimens or reports from 31 families. A single family (the HD pedigree) has had no autopsy as yet, but is related to the other Volga German pedigrees in the study and is included for that reason.

The families in this study were not ascertained at random. They were selected from more than 100 kindreds seen in our clinics and referred by colleagues in which a proband with the clinical diagnosis of AD had another family member with a history of dementia. We were specifically searching for FAD pedigrees.

Clinical criteria for probable AD were generally those suggested by Mc-Khann et al. (1983) except that complete examination and testing were usually not available for persons who died many years ago. Individuals were not considered affected unless the dementia was documented by records or confirmed by at least two separate family members. Age of onset was determined to be that age at which family members and records agreed that the individual first began showing signs of memory loss or behavioral change. Mean age at onset, age at death and duration of disease were computed for each family. Comparisons between means for the families were performed using Students t test and the Newman-Keuls multiple comparison test.

Results

Figure 1 shows the mean age of onset, range in age of onset and mean age at death for 31 families [1 additional family (EF) is included as atypical dementia Type IVA noted below]. The ethnic origin of the families were as follows: British Isles – 10, Scandinavia – 8, Volga German – 7, German – 2, Eastern Europe – 2, Japanese – 1, French Canadian – 1, and Unknown – 2.

Several observations can be made from the Figure. It is noted that some families have a wide range in age of onset, whereas the range in age of onset is quite small for other families. Duration of disease tends to be less variable than age of onset, both within and between families. Mean disease duration for the entire group was approximately 9 years with a standard deviation of about four years. Even so, the shortest

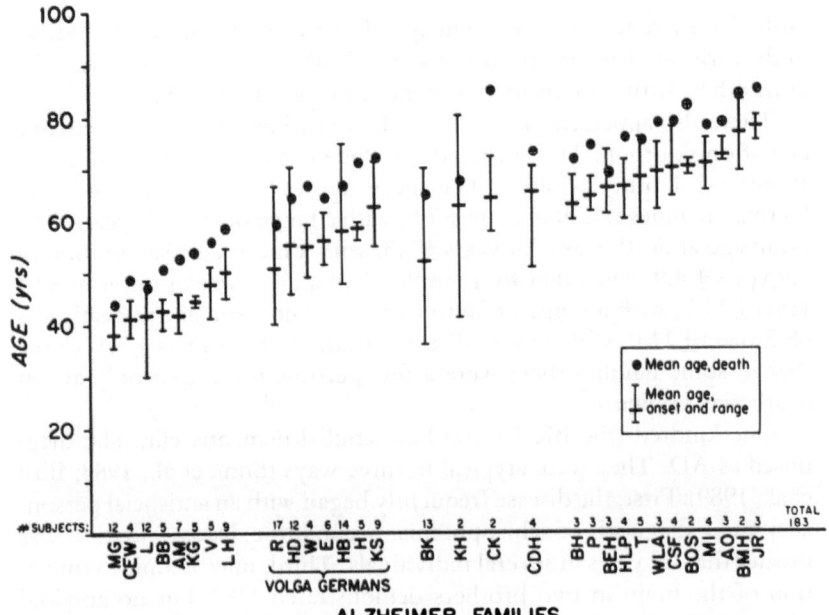

Fig. 1. Graph showing mean age of onset (small horizontal bar) with range for 31 families with FAD. Circle above each range indicates the mean age of death for that family. EF kindred not included

individual duration was one year and the longest was 23 years. Also, the total group could be divided into at least two subgroups: one with generally early onset and one with generally late onset. There also appeared to be a less well-defined intermediate or variable age of onset group. A Newman-Keuls multiple comparison test showed this grouping to be reasonable (although not unique due to overlap of the groups). Further, the differences in age of onset between families did not appear to be the result of differential longevity between the families unrelated to AD. The rank correlation between family age of onset and family age of death (excluding affected individuals) was .22 (p = .10). A more careful analysis on a subset of these families, using survival analysis methods to control for differential longevity, revealed essentially the same groupings (Cupples et al., 1989).

Eight families seemed to define an early onset FAD group. The mean age of onset was 42.9 years ± 4.9, with a range of 30 to 54 years. The mean age of death was 50.0 years ± 5.8, with a mean disease duration of 7.4 years ± 3.0.

A second group consisted of 12 families with generally late onset. The mean age at onset was 69.3 years ± 5.7, with a range of 59 to 83 years.

Note that there was no overlap in age of onset for any of the individuals in the early and late groups. Mean age at death in the late group was 77.5 years ± 5.9, with a mean disease duration of 8.7 years ± 3.9.

There also appeared to be a less well-defined intermediate or variable age of onset group. This included 7 Volga German families (Bird et al., 1988) and 4 other families. The mean age of onset for the 7 Volga German families was 56.8 years ± 7.9, with a range of 40 to 75 years. The mean age at death was 65.4 years ± 7.3, with a mean disease duration of 9.9 years ± 4.9. The other four families had a mean age of onset of 57.1 years ± 11.7, with a range of 36 to 84 years. The mean age at death was 68.5 years ± 11.0, with a mean disease duration of 9.5 years ± 5.1. (Note that in some families there were a few persons for whom only age at death was known.)

One kindred (the BK family) had familial dementia clinically diagnosed as AD. They were atypical in three ways (Sumi et al., 1988; Bird et al., 1989). First, the disease frequently began with an antisocial personality and a diagnosis of schizophrenia. Second, the disease duration was greater than 20 years in several individuals. Third, microscopic examination of the brain in two brothers demonstrated NFT but no amyloid plaques. The BK family is included in the statistical analysis of the intermediate onset families but could also be classified as a Type IV atypical dementia kindred.

The CSF family had late age of onset but also clinical and neuropathological evidence of indolent anterior horn cell disease (Bird et al., 1989). Such a family could be classified as either a Type II late onset kindred or a Type IV atypical family.

The EF family is not included in the Figure but would also be classified as a Type IV atypical dementia kindred. This family had multiple generations of individuals affected with slowly progressive dementia but there was early and prominent aphasia associated with asymmetrical brain atrophy. In addition, neuropathological examination has demonstrated amyloid plaques without NFT and neuronal degeneration in several subcortical nuclei including much of the basal ganglia. Although several members of this family had the clinical diagnosis of AD, they were atypical both clinically and pathologically.

Several small family clusters were seen in which 2 to 4 relatives had the clinical diagnosis of AD, but multigeneration involvement was not documented. That is, only siblings or cousins were demented, without clear parent to child transmission of disease.

Discussion

The results of our investigation lead us to propose the following classification of familial Alzheimer's disease:

I. Early onset FAD

A. Multigeneration involvement with mean age of onset in the family prior to 50 years with no individuals, or only rare persons, with onset after the age of 59. (Rare instances of late onset may be seen because of the coincidental occurrence of "sporadic senile AD" in any given family.) Amyloid plaques and NFT are documented in cerebral cortex in at least one person. Because these are only guidelines, a family with mean age of onset between 50 years and 55 years would probably be acceptable in this category especially if the majority of the family members had an onset prior to the age of 50.

B. Same criteria as I A but no available neuropathologic specimen.

II. Late onset FAD

A. Multigeneration involvement with mean age of onset after 60 years and no person with onset of dementia prior to age 59. Amyloid plaques and NFT are documented in cerebral cortex.

B. Same criteria as II A but no neuropathological specimen available.

III. Intermediate/variable onset FAD

A. Mean age of onset usually in the 50's but with a wide range of onset observed in the family without consistent clustering in early or late ages. Amyloid plaques and NFT are documented in cerebral cortex.

B. Same criteria as III A but no neuropathological specimen available.

IV. Familial dementia with atypical Alzheimer features or with additional system involvement

A. Cortical amyloid plaques and/or NFT associated with additional multisystem degeneration (eg. basal ganglia degeneration). Examples would be family EF noted above and the family reported by Morris et al. (1984).

B. Cortical changes of AD associated with lower motor neuron degeneration. Example includes the CSF family noted above and in Bird et al. (1989).

C. Dementia with cortical NFT but without amyloid plaques. Example includes the BK family noted above and described in Sumi et al. (1988) and Bird et al. (1989).
D. Other additional "atypical" kindreds.

V. FAD without multigeneration involvement (eg. sibs or cousins without clear documentation of affected individuals in two or more generations)

It should be noted that this proposed classification represents guidelines and not hard and fast rules. Some of the families falling within Type IV are so atypical that not all investigators would call them AD, but it should be noted that this category is carefully designated only as "familial dementia" with atypical Alzheimer features (not FAD). Excluded from category IV would be other well-recognized genetic causes of dementia such as Huntington's disease, Gerstmann-Straussler disease, familial CJD and hereditary spinocerebellar degenerations.

This classification has numerous advantages. It takes into consideration clinical, genetic and pathological criteria and is based on a systematic and detailed evaluation of 32 kindreds. Additional pedigrees and variants can be incorporated into the classification as they are reported. The classification of any given family can be changed as additional clinical or pathological material becomes available. Also, one or more categories may become irrelevant and can be incorporated into another category. For example, Type III Intermediate/variable onset FAD represents the least well-defined group, and early onset FAD and late onset FAD may eventually prove to be the only useful categories. Nevertheless, our family material presently best fits these five subgroups.

A shortcoming of this classification is that it is based on a non-randomly ascertained set of pedigrees. This may lead to various selection biases which cannot be completely quantified. For instance, there is a higher probability that families with earlier mean age at onset will be recognized and included in our sample. Consideration of the method of ascertainment is especially important in segregation analysis but may also effect estimates of age at onset (Sobel et al., 1988). For these reasons, the observed classes reported here cannot be construed as evidence for true biologic heterogeneity. Rather, the biological significance of this classification will depend upon accumulating biochemical and molecular evidence for and against heterogeneity in AD.

Acknowledgements

Supported by NIH grants #AG5136, GM15253, Veterans Administration Medical Research Funds, the French Foundation, and the American Health Assistance Foundation.

The following institutions ascertained the indicated AD kindreds: the Clinic For Alzheimer's Disease and Related Disorders, University of British Columbia Health Sciences Centre Hospital (MG and JR families); the Familial Alzheimer's Disease Research Foundation (R family); the Southern California Alzheimer's Disease Research Center Consortium, and the University of Southern California and Rancho Los Amigos Medical Centers (WLA and MI families); the Veterans Administration Medical Center Geriatric Research, Education and Clinical Center, Palo Alto (CDH family); and the Brentwood (Los Angeles) Veterans Administration Medical Center, Neurobehavior Unit (CK family).

The following individuals provided valuable records, autopsy material, or assistance in evaluating the families: H. Chui, R. Cook-Degan, J. Cummings, H. Davies, L. Forno, G. Glenner, R. Haining, E. Larson, C. Masters, G. Miner, M. Patterson, M. Pericak-Vance, E. Reiswig, A. Sadovnick, J. Tinklenberg, J. Weil, N. Zimmerer.

References

Bird TD, Lampe TH, Nemens EJ, Miner GW, Sumi SM, Schellenberg GD (1988) Familial Alzheimer's disease in American descendants of the Volga Germans: probable genetic founder effect. Ann Neurol 23:25–31

Bird TD, Sumi SM, Nemens EJ, Nochlin D, Schellenberg G, Lampe TH, Sadovnick A, Chui H, Miner GW, Tinklenberg J (1989) Phenotypic heterogeneity in familial Alzheimer's disease: a study of 24 kindreds. Ann Neurol 25:12–25

Cupples LA, Terrin NC, Meyers RH, D'Agostino RB (1989) Using survival methods to estimate age-at-onset distributions for genetic diseases with an application to Huntington disease. Genet Epidemiol 6:361–371

McKhann G, Drachman D, Folstein M, Katzman R, Price D, Stadlan EM (1984) Clinical diagnosis of Alzheimer's disease. Neurology 34:939–944

Morris JC, Cole M, Banker BQ, Wright D (1984) Hereditary dysphasic dementia and the Pick-Alzheimer spectrum. Ann Neurol 16:455–466

Sobel E, Davanipour Z, Alter M (1988) Genetic analysis of late-onset diseases using first-degree relatives. Neuroepidemiology 7:81–88

Sumi SM, Nochlin D, Bird TD (1988) Familial presenile dementia with neurofibrillary tangles but without senile (neuritic) plaques: is this familial Alzheimer's disease? Neurology 38 [Suppl 1]:266

Correspondence: T. D. Bird, M. D., Neurology (127), VA Medical Center, 1660 South Columbian Way, Seattle, WA 98108, U.S.A.

Neuropathology

Morphology of Alzheimer's disease and related disorders

K. Jellinger

Ludwig Boltzmann Institute of Clinical Neurobiology, Lainz Hospital, Vienna, Austria

Summary

The diagnosis of Alzheimer's disease (AD) can made with certainty only by histological examination of the brain, either on biopsy or at autopsy, using current criteria that are critically discussed. The morphologic diagnosis of AD is based on the finding of more than certain minimum age-related numbers of neuritic plaques (NP) and neurofibrillary tangles (NFT) that may occur independently from each other and show specific, often bilaterally symmetrical distribution patterns. Although both NFT and NP represent useful diagnostic markers and their concentrations in neocortex correlate reasonably well with certain mental status tests and some biochemical data, progressive loss of neurons and synapses in cortex and brainstem appear to be more correlated with both dementia and neuromediator changes, but their causal relationship with amyloid deposition and cytoskeletal lesions is not clear. Cortical atrophy with 40 to 60% loss of large neurons and 45 to 55% decline of synaptic density in frontal, temporal, parietal cortex and hippocampus are probable correlates of progressive dementia. Many mediator specific brainstem nuclei show progressive neuronal loss and presence of NFT with severe involvement of cholinergic nucleus basalis of Meynert (15 to 90% cell loss and atrophy of large neurons), pedunculopontine and Westphal-Edinger nuclei, serotonergic dorsal raphe nucleus (10 to 76% cell loss), noradrenergic locus ceruleus (40 to 80% cell loss in topographic relation to temporal targets) and, less, the dopaminergic striatonigral system. The morphologic heterogeneity of AD shows overlaps with Parkinson's, diffuse Lewy body disease, cerebrovascular and other brain lesions. In autopsy series, the accuracy rates for clinical diagnosis of AD are about 80 to 85%, but more standardized morphologic criteria are needed.

Introduction

There is general agreement that histopathologic evaluation of the brain, either on biopsy or at autopsy, is an essential prerequisite for the definite diagnosis of Alzheimer's disease (AD). The long accepted view as to the cerebral morphology of AD is that it is featured by extracellular deposition of amyloid forming part of the neuritic plaques (NP) or in the cerebral vasculature, preceded by deposition of A4 precursor protein (A-4P) or β-amyloid peptide (β-AP), and abnormalities of neuronal cytoskeleton producing neurofibrillary tangles (NFT) formed by paired helical and straight filaments containing phosphorylated microtubule-associated tau epitopes. Brain amyloidosis is not necessarily accompanied by NFT formation, and the development and distribution of NP and NFT are independent from each other (Moossy et al., 1988; Probst et al., 1989). The histopathologic diagnosis of AD must be based on the finding of more than certain age-related minimum quantities of NP and NFT in the neocortex and/or hippocampus (Ball, 1989), the quantitation of which, at least in part, depends on the staining methods used: The Bielschowsky and other silver techniques and immunostaining methods have been shown to be most sensitive for detecting both types of AD markers (Davies et al., 1988; Delaere et al., 1989; Wisniewski et al., 1989; Braak et al., 1989; Lamy et al., 1989), although not all silver-positive NP in neocortex contain A-4P (Ferreiro et al., 1989).

The present overview will discuss the following problems of AD:

a) Current criteria for the histological diagnosis of AD;
b) Quantitative morphology related to dementia and neurochemistry;
c) The morphologic heterogeneity or overlaps with other disorders.

a) Diagnostic criteria of AD

Although AD is featured by neocortical, hippocampal and subcortical pathologies, the criteria for the histological diagnosis of AD suggested by the NIH/AARP Working Group are based on age-related neocortical NP counts and do not include assessment of cortical NFT numbers for all age groups, since the committee noted that NFT may be found in individuals of 75 years or older fulfilling the clinical criteria of AD who had large numbers of neocortical NPs. These morphologic criteria (Khachaturian, 1985) have been widely but not universally accepted, since numerous neocortical NP with no or only very few neocortical NFT have been seen in one or two-thirds of non-demented subjects aged 75 to over 85 years (Katzman et al., 1987; Crystal et al., 1988). While Tomlinson and Henderson (1976) saw occasional NFT in the neocortex of 10 to 20% of intellectually normal elderly, they may be absent in 16 to 30% of demented DAT subjects, most being over age 75 and all showing large numbers of NFT in hippocampus (Terry et al., 1987; Morris et al., 1988; Joachim et al., 1988). These patients with "plaque-only" AD are similar to non-demented elderly with cerebral amyloidosis, but do not significantly differ in clinical, other morphological and neurochemical data from the more typical cases with both cortical NP and NFT, although the latter are associated with a tendency toward greater severity. PHF-type neuritic degeneration is accepted as a defining feature of AD (Crystal et al., 1988; Dickson et al., 1989; Delaere et al., 1989), but histopathologic criteria of AD are not uniform (Ball, 1989). Tierney et al. (1988) suggested those for diagnosis of "pure" AD excluding cerebrovascular lesions that may occur in 28 to 63% of AD cases, while concomitant Parkinson's disease (PD) pathology is seen in 15 to 85% of definite AD (Joachim et al., 1988; Jellinger, 1989; Mirra et al., 1989). Hence, more standardized histopathologic criteria for the diagnosis of definite AD are needed. Both NFT and NP that may occur independently from each other show a rather specific distribution pattern that is bilaterally symmetrical in most brain regions except for the density of NFT in hippocampus (Moossy et al., 1988), suggesting spreading of lesions from mediotemporal-hippocampus via connective pathways to neocortical midfrontal and parastriatal parietal areas, while somatosensory and visual cortex remain intact even in late stages of the disease. The density of NFT and, much less, of the NP correlates reasonably well with many mental status tests (Delaere et al., 1989; Lassmann et al., 1990) and some neurochemical data, e.g. reduction of cortical choline acetyltransferase (ChAT) or somatostatin, particularly in young AD cases, while very old SDAT patients show no or very little differences towards non-demented controls (Hansen et al., 1988; Katzman et al., 1988). Consecutive biopsy and autopsy studies show no significant changes in the density of NP and NFT in frontal and temporal cortex, but continuing loss of pyramidal cells related to progressive dementia (Mann et al., 1988). From recent data can be concluded that the demonstration and quantification of NP and NFT are useful morphological markers of AD, but may represent epiphenomena of the disease process that are of limited value as measures for the clinical or biochemical deficits. At present, not only the etiology of AD is obscure, but it is not clear whether and how the pathologic findings of amyloid, NFT and NP, or some or all of the associated neuronal losses, cause impairment of memory, cognition and behavior in patients with AD (Tomlinson, 1989).

b) Quantitative morphology changes of AD (see Table 1)

They include 1. a significant reduction of fresh brain weight by 7.5 to 10% of age-matched controls or 16 to 18% of young adults, decrease of hemispheral volume by 13–18%, with marked atrophy of medial temporal lobes and hippocampus, macroscopically prominent in about 25% of the AD cases, less frequently in frontal and parietal lobes, the latter being particularly involved in rapidly progressive AD. Ventricular volume is increased by 35 to 55%. Weight loss of cerebral cortex is largest in parietal lobe (45–58%), amygdala (45%), hippocampus (53–57%) and temporal cortex (31–45%), less in striatum (26%), while frontal and occipital cortex are not reduced. This is associated with decreased thickness of the cortex by 10–12% and loss of large neurons by 43 to 57% in hippocampus and 40–46% in midfrontal cortex in younger AD as compared to 25% in very old DAT cases (Katzman et al., 1988). Significant decrease of middle parts of corpus callosum corresponds to involvement of parietal and temporal commissural fibers (Weis et al., 1989). 2. In hippocampus, the total loss of neurons is correlated to the density of NFT (see Ball, 1989). It mainly affects the subiculum with total loss of the lamina II and III of the entorhinal cortex of the parahippocampal gyrus, the origin of the perforating pathway, the major source of synaptic terminals in the outer portion of the molecular layer of the dentate gyrus causing severe synaptic loss in this area (Hamos et al., 1989). NP formation in this area is related to the severity of neocortical NP due to deranged input from glutamate-aspartate neurons in the entorhinal cortex entering the dentate gyrus via the perforant pathway (Purohit et al., 1989), a major connection to the neocortex. Isolation of the hippocampus is considered a major cause of cognitive dysfunction in AD (Hyman et al., 1987; Kalus, 1989). 3. Cortical neuronal loss is associated with shrinkage of neurons and reduction of DNS, loss of dendrites and dendritic spines by 17 to 35%, mainly in layers III and IV (Coleman and Flood, 1987), and a fourfold increase of fibrillary astrocytes in frontal cortex replacing neuronal losses. Presynaptic axons show lesions in and around NP, which accumulation of pathologic structures. While ultrastructural studies showed no decrease in cortical synapsis numbers (Adams, 1987), recent immunostaining using synaptophysin demonstrating presynaptic vesicles displayed a constant decline of optical density of synapses in normal aging, with 45–55% loss of immunoreactivity in layers II–V of midfrontal and parietal cortex in AD (Masliah et al., 1989). The causes of synaptic loss are unknown, but Wegiel et al. (1989) suggest that, at least around NP, it may be secondary to synaptic damage due to amyloid deposition. 4. In the brainstem, many projection nuclei show progressive cell loss and presence of NFT that often are related to neuronal loss and AD lesions in cortical target areas and neuromediator changes.

a) Cholinergic system: Significant neuronal loss with frequent NP and NFT are seen in nucleus basalis of Meynert (NBM) ranging from 15 to 90% with average overall loss of 15–30% and much greater loss of large

Table 1. Quantitative morphology of AD/SDAT (for ref. see Jellinger, 1989)

Type of change		Reduction in %
Brain weight		7.5 – 18
Hemispheral volume		13 – 18
Ventricular volume		35 – 55
Weight cerebral cortex		
Inferior parietal		45 – 58
Temporal		20 – 45
Amygdala		47
Thickness of cortex		10 – 15
Numbers of neurons (> 90 μm)		
Hippocampus		43 – 57
Super. Temporal		22 – 40
Frontal		25 – 60
Dendritic spines (n)		
Frontal cortex		17
Temporal cortex		36
Synapses – midfrontal [a]		45 – 55
parietal		
Astroglia Frontal cortex		plus 400
Subcortical nuclei		
N. basalis Meynert (cholinergic) density	N neurons	15/33 – 90
Locus ceruleus (noradrenergic)	N neurons	30 – 85
Dorsal raphe nucl. (noradrenergic)	N neurons	36 – 77
Dorsal vagal nucl. (noradrenergic)	N neurons	44 ±12
Pedunculopent. nucl. (cholinergic)	N neurons	29 – 34
Supraoptic nucl.	N neurons	65
Substantia nigra (dopaminergic)	N neurons	6 – 38

[a] see Masliah et al. (1989)

neurons in intermediate and posterior parts (Ch4p). It is more severe in young AD than in old SDAT cases and similar to that in PD, where it is more severe in demented than in non-demented patients (Table 2). NBM cell loss shows significant correlation with neuronal loss, density of NP and NFT and of presynaptic M 2 muscarinic receptors in temporal cortical target areas suggesting retrograde or anterograde degeneration, while shrinkage of large NBM neurons suggests reduced enzyme synthesis or failure of transport due to neurotrophic factor dysfunction (Allen

Table 2. Neuronal loss in nucleus basalis Meynert in AD/SDAT versus age-matched controls

Disorder	N	Mean age	Neuron loss %	Max. cell density %	Authors	Year
AD	1	74.0	90.0	–	Whitehouse et al.	1981
	5	64.0	75.0	–	Whitehouse et al.	1982
	3	56.6	62.0	–	Wilcock et al.	1982
	14	60.9	70.0	54.0	Arendt et al.	1983
	4	50.5	57.3	–	Tagliavini and Pilleri	1983
	?	?	66.0	–	Nagai et al.	1983
	3	64.0	67.0	–	Rogers et al.	1985
	7	67.0	59.0	–	Jacobs and Butcher	1986
	7	67.2	55.8	–	Ichimiya et al.	1986
	15	63.9	69.5	68.0	Jellinger	1987
	?	?	64.0	–	Rinne et al.	1987
	7	70	61.0	(large)	Allen et al.	1988
SDAT	3	84.7	38.6	–	Wilcock et al.	1982
	5	81.0	35.0	–	Candy et al.	1983
	6	82.0	33.0	–	Perry et al.	1983
	5	75.6	57.3	–	Tagliavini and Pilleri	1983
	22	75.4	59.9	–	Mann et al.	1984
	5	80.0	43.0	–	Jacobs and Butcher	1986
	10	79.1	50.1	–	Mann et al.	1986
	43	80.9	46.6	39.3	Jellinger	1987
	11	?	15.0 36.0	(total) (Ch4p)	Vogels et al.	1989 [a]
Mix	9	83.8	30.0	–	Mann et al.	1986
Parkinson D.			47–77	46–77	Tagliavini and Pilleri Rogers et al.	1984 1985
PD (non demented)			32–49	34–49	Jellinger Ezrin-Waters	1986 1986
PD/SDAT	11	82.6	47.0	–	Jellinger (unpubl.)	

(for ref. see Jellinger, 1989)
[a] not included.

et al., 1988). Oyanagi et al. (1989) reported similar loss of large neurons in NBM and striatum, where many diffuse β-AP plaques are present (Joachim et al., 1989). Although the basal forebrain heavily projects to the reticular thalamic nuclei, in AD these show normal cytoarchitecture with no NP of NFT, but Alz-50 terminal staining that may be related to forebrain lesions (Tourtellotte et al., 1989). The cholinergic parabrachial *pedunculopontine nucleus,* pars compacta, in the dorsolateral part of caudal mesencephalic tegmentum having major projections from basal ganglia/thalamus to cortex, in AD shows a 30 to 40% neuronal loss with

Table 3. Dorsal raphe nucleus: quantitative changes in A.D.

Disease	Neuron loss nucleol. vol. (percent of controls)	% cells	NFT	Author (for ref. see Jellinger, 1989)	Year
AD	17.4± 6 [a]	34.4 [b]		Mann et al.	1984
	36.6 (23–48) [c]		14.6	Tabaton et al.	1985
SDAT	6.1± 1.6	22.9 [b]		Mann et al.	1984
	14.6± 4 [a]	(total)	2.25	Curcio and Kemper	1984
	27.3± 5 [a]	(large)			
	76.9±15 [c]		90.9	Yamamoto and Hirano	1985
	21.0 [b]			Ichimiya et al.	1986
	55.5± 9 [c]			Jellinger	1987
	10 restral				
	15 mid			Zweig et al.	1988
	36 caudal [c]				
	49.4 [c]			Jellinger (unpubl.)	
PD	29–43.7 [c]		4.5	Jellinger	1989

[a] $p < 0.05$; [b] $p < 0.01$; [c] $p < 0.001$

frequent involvement of cells by NFT (Mufson et al., 1988); more severe damage of this area involved in locomotor activities and regulation of sleep-waking cycles, is seen in PD and supranuclear palsy (PSP) (Jellinger, 1989; Zweig et al., 1989). The *supraoptic nucleus* in AD is relatively preserved but also has frequent occurrence of NFT that may cause effects on sleep-waking rhythm (Stopa et al., 1989).

b) In the serotonergic dorsal raphe nucleus, neuronal loss in AD ranges from 10 to 76%, most severe in the caudal parts with many NFT that may involve up to 91% of the DRN cells (Table 3).

c) The noradrenergic locus ceruleus, in AD – less than in PD – shows a significant 40–80% loss of neurons with most severe involvement and greatest density of NFT in the rostral or mid part projecting to hippocampus (Table 4), suggesting retrograde degeneration due to primary damage to its cortical projection areas (Zweig et al., 1988). Most severe lesions in LC occur in AD with major depression (Chan-Palay and Asan, 1989) and 10 to 20 fold decrease of norepinephrine (Zubenko et al., 1989).

d) The dopaminergic striatonigral system that in PD suffers cell losses from 63 to 96%, in AD is less involved with neuron losses ranging from 6 to 36% and less reduction of tyrosin hydroxylase immunoreactivity, while cell loss in ventral tegmental area (VTA) shows cell loss similar to PD ranging from 40 to 70% (Table 5). In the brainstem, there is close relationship between nerve cell losses and the distribution and density of

Table 4. Locus ceruleus: quantitative changes in Alzheimer disease

Disease	N	Mean age	Neuron loss nucleol. vol. (percent of controls)		Author	Year
AD	1	67	79 [c]		Tomonaga	1983
	7	67.2	68.6 [c]		Ichimiya et al.	1986
	1	66	68 [c]	72 central 15 ventral	Ingram et al.	1987
	22	72	70 [c] 81 [c] 67 [c]	rostral mid caudal	Zweig et al.	1988
	1		61.3 [c]		Jellinger (unpubl.)	
SDAT	15	81	51 – 56 [c]		Tomlinson	1981
	24	74 – 87	40 [c]		Vijayshankar and Brody	1979
	20	78 ± 7	80 [c]		Bondareff et al.	1982
	19	84.7	54.4 [c] 82.3 [c]		Mann et al.	1983
	22	74.5	70.3 [c] 68.5 [c]		Mann et al.	1984
	5	75.6	41.2 (30 – 86) [c]		Chui et al.	1986
	9	75.8	33 [c] (neuron atrophy) 55 [c] (mitog. activ.)		Burke et al.	1988
	18	82.4	50.3 [c]		Jellinger (unpubl.)	
	3	76	3.5 – 87.5		Chan-Palay and Asan	1989 [d]
PD			72 – 85 [c] 52 – 74 [c]		Mann et al. Jellinger	1984 1989
PD – PD/AD			26 – 94		Chan-Palay and Asan	1989 [d]

[a] $p < 0.05$; [b] $p < 0.01$; [c] $p < 0.001$; [d] (for ref. see Jellinger, 1989)

NFT that particularly involve nuclei projecting to the cortex, with following descending order of intensity (German et al., 1987; Jellinger, 1989): raphe nuclei (dorsal raphe/supratrochlear nucl.), central superior nucl., pedunculopontine nucl., locus ceruleus, nucl. paranigralis pigmentosus and nucleus paranigralis, interpeduncular, magnocellular reticular nuclei, midline thalamus and lateral hypothalamus, while substantia nigra, zona compacta, thalamic relais and motor nuclei, trochlear and oculomotor complex (except for the Westphal-Edinger n.), and medullary motor nuclei are rarely involved. These data suggest secondary involvement of most brainstem nuclei due to primary cortical AD pathology, while the distribution of NP does not always coincide with that of NFT; it shows the following descending order of intensity (Iseki et al., 1989): periaqueductal gray, superior colliculi, floor of the third and fourth ventricle,

Table 5. Substantia nigra: quantitative changes in Alzheimer disease

Disease	N	Mean age	Neuron loss nucleol. vol. (percent of controls)		Author (for ref. see Jellinger, 1989)	Year
AD	12	66	11.8±3.7 [a]	23.8±4 [c]	Mann et al.	1983
	4	63	38.1 (18–54) [c]		Tabaton et al.	1985
	1	70	47.4 [c]		Jellinger	1987
SDAT	19	85	6.3±3.6	8.4±3.9	Mann et al.	1983
SDAT+AD	22	74.5	7.5 [a]	13.0 [b]	Mann et al.	1984
SDAT+AD	5	75.6	16 (3–72) [a]		Chui et al.	1986
SDAT	16	82	28.1 (22–48) [b]		Jellinger	1987
SDAT	24	82.4	25.5 (20–40) [b]		Jellinger (unpubl.)	
SDAT	37	74.5	25.0 [a]		Gibb and Lees	1989
SDAT	22	72.7	17.0 [a]		Gibb et al.	1989
PD			63–85 [c]		Mann and Yates	1983
					Bogerts et al.	1983
					Gibb and Lees	1987

[a] $p < 0.05$; [b] $p < 0.01$; [c] $p < 0.001$

mamillary bodies and tuberomamillary nuclei, superior central nucl., less severe in reticular formation, substantia nigra, pontine tegmentum and inferior olives. The cerebellum is frequently involved by β-AP containing diffuse NP that differ from PHF-containing NP and amyloid plaques in spongiform encephalopathies (Yamaguchi et al., 1988; Braak et al., 1989; Joachim et al., 1989).

c) Morphologic heterogeneity of AD

The morphologic heterogeneity of AD is increasingly recognized, and other cytoskeletal and neurodegenerative lesions or brain pathology may occur in cases fulfilling the histologic criteria of AD. Lewy bodies (LB), the anatomical hallmark of PD, easily seen by ubiquitin immuncytochemistry (Bancher et al., 1989; Dickson et al., 1989), are found in varying numbers in both brainstem and cererbral cortex (Table 6): LB in brainstem, present in 8 to 25% of normal brains over age 60, are found in 18 to 66% of all cases of AD/DAT with or without considerable nigral damage. They may represent incidental age-related degeneration, preclinical PD (Gibb and Lees, 1988), AD with incidental LB or true combination of AD/PD. Concomitant PD pathology, i.e. nigral damage and LB in various sites, occur in 15 to 85% of definite AD cases

Table 6. Age-specific prevalence of Lewy bodies (percentage) in controls and Alzheimer disease

Age group years	Controls (1)			Alzheimer disease (2)		
	Gibb (1989) (N = 210) Sub. nigra	Jellinger (N = 274) S.N.	L.C.	Gibb-Lees (N = 106) S. nigra	Jellinger (N = 558) S.N.	L.C.
30–49	–	0	0	–	100 (3)	100
50–59	3.8	0	0	0	62.5 (24)	66.7
60–69	4.9	3.7	3.7	15.4	58.5 (94)	68.0
70–79	8.3	9.1	27.3	13.9	53.7 (205)	64.9
80–89	12.8	15.4	32.3	12.9	61.6 (211)	66.8
90–99	11.1	12.5	12.5	14.3	42.9 (21)	61.9
> 60 yr	8.2	12.7	24.0	14.0	57.3 (531)	66.1

(1) Comparison:	Lipkin (1959)	6.8% > 60 yrs
	Woodward (1962)	12% > 60 yrs
(2) Comparison:	Morris et al. (1988)	10/ 26 AD = 38%
	Joachim et al. (1988)	23/131 AD = 18%
	Crystal et al. (1988)	5/ 10 AD = 50%
	Dickson et al. (1989)	11/ 21 AD = 52%

S.N. substantia nigra; *L.C.* locus ceruleus

(Joachim et al., 1988; Mirra et al., 1989; Morris et al., 1988; Jellinger, 1989), and was present in 32% of a personal autopsy cohort of 564 AD/DAT cases (Table 7). On the other hand, in autopsy series of PD individuals, the incidence of additional AD markers ranges from 7 to 47%, and was 53.5% in a series of 306 PD cases (Table 7). Cortical LB are seen in 7–8% of all AD brains (Morris et al., 1988) and in 2.3% of aged controls (Perry et al., 1990). The close relationship between AD and PD is further confounded by diffuse Lewy body disease (DLBD) and various subsets of AD showing a variety of overlapping neuropathologic changes, those of PD, DBLD and localized spongiform changes similar to Creutzfeldt-Jakob disease (JCD) (Burkhardt et al., 1988; Hansen et al., 1989). Vacuolation in AD may be similar to that in JCD, but occurs in proximity to NP with β-AP and PHF-filled neurites, while negative transmission experiments give no evidence for a transmissible spongiform encephalopathy.

Widespread LB in both cortex and brainstem occur in DLBD, an increasingly recognized form of dementia with or without clinical signs of parkinsonism (Kosaka et al., 1988), often associated with AD lesions and similar biochemistry with decreased cortical ChAT activity and somatostatin-like immunoreactivity (Dickson et al., 1989). DLBD, in a

Table 7. Brainstem and other brain lesions in Alzheimer disease, senile dementia, Parkinson's disease, and controls

Type of lesion	Alzheimer disease N=140	SDAT N=424	Parkinson disease N=306	Controls N=286
Mean age (yr ±SEM)	65.0±0.5	81.6±0.6	74.0±0.9	75.0±0.8
S. nigra	(Numbers indicate percent of total cases)			
Neuron loss 3+	8.8	2.8	93.8	0
2+	6.1	6.6	6.2	2.1
1+	71.2	81.9	0	27.0
Lewy bodies	66.3	56.2	99.5	6.0
NFT	82.1	73.8	17.6	1.8
L. ceruleus				
Neuron loss	100	100	100	13.4
Lewy bodies	66.3	65.7	99.2	18.3
NFT	85.6	84.0	16.7	4.1
N. basalis m.	(N=70)	(N=276)	(N=94)	(N=43)
Lewy bodies	8.0	15.0	97.7	7.5
NFT	100	98.6	36.2	7.5
Westphal-Edinger nucl.	(N=22)	(N=58)	(N=30)	
Lewy bodies	10.0	11.0	96.7	N.E.
NFT	100	98.3	16.7	N.E.
Pontine tegmentum				
N. plaques	87.5	87.4	16.3	1.8
NFT	98.0	87.4	21.3	1.9
Lacunes, Basal		31.3	31.5	20.8
Ganglia	11.4	–	–	–
Old infarcts	6.0	8.1	8.2	4.6
Amyloid angiopathy	99.6	98.6	12.1	0.9
Cer. hemorrhages	7.2	10.3	5.3	1.0
Large	5.3	7.0	4.3	0
Small (multiple)	1.9	3.3	1.0	1.0
Cort. Lewy bod.	2.9	1.9	7.0	0
PD/AD	14.9/16.9			
(<70 yrs)	clin/anat.	–	18.6	0
PD/DAT		17.0/9.6		
(>70 yrs)	–	clin/anat.	35.0	0

Values given in percent of total cases; *NE* not examined; *NFT* neurofibrill. tangles; *N* plaques = neuritic plaques

recent series of 216 degenerative brain diseases, was the second most frequent cause of primary degenerative dementia after AD (Dickson et al., 1989), while it is rare in other autopsy series of dementia (Morris et al., 1988; Joachim et al., 1988; Boller et al., 1989; Jellinger, 1989). Perry et al. (1990) suggest that senile dementia of Lewy body type may comprise up to 20% of hospitalized demented old people. Most of the DLBD cases show neocortical plaques with lack of PHF-type neuritic changes in the gray matter, thus representing "plaque only" AD with extensive amyloid angiopathy but sparing of the hippocampus, while only a minority show neocortical NFT, possibly representing combination of AD and DLBD. Within the disease spectrum including AD, PD and DLBD, at one end is AD, a cortical degeneration with extensive cytoskeletal aberration of microtubule-associated tau protein, at the other end is PD, a subcortical degeneration with neurofilament aberration (Bancher et al., 1989). DLBD is the broad middle ground with some features of PD and AD, yet differing from both diseases by having widespread cortical LB and NP but lack of PHF-type neuritic changes in the neuropil and sparing of hippocampus despite widespread amyloid deposition that distinguishes DLDB from true AD/DAT.

The morphologic heterogeneity of AD is further associated with or superimposed by cerebrovascular lesions: multiple lacunes in the basal ganglia are seen in about one-third of definite cases of AD and brain infarcts are present in 8 to 35% of AD brains (Table 7; Joachim et al., 1988; Morris et al., 1988; Tierney et al., 1988; Mirra et al., 1989), but they rarely reach the extent of true combination with multiinfarct dementia (MID), representing mixed type dementia. Cerebral hemorrhages, present in 5 to 11% of AD brains (Table 7), are often caused by cerebral amyloid angiopathy, occurring in up to 100% of all AD cases (Jellinger, 1989). It may also induce cerebral infarctions and leukencephalopathy (Vinters et al., 1989).

The morphologic heterogeneity of AD, its association with other brain pathologies and the lack of uniform criteria for the diagnosis of AD and its distinct separation from "normal" brain aging are documented by varying coincidence rates for the clinical and pathological diagnosis of AD and other degenerative dementias. In various prospective or retrospective clinico-pathological studies the accuracy rates for the diagnosis of AD range from 70 to 100%, with reasonable mean values of 80 to 85% (Table 8). Retrospective clinico-pathological analysis of 675 consecutive autopsy cases of dementia disorders in the aged revealed that the histologic criteria of AD were fulfilled in 77%, only 60% representing "pure" forms, while 8% with additional cerebrovascular lesions represented "mixed" type dementia, and 9% had additional PD markers (PD/AD). MID pathology was present in 16%; 7% were other entities, and 0,3% showed nothing abnormal. The general coincidence rates for clinical and pathologic diagnosis of AD/DAT were 85.2%, for PD/AD 76%, for MID 72%, and for mixed AD/MID only 47% (Jellinger et al., 1990). These data and the results of other recent studies emphasize the need for better and more standardized clinical and morphological criteria

Table 8. Accuracy of clinical diagnosis in autopsy series of demented patients

Author (for ref. see Boller et al., 1989)[a]	Year	N/Cases	Agreement AD (Percent)
Retrospective studies			
Todorov et al.	(1975)	776 (AD)	43
Perl et al.	(1984)	26 (AD)	81
Wade et al.	(1987)	65 (AD)	85
Alafuzoff et al.	(1987)	55 (AD)	63
Kokmen et al.	(1987)	32	72
Joachim et al.	(1988)	150	87
Boller et al.	(1989)	54	63/79.7
Jellinger et al.	(1990)	675 (AD)	85.2
Prospective studies			
Sulkova et al.	(1983)	27 (AD)	82
Molsa et al.	(1985)	58 (AD)	71
Neary et al.	(1986)	24 (AD)	75
Martin et al.	(1987)	11 (AD)	100
Morris et al.	(1988)	25 (AD)	100
Crystal et al.	(1988)	19 (AD)	85
Sulkova et al.	(1989)	53 (AD)	87

AD brains with AD/SDAT alone; no other CNS lesions, i.e. "pure" AD/SDAT (Tierney et al., 1988)
[a] not included

for the classification of AD and other dementia disorder of the aged in order to increase the diagnostic accuracy rates.

References

Adams I (1987) Comparison of synaptic changes in the precentral and postcentral cerebral cortex of aging humans: a quantitative ultrastructural study. Neurobiol Aging 8:203–212

Allen SJ, Dawbarn D, Wilcock GK (1988) Morphometric immunochemical study in the nucleus basalis of Meynert in Alzheimer's disease. Brain Res 454:275–281

Ball MJ (1989) Neuropathology in the diagnosis of Alzheimer's disease. In: Hovaguimian T, Henderson S, Khachaturian Z, Orley J (eds) Classification and diagnosis of Alzheimer's disease. An international perspective. Hogrefe and Huber, Toronto, pp 135–143

Bancher C, Lassmann H, Budka H, Jellinger K, et al (1989) An antigenic profile of Lewy bodies: immunocytochemical indication for protein phosphorylation and ubiquitination. J Neuropathol Exp Neurol 48:81–93

Boller F, Lopez OL, Moossy J (1989) Diagnosis of dementia: clinicopathologic correlations. Neurology 39:76–79

Braak H (1989) Morphology of the cerebral cortex in relation to Alzheimer's dementia. J Neural Transm (P-DSect) 1:12

Braak H, Braak E, Bohl J, Lang W (1989) Alzheimer's disease: amyloid plaques in the cerebellum. J Neurol Sci 93:277–287

Burke WJ, Chung HD, Huang JS, et al (1988) Evidence for retrograde degeneration of epinephrine neurons in Alzheimer's disease. Ann Neurol 24:532–536

Burkhardt CR, Filley CM, Kleinschmidt-DeMasters BK, et al (1988) Diffuse Lewy body disease and progressive dementia. Neurology 38:1520–1528

Chan-Palay V, Asan E (1989) Alterations in catecholamine neurons of the locus ceruleus in senile dementia of the Alzheimer type and in Parkinson's disease. J Comp Neurol 287:373–392

Coleman PD, Flood DG (1987) Neuron numbers and dendritic extent in normal aging and Alzheimer's disease. Neurobiol Aging 8:521–545

Crystal H, Dickson D, Fuld P, et al (1988) Clinico-pathologic studies in dementia: nondemented subjects with pathologically confirmed Alzheimer's disease. Neurology 38:1682–1687

Davies L, Wolska B, Hilbich C, et al (1988) A 4 amyloid protein deposition and the diagnosis of Alzheimer's disease. Neurology 38:1688–1693

Delaere P, Duyckaerts C, Brion JP, et al (1989) Tau, paired helical filaments and amyloid in the neocortex: a morphometric study of 15 cases with graded intellectual status in aging and senile dementia of Alzheimer type. Acta Neuropathol (Berl) 77:645–653

Dickson DW, Crystal H, Mattiace L, et al (1989) Diffuse Lewy body disease: light and electron microscopic immunocytochemistry of senile plaques. Acta Neuropathol (Berl) 78:572–584

Ferreiro JA, Merskey H, Hachinski VC, et al (1989) Quantitative study A4 immunoreactive senile plaques and microvascular amyloid in Alzheimer's disease. J Neuropathol Exp Neurol 48:377

German DC, White CL III, Sparkman DR (1987) Alzheimer's disease: neurofibrillary tangles in nuclei that project to the cerebral cortex. Neuroscience 21:305–312

Gibb WRG (1989) The neuropathology of parkinsonian disorders. In: Jankovic J, Tolosa E (eds) Parkinson's disease and movement disorders. Urban and Schwarzenberg, Baltimore Munich, pp 205–223

Gibb WRG, Lees AJ (1988) The relevance of the Lewy body in the pathogenesis of idiopathic Parkinson's disease. J Neurol Neurosurg Psychiatry 51:745–752

Hamos JE, De Gennaro LJ, Drachman DA (1989) Synaptic loss in Alzheimer's disease and other dementias. Neurology 39:355–361

Hansen LA, DeTeresa R, Davies P, Terry RD (1988) Neocortical morphometry, lesion counts, and choline acetyltransferase levels in the age spectrums of Alzheimer's disease. Neurology 38:48–57

Hansen LA, Masliah E, Terry RD, Mirra SS (1989) A neuropathological subset of Alzheimer's disease with concomitant Lewy body disease and spongiform change. Acta Neuropathol (Berl) 78:194–201

Hyman BT, Van Hoesen GW, Damasio AR (1987) Alzheimer's disease: glutamate depletion in the hippocampal perforant pathway zone. Ann Neurol 20:472–481

Iseki E, Matsushita M, Kosaka K, et al (1989) Distribution and morphology of brain stem plaques in Alzheimer's disease. Acta Neuropathol (Berl) 78:131–136

Jellinger K (1989) Morphologie des alternden Gehirns und der (prä)senilen Demenzen. In: Platt D, Österreich K (Hrsg) Handbuch der Gerontologie, Bd 5. Fischer, Stuttgart New York, S 3–56

Jellinger K, Danielczyk W, Gabriel E (1989) Clinicopathological analysis of dementia disorders in the aged. J Neuropathol Exp Neurol 48:379

Jellinger K, Danielczyk W, Fischer P, Gabriel E (1990) Clinicopathological analysis of dementia disorders in the elderly. J Neurol Sci 95 (in press)

Joachim CL, Morris JH, Selkoe DJ (1988) Clinically diagnosed Alzheimer's disease: autopsy results in 150 cases. Ann Neurol 24:50–56

Joachim CL, Morris JM, Selkoe DJ (1989) Diffuse senile plaques occur commonly in the cerebellum of Alzheimer's disease. Am J Pathol 135:309–320

Kalus P, Braak H, Braak E, Bohl J (1989) The presubicular region in Alzheimer's disease. Topography of amyloid deposits and neurofibrillary changes. Brain Res 494:198–203

Katzman R, Terry R, DeTeresa R, et al (1988) Clinical, pathological, and neurochemical changes in dementia: a subgroup with preserved mental status and numerous neocortical plaques. Ann Neurol 23:138–144

Khachaturian ZS (1985) Diagnosis of Alzheimer's disease. Arch Neurol 42:1097–1105

Kosaka K, Tsuchiya K, Yoshimura M (1988) Lewy body disease with and without dementia. A clinicopathologic study of 35 cases. Clin Neuropathol 7:299–305

Lamy C, Duyckaerts C, Delaere P, et al (1989) Comparison of seven staining methods for senile plaques and neurofibrillary tangles in a prospective series of 15 elderly patients. Neuropathol Appl Neurobiol 15:563–578

Lassmann H, Fischer P, Bancher C, Jellinger K (1990) Immunocytochemical and ultrastructural pathology of nerve cells in Alzheimer's disease and related disorders. In: Maurer K, Riederer P, Beckmann H (eds) Alzheimer's disease. Epidemiology, Neuropathology, Neurochemistry, and Clinics. Springer, Wien New York, pp 171–179 (Key Topics in Brain Research)

Mann DMA, Marcyniuk B, Yates PI, et al (1988) The progression of the pathological changes of Alzheimer's disease in frontal and temporal neocortex examined both at biopsy and at autopsy. Neuropathol Appl Neurobiol 14:177–195

Masliah E, Terry RD, DeTeresa R, et al (1989) Morphometric quantification of a synaptic marker in neocortex of Alzheimer and Pick disease. J Neuropathol Exp Neurol 48:333

Mirra SS, Brownlee LM, Sumi SM, et al (1989) The CERAD neuropathology protocol: observations on cases clinically diagnosed as probable Alzheimer's disease. J Neuropathol Exp Neurol 48:334

Moossy J, Zubenko GS, Martinez AJ, Rao GR (1988) Bilateral symmetry of morphologic lesions in Alzheimer's disease. Arch Neurol 45:251–254

Morris JC, McKeel DW, Fulling K, et al (1988) Validation of clinical diagnostic criteria for Alzheimer disease. Ann Neurol 24:17–22

Mufson EJ, Mash DC, Hersh LB (1988) Neurofibrillary tangles in cholinergic pedunculopontine neurons in Alzheimer's disease. Ann Neurol 24:623–629

Oyanagi K, Takahashi H, Wakabayashi K, Ikuta F (1989) Correlative decrease of large neurons in the neostriatum and basal nucleus of Meynert in Alzheimer disease. J Neuropathol Exp Neurol 48:336

Perry RH, Irving D, Blessed G, et al (1990) Senile dementia of Lewy body type. A clinically and neuropathologically distinct form of Lewy body dementia in the elderly. J Neurol Sci 95:119–139

Probst A, Anderton BH, Brion JP, Ulrich JU (1989) Senile plaque neurites fail to demonstrate anti-paired helical filament and anti-microtubule-associated protein-tau immunoreactive proteins in the absence of neurofibrillary tangles in the neocortex. Acta Neuropathol (Berl) 77:430–436

Purohit C, Perry RH, Irvinb D (1989) Alzheimer-type pathology in dentate gyrus of the hippocampus. J Neuropathol Exp Neurol 48:340

Stopa EG, Tate-Ostroff B, Walcott EC (1989) Human suprachiasmatic nuclei in Alzheimer disease. J Neuropathol Exp Neurol 48:327

Sulkova R, Erkinjuntti T, Haltia M, et al (1989) Non-Alzheimer dementias fulfilling the NINCDS-ADRDA criteria for probable Alzheimer's disease. Aging of the brain and dementia: ten years later. Florence, May 31–June 3, 1989, p 125 (Abstr)

Terry RD, Hansen LA, DeTeresa R, et al (1987) Senile dementia of the Alzheimer type without neocortical neurofibrillary tangles. J Neuropathol Exp Neurol 46:262–268

Tierney MC, Fisher RH, Lewis AJ, et al (1988) The NINCDS-ADRDA Work Group criteria for the clinical diagnosis of probable Alzheimer disease: clinicopathologic study of 57 cases. Neurology 38:359–364

Tomlinson BE (1989) The neuropathology of Alzheimer's disease – Issues in need of resolution. Neuropathol Appl Neurobiol 15:491–512

Tomlinson BE, Henderson G (1976) Some quantitative cerebral findings in normal and demented old people. In: Terry RD, Gershon S (eds) Neurobiology of aging. Raven Press, New York, pp 183–209

Tourtellotte WG, Van Hoesen GW, Hyman BT, et al (1989) Afferents of the thalamic reticular nucleus are pathologically altered in Alzheimer's disease. J Neuropathol Exp Neurol 48:336

Vinters HV, Gray F, Partridge WM, et al (1989) Cerebral amyloid angiopathy with leukoencephalopathy. J Neuropathol Exp Neurol 48:379

Vogels OJM, Broere CAJ, Renkawek K (1989) Neuron numbers and sizes of nucleus basalis Meynert in Alzheimer's disease. Aging of the brain and dementia: ten years later. Florence, May 31–June 3, 1989, p 77 (Abstr)

Wegiel J, Wisniewski HM, Wang KC (1989) Neuronal reaction on plaque amyloid deposits. J Neuropathol Exp Neurol 48:325

Weis S, Wenger E, Jellinger K (1989) The corpus callosum in normal aging and Alzheimer disease. Aging of the brain and dementia: ten years later. Florence, May 31–June 3, 1989, p 79 (Abstr)

Wisniewski HM, Wen GY, Kim LS (1989) Comparison of four staining methods on the detection of neuritic plaques. Acta Neuropathol (Berl) 78:22–27

Yamaguchi H, Hirai S, Morimatsu M, et al (1988) Diffuse type of senile plaques in the cerebellum of Alzheimer type dementia demonstrated by β-protein immunostaining. Acta Neuropathol 77:113–114

Zubenko GS, Moossy J, Claassen AJ, et al (1989) Neurochemical and morphological correlates of major depression in primary dementia. J Neuropathol Exp Neurol 48:333

Zweig RM, Ross CA, Hedreen JC, et al (1988) The neuropathology of aminergic nuclei in Alzheimer's disease. Ann Neurol 24:233–242

Zweig RM, Jankel WR, Hedreen JC, et al (1989) The pedunculopontine nucleus in Parkinson's disease. Ann Neurol 26:41–46

Correspondence: Dr. K. Jellinger, Ludwig Boltzmann Institute of Clinical Neurobiology, Lainz Hospital, Wolkersbergenstrasse 1, A-1130 Wien, Austria.

Morphology of white matter, subcortical, dementia in Alzheimer's disease

A. Brun[1], L. Gustafson[2], and E. Englund[1]

[1] Institute of Pathology, Department of Neuropathology, and
[2] Department of Psychogeriatrics, University Hospital, Lund, Sweden

Summary

Dementing disorders are not only caused by grey matter alterations but also by white matter lesions, either together with grey matter changes or as the sole change. In dementia of Alzheimer's type white matter disease in the nature of a selective incomplete white matter infarction was found in around 60% of cases. It was regarded as being most likely due to regional hypoperfusion of the white matter and might therefore be prevented, halted or even partly reversed by therapy.

Introduction

Dementias are usually thought of as cortical diseases, with a small group of subcortical dementias affecting basal ganglia or brain stem. The role of rarely considered or little known white matter diseases remains disregarded. They seem, however, to be common and may be an important component of a syndrome or even the sole disease behind a dementia. However, they are submerged among the rather poorly defined subcortical dementias. Nevertheless, on the basis of their frequency and importance they deserve an anatomical plateau of their own, the resulting disease logically termed white matter dementia, and are to be regarded as a distinctive part of the subcortical dementias. This definition is used in the following. Subcortical white matter lesions of the type under consideration here is the white matter disease encountered in Alzheimer's disease.

Material

The material consists of 50 cases of presenile and senile dementia of Alzheimer's type (AD and SDAT respectively) aged 52 to 90 years. For a comparison were used normal age-matched cases, cases of frontal lobe dementia of non-Alzheimer type (FLD), (Brun et al., 1987), Pick's disease, multi-infarct dementia (MID), Binswanger's disease, incomplete white matter infarction as well as occasional cases of Jacob Creutsfeldt's disease (JC), AIDS and multiple sclerosis (MS).

Method

In the neuropathological study whole brain semiserial frontal sectioning was consistently employed. 5-micra thick sections from paraffin-embedded slices were stained according to a wide variety of methods including hematoxylin and eosin, Luxol fast Blue for myelin, Naoumenko and Bielschowsky for plaques and tangles, Kongo red and thioflavine S for amyloid and gliofibrillar acid protein with PAP-technique for astroglial protein. On templates of such sections the grey and white matter changes were recorded as to type, distribution and severity. In this way also non-relevant additional important lesions could be detected, sometimes disqualifying for inclusion in respective groups. This method also allowed a close comparison between topography of the various lesions. As reported earlier (Englund et al., 1987, 1988) 50 small samples were also taken from the white matter, both normal and pathological, for a small piece in vitro MR study, analysis of water and myelin lipids as well as microscopical assessment of type and severity of lesions. The samples where identical for all these studies which allowed a close correlation between the results.

The primary dementia cases all belong to a large prospective study why relevant clinical features such as blood pressure variations, dementia symptoms, regional cerebral blood flow changes and other signs and symptoms could be derived from the clinical records.

Results

Reported here are only the findings in the white matter of cases with primary degenerative dementia of the Alzheimer type.

White matter changes were found in 30 out of 50 DAT cases. They were located in the deep white matter periventricularly, sometimes spreading to more superficial strata. They were most common frontally and seen less often parietally and centrally.

The changes were not circumscribed but irregularly and diffusely outlined with a gradual transition between damaged and non-damaged areas. On myelin-stained sections they appeared as pale areas. On parallel sections stained differently these same areas were checked for cellular reactions: If not present the myelin paling was regarded as an artefact.

Cell changes found consisted of an attenuation of the tissues with a partial loss of axons and myelin together with oligodendroglial cells, accompanied by mild astrocytic gliosis and a few macrophages. There was no swelling or water increase, no inflammatory changes and no cavitation or complete loss of axons or myelin sheaths. Vessels within the changed areas showed fibrohyaline wall thickening with severe narrowing of the lumen but no hypertensive changes. The basomeningeal arteries displayed only mild arteriosclerosis and amyloid angiopathy was limited to meningocortical vessels.

The white matter changes were graded with regard to the severity of loss of normal constituents and appearance of reactive changes. The severest degree of white matter change was found in $\frac{1}{5}$ of the material, slight to moderate degree in $\frac{2}{5}$ and no changes in the remaining $\frac{2}{5}$. There was no regional correspondence between the cortical encephalopathy with plaques and tangles etc. on the one hand and the white matter changes on the other hand, neither with regard to location nor to severity.

The small sample study showed an MR T-1 and T-2 time prolongation and reduction of cerebroside and ganglioside by up to 50%. These changes parallelled each other and also the microscopical grade of severity. Morphological, MR and biochemical changes were found only within these white matter areas, whereas outside the morphology, MR and lipid values were normal. Thus there was never any diffuse involvement of the white matter.

The clinical records showed cardiovascular disorders and hypotensive attacks to be 3 times as common in cases of DAT with white matter changes as in cases without. Hypertension was rare. From a symptomatological point of view DAT with white matter changes had more symptoms such as lack of insight and disinhibition and also more prominent symptoms of a cerebrovascular type such as vertigo, fainting spells or syncope and a fluctuating course, as compared with DAT without white matter changes. Classical DAT symptoms were rather subdued to some extent.

Discussion

The white matter changes described may have a component of Wallerian degeneration. They are, however, not mainly of this type since white matter was normal outside the lesions and the changes may in non-DAT cases exist as the sole brain disorder and also since there was no correlation between cortical and white matter changes (Brun et al., 1986). Neither do the latter appear as part of a generalized metabolic oligodendroglial disorder since they are focal. They are also clearly different from those in multiple sclerosis and from regular complete infarctions. From a morphological point of view they appear as incomplete infarctions with varying attenuation of the tissue and mild cellular reaction. Against this background they are most likely due to hypoperfusion caused by the local vascular changes in combination with general cardiovascular factors inducing hypotensive attacks. This may be supported by the partly cerebrovascular type of symptomatology.

The changes are called selective incomplete white matter infarction (SIWI) due to the microscopical appearance of the changes, the proposed etiology and the fact that the changes selectively engage the white matter.

The frequent association between DAT and SIWI may indicate a connection between the two disorders. This connection could be in the nature of a wide-spread transmitter deficit affecting also the cerebral vascular innervation or the common meningocortical amyloid angiopathy with consequences for the perfusion of the underlying white matter.

In DAT the white matter lesions may be expected to block the transport routes for acetylcholine between nucleus basalis Meynert and the cortex, and might hypothetically in a retrograde fashion cause degenerative changes in this nucleus.

The ventricular widening most likely is mainly due to this white matter disease. A tendency to a frontal accent of the white matter changes appears to coincide with the frontal component of the symptomatology. Probably only severe changes produce symptoms. Symptoms related to the white matter by other authors (Filley et al., 1989) include inattention, forgetfulness and emotional changes whereas amnesia, aphasia and movement disorders are more characteristic of cortical disorders. The histologic type of the lesions and the reflection in MR relaxation times may indicate that the changes are visible on MRI as also proposed by some authors. This is in agreement with a preliminary MR study of later series of cases in this material (unpublished results), the frontal white matter bearing the brunt of the SIWI.

A few other groups have since presented similar findings e.g. Gottfries et al. (1985) and George et al. (1981). White matter disease is not unique to Alzheimer's dementia. Changes, though of a different type, are found also in Pick's disease and a leucoencephalopathic form of Jacob Creutsfeldts's disease and are also a part of multi-infarct dementia. They are even more prominent in several other disorders and then in some instances the main change such as in Binswanger's disease and progressive subcortical gliosis (PSG). In still other disorders they are the only change such as in AIDS, multiple sclerosis and some cases affected by SIWI alone. They may also be the sole or main lesion in alcoholic dementia, normal pressure hydrocephalus and toluene intoxication (Filley et al., 1989). White matter alterations may also be the dominant change in the ageing brain.

Thus in dementia white matter lesions may arise as an important field of research and an important correlate to dementing symptoms. If the white matter lesions in DAT are of a hypoperfusive etiology a preventive or even partly reversing therapy may be considered.

References

Brun A, Englund E (1986) A white matter disorder in dementia of the Alzheimer type: a pathoanatomical study. Ann Neurol 19:253–262

Brun A (1987) Frontal lobe degeneration of non Alzheimer type. I. Neuropathology. Arch Gerontol Geriatr 6:193–208

Englund E, Brun A, Persson B (1987) Correlations between histopathologic white matter changes and proton MR relaxation times in dementia. Alzheimer Disease and Associated Disorders 1(3):156–170

Englund E, Brun A, Alling C (1988) White matter changes in dementia of Alzheimer's type. Brain 3:1425–1539

Filley CM, Franklin GM, Heaton RK, Rosenberg NL (1989) White matter dementia. Clinical disorders and implications. Neuropsych Neuropsychol Behav Neurol 1(4):239–254

George AE, deLeon MJ, Ferris SH, Kricheff II (1981) Parenchymal CT correlates of senile dementia (Alzheimer's disease): loss of grey-white matter discriminability. AJNR 2:205–211

Gottfries CG, Karlsson I, Svennerholm L (1985) Senile dementia – a white matter disease? In: Gottfries CG (ed) Normal ageing, Alzheimer's disease and senile dementia. Editions de'l Université, Bruxelles, pp 111–118

Correspondence: Prof. A. Brun, M. D., Ph. D., Department of Neuropathology, University Hospital, S-22185 Lund, Sweden.

Morphology of the cerebral cortex in relation to Alzheimer's dementia

H. Braak and E. Braak

Zentrum der Morphologie, Johann Wolfgang Goethe Universität,
Frankfurt/Main, Federal Republic of Germany

Summary

The hallmarks of Alzheimer's disease are extracellular amyloid deposits and intraneuronal neurofibrillary changes. The distribution pattern of amyloid is different from that of neurofibrillary changes. Plaque-like deposits of amyloid should carefully be distinguished from the well known neuritic plaques. The brunt of the cortical pathology is borne by the entorhinal region, the hippocampal formation, and the isocortical association areas. The early occurring destruction of the entorhinal cortex disrupts the transport of information from isocortical association areas to the hippocampal formation.

The cerebral cortex represents the dominating structure of the human brain. It is divided into the more or less uniformly built isocortex and the heterogeneously composed allocortex (Braak, 1980, 1984). The phylo- and ontogenetically late developing association areas occupy most of the total area of the isocortex (Fig. 1 a, white areas). The allocortex is small in comparison to the isocortex. Its central portion is the hippocampal formation. A further component is the entorhinal region that is located in the anterior portions of the parahippocampal gyrus and the gyrus ambiens. The entorhinal cortex exhibits a particularly complex lamination pattern (Stephan, 1975; Braak and Braak, 1985).

Recent investigations in the primate brain have shown that stout fibre bundles originating from all isocortical association areas converge on the entorhinal region hereby providing it with abundant somato-motor, somato-sensory, acoustic, and visual information (Fig. 1 a; van Hoesen, 1982; van Hoesen et al., 1972). The outer cellular layers of the entorhinal region generate the perforant tract which in turn transmits the information to the hippocampus (Fig. 1 b). This information transfer from isocortical association areas – via the entorhinal region – to the hippocampal formation is of significance for maintaining mnestic functions.

All of the three territories under consideration: the isocortical association areas, the entorhinal region, and the hippocampal formation exhibit characteristic and thoroughgoing structural changes during the course of Alzheimer's disease.

A large amount of unsoluble fibrous material that normally does not occur in the central nervous system is both extracellularly and intraneuronally deposited. The extracellularly located amyloid deposits should carefully be distinguished from the intraneuronal components which make up the neurofibrillary changes.

Modern silver techniques which take advantage of physical development of the latent image are available for the demonstration of both the amyloid (Campbell et al., 1987) and neurofibrillary changes (Gallyas and Wolff, 1986). The sensitivity and specificity of these advanced silver stains is comparable to that seen in immunostained preparations (Braak et al., 1989).

Amyloid deposits

In Alzheimer's disease, large numbers of plaque-like amyloid deposits are more or less evenly distributed throughout all layers and all areas of the isocortex. Most of the amyloid plaques remain devoid of pathologically changed argyrophilic neurites. Furthermore, the tissue within the range of many of these plaques does not reveal any conspicuous neuropil distortions in Nissl stained material. It has to be emphasized that large numbers of amyloid plaques occur in layers that never or only occasionally harbour neuritic plaques, among these are the molecular layer and the multiform layer (Fig. 1 c). In addition to that, many amyloid deposits are also present in the white matter and there even at a considerable distance to the cortex. Thus, it appears inappropriate to consider the amyloid plaques as forerunners of neuritic plaques or as structures that

---

Fig. 1. a, b Fibre bundles from isocortical association areas converge upon the entorhinal cortex which in turn projects via the perforant path to the hippocampal formation. c Schematic lamination pattern of an isocortical area (striate area, Brodmann field 17) and its Alzheimer-related pathology as seen in preparations stained for amyloid and neurofibrillary changes. Note the difference in the distribution pattern of amyloid plaques and neuritic plaques. d Distribution of Alzheimer-related cortical changes. The black area represents the most severely affected region. e Neurofibrillary changes consist of three kinds of lesions, namely the neuritic plaques, neurofibrillary tangles, and neuropil threads. f The transport of information from isocortical association areas to the hippocampal formation is bilaterally interrupted by destruction of the cells of origin of the perforant tract

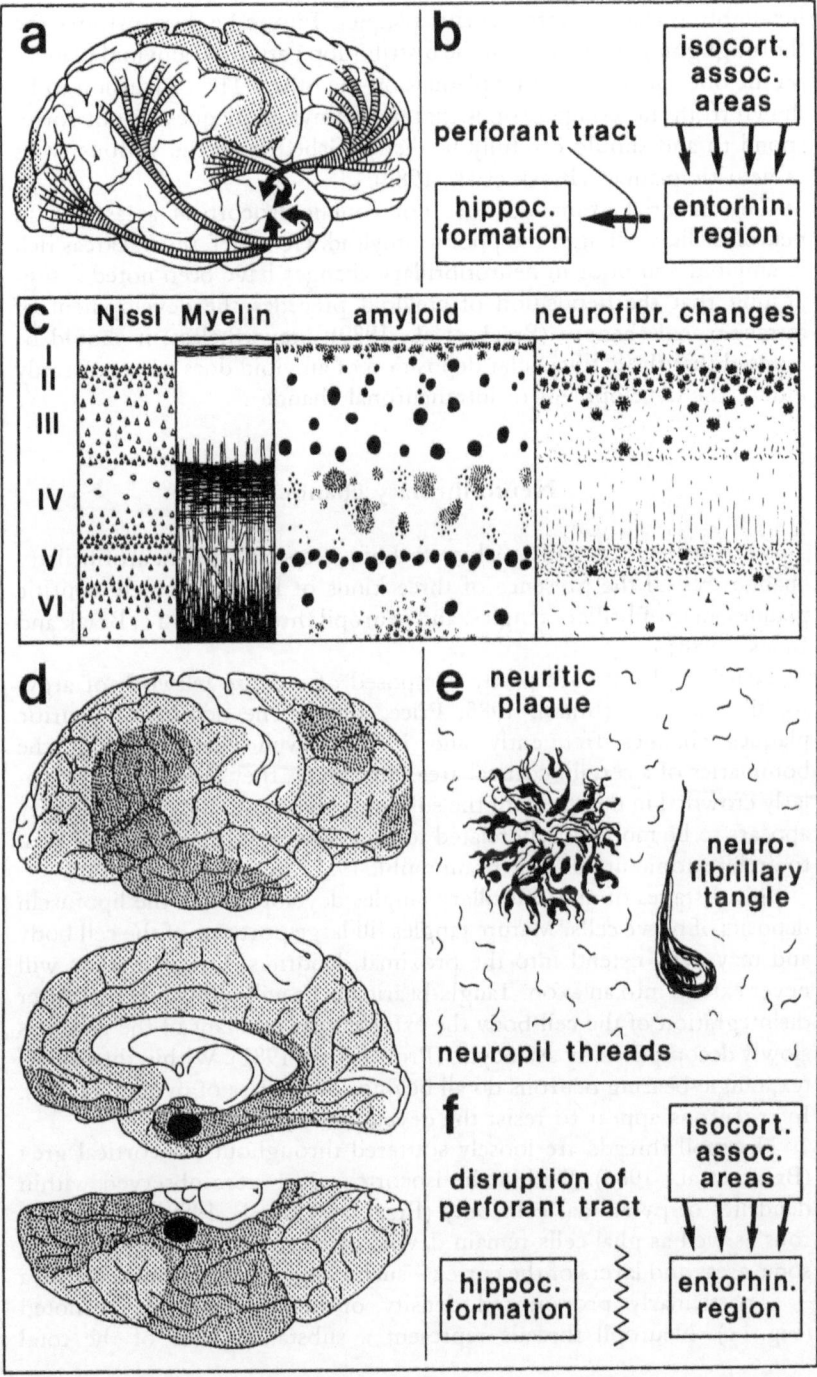

inevitably transform into neuritic plaques. Figure 1 c demonstrates the thoroughgoing differences in the distribution pattern of amyloid plaques on the one side and neuritic plaques on the other. The conclusion to be drawn from this comparison is that most amyloid plaques do not correspond to and should carefully be distinguished from the various types of neuritic plaques (Braak et al., 1989).

The material examined does not contain isocortical areas rich in neurofibrillary changes and poor in amyloid. However, several areas rich in amyloid and poor in neurofibrillary changes have been noted – suggesting that the deposition of amyloid precedes the development of intraneuronal changes (Braak et al., 1989). Nevertheless, it should be stressed that the extracellular deposition of amyloid does not necessarily induce the development of intraneuronal changes.

Neurofibrillary changes

Sections stained with a silver technique specific for neurofibrillary changes reveal the presence of three kinds of lesions, namely neuritic plaques, neurofibrillary tangles, and neuropil threads (Fig. 1 e; Braak and Braak, 1988 a).

Neuritic plaques are partly composed of a dense feltwork of argyrophilic neurites (Mann, 1985; Price, 1986). The density of neuritic plaques changes frequently and it may even change within the boundaries of a certain cortical area. Moreover, they tend to be particularly crowded in the depth of the sulci and, therefore, the plaque density appears to be more closely related to the configuration of the sulci than to architectonic units (von Braunmühl, 1957; Braak et al., 1989).

Initial stages of neurofibrillary tangles develop within the lipofuscin deposits of nerve cells. Mature tangles fill large portions of the cell body and may even extend into the proximal dendrites, however, they will never extend into an axon. Tangle-bearing cells will ultimately die. After disintegration of the cell body the extracellular remnant of the tangle is slowly decomposed by astrocytes (Probst et al., 1982). Within the isocortex, tangle-bearing neurons do all belong to the class of pyramidal cells. Interneurons appear to resist the development of tangles.

Neuropil threads are loosely scattered throughout the cortical grey (Braak et al., 1986). Within the isocortex, they were observed within dendrites of pyramidal cells only (Braak and Braak, 1988 b). Interneurons as well as glial cells remain devoid of these specific alterations. In some areas and layers of the cortex – such as in layer V of the striate area – a particularly pronounced density of neuropil threads is noted (Fig. 1 c). Neuropil threads represent a substantial part of the total amount of neurofibrillary changes and should thus justify more efforts to elucidate the conditions responsible for their development and pattern of distribution (Braak et al., 1989).

Distribution pattern

Amyloid deposits, neuritic plaques, neurofibrillary tangles, and neuropil threads are by no means distinguishing features of Alzheimer's disease, since they also occur – although in considerably lesser number – in the brain of the aged and non-demented individual (von Braunmühl, 1957; Mann, 1985; Price, 1986; Braak and Braak, 1988a). This is also why the neuropathological diagnosis is only an easy one in severe cases. Recognition of early stages of the disease is extremely difficult and requires extended and time-consuming quantitative analyses of a large number of cortical areas. In this context, it should be emphasized that the advanced and highly reliable silver techniques mentioned above permit the processing of numerous preparations such as sections running through entire hemispheres. All structures containing the pathological material are specifically marked. They are clearly visible against an almost unstained background and thereby enable the immediate recognition – even by the unaided eye – of the pathology and of the striking differences in the severity of the change.

Figure 1d shows the distribution of the pathological changes, as compiled from data available in the literature (Brun and Gustafson, 1976; Brun, 1983). Isocortical association areas are particularly heavily involved in Alzheimer's disease, nevertheless the brunt of the pathology is borne by the allocortex with severe destruction of the entorhinal cortex and hippocampal formation. The pyramidal cells within the superficial layers of the entorhinal cortex are extremely susceptible to the development of tangles (Braak and Braak, 1985). Advanced stages of Alzheimer's disease show that all projection cells of the second layer are affected. The projection cells within this layer are also the first cortical neurons affected by the pathology. When focusing on the initial stages of the disease, neither the temporal isocortex nor the hippocampal formation show the presence of large numbers of neurofibrillary tangles and neuropil threads while the second layer of the entorhinal region is already severely and specifically affected. The projection cells of this layer generate important portions of the entorhinal-hippocampal connection, the perforant path. Therefore, the transport of isocortical information via the entorhinal region to the hippocampal formation becomes bilaterally hampered or interrupted (Fig. 1f). The lamina-specific destructive process leads to a "disconnection syndrome" in the sense of the late american neurologist Geschwind; isocortex and allocortex become disconnected from each other (Hyman et al., 1984, 1986).

This disruptive process which already occurs in the initial stages of Alzheimer's disease might serve as an explanation for the early impairment of intellectual capabilities of individuals suffering from this disorder. In the next stage, the hippocampal formation itself as well as the association areas of the isocortex become gradually destroyed, so that ultimately, the three cortical regions necessary for maintaining mnestic functions are subject to thoroughgoing pathological changes.

Acknowledgements

This work was supported by the Deutsche Forschungsgemeinschaft, the Friedrich Merz Stiftung, Frankfurt, and Degussa, Hanau. The authors are indebted to Profs. Drs. Hübner and Stutte (Department of Pathology, Frankfurt) and Profs. Drs. Schlote, Goebel, Bohl, Jakob, Gulotta, Peiffer, Ulrich, Probst, Kleihues, Lang (Department of Neuropathology, Frankfurt, Mainz, Wiesloch, Münster, Tübingen, Basel, Zürich).

References

Braak H (1980) Architectonics of the human telencephalic cortex. In: Braitenberg V, Barlow HB, Bizzi E, Florey E, Grüsser OJ, van der Loos H (eds) Studies of brain function, vol 4. Springer, Berlin Heidelberg New York, pp 1–147

Braak H (1984) Architectonics as seen by lipofuscin stains. In: Peters A, Jones EG (eds) Cerebral cortex, vol 1. Plenum Press, New York, pp 59–104

Braak H, Braak E (1985) On areas of transition between entorhinal allocortex and temporal isocortex in the human brain. Normal morphology and lamina-specific pathology in Alzheimer's disease. Acta Neuropathol 68: 325–332

Braak H, Braak E (1988a) Morphology of the human isocortex in young and aged individuals: qualitative and quantitative findings. In: von Hahn HP (ed) Interdisciplinary topics in gerontology, vol 25. Karger, Basel, pp 1–15

Braak H, Braak E (1988b) Neuropil threads occur in dendrites of tangle-bearing nerve cells. Neuropathol Appl Neurobiol 14: 39–44

Braak H, Braak E, Grundke-Iqbal I, Iqbal K (1986) Occurrence of neuropil threads in the senile human brain and in Alzheimer's disease: a third location of paired helical filaments outside of neurofibrillary tangles and neuritic plaques. Neurosci Lett 65: 351–355

Braak H, Braak E, Kalus P (1989) Alzheimer's disease: areal and laminar pathology in the occipital isocortex. Acta Neuropathol 77: 494–506

Brun A (1983) An overview of light and electron microscopic changes. In: Reisberg B (ed) Alzheimer's disease. The standard reference. Free Press, New York, pp 37–47

Brun A, Gustafson L (1976) Distribution of cerebral degeneration of Alzheimer's disease. Arch Psychiat Nervenkr 223:15–33

Campbell SK, Switzer RC, Martin TL (1987) Alzheimer's plaques and tangles: a controlled and enhanced silver-staining method. Soc Neurosci Abstr 13:678

Gallyas F, Wolff JR (1986) Metal-catalyzed oxidation renders silver intensification selective. Applications for the histochemistry of diaminobenzidine and neurofibrillary changes. J Histochem Cytochem 34:1667–1672

Hyman BT, van Hoesen GW, Damasio AR, Barnes CL (1984) Alzheimer's disease: cell-specific pathology isolates the hippocampal formation. Science 225:1168–1170

Hyman BT, van Hoesen GW, Kromer LJ, Damasio AR (1986) Perforant pathway changes and the memory impairment of Alzheimer's disease. Ann Neurol 20:472–481

Mann DMA (1985) The neuropathology of Alzheimer's disease: a review with pathogenetic, aetiological and therapeutic considerations. Mech Ageing Dev 31:213–255

Price DL (1986) New perspectives of Alzheimer's disease. Ann Rev Neurosci 9:489–512

Probst A, Ulrich J, Heitz PU (1982) Senile dementia of Alzheimer type: astroglial reaction to extracellular neurofibrillary tangles in the hippocampus. An immunocytochemical and electron-microscopic study. Acta Neuropathol 57:75–79

Stephan H (1975) Allocortex. In: Bargmann W (Hrsg) Handbuch der mikroskopischen Anatomie des Menschen, Bd 4/9. Springer, Berlin Heidelberg New York

Van Hoesen GW (1982) The parahippocampal gyrus. New observations regarding its cortical connections in the monkey. Trends Neurosci 5:345–350

Van Hoesen GW, Pandya DN, Butters N (1972) Cortical afferents to the entorhinal cortex of the rhesus monkey. Science 175:1471–1473

Von Braunmühl A (1957) Alterserkrankungen des Zentralnervensystems. Senile Involution. Senile Demenz. Alzheimersche Krankheit. In: Lubarsch O, Henke F, Rössle R (Hrsg) Handbuch der speziellen pathologischen Anatomie und Histologie, Bd 13/1A. Springer, Berlin, S 337–539

Correspondence: Prof. Dr. med. H. Braak, Zentrum der Morphologie, Klinikum der Johann Wolfgang Goethe Universität, Theodor-Stern-Kai 7, D-6000 Frankfurt/Main 70, Federal Republic of Germany.

Quantitative investigations of presenile and senile changes of the human entorhinal region

H. Heinsen[1], **L. Abel**[1], **S. Heckers**[1], **Y. L. Heinşen**[1], **H. Beckmann**[1], **P. Riederer**[1], and **G. Hebenstreit**[2]

[1] Department of Psychiatry, University of Würzburg, Würzburg, Federal Republic of Germany
[2] Department of Psychiatry, Landeskrankenhaus, Amstetten/Mauer, Austria

Summary

In a quantitative study on thick frozen sections stained by the Gallyas' silver impregnation method neurofibrillary tangles (NFT), neuritic plaques (NP) and neuropil threads (NT) in the entorhinal as well as transentorhinal area (Braak and Braak, 1985) have been evaluated. The amount of NT exceeded by far that of NFT and NP.

Introduction

Since the first description by Alois Alzheimer (1907) there has been much debate on the significance of neurofibrillary tangles as well as neuritic plaques. Alzheimer stressed that "within an otherwise normal looking cell you first find one or more fibrils which impress by their thickness and impregnability". – "Miliar cores" were mentioned only as concomitant changes throughout the cortex. Simchowicz (1911) recognized the ammonshorn and the frontal cortex as the predilection areas for neurofibrillary tangles. Mutrux (1947) estimated the neuritic plaques to be less significant. Hirano and Zimmerman (1962) noted a special topographic distribution of neurofibrillary tangles in allocortical areas. Kemper (1978) stressed the heavy involvement of the lateral entorhinal region, more precisely described later by Braak and Braak (1985). Blessed, Tomlinson and Roth (1968) tried to set up a correlation of an overall plaque count with an overall dementia score (1968) being heavily criticized by Dayan (1970). Recently Braak et al. (1986) described numerous tubular

argyrophilic profiles in histologic sections from patients with AD or
SDAT using Gallyas' silver impregnation. These thread-like processes
representing either fine dendrites or axons were designated neuropil
threads by Braak et al. (1986). The intention in the present study is to
quantify the presence of NFT, NP and NT using a novel method based
on thick sections (500 mikrometer) and a modified Gallyas' silver impregnation for the demonstration of pathologic fibrils.

Materials and methods

Twenty brains from patients older than 60 years at death were fixed in 4%
formalin, embedded in gelatine, frozen, cut into consecutive series of 0.5 mm
thick sections, and stained following the procedure described by Gallyas (1971).
Point-counting was performed, using an ocular grid at 1000 × magnification, on
microscopically identified sections through the Area entorhinalis and Area
transentorhinalis.

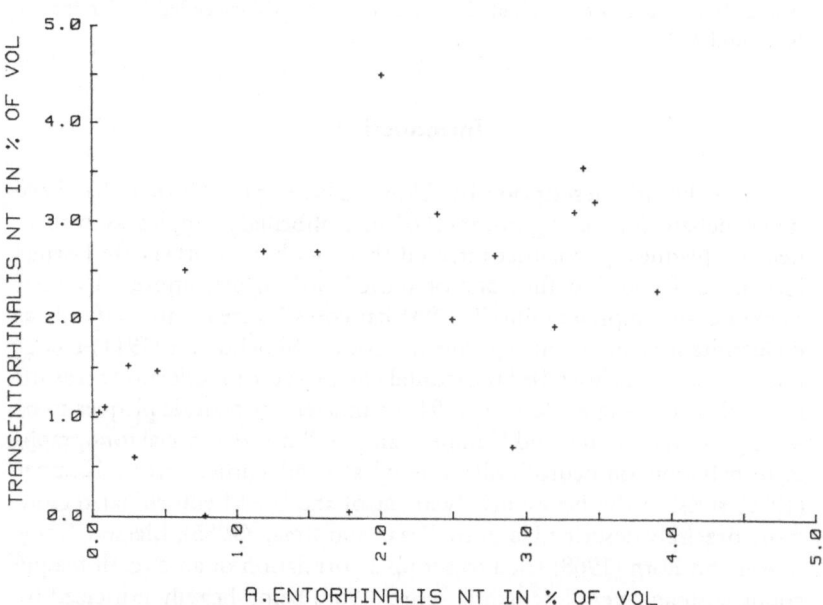

Fig. 1. The volume density of neuropil threads (NT) in the transentorhinal area is plotted
against the entorhinal area. Each cross signifies an individual case. Note that even in the
most severe cases the sum of the quantitative changes of NFT and NT comprise only 7%
of nervous tissue. The quantitative distribution of NP was negligible in the entorhinal
region

Results

In general, neuropil threads were more frequently encountered than neurofibrillary tangles or neuritic plaques (cf. Figs. 1, 2) yielding together maximal values of 7%. With NT density up to 2.5% in both regions (Fig. 1) the density of NT in the transentorhinal region generally exceeds that in the entorhinal region. NT densities higher than 2.5% appear to be distributed equally in both regions. Neuritic plaques have been encountered in the two regions under investigation only in cases with the most massive histological changes. The rostral parts of entorhinal region which cover the amygdaloid body were crowded with neuritic plaques in these latter cases. Similar changes were also observed in the adjacent isocortex, yet not estimated. Two of the control cases showed lesions confined to the transentorhinal area. Neuropil thread density exceeded in both cases the density of neurofibrillary tangles. Neuritic plaques were completely absent.

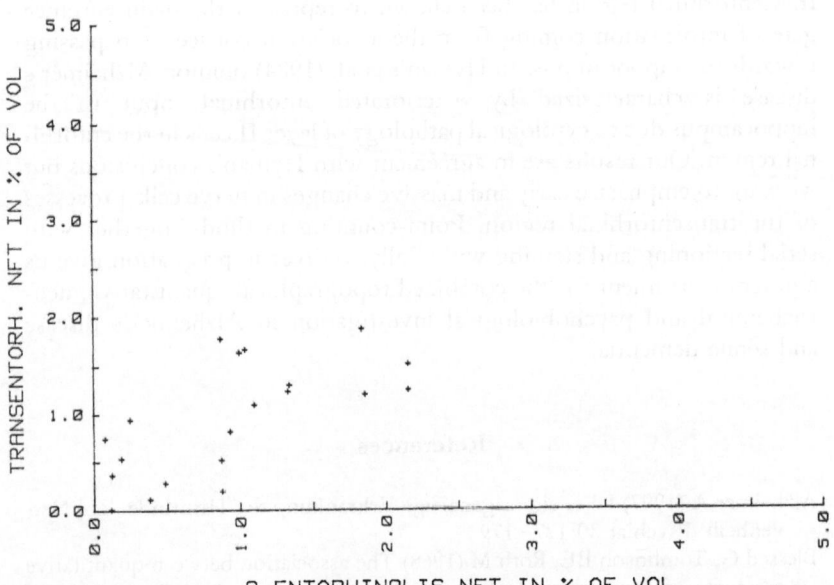

Fig. 2. Volume density of neurofibrillary tangles (NFT) in transentorhinal area versus entorhinal area. Each cross represents an individual case. The overall quantitative changes in the individual cases are obtained by summing the data on the abscissa as well as ordinate

Discussion

The present point-counting estimations of argyrophilic tissue in cases with Alzheimer's disease, senile dementia and control cases show an hitherto unexpected excess of neuropil threads. Interpreting our results one is inclined to designate both forms of dementia principally as a disease of neuropil threads. Previous quantitative studies emphasized the number of neurofibrillary tangles and neuritic plaques. By means of new stereological methods (reviewed by Gundersen et al., 1988) it has become easy to estimate the volume density of small structures like neuropil threads. The second uncommon feature which so far has only been described by Matsuyama et al. (1978) is the early affliction by senile changes of the transentorhinal area in subjects with no clinical known history of dementia. Our results represent an extension of Kemper's (1978) description with generally higher concentrations of senile changes in this latter area in demented subjects. These two observations, the prevalence and excess of neuropil threads and the time course and topographical distribution of argyophilic material, are of special interest. In a previous HRP study in the cat (Room and Groenewegen, 1986) the transentorhinal region has been shown to represent the main entrance gate of information coming from the association cortices and passing towards the hippocampus. In Hyman's et al. (1984) opinion Alzheimer's disease is characterized by deteriorated entorhinal input to the hippocampus due to cytological pathology of layer II cells in the entorhinal region. Our results are in agreement with Hyman's conclusions but we want to emphasize early and massive changes in nerve cells processes of the transentorhinal region. Point-counting methods together with serial sectioning and staining with Gallyas' silver impregnation give us a potent instrument for the combined topographical, quantitative, neurochemical and psychobiological investigation in Alzheimer's disease and senile dementia.

References

Alzheimer A (1907) Über eine eigenartige Erkrankung der Hirnrinde. Zbl Nervenheilk Psychiat 30:177–179

Blessed G, Tomlinson BE, Roth M (1968) The association between quantitative measures of dementia and of senile change in the cerebral grey matter of elderly subjects. Br J Psychiatry 114:797–811

Braak H, Braak E (1985) On areas of transition between entorhinal allocortex and temporal isocortex in the human brain. Normal morphology and lamina-specific pathology in Alzheimer's disease. Acta Neuropathol 68:325–332

Braak H, Braak E, Grundke-Iqbal I, Iqbal K (1986) Occurrence of neuropil threads in the senile human brain and in Alzheimer's disease: a third location of paired helical filaments outside of neurofibrillary tangles and neuritic plaques. Neurosci Lett 65:351–355

Dayan DA (1970) Quantitative histological studies on the aged human brain I and II. Acta Neuropathol 16:85–102

Gallyas F (1971) Silver staining of Alzheimer's neurofibrillary changes by means of physical development. Acta Morphol Acad Sci Hung 19:1–8

Gundersen HJG, et al (1988) Some new, simple and efficient stereological methods and their use in pathological research and diagnosis. APMIS 96:379–394

Hirano A, Zimmerman HM (1962) Alzheimer's neurofibrillary changes. Arch Neurol 7:73–88

Hyman B, van Hoesen G, Damasio A, Barnes C (1984) Alzheimer's disease: cell specific pathology isolates the hippocampal formation. Science 225:1168–1170

Kemper T (1978) Senile dementia: a focal disease of the temporal lobe. In: Nandy K (ed) Senile dementia: a biomedical approach. Elsevier, pp 105–113

Matsuyama H, Nakamura S (1978) Senile changes in the brain in the Japanese: incidence of Alzheimer's neurofibrillary change and senile plaques. In: Katzman R, Terry RD, Blick KL (eds) Alzheimer's disease: senile dementia and related disorders. Raven Press, New York, pp 287–297

Mutrux S (1947) Diagnostic différentiel histologique de la maladie d'Alzheimer et de la démence sénile: Pathophobie de la zone de projection corticale. Mschr Psychiat Neurol 113:100–117

Room P, Gronewegen HJ (1986) Connections of the parahippocampal cortex. I. Cortical afferents. J Comp Neurol 254:415–450

Simchowicz T (1911) Histologische Studien über die senile Demenz. Histologische und Histopathologische Arbeiten der Großhirnrinde, Bd. 4, S 267–462

Correspondence: Dr. H. Heinsen, Department of Psychiatry, University of Würzburg, Füchsleinstrasse 15, D-8700 Würzburg, Federal Republic of Germany.

Neuronal plasticity of the septo-hippocampal pathway in patients suffering from dementia of Alzheimer type

H.-J. Gertz and J. Cervos-Navarro

Departments of Gerontopsychiatry, and Neuropathology, Free University Berlin, Berlin

Summary

Quantitative morphological examination of nerve cell number of the medial septal nuclei (area CH_1) and of the vertical limb of the diagonal band of Broca (area CH_2) and of spine density of fascia dentata dendrites was performed in DAT and control cases. The data revealed that compensatory axonal sprouting of surviving axons seems to replace lost cholinergic afference of fascia dentata cells after degeneration of the basal forebrain nuclei. Changes of fascia dentata innervation may be an additional pathogenetic factor in dementia of Alzheimer's type.

Introduction

The loss of acetylcholinesterase and choline acetyl transferase in the cerebral cortex and the loss of cholinergic source neurons within the basal forebrain nuclei (bfn) ($CH_1 - CH_4$) are consistent features in cases of dementia of Alzheimer's type (DAT). In order to elucidate the influence of the loss of subcortical cholinergic neurons on postsynaptic morphology of the cholinergic innervated cortical nerve cells we investigated the septo-hippocampal pathway (SHP). The fibers originate from the cholinergic neurons of the nuclei of the medial septum (CH_1) and of the vertical limb of the diagonal band of Broca (CH_2) and are believed to provide a laminar organized cholinergic innervation of the hippocampus and the fascia dentata (FD).

Material and methods

Fourteen brains of autopsied patients were examined. Seven patients had clinically assessed severe dementia, their age being 91.1 ± 1.95 years (mean $\pm$ S.D.).

Seven age matched controls (age 91.5 ± 3.60) had been mentally unimpaired. All had died from internal diseases. Postmortem delay ranged between 2 and 14 h (mean 7.4 ± 4.9) in DAT and between 3 and 6 h (mean 7.7 ± 4.63) in controls. The brains were removed and fixed in formaldehyde 4% for more than half a year.

Tissue blocks containing the area CH_1 and CH_2 were embedded in celloidin, cut in serial coronal sections of 20 µm und stained with Cresyl-fast-violet. At intervals of 400 µm cell counts were performed. In area CH_1 all nucleated nerve cells measuring more than 20 µm and in area CH_2 all nerve cells measuring more than 30 µm were counted. The total cell number of both nuclei was calculated by the simple addition of the cell numbers of the separate levels, multiplied by 20, because every 20th section had been counted. Tissue blocks of the hippocampus were stained by a modified Golgi impregnation technique. After staining the blocks were embedded in paraffin and cut in 75-µm sections.

The total length of the apical dendrites of the granular cells of the FD was measured. The dendrites were subdivided into 10-µm segments starting from the soma, and the spines were counted. Twenty 10-µm segments were studied so that a total length of 200 µm was evaluated. Fifteen granular cells of FD per brain, i.e. 105 cells per group, were examined.

Results

The mean number of neurons in areas CH_1 and CH_2 differed significantly between the two groups. The total length of the dendrites of FD granular cells was 229 ± 25 µm in DAT and 238 ± 32 µm in controls. It did not differ significantly between the two groups (Mann-Whitney U-test, $P > 0.5$). Figure 1 shows the spine density per group ($n = 105$ cells) along the apical dendrites of the granular cells of FD as a function

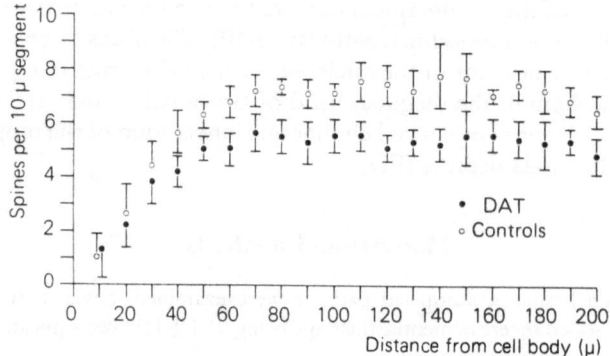

Fig. 1. Graphic representation of spine number of dendrites of FD granular cells as a function of the distance from the cell body. *Filled circles:* mean spine number of DAT cases; *open circles:* mean spine number of controls. Vertical bars show SD

of the distance from the cell body. In the most proximal part, spine density was not significantly different between the two groups (Mann-Whitney U-test, $P > 0.05$). Starting from the 4th segment, there was, however, a statistically significant decrease in spine density (by 25.4% on the average) in the DAT group as compared to controls (Mann-Whitney U-test, $P < 0.001$). Spine density was not significantly correlated to nerve cell number in CH_1 and CH_2 (Spearman rank correlation; $P > 0.05$).

Discussion

Our findings of nerve cell loss of CH_1 and CH_2 are in agreement with previous reports on the bFN (Whitehouse et al., 1981; Wilcock et al., 1983; Lowes-Hummel et al., 1989).

Our data on the total dendritic length of FD granular cells are in part in agreement with those of Flood et al. (1987) and De Ruiter et al. (1987) who found a slight tendency for DAT cases to have shorter FD dendrites than age matched controls. The significant reduction of spine density in the distal part of the dendrites is probably due to the fact that the source neurons of the perforant path, i.d. neurons of lamina II of the area entorhinalis are strongly affected and decreased in SDAT (Hyman et al., 1984). The perforant path innervates the outer ⅔ of FD molecular layer.

It is of particular interest that, in the proximal segments, the differences between DAT and control cases were not statistically significant.

Septo-hippocampal fibers from a dense supragranular innervation immediately above the granular cell layer (Mosko et al., 1973). The cholinergic synapses of the hippocampus and FD have shown to be axodendritic (Shute et al., 1966). There is sufficient evidence that granular cells are excited by cholinergic boutons mainly located on apical dendrites near the soma (Fonnum, 1970).

Disappearance of dendritic spines is shown to follow presynaptic interruption. In cases of specific afferents, spines along only a portion of the dendrites are affected (Paranavelas et al., 1974). One might therefore have expected that the reduction of nerve cells in CH_1 and CH_2 should lead to a pronounced reduction of spines in that part of the dendrites where cholinergic afferents form synapses, i.e. in the most proximal part. But in this dendritic region, spine density actually remained unchanged.

A possible explanation for our findings might be found in compensatory axonal sprouting of surviving axons (Gertz et al., 1987). It is known that terminals lost to a structure after injury to one of its afferents are replaced in part or wholly by sprouting of undamaged input (Cotman et al., 1981). Our data are consistent with replacement of the lost cholinergic afferents in DAT. Two possible mechanisms are suggested: (1) The remaining cholinergic neurons of CH_1 and CH_2 sprout collaterally to maintain the number of synapses with the granular cells to the FD. This is in agreement with the findings that cholinergic muscarinic receptors do not change in number in DAT (Lang and Henke, 1983). On the other hand, the loss of the presynaptic marker ChAT in DAT (Henke and Lang, 1983) may indicate a qualitative change in the axon terminals rather than a loss of synapses. (2) Decreased cholinergic input to the hippocampus might have been substituted by a monoaminergic one as in experimental animals (Crutcher et al., 1981). This suggestion is supported by the finding that in DAT MAO-B activity is significantly increased in the hippocampus (Adolfson et al., 1980). MAO-B is accumulated in synaptosomal fractions (Jellinger and Riederer, 1984).

Our findings indicate that neuronal plasticity plays a role in SDAT.

At present it is unclear if collateral sprouting provides additional stability of neuronal circuits, and leads at least in part to functional recovery of damaged neurone function.

In case transmitter changes actually occur it cannot be excluded that neurone plasticity causes additional irritation within the neuronal network and may be an important factor in the pathogenesis of dementia.

References

Adolfsson R, Gottfries CG, Orland L, Wiberg A, Winblad B (1980) Increased activity of brain and platelet monoamine oxidase in dementia of Alzheimer type. Life Sci 27:1023–1034

Cotman CW, Lewis ER, Hand D (1981) The critical afferent theory: a mechanism to account for septohippocampal development and plasticity. In: Flohr H, Precht W (eds) Lesion-induced neuronal plasticity in sensorimotor systems. Springer, Berlin Heidelberg New York, pp 13–26

Crutcher KA, Brothers L, Davis JN (1981) Sympathetic noradrenergic sprouting in response to central cholinergic denervation: a histochemical study of neuronal sprouting in the rat hippocampal formation. Brain Res 210:115–128

DeRuiter JP, Uylings HBM (1987) Morphometric and dendritic analysis of fascia dentata granule cells in human aging and senile dementia. Brain Res 402:217–229

Flood DG, Buell SJ, Horwitz GJ, Coleman PD (1987) Dendritic extent in human dentate gyrus granule cells in normal aging and senile dementia. Brain Res 402:205–216

Fonnum F (1970) Topographical and subcellular localization of choline acetyltransferase in rat hippocampal region. J Neurochem 17:1029–1037

Gertz H-J, Cervos-Navarro J, Ewald V (1987) The septohippocampal pathway in patients suffering from senile dementia of Alzheimer's type. Evidence for neuronal plasticity? Neurosci Lett 76:228–232

Henke H, Lang W (1983) Cholinergic enzymes in neocortex, hippocampus and basal forebrain of non-neurological and senile dementia of Alzheimer-type patients. Brain Res 267:271–280

Hyman BT, Van Hoesen GW, Damasio AR, Barnes CL (1984) Alzheimer's disease: cell-specific pathology isolates the hippocampal formation. Science 225:1168–1170

Jellinger K, Riederer P (1984) Dementia in Parkinson's disease and (pre)senile dementia of Alzheimer type: morphological aspects and changes in the intracerebral MAO activity. Adv Neurol 40:199–210

Lang W, Henke H (1983) Cholinergic receptor binding and autoradiography in brains of non-neurological and senile dementia of Alzheimer-type patients. Brain Res 267:271–280

Lowes-Hummel P, Gertz H-J, Ferszt R, Cervos-Navarro J (1989) The basal nucleus of Meynert revised: the nerve cell number decreases with age. Arch Gerontol Geriatr 8:21–27

Mosko S, Lynch G, Cotman CW (1973) The distribution of septal projections to the hippocampus of the rat. J Comp Neurol 152:163–174

Paranavelas IG, Lynch G, Brecha N, Cotman CW, Globus A (1974) Spine loss and regrowth in hippocampus following deafferentiation. Nature (London) 248:71–73

Shute CCD, Lewis PR (1966) Electron microscopy of cholinergic terminals and acetylcholinesterase containing neurons in the hippocampal formation of the rat. Z Zellforsch 69:334–343

Whitehouse PJ, Price DL, Clark AW, Coyle JP, DeLong MR (1981) Alzheimer's disease: evidence for selective loss of cholinergic neurons in the nucleus basalis. Ann Neurol 10:122–126

Wilcock GK, Esiri MM, Bowen DM, Smith CCT (1983) The nucleus basalis in Alzheimer's disease: cell counts and cortical biochemistry. Neuropathol Appl Neurobiol 9:175–179

Correspondence: Department of Gerontopsychiatry, Freie Universität Berlin, Eschenallee 3, D-1000 Berlin 19.

Morphology of neurofibrillary tangles and senile plaques

J. Flament-Durand and J. P. Brion

Department of Pathology and Electron Microscopy, Erasmus University Hospital,
Université Libre de Bruxelles, Bruxelles, Belgium

Summary

Abnormal filaments (PHF) accumulate in neurons in Alzheimer's disease in target areas. They fill the pericaryon but are also found in dendrites and axons. Their presence is associated with a disappearance of microtubules and neurofilaments, and an accumulation of dense bodies, altered mitochondria and smooth endoplasmic reticulum. Their ultrastructure differs from normal cytoskeletal fibers but one of their main components are the tau proteins. Senile plaques are composed of a core of amyloid fibers surrounded by abnormal neurites containing PHF and accumulations of organelles, reactive astrocytes and microglia.

Introduction

Neurofibrillary tangles and senile plaques are the hallmark neuropathological lesions of Alzheimer's disease. In quantitative studies, their number has been repeatedly correlated with the severity of dementia (Blessed et al., 1968; Wilcock and Esiri, 1982; Duyckaerts et al., 1987). Although their primary role as the causal factors leading to the dementia syndrome has been questioned (Katzmann et al., 1988) great numbers of neurofibrillary tangles and plaques are never observed in normal people. Unraveling the mechanisms leading to their formation will undoubtedly provide insight into the etiology of this disease.

Although neurofibrillary tangles and senile plaques were first described early this century (Alzheimer, 1907; Fisher, 1907), the mapping of their distribution in human brain in Alzheimer's disease was not investigated in detail until relatively recently. Initial electron microscopy studies (Kidd, 1963; Terry, 1963) demonstrated especially the unique ultrastructure of abnormal filaments in tangles; the ultrastructure of tangles and plaques has since been investigated by many authors. Numerous biochemical data concerning the composition of neurofibrillary tangles and the amyloid of senile plaques has also been obtained in the last years.

Morphological studies, especially in electron microscopy, of Alzheimer's brain tissue and immunocytochemical studies of neurofibrillary tangles and plaques have been performed in our laboratory and we present some of these data in this paper.

Material and methods

Photonic microscopy

Brain tissue samples obtained post-mortem are fixed in 4% formaldehyde and embedded in paraffin. Congo red, Holmes and Bodian's staining methods are used to demonstrate neurofibrillary tangles and senile plaques in tissue sections.

Brain biopsies

Six brain biopsies from the frontal lobes of patients with senile dementia of the Alzheimer type were fixed in 4% glutaraldehyde, post-fixed in 1% osmium and embedded in Epon. Ultrathin sections were counterstained with uranyl acetate and lead citrate, and studied by transmission electron microscopy.

Isolation of neurofibrillary tangles

Neurofibrillary tangles insoluble in sodium dodecyl sulfate were isolated from post-mortem brain tissue by the method of Ihara et al. (1983). Isolated abnormal filaments were adsorbed on carbon-formvar coated grids and negatively stained with 1% phosphotungstic acid. Abnormal filaments adsorbed on formvar coated grids were rotatory shadowed with platinium:iridium.

Preparation of antisera

An antiserum was raised using the enriched fraction of neurofibrillary tangles previously prepared (anti-PHF serum). Other antisera were raised against several cytoskeletal proteins: the microtubule-associated proteins tau and MAP2, prepared by thermodenaturation of microtubules polymerized in vitro, and separated on SDS-polyacrylamide gels. A serum was also raised against a β-galactosidase fusion protein corresponding to a MAP2 cDNA. The anti-ubiquitin serum was prepared in Dr. B. H. Anderton's laboratory.

Immunocytochemistry was performed with these sera on paraffin sections using the PAP method (Sternberger, 1979). The immunogold labeling method (Demey et al., 1981) was used for labeling isolated PHF for electron microscopy.

Results

Photonic microscopy

Neurofibrillary tangles are observed in the neocortex, hippocampus, amygdala, Meynert nucleus basalis, locus coeruleus and raphé's nuclei (Tomlinson et al., 1981; Mann, 1985). Senile plaques are observed chiefly in neocortex and hippocampus and occasionally in subcortical areas and cerebellum. Silver staining (Fig. 1) demonstrates the greatest number of lesions, but after Congo red staining neurofibrillary tangles and some plaques can also be observed due to their birefringence when observed under crossed Nicolls filters.

Ultrastructural observations

The neurofibrillary tangles are made of bundles of 20–25 nm wide abnormal filaments showing regular constrictions up to 10 nm wide every 80 nm and are known as "paired helical filaments" (PHF). Electron microscopy indicates that they are located in the neuronal pericarya, in axons and in dendrites. They are more frequent in non-myelinated processes but are also present in myelinated fibers (Fig. 2). Morphologically normal synapses make contact with these fibers (Fig. 3). A striking observation in these biopsies is that in the neurites full of PHF, there is almost a complete disappearance of the normal microtubules (Flament-Durand and Couck, 1979; Dustin and Flament-Durand, 1982) associated with accumulation of osmiophilic dense bodies, altered mitochondria (Figs. 4 and 5), and smooth endoplasmic reticulum (Richard et al., 1989).

After platinium : iridium shadowing of isolated PHF, a regular right-handed twisting is clearly visible (Fig. 7 A) (Brion et al., 1984).

The "classical" senile plaque is composed of an extracellular core of amyloid surrounded by reactive glial cells, microglia and neurites filled with PHF (Fig. 6). Here also, electron microscopy clearly indicates that these abnormal neurites contain accumulation of dense bodies, altered mitochondria and smooth endoplasmic reticulum. The amyloid core is made of packed 10 nm wide amyloid fibrils which do not show the regular constrictions of PHF.

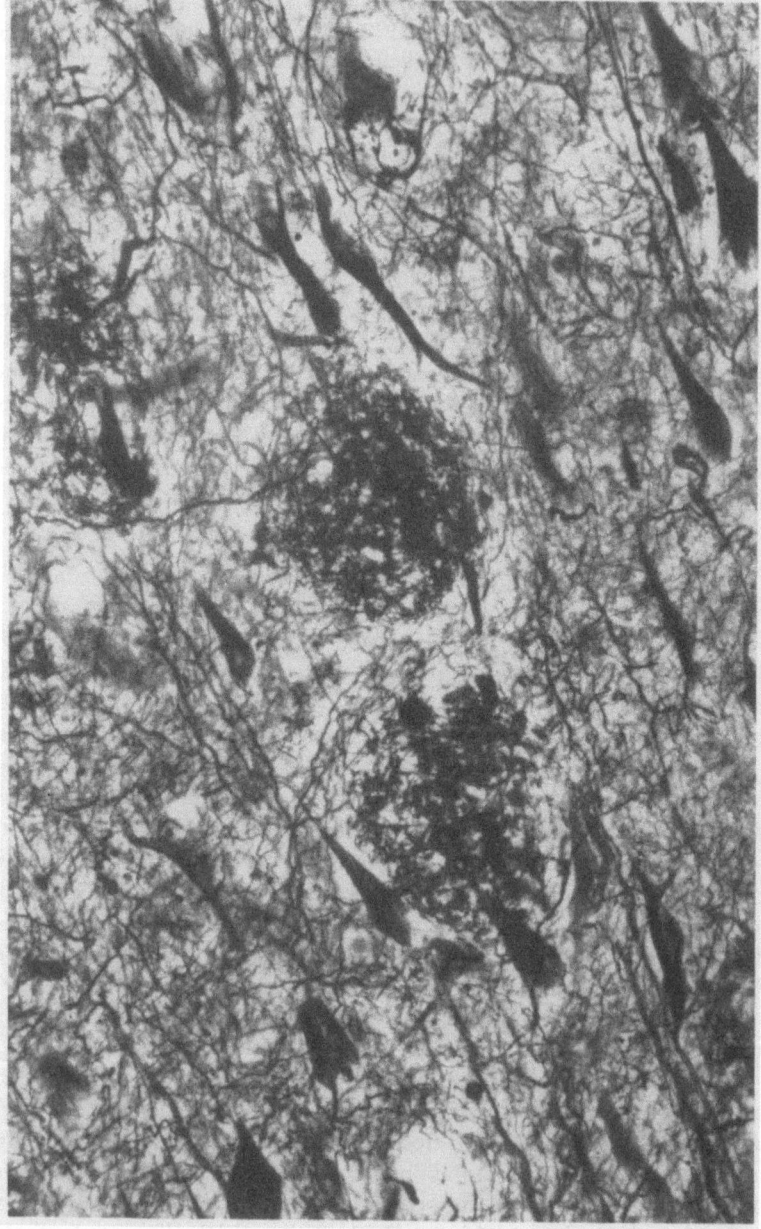

Fig. 1. Paraffin section. Silver staining demonstrating numerous intraneuronal tangles and neuritic plaques. × 650

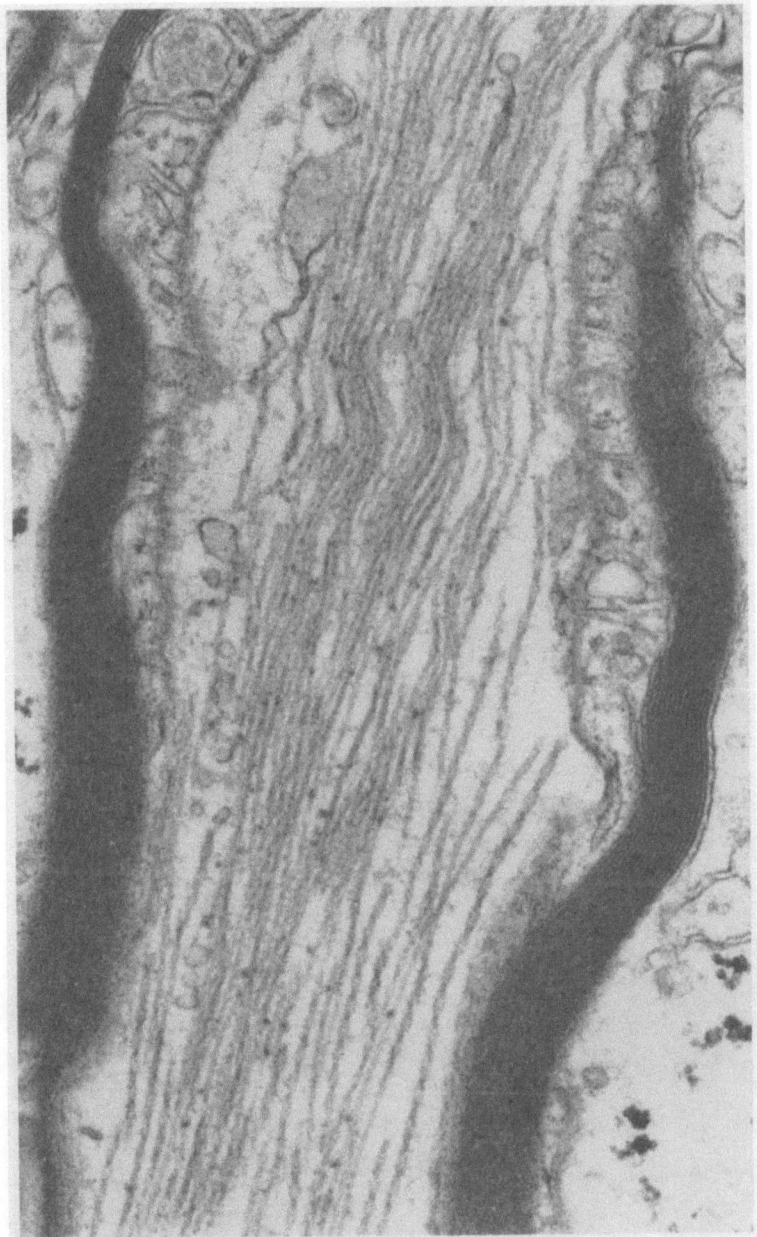

Fig. 2. Electron microscopy. Myelinated axon filled by paired helical filaments. Disappearance of normal microtubules and neurofilaments. ×48 000

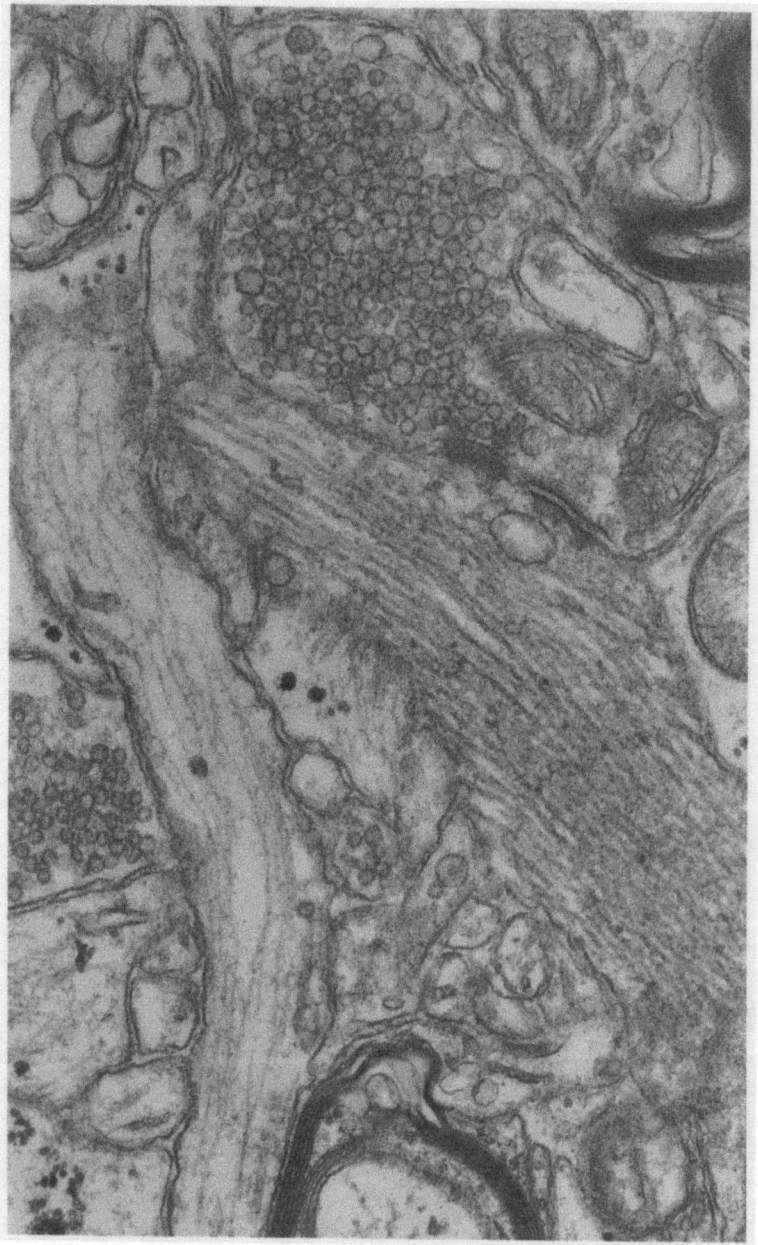

Fig. 3. Electron microscopy. A morphologically normal synapse makes contact with a dendrite full of paired helical filaments. × 48 000

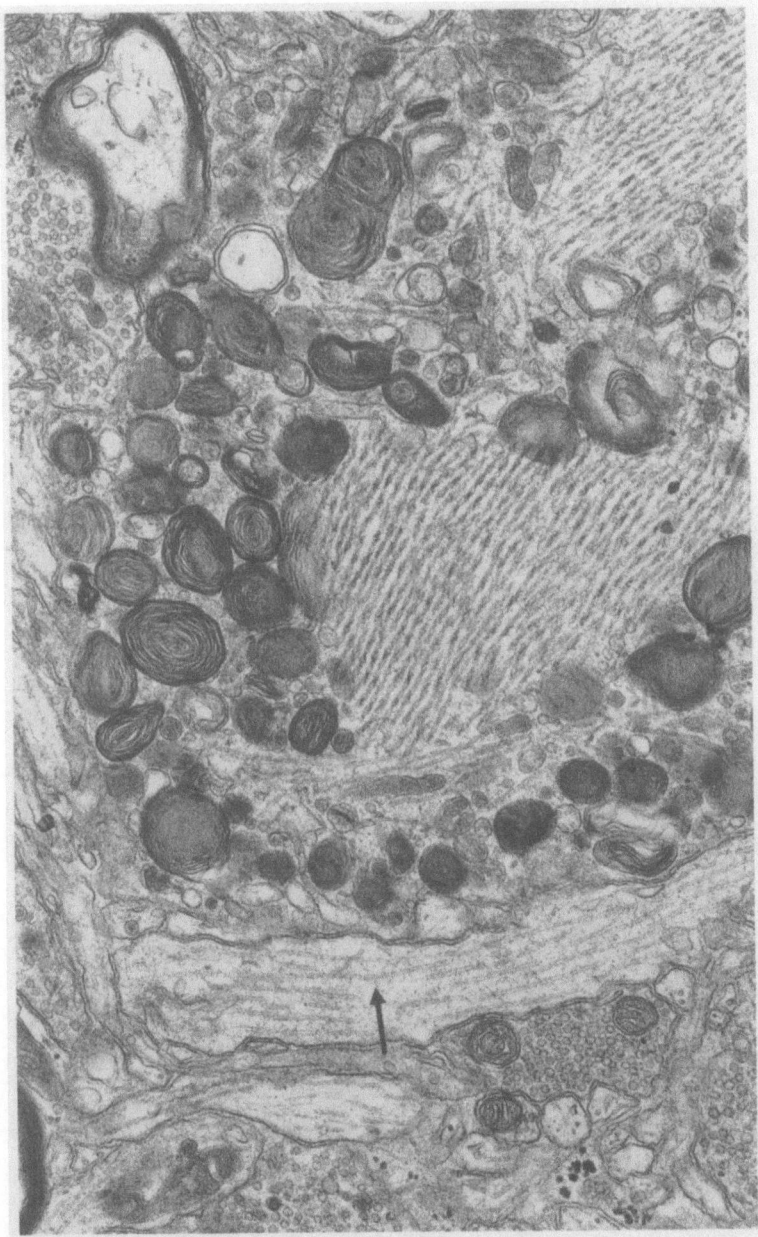

Fig. 4. Electron microscopy. Close to a neurite full of normal microtubules (↓) an en-larged neurite is seen full of paired helical filaments and accumulation of multilamellar osmiophilic bodies. × 28 800

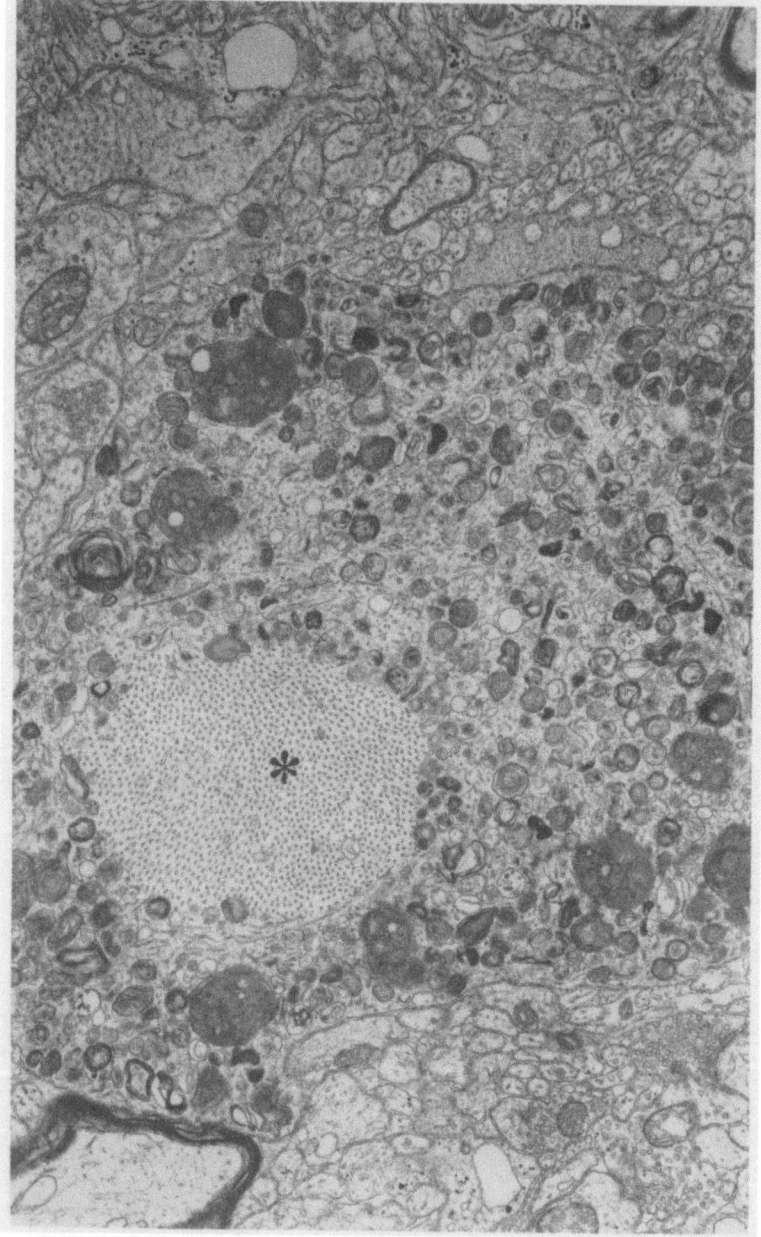

Fig. 5. Electron microscopy. Enlarged neurite containing paired helical filaments cut transversally (∗) surrounded by numerous multilamellar osmiophilic bodies and abnormal mitochondria. ×7650

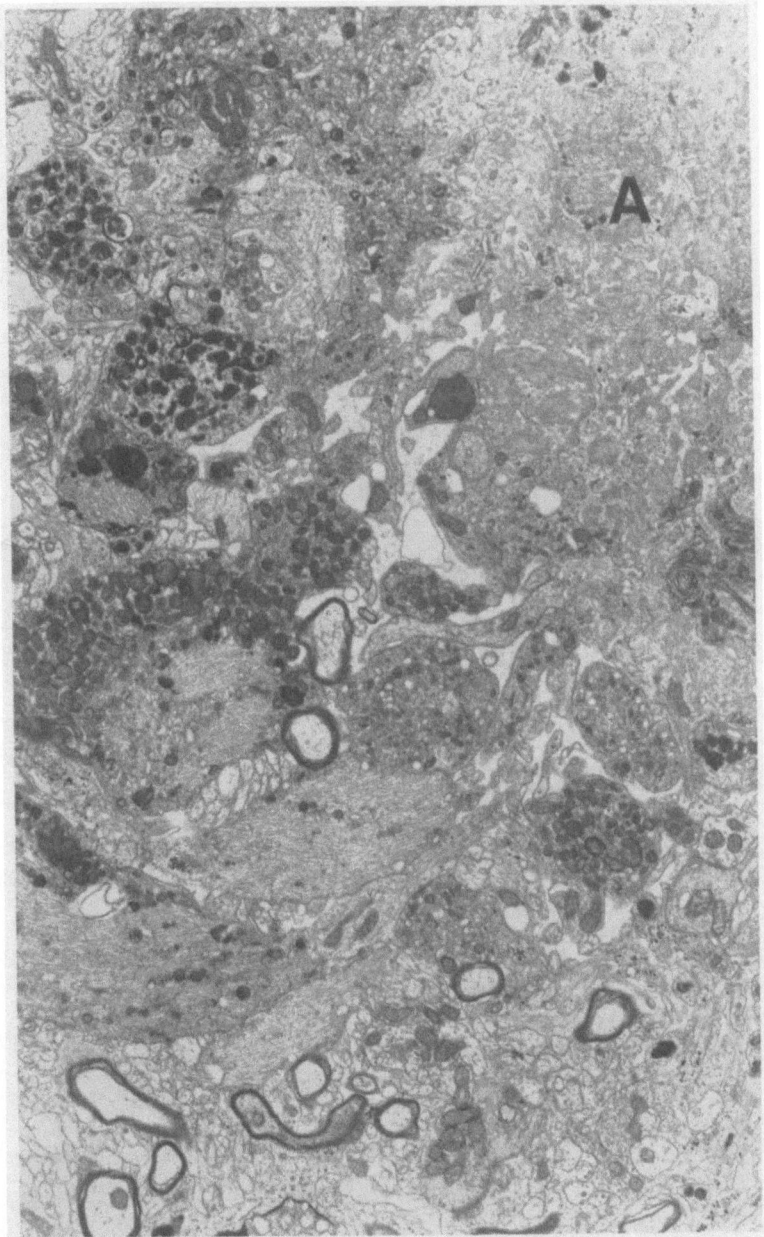

Fig. 6. Electron microscopy. Part of a neuritic plaque showing the core of amyloid (A) surrounded by abnormal neurites containing paired helical filaments and accumulation of dense osmiophilic bodies. × 5520

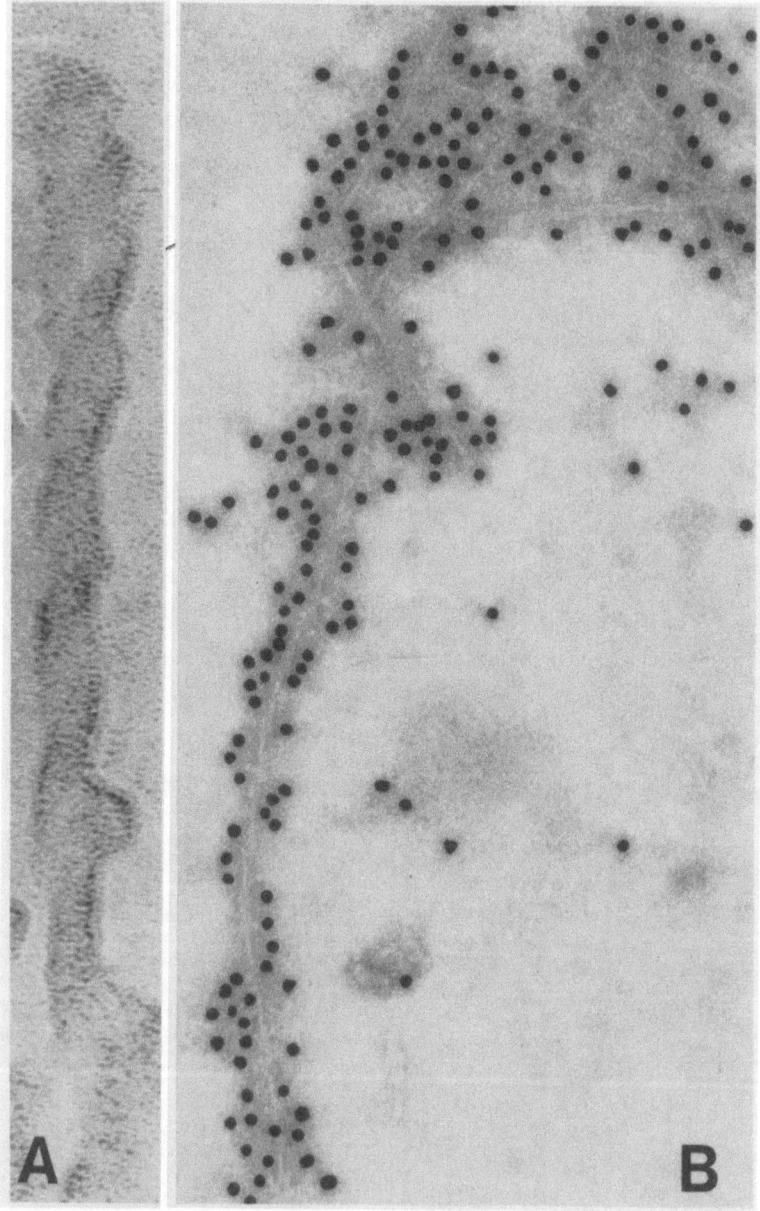

Fig. 7. Electron micrographs of isolated paired helical filaments. **A** After shadowing with platinium:iridium, the right-handed helix with a periodic torsion every 80 nm is clearly demonstrated. × 360 000. **B** Immunogold labeling with an anti-tau serum. × 120 000

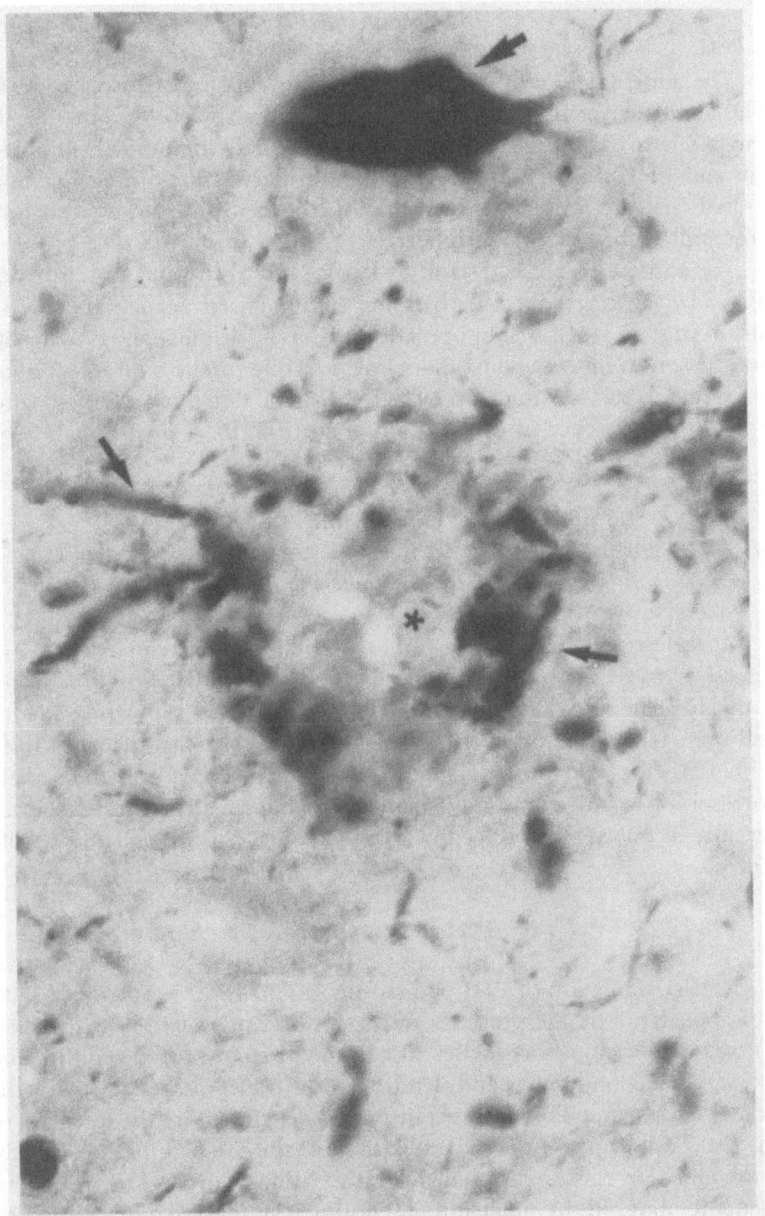

Fig. 8. Paraffin section. Immunolabelling of neurofibrillary tangles (⬇) and plaque neu-rites (⬇) with the anti-PHF serum. A Congo red counterstaining shows the unlabeled amyloid core (★) of the plaque. ×1250 (Reprinted from Flament-Durand J, Brion JP. In: Interdisciplinary topics in gerontology, vol 26, pp 56–62, with permission of Karger, Basel)

Immunohistochemistry

The anti-PHF serum strongly labels NFT in situ and isolated PHF. The abnormal neurites of many plaques are also labeled but not the amyloid core (Brion et al., 1985a) (Fig. 8). Double-immunolabeling with this anti-PHF and an anti-GFAP (glial fibrillary acidic protein) demonstrated that astrocytes are surrounding the neuritic plaque with long processes embracing its periphery.

The anti-tau serum also strongly labels NFT in situ and after isolation (Fig. 7B) (Brion et al., 1985b). The serum raised against the MAP2 fusion protein labels a subpopulation of NFT (Brion et al., 1989a), but not the serum raised against the whole MAP2 protein. The anti-ubiquitin serum labels a subgroup of NFT in situ but most of them after isolation in presence of SDS (Brion et al., 1989b).

Discussion

Subcellular distribution of PHF

Our electron microscopy studies of brain biopsies in Alzheimer's disease have emphasized the wide distribution of PHF in all neuronal compartments, i.e. the pericaryon, the dendrites and axons (myelinated and unmyelinated), and even at the synaptic level. Small neurites filled with PHF are also observed in electron microscopy, and they could correspond to the numerous small-sized abnormal neurites described recently in photonic microscopy (Braak et al., 1986; Kowall and Kosik, 1987).

Disturbances of axoplasmic transports

The accumulation of PHF in neurons and neurites is associated with a disappearance of microtubules and accumulation of various organelles. These membranous organelles are normally transported antero- and retrogradely along axons and dendrites, this transport belonging to the "quick axoplasmic flow"; this transport is dependent on the integrity of the microtubule network. Accumulations of smooth endoplasmic reticulum similar to those observed in PHF-bearing neurons can be experimentally induced by microtubule poisons and acrylamide (Chretien et al., 1981). Thus these observations strongly suggest that the presence of PHF is associated with disturbances of axoplasmic flows (Flament-Durand and Couck, 1979; Dustin and Flament-Durand, 1982; Gray et al., 1987; Richard et al., 1989), especially the transport of membranous organelles. If the formation of PHF directly induces these disturbances of axoplasmic flows, this would primarily impair neuron metabolism.

The senile plaque

In electron microscopy the abnormal neurites surrounding the amyloid core of senile plaques show the same accumulation of organelles in association with PHF and have been observed by other authors (Gonatas et al., 1967); they suggest the existence of flow disturbances also at the level of these neurites. According to the classical view some plaques do not contain amyloid core and others would consist of amyloid without or with very few abnormal neurites; very "immature plaques" have also been described recently in light microscopy (Probst et al., 1987). The exact evolutionary relationship, if any, between these different aspects of plaques is still hypothetic. We thus do not yet know whether these abnormal neurites in plaques develop primarily, after amyloid fiber deposition or simultaneously. Although some neurites cross senile plaque apparently without being affected, others show morphological changes which could correspond to "sprouting" phenomena (Probst et al., 1983).

Ultrastructurally amyloid fibers (6–10 nm wide filaments) are clearly different from PHF, nor do they share the same antigenic cross-reactivities (this volume) and are essentially composed of a 42 aminoacids polypeptide (A4 polypeptide) (Masters et al., 1985). The same polypeptide accumulates in the walls of small cerebral blood vessels (Congophilic angiopathy) (Glenner and Wong, 1984). Amyloid deposits in plaques could be often in close relationship with vascular amyloid (Miyakawa et al., 1982).

Substructure of PHF

Initial studies by Kidd (1963) and Terry (1963) indicated that abnormal filaments constituting the neurofibrillary tangle are made of two filaments intertwined around each other or of a "twisted tubule" or ribbon (Hirano et al., 1968). Using a goniometric stage, Wisniewski et al. (1976) confirmed the helicoidal nature of these abnormal filaments. Differences in the periodicity of PHF constriction and the presence of straight filaments in Alzheimer's disease and in other diseases where neurofibrillary tangles develop (supranuclear palsy, Down's syndrome) have been reported by other authors (Monpetit et al., 1985). These different types of filaments however keep the same antigenic cross-reactivities.

A protofilamentous substructure of PHF has been described by some authors (Brion et al., 1984; Wisniewski et al., 1984). Shadowing of PHF allowed to demonstrate that these filaments exhibit mainly a right-handed helix, although a mixture of right- and left-handed PHF has been reported. In a study using image analysis and reconstruction, Wischik et al. (1985) and Crowther and Wischik (1985) have proposed a high-definition structural model of PHF: they would be made by the stacking of globular units aligned perpendicularly to the long axis of the filament and would give a general shape evoking rather a twisted ribbon. The volume of the subunit would account for a 100 kDa polypeptide.

All these ultrastructural studies point to the morphological differences between PHF and the normal microtubules and neurofilaments.

Immunolabeling of neurofibrillary tangles

Neurofibrillary tangles are labeled by antibodies against several cytoskeletal proteins. A detailed account of these immunoreactivities is presented elsewhere in this volume. The more consistent finding is the labeling of neurofibrillary tangles, in situ as well as after isolation with strong detergents, by all antibodies raised against the microtubule-associated tau proteins (Brion et al., 1985 b; Kosik et al., 1986; Delacourte and Defossez, 1986). Tau proteins have also been identified as a component of PHF by aminoacid sequencing of peptides obtained from pure preparation of PHF (Goedert et al., 1988). PHF share also epitopes with neurofilament (Miller et al., 1986) and MAP2 (Kosik et al., 1984; Brion et al., 1989 b). Tau proteins incorporated in PHF could be abnormal; since these proteins are involved in the polymerization and the stability of the microtubules, they could be unable in PHF-bearing neurons to participate in microtubule polymerization and turn-over (Iqbal et al., 1986). This would lead to the morphological changes of the cytoskeleton observed in these neurons.

There is a significant correlation between the number of neurofibrillary tangles detected by the anti-PHF and the anti-tau antibodies and the gravity of dementia (Duyckaerts et al., 1987; Delaère et al., 1989).

In our experience there is no antigenic cross-reactivity between the amyloid component of plaques and neurofibrillary tangles. Their common property of birefringence after Congo red staining is related to the presence of polypeptide chains in β-pleated sheet conformation.

Ubiquitin is also a component of PHF, at least a proportion of them (Mori et al., 1987; Perry et al., 1987; Brion et al., 1989 a). Since ubiquitin is involved in a non-lysosomial proteolytic pathway, the presence of this molecule on PHF could witness an attempt by the affected neurons to remove these abnormal fibers.

Acknowledgements

This work was supported by grants from the Belgian FNRS, FRFC, FRSM (no 96 048 082, 90 548 082, 90 148 082).

References

Alzheimer A (1907) Über eine eigenartige Erkrankung der Hirnrinde. Allgemeine Zeitschrift für Psychiatrie 64:146–148

Blessed G, Tomlinson BE, Roth M (1968) The association between quantitative measurements of dementia and of senile changes in the cerebral gray matter of elderly subjects. J Psychiat 114:797–805

Braak H, Braak E, Grundke-Iqbal I, Iqbal K (1986) Occurrence of neuropil threads in the senile human brain and in Alzheimer's disease: a third location of paired helical filaments outside of neurofibrillary tangles and neuritic plaques. Neurosci Lett 65:251–355

Brion JP, Couck AM, Flament-Durand J (1984) Ultrastructural study of enriched fractions of "tangles" from human patients with senile dementia of the Alzheimer type. Acta Neuropathol (Berl) 64:148–152

Brion JP, Couck AM, Passareiro H, Flament-Durand J (1985a) Neurofibrillary tangles of Alzheimer's disease: an immunocytochemical study. J Submicrosc Cytol 17:89–96

Brion JP, Passareiro H, Nunez J, Flament-Durand J (1985b) Mise en évidence immunologique de la protéine tau au niveau des lésions de dégénérescence neurofibrillaire de la maladie d'Alzheimer. Arch Biol (Brux) 95:229–235

Brion JP, Power D, Hue D, Couck AM, Anderton BH, Flament-Durand J (1989a) Heterogeneity of ubiquitin immunoreactivity in neurofibrillary tangles of Alzheimer's disease. Neurochem Int 14:121–128

Brion JP, Couck AM, Cheetham ME, Hanger D, Anderton BH, Flament-Durand J (1989b) A cDNA encodes epitopes shared between microtubule-associated protein MAP2 and Alzheimer neurofibrillary tangles: in situ hybridization and immunocytochemistry. In: Christen Y (ed) Biological markers of Alzheimer's disease. Research and perspectives in Alzheimer's disease. Springer, Berlin Heidelberg New York Tokyo, pp 56–65

Chretien M, Patey G, Souyri F, Droz B (1981) "Acrylamide-induced" neuropathy and impairment of axonal proteins. II. Abnormal accumulations of smooth endoplasmic reticulum at sites of focal retention of fast transported proteins. Electron microscope radioautographic study. Brain Res 205:15–28

Crowther RA, Wischik CM (1985) Image reconstruction of the Alzheimer paired helical filament. EMBO J 4:3661–3665

Delacourte A, Defossez A (1986) Alzheimer's disease: tau proteins, the promoting factors of microtubule assembly, are major components of paired helical filaments. J Neurol Sci 76:173–186

De Mey J, Moeremans M, Geuens G, Nuydens R, De Brabander M (1981) High-resolution light and electron-microscopic localization of tubulin with the IGS (immunogold staining) method. Cell Biol Int Rep 5:889–899

120   J. Flament-Durand and J. P. Brion

Dustin P, Flament-Durand J (1982) Disturbances of axoplasmic transport in Alzheimer's disease. In: Weiss DG, Gorio A (eds) Axoplasmic transport in physiology and pathology. Springer, Berlin Heidelberg New York, pp 131–136

Delaere P, Duyckaerts C, Brion JP, Poulain V, Hauw JJ (1989) Tau, paired helical filaments and amyloid in the neocortex: a morphometric study of 15 cases with graded intellectual status in aging and senile dementia of Alzheimer type. Acta Neuropathol (Berl) 77:645–653

Duyckaerts C, Brion JP, Hauw JJ, Flament-Durand J (1987) Comparison of immunocytochemistry with a specific antibody and Bodian's protargol method. Quantitative assessment of the density of neurofibrillary tangles and senile plaques in senile dementia of the Alzheimer type. Acta Neuropathol (Berl) 73:167–170

Fischer O (1907) Miliare Nekrosen mit drusigen Wucherungen der Neurofibrillen, eine regelmäßige Veränderung der Hirnrinde bei seniler Demenz. Monatsschrift für Psychiatrie und Neurologie 22:361–372

Flament-Durand J, Couck AM (1979) Spongiform alterations in brain biopsies of presenile dementia. Acta Neuropathol (Berl) 46:159–162

Glenner GG, Wong CW (1984) Alzheimer's disease and Down's syndrome: sharing of a unique cerebrovascular amyloid fibril protein. Biochem Biophys Res Comm 122:1131–1135

Goedert M, Wischik CM, Crowther RA, Walker JE, Klug A (1988) Cloning and sequencing of the cDNA encoding a core protein of the paired helical filament of Alzheimer's disease: identification as the microtubule-associated protein tau. Proc Natl Acad Sci USA 85:4051–4055

Gonatas NK, Anderson W, Evangelistica I (1967) The contribution of altered synapses in the senile plaque: an electron microscope study in Alzheimer's dementia. J Neuropathol Exp Neurol 26:25–39

Gray EG, Paula-Barbosa M, Rober A (1987) Alzheimer's disease: paired helical filaments and cytomembranes. Neuropathol Appl Neurobiol 13:91–110

Hirano A, Dembitzer HM, Kurland LT (1968) The fine structure of some intraganglionic alterations. J Neuropathol Exp Neurol 27:167–182

Ihara Y, Abraham C, Selkoe DJ (1983) Antibodies to paired helical filaments in Alzheimer's disease do not recognize normal brain proteins. Nature 304:727–730

Iqbal K, Grundke-Iqbal I, Zaidi T, Merz PA, Wen GY, Shaikh SS, Wisniewski HM (1986) Defective brain microtubule assembly in Alzheimer's disease. Lancet i:421–426

Katzman R, Terry RD, De Teresa R, Brown T, Davies P, Fuld P, Renbing X, Peck A (1988) Clinical, pathological and neurochemical changes in dementia: a subgroup with preserved mental status and numerous neocortical plaques. Ann Neurol 23:138–144

Kidd M (1963) Paired helical filaments in electron microscopy of Alzheimer's disease. Nature 197:192–193

Kosik KS, Duffy LK, Dowling MM, Abraham C, McCluskey A, Selkoe DJ (1984) Microtubule-associated protein 2: monoclonal antibodies demonstrate the selective incorporation of certain epitopes into Alzheimer neurofibrillary tangles. Proc Natl Acad Sci USA 81:7941–7945

Kosik KS, Joachim CL, Selkoe DJ (1986) The microtubule-associated protein, tau, is a major antigenic component of paired helical filaments in Alzheimer's disease. Proc Natl Acad Sci USA 83:4044–4048

Kowall NW, Kosik KS (1987) Axonal disruption and aberrant localization of tau protein characterize the neuropil pathology of Alzheimer's disease. Ann Neurol 22:639–643

Mann DMA (1985) The neuropathology of Alzheimer's disease: a review with pathogenetic, aetiological and therapeutic considerations. Mech Ageing Dev 31:213–255

Masters CL, Simms G, Weinman NA, Multhaup G, McDonald BL, Beyreuther K (1985) Amyloid plaque core protein in Alzheimer disease and Down syndrome. Proc Natl Acad Sci USA 82:4245–4249

Miller CCJ, Brion JP, Calvert R, Chin TK, Eagles PAM, Downes MJ, Flament-Durand J, Haugh M, Kahn J, Probst A, Ulrich J, Anderton BH (1986) Alzheimer's paired helical filaments share epitopes with neurofilament side arms. EMBO J 5:269–276

Miyakawa T, Shimuji A, Kuramoto R, Higuchi Y (1982) The relationship between senile plaques and cerebral blood vessel in Alzheimer's disease and senile dementia. Virchows Arch (Cell Pathol) 40:121–129

Monpetit V, Clapin DF, Guberman A (1985) Substructure of 20 nm filaments of progressive supranuclear palsy. Acta Neuropathol (Berl) 68:311–318

Mori H, Kondo J, Ihara Y (1987) Ubiquitin is a component of paired helical filaments in Alzheimer's disease. Science 325:1641–1644

Perry G, Friedman R, Shaw G, Chau V (1987) Ubiquitin is detected in neurofibrillary tangles and senile plaque neurites of Alzheimer's disease brains. Proc Natl Acad Sci USA 84:3033–3036

Probst A, Basler V, Bron B, Ulrich J (1983) Neuritic plaques in senile dementia of the Alzheimer type: a Golgi analysis in the hippocampal region. Brain Res 268:249–254

Probst A, Brunnschweiler H, Lautenschlager C, Ulrich J (1987) A special type of senile plaque, possibly an initial stage. Acta Neuropathol (Berl) 74:133–141

Richard S, Brion JP, Couck AM, Flament-Durand J (1989) Accumulation of smooth endoplasmic reticulum in Alzheimer's disease: new morphological evidence of axoplasmic flow disturbances. J Submicrosc Cytol 21:461–467

Sternberger LA (1979) Immunocytochemistry. Wiley, New York

Terry RD (1963) The fine structure of neurofibrillary tangles in Alzheimer's disease. J Neuropathol Exp Neurol 22:629–642

Tomlinson BE, Irving D, Blessed G (1981) Cell loss in the locus caeruleus in senile dementia of Alzheimer type. J Neurol Sci 49:419–428

Wilcock GK, Esiri MM (1982) Plaques, tangles and dementia. A quantitative study. J Neurol Sci 56:343–356

Wischik CM, Crowther RA, Stewart M, Roth M (1985) Subunit structure of paired helical filaments in Alzheimer's disease. J Cell Biol 100:1905–1912

Wisniewski HM, Narang HK, Terry RD (1976) Neurofibrillary tangles of paired helical filaments. J Neurol Sci 27:173–181

Wisniewski HM, Merz PA, Iqbal K (1984) Ultrastructure of paired helical filaments of Alzheimer's neurofibrillary tangles. J Neuropathol Exp Neurol 43:643–656

Correspondence: Prof. J. Flament-Durand, Department of Pathology, Erasmus University Hospital, Université Libre de Bruxelles, 808, route de Lennik, B-1070 Brussels, Belgium

An in vitro model for the study of the neurofibrillary degeneration of the Alzheimer type

S. Flament, A. Delacourte, and A. Défossez

Unité INSERM n° 16, Laboratoire de Neurosciences, Faculté de médecine, Lille, France

Summary

We have recently reported that two abnormally phosphorylated Tau proteins (Tau 64 and 69) were systematically present in Alzheimer brain areas which are affected by the neurofibrillary degeneration (NFD). We suggest here that these likely early markers of the NFD might define an in vitro model for the study of the NFD. A such model might be useful to verify the different etiopathogenic hypothesis which have been proposed up to here.

Introduction

Alzheimer's disease (AD) is characterized histopathologically by the presence of numerous neurofibrillary tangles (NFT) and senile plaques (SP) in the brain of affected individuals (Hansen et al., 1988). NFT consist ultrastructurally of the accumulation of bundles of paired helical filaments (PHF) in the cytoplasm of neurons (Kidd, 1964). Tau proteins, the promoting factor of microtubule assembly, have been shown to be the major antigenic component of these PHF (Brion et al., 1985; Delacourte and Defossez, 1986; Grundke-Iqbal et al., 1986a; Kosik et al., 1986; Nukina and Ihara, 1986; Wood et al., 1986). Nevertheless, the reason for the incorporation of these normal cytoskeletal components in such insoluble structures remains unknown. Indeed, even if an abnormal phosphorylation of Tau proteins has been suggested (Grundke-Iqbal et al., 1986b; Ihara et al., 1986; Wood et al., 1986), this theory is still discussed (Ksiezak-Reding et al., 1988).

We have reported recently (Flament and Delacourte, 1989), that two abnormally phosphorylated Tau proteins (Tau 64 and 69) are always found in Alzheimer brain areas containing NFT and SP. These proteins might appear at early stages of the disease and the phosphorylation of Tau proteins might be the trigger that enhance their incorporation into insoluble PHF structures (Flament and Delacourte, 1989).

We suggest here that using such early markers of the NFD, experimental models for the study of the NFD might be established and that the different hypothesis about the etiopathogenesis of AD might be verified.

Material and methods

Clinical datas about the 10 patients with AD and the 10 age-matched controls have been published (Flament and Delacourte, 1989).

1) Immuno-histochemical studies: one cerebral hemisphere (randomly left or right) was fixed in formalin and different brain areas were then dissected. After paraffin embedding, serial sections were cut at a thickness of 7,5 µm and were examined immunohistochemically for the presence of NFT and SP.

2) Immuno-blot studies: the other cerebral hemisphere was frozen. Brain areas corresponding to those previously described for histology were dissected and treated for the analysis by western-blotting as in Flament and Delacourte (1989).

3) Antisera: the anti-PHF was raised against PHF extracted from the brain of a patient with AD (Delacourte and Défossez, 1986). For the anti-human Tau antiserum, the immunogen was a heat-stable preparation of Tau proteins obtained from a control brain and further purified by preparative SDS-PAGE (Defossez et al., 1988). These two antisera have identical biochemical and histochemical properties as shown by numerous studies (Delacourte and Defossez, 1986; Defossez et al., 1988).

4) Alkaline phosphatase treatment: a heat-stable preparation of Tau proteins extracted from the temporal cortex of a patient with AD was dialyzed against adequate buffer and mixed with calf intestine alkaline phosphatase as described in Flament and Delacourte (1989). A control was done by addition of sodium pyrophosphate at 100 mg/ml.

Results

On immunoblots of control brain homogenates, only the normal set of Tau proteins with molecular weight (MW) ranging from 45 to 62 kDa was detected by our two antisera. This pattern was found whatever the area (for example see lane 1 on Fig. 1: temporal cortex), except in the hippocampus of an eighty years old control where two additional Tau variants were detected at 64 and 69 kDa. On immunoblots of AD brain homogenates corresponding to areas such as thalamus, cerebellum or caudate nucleus (lane 3 in Fig. 1), the Tau staining pattern was quite

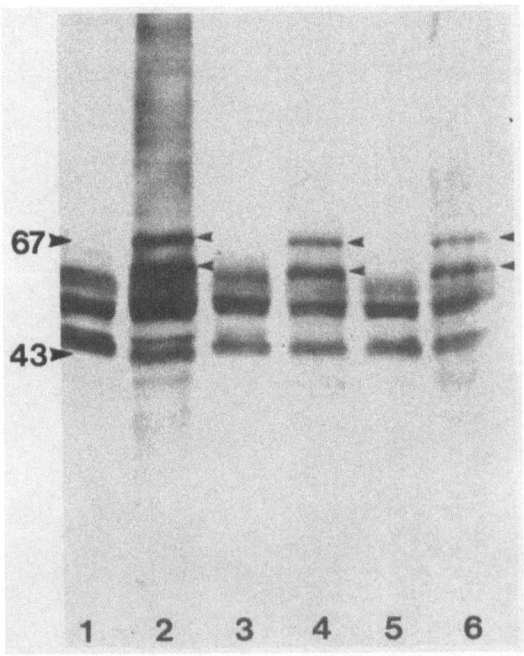

Fig. 1. Immunodetection and characterization of Tau 64,69. Lane 1: temporal cortex from a control patient; lane 2: temporal cortex from a patient with AD; lane 3: caudate nucleus from a patient with AD; lane 4: frontal cortex from a patient with AD; lane 5: the same sample than in lane 4 after four hours of incubation with the alkaline phosphatase; lane 6: same sample where sodium pyrophosphate was added. In homogenates of Alzheimer brain areas that are affected by the process of neurofibrillary degeneration, two additional Tau variants were detected by the anti-PHF at 64 and 69 kDa (arrowheads). After alkaline phosphatase treatment, Tau 64,69 disappeared whereas the immunostaining profile did not change when sodium pyrophosphate was added to the sample; their heavy MW is due to an abnormal phosphorylation. Standard MW markers are expressed in kDa

similar to those usually seen in control brains. In areas such as hippocampus, frontal (lane 4 in Fig. 1) or temporal cortex (lane 2 in Fig. 1), two additional Tau variants were detected at 64 and 69 kDa.

In order to know if the heavy MW of these pathological Tau variants was the result of an abnormal phosphorylation (Baudier and Cole, 1987), we have dephosphorylated a heat-stable preparation of Tau proteins extracted from the brain of a patient with AD where Tau 64 and 69 had already been detected (Fig. 1 lane 4). After four hours of incubation with the calf intestine alkaline phosphatase, the two Tau variants disappeared while the bands between 45 and 62 kDa were more strongly stained (Fig. 1 lane 5). In the sample where sodium pyrophosphate had been added, Tau 64 and 69 were still detected at the end of the incubation time

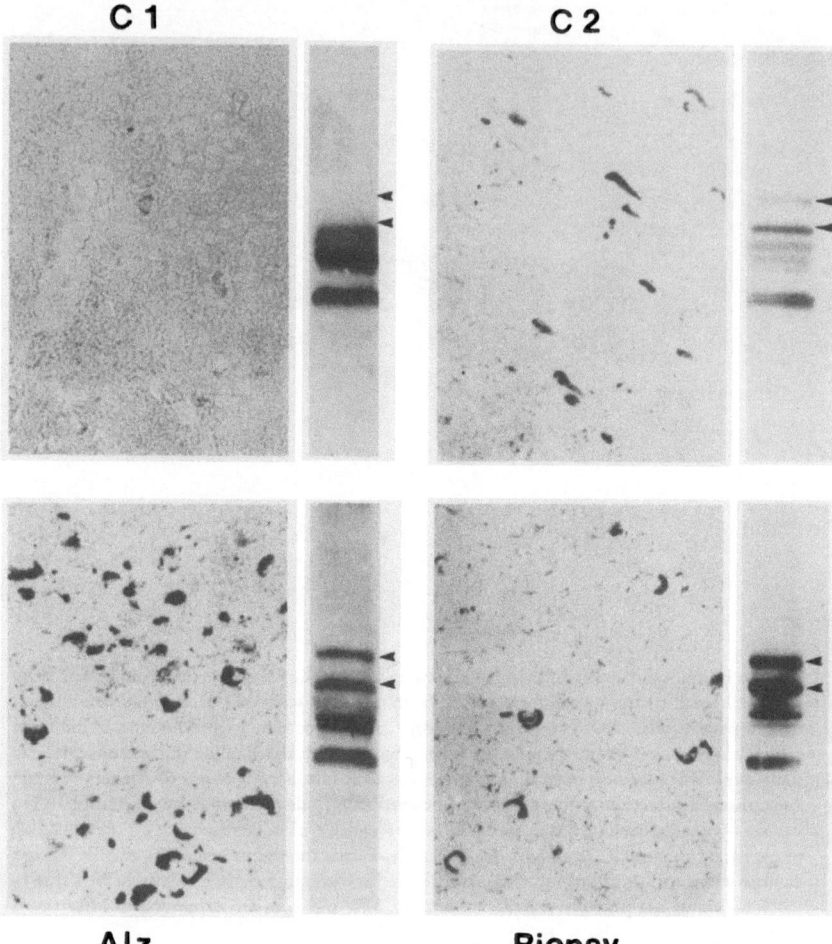

Fig. 2. Tau 64,69 and neurofibrillary degeneration. Tissue sections and corresponding immunoblots stained with the anti-PHF. C1: temporal cortex from a 52 years old control; C2: hippocampus from an 80 years old control; Alz: frontal cortex from a 73 years old Alzheimer patient; Biopsy: frontal biopsy from a 58 years old patient who began to develop an unusual dementia. In autopsic pieces, Tau 64 and 69 (arrowheads) are more strongly detected in areas which contain a much higher density of NFT and SP. In the frontal biopsic tissue, Tau 64 and 69 are more strongly detected than in the frontal autopsic tissue though the density of NFT and SP is not higher (compare Alz and Biopsy). Therefore, it seems that the phosphorylation of Tau proteins might appear before their incorporation in PHF and that Tau 64 and 69 are early reliable markers of the neurofibrillary degeneration

(Fig. 1 lane 6). These results demonstrated that the decrease of the MW of Tau 64 and 69 was due to their dephosphorylation rather than to proteolysis (proteases often contaminate phosphatase preparations).

When the results of the immuno-histochemical study were correlated with those of the immuno-blot study, it appeared that the detection of Tau 64 and 69 closely followed the presence of the neurofibrillary degeneration. Indeed, Tau 64 and 69 were only found in areas showing NFT and SP and were more strongly detected in the regions that are more seriously affected during the disease (compare C1, C2 and Alz in Fig. 2).

We had the opportunity to analyze a frontal biopsy from a 52 years old patient who began to develop an unusual dementia with atypical clinical signs. Tau 64 and 69 were present at a much higher level than in all the autopsic brain tissue which were studied up to here (Fig. 2 biopsy). We concluded to the presence of NFD in this area, which was later confirmed by the histopathologist (Fig. 2 biopsy). Therefore, Tau 64 and 69 have a reliable diagnostic value for the presence of the NFD in brain tissue.

Discussion

Progress in the etiology of AD are hampered by the lack of either in vitro or animal models. Indeed, most of the investigations are done on post-mortem material which reveals only the ultimate stages of the pathological changes. The discovery of Tau 64 and 69 that are likely early markers of the NFD (they are always detected in conditions that are not able to solubilize the PHF – low amounts of SDS – and besides they are present in higher amounts in the biopsy than in autopsic tissue), opens up the possibility that experimental models for the study of the NFD might be established.

At first, the phosphorylation of heat-stable preparations of human Tau proteins from control brains might be attempted using different kinases. This further characterization of Tau 64 and 69 will get clues about the enzymes that are involved in the process of neurofibrillary degeneration.

Given such data, it will be attempted to induce as precisely as possible the formation of Tau 64 and 69 in nerve cell cultures by an activation of the kinase that is responsible for the decrease of the electrophoretic mobility of Tau proteins.

Moreover, it will be easy to verify the different etiopathogenic hypothesis that have already been presented for AD.

Glutamate, an excitatory amino acid neurotransmitter that has been shown to be implicated in learning and memory (Lynch and Baudry,

1984), has neurotoxic properties and might be involved in the patho-physiology of AD (Maragos et al., 1987).

Aluminosilicates are found in SP (Candy et al., 1986) and NFT (Perl and Brody, 1980) in the brain of patients with AD; hence, it has been proposed that Aluminium might play a role in the etiopathogenesis of AD.

Therefore, it will be possible to see if Tau 64 and 69 are produced in nerve cells growing in a medium containing Glutamate or Aluminium salts.

All persons with Down's syndrome who live past the age of 35 years show within their brains NFT and SP (Mann, 1988). Therefore it is likely that protein or enzyme changes associated with the additional chromoso-mal material might be involved in the formation of SP and NFT both in adult DS and possibly in AD.

For example it has been shown that the genetic locus for the Cu/Zn dependent superoxide dismutase is situated on the long arm of chromo-some 21 which is duplicated in DS. Increases in oxidative reactions catalysed by this enzyme may lead to neuronal damages (Delacourte et al., 1988).

It has also been shown that the gene coding for the β amyloid A4 precursor was localized on chromosome 21 (Kang et al., 1987). It seems that amyloid deposits precede the NFD (Duyckaerts et al., 1988). There-fore, it will be interesting to know if Tau 64, 69 may be produced in transfected cells with the gene of superoxide dismutase or with those of protein A4 precursor. Another approach would be to obtain the produc-tion of Tau 64 and 69 in the brain of transgenic animals for these genes. A such model might help to spot the first biochemical events that enhance the abnormal phosphorylation of Tau proteins and later their accumulation into insoluble PHF structures leading to neuronal death.

References

Baudier J, Cole RD (1987) Phosphorylation of Tau proteins to a state like that in Alzheimer's brain is catalized by a calcium/calmodulin dependent kinase and modulated by phospholipids. J Biol Chem 262:17577–17583

Brion JP, Passareiro H, Nunez J, Flament-Durand J (1985) Mise en évidence immunologique de la proteine tau au niveau des lésions de dégénérescence neurofibrillaire de la maladie d'Alzheimer. Arch Biol 95:229–235

Candy JM, Klinowski J, Perry RH, Perry EK, Fairbrain A, Oakley AE, Carpen-ter TA, Atack JR, Blessed G, Edwardson JA (1986) Aluminosilicates and senile plaque formation in Alzheimer's disease. Lancet i:354–356

Defossez A, Beauvillain J-C, Delacourte A, Mazzuca M (1988) Alzheimer's disease: a new evidence for common epitopes between microtubule associated protein Tau and paired helical filaments (PHF): demonstration at the electron microscope by a double immunogold labelling. Virchows Arch 413:141–145

Delacourte A, Défossez A (1986) Alzheimer's disease: tau proteins, the promoting factors of microtubule assembly, are major antigenic components of paired helical filaments. J Neurol Sci 76:173–186

Delacourte A, Défossez A, Céballos I, Nicole A, Sinet PM (1988) Preferential localization of copper zinc superoxide dismutase in the vulnerable cortical neurons in Alzheimer's disease. Neurosci Lett 92:247–253

Duyckaerts C, Delaère P, Poulain V, Brion J-P, Hauw JJ (1988) Does amyloid precede "PHF" in the senile plaque? A study of 15 cases with graded intellectual status in aging and AD. Neurosci Lett 91:354–359

Flament S, Delacourte A (1989) Abnormal Tau species are produced during Alzheimer's disease neurodegenerating process. FEBS Lett 247:213–216

Grundke-Iqbal J, Iqbal K, Quinlan M, Tung Y-C, Zaidi MS, Wisniewski HM (1986a) Microtubule-associated protein tau a component of alzheimer paired helical filaments. J Biol Chem 261:6084–6089

Grundke-Iqbal I, Iqbal K, Tung Y-C, Quinlan M, Wisniewski HM, Binder LI (1986b) Abnormal phosphorylation of the microtubule-associated protein tau in Alzheimer cytoskeletal pathology. Proc Natl Acad Sci USA 83:4913–4917

Hansen LA, De Teresa R, Davies P, Terry RD (1988) Neocortical morphometry, lesion counts, and choline acetyltransferase levels in the age spectrum of Alzheimer's disease. Neurology 38:48–54

Ihara Y, Nukina N, Miura R, Ogawara M (1986) Phosphorylated tau protein is integrated into paired helical filaments in Alzheimer's disease. J Biochem (Tokyo) 99:1807–1810

Kang J, Lemaire H-G, Unterbeck A, Salbaum JM, Masters CL, Grzeschik K-H, Multhaup G, Beyreuther K, Muller-Hill B (1987) The precursor of Alzheimer's disease amyloid A4 protein resembles a cell surface receptor. Nature 325:733–736

Kidd M (1964) Alzheimer's disease, an electron microscopical study. Brain 87:309–320

Ksiezak-Reding H, Binder LI, Yen S-H (1988) Immunochemical and biochemical characterization of tau proteins in normal and Alzheimer's disease brains with Alz 50 and Tau-1. J Biol Chem 263:7947–7953

Kosik KS, Joachim CL, Selkoe DJ (1986) Microtubule-associated protein tau is a major antigenic component of paired helical filaments in Alzheimer disease. Proc Natl Acad Sci USA 83:4044–4048

Lynch G, Baudry M (1984) The biochemistry of memory: a new and specific hypothesis. Science 224:1057–1063

Mann DMA (1988) Alzheimer's disease and Down's syndrome. Histopathology 13:125–137

Maragos WF, Greenamyre JT, Penney JB, Young AB (1987) Glutamate dysfunction in Alzheimer's disease: an hypothesis. Trends Neurosci 10:65–68

Nukina N, Ihara Y (1986) One of the antigenic determinants of paired helical filaments is related to tau protein. J Biochem (Tokyo) 99:1541–1544

Perl DP, Brody AR (1980) Alzheimer's disease: X-ray spectrometric evidence of aluminium accumulation in neurofibrillary tangle-bearing neurones. Science 208:297–299

Wood JG, Mirra SS, Pollock NJ, Binder LI (1986) Neurofibrillary tangles of Alzheimer's disease share antigenic determinants with the axonal micro-tubule-associated protein tau. Proc Natl Acad Sci USA 83:4040–4043

Correspondence: Dr. A. Delacourte, Laboratoire des neurosciences, Unité IN-SERM n° 16, Faculté de médecine, Place de Verdun, F-59045 Lille, France.

Molecular and cellular changes associated with neurofibrillary tangles and senile plaques

J-P. Brion[1], M. E. Cheetham[2], M. Coleman[2], G. Dale[2], J-M. Gallo[2], D. P. Hanger[2], A. Probst[3], D. C. Smith[2], and B. H. Anderton[2,4]

[1] Université Libre de Bruxelles, Laboratoire d'Anatomie Pathologique et de Microscopie Electronique, Faculté de Médecine, Bruxelles, Belgium
[2] Department of Immunology, St George's Hospital Medical School, London, United Kingdom
[3] Institut für Pathologie, Universität Basel, Basel, Switzerland
[4] Department of Neuroscience, Institute of Psychiatry, London, United Kingdom

Summary

Immunocytochemical studies have shown that changes in amyloid processing and in proteoglycan core proteins occur in senile plaques. Western blotting studies have shown that an extra tau species is present in Alzheimer brain and tangle-reactive neurofilament antibodies were found to be more selective towards phosphorylated neurofilaments than those that do not recognise tangles. A fragment of MAP2 was found to be present in tangles.

Introduction

It is now accepted that the microtubule-associated protein, tau, and the protein ubiquitin are integral components of PHF (Brion et al., 1985; Grundke-Iqbal et al., 1986; Wood et al., 1986; Kosik et al., 1986; Mori et al., 1987; Perry et al., 1987; Cole and Timiras, 1987; Shaw and Chau, 1988). The presence of other proteins in PHF, particularly neurofilaments and MAP2, is still controversial (Anderton et al., 1982; Kosik et al., 1984; Miller et al., 1986; Haugh et al., 1986; Lee et al., 1988; Ksiezak-Reding et al., 1987; Nukina et al., 1987). However, this suggests that there is a cytoskeleton – PHF transition which would have serious consequences for affected neurones.

Senile plaques display apparent maturational stages. One of us has described a new type of plaque, possibly representing an earlier stage than previously recognised, which appears as an alteration in neuropil texture with a central putative microglial cell (Probst et al., 1987). These plaques, termed plaques A, may be similar to diffuse plaques which are stained with antibodies to the amyloid A4 or β-protein following pre-treatment of brain material with formic acid (Tagliarini et al., 1988; Yamaguchi et al., 1989). Plaque cores also have an inorganic component of aluminosilicate (Candy et al., 1986) and other proteinacous con-stituents have been described, including alpha-1-anti-chymotrypsin and basement membrane heparan sulfate proteoglycan core protein (Abra-ham et al., 1988; Snow et al., 1988; Dale et al., 1989).

In this article, we describe recent findings concerning cytoskeletal involvement in neurofibrillary tangles and the immunohistochemical characterisation of plaques A.

Materials and methods

Immunohistochemical investigation of plaques A

Plaque A-rich tissue sections were stained with a rabbit antiserum to A4 protein (residues 1–42) (kindly provided by C Masters) and a monoclonal antibody to heparan sulfate proteoglycan core protein. Immunohistochemistry was performed using the streptavidin-biotin-horseradish peroxidase procedure.

Investigation of phosphate-dependent NF and tangle epitopes

A synthetic peptide of NF-H was conjugated to keyhole limpet haemocyanin (KLH) by the method of Liu et al. (1979) and chemically phosphorylated as previously described (Otvos et al., 1988). Antibody binding was assayed by ELISA, using phosphorylated KLH as a control (Coleman and Anderton, sub-mitted). To investigate phosphate-dependence of the antibodies some purified NF-H was dephosphorylated by alkaline phosphatase as in Carden et al. (1985) and antibody reactivities towards phosphorylated and dephosphorylated NF-H were compared by ELISA.

Western blotting

Brain homogenates (in PBS containing 1% w/v SDS, 1 mM DTT) were heated in a boiling water bath for ten minutes and then spun at 6,500 rpm (4000 g_{av}) for 30 minutes (Sorval SS34 rotor). The resultant supernatants were then further spun at 100,000 rpm (300,000 g_{av}) for one hour (Beckman TL100.3 rotor). These final supernatants were used for Western blotting experiments from 12.5% (w/v) polyacrylamide gels. Nitrocellulose filters were blocked in

TBS buffer containing skimmed milk powder (6% w/v) and Tween 20 (0.2% w/v) for eighteen hours at 4 °C before probing with a polyclonal antiserum to tau.

Results and discussion

Neurofilaments and tangles

It is well established that only some antibodies to neurofilaments label Alzheimer neurofibrillary tangles in situ; those that do label tangles all preferentially recognise phosphorylated forms of either NF-H and/or NF-M, the two larger neurofilament triplet polypeptides (Anderton et al., 1982; Kahn et al., 1987; Sternberger et al., 1984; Miller et al., 1986; Lee et al., 1988; Haugh et al., 1986). It has been reported that tangle-reactive neurofilament antibodies also recognise tau and that their staining of tangles is probably a result of this neurofilament-tau cross-reactivity (Ksiezak-Reding et al., 1987; Nukina et al., 1987), tau being an undisputed component of PHF (Brion et al., 1985; Grundke-Iqbal et al., 1986; Wood et al., 1986; Wischik et al., 1988). We therefore undertook a study of the epitope specificity of a panel of neurofilament antibodies some of which label tangles.

Four monoclonal antibodies which label tangles (RT97, 1215, 8D8 and SMI-31) bound strongly to a serine-phosphorylated synthetic peptide Glu-Ala-Lys-Ser-Pro-Ala (Coleman and Anderton, submitted) which corresponds to a multiple phosphorylation site in the C-terminal tail domain of NF-H (Breen et al., 1988; Geisler et al., 1987). This suggests that in tangles the antibodies bind either to an identical sequence or to similar Lys-Ser-Pro containing sequences, such as those occurring in tau (twice) and elsewhere in NF-H (34 times).

Two monoclonal antibodies which do not label tangles (147 and RS18) also bound strongly to the same phosphorylated synthetic peptide, but there seems to be a subtle difference between how they and the tangle-reactive antibodies recognise this sequence, since 147 and RS18 bind both the synthetic peptide and NF-H in a less phosphate-dependent manner than RT97, 1215, 8D8 and SMI-31.

Our results show that of the antibodies investigated, those which label tangles are more phosphate-dependent than those which do not, although all antibodies studied are capable of recognising the same neurofilament epitope. Unfortunately, this does not elucidate why only some of these antibodies label tangles in situ.

Western blotting with tau antibodies

Western blots of brain homogenates labelled with a polyclonal anti-tau serum is seen in Fig. 1. Labelling was predominantly in the 50–60 kD range. Intensity of tau staining appeared to vary from case to case but in three of the four Alzheimer cases studied, an extra band of molecular weight 68 kD was seen (marked with a small arrow in Fig. 1); this band was not seen in any of the old controls used in the study. A 68 kD band was seen, however, in the young control case (marked YC in Fig. 1). This extra band may correspond to A68 described by Wolozin et al. (1986) using the monoclonal antibody Alz 50. These authors also reported some staining of the antigen in a juvenile brain (Wolozin et al., 1988). Our results are similar to those of Flament and Delacourte (1989).

MAP 2

Certain antibodies to MAP2 have previously been reported to label neurofibrillary tangles in situ (Kosik et al., 1984). It has been shown that MAP2 and tau cross-react and indeed they share sequence homology in

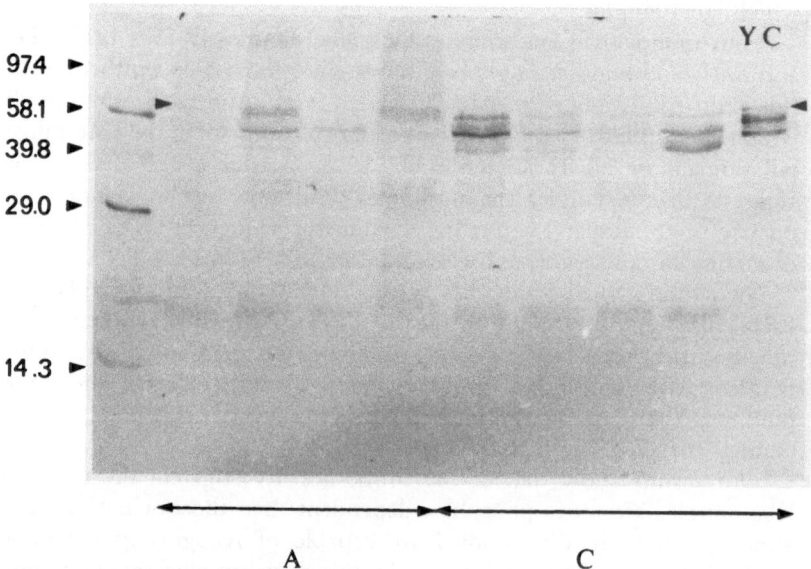

Fig. 1. Western blot of Alzheimer (A) and control (C) samples using 300,000 g_{av} supernatants. The blot was stained using a polyclonal anti-tau antibody. An additional 68 kD species in 3 of the Alzheimer cases studied as well as in the young control (YC) is marked with an arrow head. Molecular weight markers are shown on the left of the blot

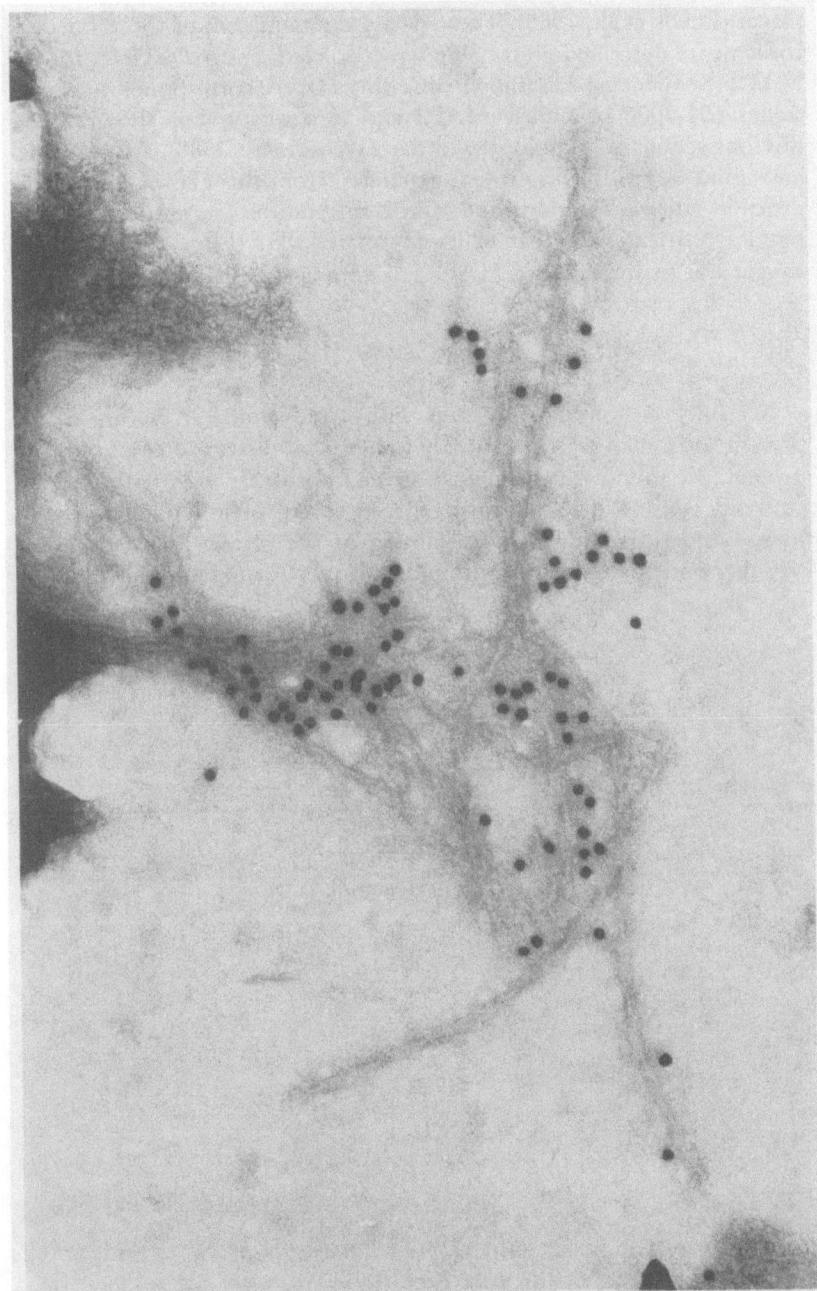

Fig. 2. Labelling of isolated, SDS extracted PHF with antiserum to MAP2 fusion protein. × 48,000

places (Lewis et al., 1988). These data are reminiscent of those for neurofilaments described above. We have isolated a partial cDNA for rat
MAP2. Sequencing has shown that this cDNA corresponds to nucleotides 691-3006 of mouse MAP2 and is a region not showing any
obvious sequence homology with tau (Wang et al., 1988). A polyclonal
antiserum to the fusion protein generated from this cDNA labels dendrites in rat brain sections like MAP2 antibodies. The antiserum labels
tangles in situ and about 15–20% of isolated PHF (Fig. 2). We conclude
that at least a fragment of MAP2 is a constituent of PHF.

Senile plaques A

We have examined plaques A with a polyclonal antiserum to A4
protein and a monoclonal antibody to heparan sulfate proteoglycan core
protein. We found that plaques A stained positively with both types of
antibody (Fig. 3). The anti-proteoglycan core protein monoclonal antibody besides showing diffuse staining of the plaques A, also gave a
stronger punctuate staining within the plaques A and in the surrounding

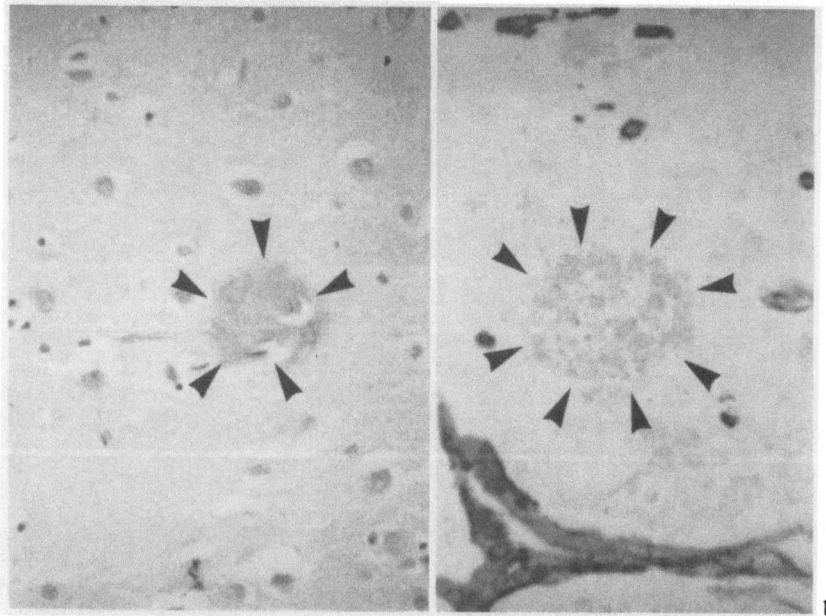

Fig. 3. a Plaque A stained with antiserum to A4 protein × 240 (outlined by arrows).
b Plaque A stained with antibody to heparan sulfate proteoglycan core protein × 240
(outlined by arrows)

neuropil. The latter may represent abnormal synaptic elements and im-munoelectron microscopic studies are in progress to investigate this possibility.

The labelling of classical senile plaques by heparan sulfate proteogly-can core protein antibodies has been described by others (Snow et al., 1988). Our results suggest that if plaques A represent very early stage in senile plaque development other changes in the extracellular matrix, including cell membrane surfaces, may occur in addition to changes in the processing of the A4 precursor protein.

It may therefore be more fruitful to look for extraneuronal factors which may play a role in plaque formation rather than focusing on neuronal synthesis and processing of A4 precursor. The obvious candi-date for investigation is the central microglial cell in the plaques A.

Acknowledgements

This work was supported by the Medical Research Council, the Wellcome Trust and NATO.

References

Abraham CR, Selkoe DJ, Potter H (1988) Immunochemical identification of the serine brain inhibitor, α_1-antichymotrypsin in the brain amyloid deposits of Alzheimer's disease. J Neuropathol Exp Neurol 47:354 (Abstract)

Anderton BH, Breinberg D, Downes MJ, Green PJ, Tomlinson B, Ulrich J, Wood JN, Kahn J (1982) Monoclonal antibodies show that neurofibrillary tangles and neurofilaments share antigenic determinants. Nature 298:84–86

Breen KC, Robinson PA, Wion D, Anderton BH (1988) Partial sequence of the rat heavy neurofilament polypeptide. FEBS Lett 241:213–218

Brion J-P, Van Der Bosch de Aguilar P, Flament-Durand J (1985) Morpholog-ical and immunocytochemical studies. In: Traber J, Gispen WH (eds) Senile dementia of the Alzheimer type. Springer, Berlin Heidelberg New York Tokyo, pp 164–175

Candy JM, Oakley AE, Klinowski J, Carpenter TA, Perry RH, Atack JR, Perry EK, Blessed G, Fairbairn A, Edwarson JA (1986) Aluminosilicates and senile plaque formation in Alzheimer's disease. Lancet i:354–356

Carden MJ, Schlaepfer WW, Lee VM-Y (1985) The structure, biochemical prop-erties, and immunogenicity of neurofilament peripheral regions are deter-mined by phosphorylation state. J Biol Chem 260:9805–9817

Cole GM, Timiras P (1987) Ubiquitin-protein conjugates in Alzheimer's lesions. Neurosci Lett 79:207–212

Coleman MP, Anderton BH. Phosphate dependent monoclonal antibodies to neurofilaments and Alzheimer neurofibrillary tangles recognise a synthetic peptide. (Submitted for publication)

Dale G, Probst A, Anderton BH, Ulrich J (1989) Very early senile plaques contain pre-amyloid A4 deposits. Neuropathol Appl Neurobiol 15:272

Flament S, Delacourte A (1989) Abnormal tau species are produced during Alzheimer's disease neurodegenerating process. FEBS Lett 247:213–216

Geisler N, Vandekerckhove J, Weber K (1987) Location and sequence characterization of the major phosphorylation sites of the high molecular mass neurofilament proteins M and H. FEBS Lett 221:403–407

Grundke-Iqbal I, Iqbal K, Yunn-Chyn T, Quinlan M, Wisniewski HM, Binder LI (1986) Abnormal phosphorylation of the microtubule-associated protein τ (tau) in Alzheimer cytoskeletal pathology. Proc Natl Acad Sci USA 83:4913–4917

Haugh MC, Probst A, Ulrich J, Kahn J, Anderton BH (1986) Alzheimer neurofibrillary tangles contain phosphorylated and hidden neurofilament epitopes. J Neurol Neurosurg Psychiatry 49:1213–1220

Kahn J, Anderton BH, Miller CC, Wood JN, Esiri MM (1987) Staining with monoclonal antibodies to neurofilaments distinguishes between subpopulations of neurofibrillary tangles, between groups of axons and between groups of dendrites. J Neurol 234:241–246

Kosik KS, Duffy LK, Dowling MM, Abraham C, McCluskey A, Selkoe DJ (1984) Microtubule-associated-protein 2: monoclonal antibodies demonstrate the selective incorporation of certain epitopes into Alzheimer neurofibrillary tangles. Proc Natl Acad Sci USA 81:7941–7945

Kosik KS, Joachim CL, Selkoe DJ (1986) Microtubule-associated protein tau (τ) is a major antigenic component of paired helical filaments in Alzheimer's disease. Proc Natl Acad Sci USA 83:4044–4048

Ksiezak-Reding H, Dickson DW, Davies P, Yen S-H (1987) Recognition of tau epitopes by anti-neurofilament antibodies that bind to Alzheimer neurofibrillary tangles. Proc Natl Acad Sci USA 84:3410–3414

Lee VM-Y, Otros L Jr, Schmidt ML, Trojanowski JQ (1988) Alzheimer disease tangles share immunological similarities with multiphosphorylation repeats in the two large neurofilament proteins. Proc Natl Acad Sci USA 85:7384–7388

Lewis SA, Wang D, Cowan NJ (1988) Microtubule-associated protein MAP2 shares a microtubule binding motif with tau protein. Science 242:936–939

Liu F-T, Zinnecker M, Hamaoka T, Katz DH (1979) New procedures for preparation and isolation of conjugates of proteins and a synthetic copolymer of D-amino acids and immunochemical characterization of such conjugates. Biochemistry 18:690–697

Miller CCJ, Brion J-P, Calvert R, Chin TK, Eagles PAM, Downes MJ, Flament-Durand J, Haugh M, Kahn J, Probst A, Ulrich J, Anderton BH (1986) Alzheimer's paired helical filaments share epitopes with neurofilament side arms. EMBO J 5:269–276

Mori H, Kondo J, Ihara Y (1987) Ubiquitin is a component of paired helical filaments in Alzheimer's disease. Science 235:1641–1644

Nukina N, Kosik KS, Selkoe DJ (1987) Recognition of Alzheimer paired helical filaments by monoclonal neurofilament antibodies is due to cross reaction with tau protein. Proc Natl Acad Sci USA 84:3415–3419

Otvos L, Hollosi M, Perczel A, Dietzschold B, Fasman GD (1988) Phosphorylation loops in synthetic peptides of the human neurofilament protein middle-sized subunit. J Prot Chem 7:365–376

Perry G, Friedman R, Shaw G, Chau V (1987) Ubiquitin is detected in neurofibrillary tangles and senile plaque neurites of Alzheimer's disease brains. Proc Natl Acad Sci USA 84:3033–3036

Probst A, Brunnschweiler H, Lautenschlager C, Ulrich J (1987) A special type of senile plaque, possibly an initial stage. Acta Neuropathol 74:133–141

Shaw G, Chau V (1988) Ubiquitin and microtubule-associated protein tau immunoreactivity each define distinct structures with differing distributions and solubility properties in Alzheimer brain. Proc Natl Acad Sci USA 85:2854–2858

Snow AD, Mar H, Hochlin D, Wight TN (1988) Ultrastructural immunolocalisation of heparan sulfate proteoglycans to amyloid fibrils in neuritic plaques of Alzheimer's disease. J Neuropathol Exp Neurol 47:355 (Abstract)

Sternberger NH, Sternberger LA, Ulrich J (1985) Aberrant neurofilament phosphorylation in Alzheimer's disease. Proc Natl Acad Sci USA 82:4274–4276

Tagliarini F, Giaccone G, Frangione B, Bugiani O (1988) Preamyloid deposits in the cerebral cortex of patients with Alzheimer's disease and non-demented individuals. Neurosci Lett 93:191–196

Wang D, Lewis SA, Cowan NJ (1988) Complete sequence of a cDNA encoding mouse MAP2. Nucleic Acids Res 16:11369–11370

Wischik CM, Novak M, Thogersen HC, Edwards PC, Runswick MJ, Jakes R, Walker JE, Milsten C, Roth M, Klug A (1988) Isolation of a fragment of tau derived from the core of the paired helical filament of Alzheimer disease. Proc Natl Acad Sci USA 85:4506–4510

Wolozin BL, Pruchnicki A, Dickson DW, Davies P (1986) A neuronal antigen in the brains of Alzheimer patients. Science 232:648–651

Wolozin BL, Scicutella A, Davies P (1988) Re-expression of a developmentally regulated antigen in Down syndrome and Alzheimer's disease. Proc Natl Acad Sci USA 85:6202–6206

Wood JG, Mirra SS, Pollock NJ, Binder LI (1986) Neurofibrillary tangles of Alzheimer's disease share antigenic determinants with the axonal microtubule-associated protein tau. Proc Natl Acad Sci USA 83:4040–4043

Yamaguchi H, Hirai S, Morimatsu M, Shoji M, Nakazato Y (1989) Diffuse type of senile plaques in the cerebellum of Alzheimer-type dementia demonstrated by β protein immunostain. Acta Neuropathol 77:314–319

Correspondence: Prof. B. H. Anderton, Department of Neuroscience, Institute of Psychiatry, De Crespigny Park, Denmark Hill, London SE5 8AF, United Kingdom.

Brain abnormalities in aged monkeys: a model sharing features with Alzheimer's disease

D. L. Price[1,2,3,5], E. H. Koo[1,2,5], S. S. Sisodia[2,5], L. J. Martin[2,5], L. C. Walker[2,5], C. A. Kitt[2,5], and L. C. Cork[2,4,5]

[1] Departments of Neurology, [2] Pathology, and [3] Neuroscience, [4] Division of Comparative Medicine (LCC), and [5] Neuropathology Laboratory, The Johns Hopkins University School of Medicine, Baltimore, Maryland, U.S.A.

Summary

Macaques develop age-associated cognitive/memory impairments as well as brain abnormalities (particularly senile plaques) similar to those occurring in the brains of older humans and individuals with Alzheimer's disease (AD). Some of these abnormalities can be clarified by investigations of aged nonhuman primates. Brain tissues from animals (ranging in age from 4–35 years) were examined by RNA blotting techniques, in situ hybridization, immunoblotting, and immunocytochemistry. These approaches allowed analysis of the evolution of some of the structural/chemical alterations (formation of neurites, amyloid deposition, and neurofibrillary tangles [NFT]) that occur in these animals. These investigations have relevance for understanding some of the behavioral, neuropathological, and neurochemical abnormalities that occur in older humans and in individuals with AD.

Introduction

Characterized by the presence of NFT and senile plaques in brain, AD preferentially affects specific brain regions and certain populations of nerve cells in the brainstem, basal forebrain, amygdala, hippocampus, and neocortex (Kemper, 1983; Price et al., 1986). Affected neurons develop two types of neurofibrillary pathology: enlarged distal axons, dendrites, and nerve terminals (neurites in plaques); and NFT (Price et al., 1986; Selkoe, 1989; Kosik, 1989). These two types of neurofibrillary abnormalities are associated with intracytoplasmic accumulations of filaments in neuronal processes (neurites) and perikarya (NFT), which appear to be related to perturbations of the neuronal cytoskeleton, particularly tau, and, possibly, neurofilaments (Anderton et al., 1982; Brion et al., 1985; Cork et al., 1986; Grundke-Iqbal et al., 1986; Wischik et al., 1988a, b; Kosik, 1989). Associated with neurites in senile plaques are extracellular deposits of amyloid that contain $\beta/A4$, a 42–43 amino acid polypeptide (Glenner, 1983; Masters et al., 1985; Wong et al., 1985) derived from an amyloid precursor protein (APP) (Kang et al., 1987; Weidemann et al., 1989) coded for by a gene on chromosome 21 (Goldgaber et al., 1987; Kang et al., 1987; Robakis et al., 1987b; Tanzi et al., 1987). At least three APP transcripts exist: APP-695 mRNA codes for a membrane-spanning glycoprotein (Goldgaber et al., 1987; Kang et al., 1987; Tanzi et al., 1987; Robakis et al., 1987a, b); and APP-751 and APP-770 mRNAs code for glycoproteins with a protease inhibitor domain (Kitaguchi et al., 1988; Ponte et al., 1988; Tanzi et al., 1988). APP-695 mRNA is enriched in brain (Ponte et al., 1988); APP-751/770 mRNAs are present in brain and in a variety of other organs (Kitaguchi et al., 1988; Ponte et al., 1988; Tanzi et al., 1988). All APP transcripts are present in neurons (Bahmanyar et al., 1987; Goedert, 1987; Allsop et al., 1988; Cohen et al., 1988; Higgins et al., 1988; Koo et al., 1988; Palmert et al., 1988; Shivers et al., 1988; Zimmerman et al., 1988), and APP isoforms have been identified in nerve cells (see below). Although much has been learned concerning the brain abnormalities that occur in AD, the evolution of the pathology (i.e., neurite formation, deposition of amyloid, and development of plaques and NFT) is not yet well understood.

Because aged nonhuman primates develop some of the brain abnormalities that occur in aged humans and in individuals with AD (Wisniewski et al., 1973; Schlaepfer and Freeman, 1978; Kurucz et al., 1981; Struble et al., 1982; Struble et al., 1984; Kitt et al., 1985a, b; Selkoe et al., 1987; Walker et al., 1988a, b; Abraham et al., 1989; Cork et al., 1989), these animals can be used to study the temporal and spatial evolution of these age-associated brain lesions. To illustrate this approach, we outline our studies of the development of amyloid deposits in senile plaques. First, we review the results of our investigations of the normal biology of APP in brain and then we discuss studies of abnormalities of APP that occur in the cortices of older animals.

Materials and methods

Brain tissues from rhesus monkeys (ranging from 4–35 years of age) were examined with classical histological techniques, Northern blotting methods,

RNase S$_1$ protection assays, immunoblotting, in situ hybridization, and immunocytochemistry (Struble et al., 1982; Casella et al., 1983; Kitt et al., 1984; Struble
et al., 1984; Kitt et al., 1985a, b; Struble et al., 1985; Walker et al., 1985; Walker
et al., 1987; Walker et al., 1988a; Martin et al., 1989; Sisodia et al., 1989; Koo
et al., 1990, in press).

Results

Cortices of young controls

APP-751/770 and APP-695 transcripts were present in cerebral cortices, and the ratio of APP-695 mRNA to APP-751/770 mRNA was ca.
one (Koo et al., 1990). Neurons expressed APP-695 and APP-751/770
mRNAs, and the distribution of different transcripts in cortex was
similar. Immunoblots showed APP-like immunoreactivity in cerebral
cortex, and APP isoforms in brain tissue comigrated with APP synthesized in vitro (i.e., in APP minigene transfected cell lines) (Sisodia et al.,
1989). APP-like immunoreactivity was present in perikarya, proximal
dendrites, and axons of cortical neurons; large pyramidal neurons (layers
III and IV) showed the highest levels of immunoreactivity. The localization of APP within axons is consistent with our recent investigations
demonstrating that APP is carried by rapid anterograde transport in
axons of rat dorsal root ganglia (Koo et al., in press).

Cortices of older animals

Older animals exhibited a variety of age-associated abnormalities of
brain. Abnormal, irregular varicose axons/neurites were visualized by
silver-impregnation methods and by immunocytochemistry with antibodies that recognized several transmitter markers, synaptophysin,
phosphorylated neurofilaments, tau, A68, and APP. Neurites were often
present in proximity to β/A4-positive plaques. β/A4-positive deposits
appeared to cap some of these neurites. The demonstration of APP-like
immunoreactivity in neurites and the presence of β/A4 in immediate
proximity to these neurites are consistent with the concept that neurite-
derived APP may be abnormally processed to form some of the β/A4
deposited in the neuropil. Amyloid was also present in the walls of some
blood vessels but, in aged Macaca mulatta, there is not a striking correlation between the presence of amyloid in plaques and blood vessels. In
addition, older animals showed patterns of abnormal immunoreactivities
in cortical perikarya. For example, a number of these animals exhibited
A68 immunoreactivity in some cells. One old animal developed argy-

rophilic NFT associated with the presence of tau, A68, phosphorylated neurofilaments, and epitopes of paired helical filaments (PHF) (Cork et al., 1989).

Discussion

Rhesus monkeys have a potential lifespan of > 30 years, and some older animals are impaired on a range of tasks including visual recognition memory, spatial memory, habit formation, and visuospatial orientation (Presty et al., 1987; Walker et al., 1988 b). Among animals of the same age, there were some differences in performance on specific tasks (Struble et al., 1982; Casella et al., 1983; Kitt et al., 1984; Struble et al., 1984; Kitt et al., 1985 a, b; Struble et al., 1985; Walker et al., 1985; Walker et al., 1987; Walker et al., 1988 a; Martin et al., 1989; Sisodia et al., 1989; Koo et al., 1990, in press).

On the basis of our studies of nonhuman primates (Martin et al., 1989), we have postulated a possible sequence for the development of some of the structural abnormalities that appear in the brains of these animals (Fig. 1). In normal neurons (A), transcription and translation occur in perikarya; a variety of proteins are then translocated to target sites by fast and slow transport. For example, APP is rapidly transported in axons (Koo et al., in press), whereas cytoskeletal elements (neurofilament proteins, tau, tubulin, etc.) are transported slowly. The initial age-associated structural abnormalities are still uncertain (B). Early manifestations of disease appear to be the formation of neurites (i.e., enlarged distal axons and nerve terminals) and the appearance of preamyloid deposits, detected with newly available antibodies to $\beta/A4$. The cellular source of amyloid is not known: some investigators believe that $\beta/A4$ is derived from APP synthesized in neurons (Price, 1986; Price et al., 1989); others suggest that APP and $\beta/A4$ are derived from the blood stream or from endothelial cells (Glenner, 1983). Early plaques (C) show both neurites and amyloid. APP- and phosphorylated neurofilament-immunoreactive neurites are observed in a subset of plaques that exhibit $\beta/A4$ immunoreactivity. Some APP-immunoreactive neurites were also capped by deposits of $\beta/A4$ (Martin et al., 1989). The presence of APP-like immunoreactivity in some neurites within $\beta/A4$-containing classical plaques (D) provides indirect evidence for a possible neuronal origin of some of the amyloid deposited in the brains of these aged animals. One source of parenchymal $\beta/A4$ may be abnormally processed neurite-derived APP. Alterations in levels of proteases or protease inhibitors in proximity to the surface of neurons in the microenvironment of the neuropil (associated with nonneuronal cells [i.e., microglia, astrocytes, etc.]) could play roles in the processing of APP. $\beta/A4$ associated with the

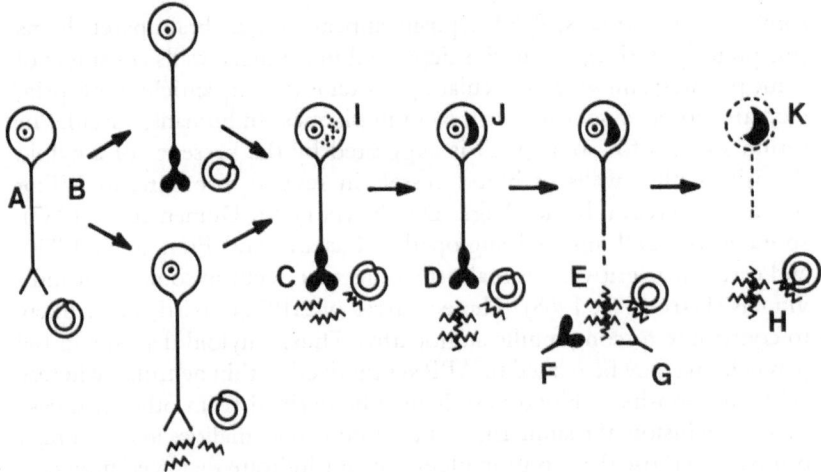

Fig. 1. Hypothetical sequence for the development of some of the structural abnormalities that appear in the brains of aged nonhuman primates (Figure described in text)

membrane does not appear to self-assemble into fibrils, and it is likely that self-assembly of amyloid fibrils will only occur when the $\beta/A4$ fragment is liberated from neuronal membranes. A neuronal origin for parenchymal deposits of $\beta/A4$ is favored by several lines of evidence: APP mRNAs and isoforms are present in neurons; APP is rapidly transported to distal nerve terminals; $\beta/A4$ is deposited in brain (gray matter, not white matter); amyloid in cortical parenchyma has an anatomical (laminar) distribution; APP-positive neurites are present in proximity to $\beta/A4$ deposits (plaques); and there is only a weak correlation between the presence of plaques and the presence of congophilic angiopathy.

As the degenerative process continues, axons may degenerate (E). Some plaque neurites are degenerating elements (F), whereas others may be reactive sprouts (G). End-stage plaques show amyloid cores (H). Some neurons (I) develop abnormalities in perikarya, including the presence of tau, phosphorylated neurofilaments, and A68 immunoreactivities. Some cells may become atrophic, and NFT develop in a few cells (J). The final stage of disease is marked by evidence of neuronal degeneration, leaving behind "ghost" tangles (K) and end-stage plaques (H).

It should be emphasized that the evidence for a neuronal origin for $\beta/A4$ is indirect and does not exclude other sources of amyloid (i.e., APP) derived, in part, from other cells in brain or from the serum. A systemic (or endothelial) origin for vascular amyloid is suggested by several mechanisms: APP mRNA and isoforms are present in vessels and some systemic organs; $\beta/A4$ deposits appear outside brain parenchyma (meningeal arteries); amyloid is deposited in capillary walls at centers of some plaques; and several circulating proteins (i.e., α_1-antichymotrypsin, etc.) are closely associated with $\beta/A4$ in plaques. In humans, vascular or serum sources for some $\beta/A4$ is suggested by the presence of amyloid deposits within walls of blood vessels in several disorders, including hereditary cerebral hemorrhage (Dutch type) (van Duinen et al., 1987), sporadic cerebral amyloid angiopathy (Castaño and Frangione, 1988), and cerebral arteriovenous malformations that occur in some older individuals (Hart et al., 1988). These sources of APP are particularly likely to contribute to congophilic angiopathy. Thus, amyloid deposits in the parenchyma may be related to APP synthesized within neurons, whereas amyloid deposits in blood vessels may be derived from other sources.

In conclusion, the similarities of the brain abnormalities in nonhuman primates to those that occur in older humans indicate that aged monkeys can serve as a useful model for future investigations of a variety of age-associated abnormalities in brain.

Acknowledgements

The authors thank Drs. Axel Unterbeck, Konrad Beyreuther, Robert G. Struble, Dennis J. Selkoe, Carmela R. Abraham, and Huntington Potter for helpful discussions.

This work was supported by grants from the U.S. Public Health Service (NIH NS 20471, NS 17079, AG 05146, AG 03359) and funds from The Robert L. & Clara G. Patterson Trust and the American Health Assistance Foundation. Drs. Price and Koo are recipients of a Leadership and Excellence in Alzheimer's Disease (LEAD) award (NIA AG 07914). Dr. Price is the recipient of a Javits Neuroscience Investigator Award (NIH NS 10580).

References

Abraham CR, Selkoe DJ, Potter H, Price DL, Cork LC (1989) α_1-Antichymotrypsin is present together with the β-protein in monkey brain amyloid deposits. Neuroscience 32:715–720

Allsop D, Wong CW, Ikeda S-I, Landon M, Kidd M, Glenner GG (1988) Immunohistochemical evidence for the derivation of a peptide ligand from the amyloid β-protein precursor of Alzheimer disease. Proc Natl Acad Sci USA 85:2790–2794

Anderton BH, Breinburg D, Downes MJ, Green PJ, Tomlinson BE, Ulrich J, Wood JN, Kahn J (1982) Monoclonal antibodies show that neurofibrillary tangles and neurofilaments share antigenic determinants. Nature 298:84–86

Bahmanyar S, Higgins GA, Goldgaber D, Lewis DA, Morrison JH, Wilson MC, Shankar SK, Gajdusek DC (1987) Localization of amyloid β-protein messenger RNA in brains from patients with Alzheimer's disease. Science 237:77–79

Brion JP, van den Bosch de Aguilar P, Flament-Durand J (1985) Senile dementia of the Alzheimer type: morphological and immunocytochemical studies. In: Traber J, Gispen WH (eds) Senile dementia of the Alzheimer type. Springer, Berlin Heidelberg New York Tokyo, pp 164–174

Casella JF, Flanagan MD, Lin S (1983) Cytochalasins: their use as tools in the investigation of actin polymerization in vivo and in vitro. In: DosRemedios CG, Barden JA (eds) Actin: structure and function in muscle and non-muscle cells. Academic Press, Sydney, pp 227–240

Castaño EM, Frangione B (1988) Biology of disease. Human amyloidosis, Alzheimer disease and related disorders. Lab Invest 58:122–132

Cohen ML, Golde TE, Usiak MF, Younkin LH, Younkin SG (1988) In situ hybridization of nucleus basalis neurons shows increased β-amyloid mRNA in Alzheimer disease. Proc Natl Acad Sci USA 85:1227–1231

Cork LC, Sternberger NH, Sternberger LA, Casanova MF, Struble RG, Price DL (1986) Phosphorylated neurofilament antigens in neurofibrillary tangles in Alzheimer's disease. J Neuropathol Exp Neurol 45:56–64

Cork LC, Walker LC, Price DL (1989) Neurofibrillary tangles and senile plaques in a cognitively impaired, aged nonhuman primate. J Neuropathol Exp Neurol 48:378

Glenner GG (1983) Alzheimer's disease: multiple cerebral amyloidosis. Biological aspects of Alzheimer's disease. Banbury Rep 15:137–144

Goedert M (1987) Neuronal localization of amyloid beta protein precursor mRNA in normal human brain and in Alzheimer's disease. EMBO J 6:3627–3632

Goldgaber D, Lerman MI, McBride OW, Saffiotti U, Gajdusek DC (1987) Characterization and chromosomal localization of a cDNA encoding brain amyloid of Alzheimer's disease. Science 235:877–880

Grundke-Iqbal I, Iqbal K, Quinlan M, Tung Y-C, Zaidi MS, Wisniewski HM (1986) Microtubule-associated protein tau. A component of Alzheimer paired helical filaments. J Biol Chem 261:6084–6089

Hart MN, Merz P, Bennett-Gray J, Menezes AH, Goeken JA, Schelper RL, Wisniewski HM (1988) β-Amyloid protein of Alzheimer's disease is found in cerebral and spinal cord of vascular malformations. Am J Pathol 132:167–172

Higgins GA, Lewis DA, Bahmanyar S, Goldgaber D, Gajdusek DC, Young WG, Morrison JH, Wilson MC (1988) Differential regulation of amyloid-β-protein mRNA expression within hippocampal neuronal subpopulations in Alzheimer's disease. Proc Natl Acad Sci USA 85:1297–1301

Kang J, Lemaire H-G, Unterbeck A, Salbaum JM, Masters CL, Grzeschik K-H, Multhaup G, Beyreuther K, Müller-Hill B (1987) The precursor of Alzheimer's disease amyloid A4 protein resembles a cell-surface receptor. Nature 325:733–736

Kemper TL (1983) Organization of the neuropathology of the amygdala in Alzheimer's disease. Biological aspects of Alzheimer's disease. Banbury Rep 15:31–35

Kitaguchi N, Takahashi Y, Tokushima Y, Shiojiri S, Ito H (1988) Novel precursor of Alzheimer's disease amyloid protein shows protease inhibitory activity. Nature 331:530–532

Kitt CA, Price DL, Struble RG, Cork LC, Wainer BH, Becher MW, Mobley WC (1984) Evidence for cholinergic neurites in senile plaques. Science 226: 1443–1445

Kitt CA, Struble RG, Cork LC, Mobley WC, Walker LC, Joh TH, Price DL (1985a) Catecholaminergic neurites in senile plaques in prefrontal cortex of aged nonhuman primates. Neuroscience 16:691–699

Kitt CA, Walker LC, Molliver ME, Price DL (1985b) Serotonergic neurites in senile plaques of aged nonhuman primates. Anat Rec 211:98A

Koo EH, Goldgaber D, Sisodia SS, Applegate MD, Gajdusek DC, Price DL (1988) Studies of β-amyloid precursor gene expression in brains of aged monkeys. Soc Neurosci Abstr 14:637

Koo EH, Sisodia SS, Archer DR, Martin LJ, Weidemann A, Beyreuther K, Price DL (1990a) Amyloid precursor protein undergoes fast anterograde axonal transport. Proc Natl Acad Sci (in press)

Koo EH, Sisodia SS, Cork LC, Unterbeck A, Bayney RM, Price DL (1990b) Differential expression of amyloid precursor protein mRNAs in cases of Alzheimer's disease and in aged nonhuman primates. Neuron 2:97–104

Kosik KS (1989) Minireview: the molecular and cellular pathology of Alzheimer neurofibrillary lesions. J Gerontol Biol Sci 44:B55–B58

Kurucz J, Charbonneau R, Kurucz A, Ramsey P (1981) Quantitative clinicopathologic study of cerebral amyloid angiopathy. J Am Geriatr Soc 29:61–69

Martin LJ, Cork LC, Koo EH, Sisodia SS, Weidemann A, Beyreuther K, Masters C, Price DL (1989) Localization of amyloid precursor protein (APP) in brains of young and aged monkeys. Soc Neurosci Abstr 15:23

Masters CL, Simms G, Weinman NA, Multhaup G, McDonald BL, Beyreuther K (1985) Amyloid plaque core protein in Alzheimer's disease and Down syndrome. Proc Natl Acad Sci USA 82:4245–4249

Palmert MR, Golde TE, Cohen ML, Kovacs DM, Tanzi RE, Gusella JF, Usiak MF, Younkin LH, Younkin SG (1988) Amyloid protein precursor messenger RNAs: differential expression in Alzheimer's disease. Science 241:1080–1084

Ponte P, Gonzalez-DeWhitt P, Schilling J, Miller J, Hsu D, Greenberg B, Davis K, Wallace W, Lieberburg I, Fuller F, Cordell B (1988) A new A4 amyloid mRNA contains a domain homologous to serine proteinase inhibitors. Nature 331:525–527

Presty SK, Bachevalier J, Walker LC, Struble RG, Price DL, Mishkin M, Cork LC (1987) Age differences in recognition memory of the rhesus monkey (Macaca mulatta). Neurobiol Aging 8:435–440

Price DL, Koo EH, Unterbeck A (1989) Cellular and molecular biology of Alzheimer's disease. BioEssays 10:69–74

Price DL (1986) New perspectives on Alzheimer's disease. Annu Rev Neurosci 9:489–512

Robakis NK, Ramakrishna N, Wolfe G, Wisniewski HM (1987a) Molecular cloning and characterization of a cDNA encoding the cerebrovascular and the neuritic plaque amyloid peptides. Proc Natl Acad Sci USA 84:4190–4194

Robakis NK, Wisniewski HM, Jenkins EC, Devine-Gage EA, Houck GE, Yau X-L, Ramakrishna N, Wolfe G, Silverman WP, Brown WT (1987b) Chromosome 21q21 sublocalisation of gene encoding beta-amyloid peptide in cerebral vessels and neuritic (senile) plaques of people with Alzheimer's disease and Down syndrome. Lancet i:384–385

Schlaepfer WW, Freeman LA (1978) Neurofilament proteins of rat peripheral nerve and spinal cord. J Cell Biol 78:653–662

Selkoe DJ, Bell DS, Podlisny MB, Price DL, Cork LC (1987) Conservation of brain amyloid proteins in aged mammals and humans with Alzheimer's disease. Science 235:873–877

Selkoe DJ (1989) Biochemistry of altered brain proteins in Alzheimer's disease. Annu Rev Neurosci 12:463–490

Shivers BD, Hilbich C, Multhaup G, Salbaum M, Beyreuther K, Seeburg PH (1988) Alzheimer's disease amyloidogenic glycoprotein: expression pattern in rat brain suggests a role in cell contact. EMBO J 7:1365–1370

Sisodia SS, Koo EH, Martin LJ, Unterbeck AJ, Beyreuther K, Weidemann A, Price DL (1989) Biosynthesis and processing of amyloid precursor protein (APP) in vitro. Soc Neurosci Abstr 15:23

Struble RG, Cork LC, Whitehouse PJ, Price DL (1982) Cholinergic innervation in neuritic plaques. Science 216:413–415

Struble RG, Hedreen JC, Cork LC, Price DL (1984) Acetylcholinesterase activity in senile plaques of aged macaques. Neurobiol Aging 5:191–198

Struble RG, Price DL Jr, Cork LC, Price DL (1985) Senile plaques in cortex of aged normal monkeys. Brain Res 361:267–275

Tanzi RE, Gusella JF, Watkins PC, Bruns GAP, St George-Hyslop P, Van Keuren ML, Patterson D, Pagan S, Kurnit DM, Neve RL (1987) Amyloid β-protein gene: cDNA, mRNA distribution, and genetic linkage near the Alzheimer locus. Science 235:880–884

Tanzi RE, McClatchey AI, Lampert ED, Villa-Komaroff L, Gusella JF, Neve RL (1988) Protease inhibitor domain encoded by an amyloid protein precursor mRNA associated with Alzheimer's disease. Nature 331:528–530

van Duinen SG, Castaño EM, Prelli F, Bots GTAB, Luyendijk W, Frangione B (1987) Hereditary cerebral hemorrhage with amyloidosis in patients of Dutch origin is related to Alzheimer's disease. Proc Natl Acad Sci USA 84:5991–5994

Walker LC, Kitt CA, Cork LC, Struble RG, Dellovade TL, Price DL (1988a) Multiple transmitter systems contribute neurites to individual senile plaques. J Neuropathol Exp Neurol 47:138–144

Walker LC, Kitt CA, Schwam E, Buckwald B, Garcia F, Sepinwall J, Price DL (1987) Senile plaques in aged squirrel monkeys. Neurobiol Aging 8:291–296

Walker LC, Kitt CA, Struble RG, Schmechel DE, Oertel WH, Cork LC, Price DL (1985) Glutamic acid decarboxylase-like immunoreactive neurites in senile plaques. Neurosci Lett 59:165–169

Walker LC, Kitt CA, Struble RG, Wagster MV, Price DL, Cork LC (1988b) The neural basis of memory decline in aged monkeys. Neurobiol Aging 9:657–666

Weidemann A, König G, Bunke D, Fischer P, Salbaum JM, Masters CL, Beyreuther K (1989) Identification, biogenesis, and localization of precursors of Alzheimer's disease A4 amyloid protein. Cell 57:115–126

Wischik CM, Novak M, Edwards PC, Klug A, Tichelaar W, Crowther RA (1988a) Structural characterization of the core of the paired helical filament of Alzheimer disease. Proc Natl Acad Sci USA 85:4884–4888

Wischik CM, Novak M, Thogersen HC, Edwards PC, Runswick MJ, Jakes R, Walker JE, Milstein C, Roth M, Klug A (1988b) Isolation of a fragment of tau derived from the core of the paired helical filament of Alzheimer disease. Proc Natl Acad Sci USA 85:4506–4510

Wisniewski HM, Ghetti B, Terry RD (1973) Neuritic (senile) plaques and filamentous changes in aged rhesus monkeys. J Neuropathol Exp Neurol 32:566–584

Wong CW, Quaranta V, Glenner GG (1985) Neuritic plaques and cerebrovascular amyloid in Alzheimer disease are antigenically related. Proc Natl Acad Sci USA 82:8729–8732

Zimmerman Z, Herget T, Salbaum JM, Schubert W, Hilbich C, Cramer M, Masters CL, Multhaup G, Kang J, Lemaire H-G, Beyreuther K, Starzinski-Powitz A (1988) Localization of the putative precursor of Alzheimer's disease-specific amyloid at nuclear envelopes of adult human muscle. EMBO J 7:367–373

Correspondence: D. L. Price, M. D., Neuropathology Laboratory, 509 Pathology Building, The Johns Hopkins University School of Medicine, 600 North Wolfe Street, Baltimore, MD 21205-2181, U.S.A.

Aged dogs: an animal model to study beta-protein amyloidogenesis

H. M. Wisniewski, J. Wegiel, J. Morys*, C. Bancher*, Z. Soltysiak*, and K. S. Kim

Office of Mental Retardation and Developmental Disabilities, New York State Institute for Basic Research in Developmental Disabilities, Staten Island, New York, U.S.A.

Summary

The morphology, incidence and distribution of plaques and diffuse amyloid deposits in the brains of seven old dogs (18.5–26.5 years of age) were examined on tissue sections immunocytochemically stained with two monoclonal antibodies to two distinct epitopes of the beta-protein. Amyloid deposits were found in all seven brains examined. Amyloid occurred in three morphological forms: 1. focal amyloid deposits (plaques), 2. large diffuse amyloid deposits and 3. amyloid angiopathy. The number of these deposits was comparable to the numbers of all three types of amyloid deposits seen in the brains of people with Alzheimer's disease. The number and type of morphological forms of the amyloid deposits depends on topography and the age of the animals. The number of plaques was highest in the brains of the animals 18.5–21 years of age. The oldest animals (21.5, 24 and 26.5 years of age) had a smaller number of amyloid deposits. With age, the number of plaques decreased in superficial layers of the cerebral cortex (II–III) and increased in the deeper layers (IV–VI). In the oldest animals, diffuse amyloid deposits in the deeper layers of the cortex predominated. Our studies suggest that the frequency and the extent of amyloid deposits in the brains of aged dogs are much wider than so far appreciated. It thus appears that aged dogs may be suitable as an animal model for the study of pathomechanisms involved in beta-protein amyloidogenesis.

---

* J. M. is a visiting scientist from the Department of Anatomy, Medical Academy of Gdansk, Poland. C. B. is a visiting scientist from the Neurological Institute, University of Vienna, Austria. Z. S. Veterinary Faculty of Higher Agricultural School, Wroclaw, Poland.

Introduction

Studies on the sequence of changes which lead to neuritic (senile) plaque (NP) formation in scrapie, Down syndrome and Alzheimer's disease (AD) lead us to the conclusion that amyloid deposits initiate the neuropil pathology and are the nidus of the plaque formation, the cause of neuronal loss and, in AD, probably also the cause of neurofibrillary changes. Whether the amyloid deposits in AD destroy the brain in such a way that the liver and kidneys are destroyed in systemic amyloidosis or whether beta-protein pathology induces other changes in neuronal functions remains to be determined. However, in our opinion understanding of the sequence of events leading to brain amyloidosis in AD is of critical importance for future therapy and prevention of this devastating disease. Animal models of human diseases usually bring closer the day when we can help affected people. The need for an animal model for AD is particularly pressing considering the number of people who are suffering or will suffer from this disease. Amyloid deposits similar to those observed in human brains have been described in aged non-human primates (Dayan, 1971; Kitt et al., 1984, 1985; Price et al., 1985; Struble et al., 1985; Walker et al., 1986, 1987; Wisniewski et al., 1973; Wisniewski and Terry, 1973). Numerous plaques were also observed in the brains of aged bears (Cork et al., 1988). Vaughan and Peters (1981) showed plaques in the cerebral cortex of three aged rats.

Senile plaques and cerebrovascular amyloidosis in the brains of dogs were observed by von Braunmuhl (1956) and Osetowska (1966). Von Braunmuhl described structures resembling NPs in the brains of three out of 20 old dogs between 14 and 20 years of age. In five aged dogs studied by Wisniewski et al. (1970), occasional NPs were noted. Their ultrastructure was similar to that of the human plaques. However in detailed morphological studies of the brains of 25 beagle dogs, varying in age from one year to over 16 years, Ball et al. (1983) did not find NP.

An immunocytochemical study revealed the presence of NPs in five species of aged mammals, including dogs, and showed that the amyloid deposits that are part of these NPs consist of the beta-protein (Selkoe et al., 1987).

In this paper, we present data on the spectrum of morphological appearance of amyloid deposits in the brains of aged dogs. This study shows that the extent of brain amyloidosis in aged dogs is much greater than previously reported and that the number of amyloid deposits is comparable to the number of NPs found in the brains of people with AD. Aged dogs, therefore, may become the animal of choice for studying the genesis of beta-protein amyloidosis, the type of brain amyloidosis that affects the victims of AD.

Material and methods

The brains of seven dogs 18.5 to 26.5 years of age were used in this study. According to information obtained from the owners, in the last months of life, all dogs had been inactive and the three oldest were almost completely immobilized. All dogs were unable to control stool and urination. Loss of sight was observed in all animals, and three dogs were completely blind as a result of bilateral cataracts. Hearing difficulties ranging from poor audition to complete deafness were also reported.

The brains were fixed in 10% formalin for four to six weeks. Brains were cut in a coronal plane into 11–12 five-mm thick slabs. The slabs were embedded in paraffin. Six-μm-thick sections were prepared and stained with cresyl violet, Bielschowsky's silver impregnation method and by immunocytochemistry using monoclonal antibodies (mAbs) raised against the following antigens: (1) A synthetic peptide corresponding to amino acids 1 to 24 of the amyloid beta-protein (4G8, IgG_{2b}, 1:4000, Kim et al., 1988; Bancher et al., 1989); the epitope recognized by this mAb resides of amino acids 17–24. (2) A synthetic peptide corresponding to amino acids 1 to 17 of the amyloid beta-protein (6E10, IgG_1, 1:4000, Kim et al., unpublished); the epitope recognized by mAb 6E10 is likely to be located in the vicinity of amino acid 11. (3) Bovine tau (Tau-1, IgG_{2a}, 1:50000, Binder et al., 1985; Grundke-Iqbal et al., 1986); the epitope recognized by mAb Tau-1 is located between amino acids 131 and 149 of the human tau sequence, i.e., in the central part of the molecule (Kosik et al., 1988). This mAb was used to assess the presence of neurofibrillary tangles and PHF-containing neurites.

Immunocytochemistry was performed using the avidin-biotin method. Sections were deparaffinized in xylene/graded alcohols, and endogenous peroxidase was inactivated by 0.2% hydrogen peroxide in methanol for 30 min. Sections were then incubated in 10% fetal calf serum in PBS/TBS for 30 min, followed by the primary mAb at the above-mentioned dilutions. As a detection system, biotinylated species-specific anti-mouse immunoglobulins (Amersham, 1:200) and horseradish peroxidase-labelled avidin (Extravidin, Sigma, 10 μg/ml) were employed. Primary antibody incubation was performed at 4 °C overnight. All other incubations were done for 1 hour at room temperature. Peroxidase was visualized with diaminobenzidine (500 μg/ml in PBS, pH 7.4, 0.01% H_2O_2). Sections were counterstained with hematoxylin.

To enhance the immunoreactivity of amyloid deposits on paraffin sections, the tissue was treated with concentrated formic acid before immunostaining (Kitamoto et al., 1987). To obtain staining of PHF with the mAb to tau, sections had to be treated with alkaline phosphatase (Sigma, Type VII-L, 400 μg/ml in 0.1 M Tris-HCl, pH 8, for 2.5 hrs. at 37 °C) before antibody incubation (Grundke-Iqbal et al., 1986).

Hemispheric sections (11–13 from one brain) stained for amyloid beta-protein using mAb 4G8 were used to determine the number of plaques per mm^2. Measurements were performed with a final magnification of 89 × in a projecting microscope (Pictoval, C. Zeiss). The cortex of one hemisphere was represented by 260–300 test areas.

Results

On sections immunostained with the mAb to the beta-protein, amyloid appeared in three morphological forms: focal amyloid deposits (plaques), diffuse amyloid deposits and amyloid angiopathy. Because immunohistochemistry for the beta-protein selectively visualizes amyloid deposits, comparison with the various types of NP seen with classical histopathological methods is not always possible. The immunohistochemical staining method used in this study does not label all the structures that participate in the plaque formation. Thus, it is necessary to define the terms "focal amyloid deposits" (plaques) and "diffuse amyloid deposits", which will be used in the following description.

Focal amyloid deposits (plaques) are usually round or oval circumscribed lesions with a diameter ranging from 10–140 μm. They stain darkly by immunohistochemistry, and the intensity of the reaction decreases gradually from the center to the periphery of the plaque. The morphology of these lesions, as seen by beta-protein immunostaining, is reminiscent of the primitive and classical type of plaques described in humans (Wisniewski and Terry, 1973).

In contrast to the focal amyloid deposits (plaques), diffuse amyloid deposits are usually irregular in shape and much larger in size. The staining intensity within an individual lesion is even, but it is lower than that of the plaques.

The number, type (focal or diffuse) and topography of the amyloid deposits are subject to extensive interindividual variability.

All types of amyloid deposits (plaques, diffuse deposits and amyloid angiopathy) were visualized with both beta-protein antibodies. The following description is based on observations made on sections immunostained with mAb 4G8. Similar results were obtained with mAb 6E10. MAb 4G8, but not mAb 6E10, also reacted with neuronal lipofuscin (Grundke-Iqbal et al., 1989; Bancher et al., 1989).

Cerebral cortex

Immunohistochemistry with mAb 4G8 revealed amyloid deposits in the cerebral cortex of all seven dogs investigated. The most common form of lesions were focal amyloid deposits (plaques). In every brain studied, it was possible to find test areas with the number of plaques exceeding 10/mm². In four brains, the number of plaques exceeded 15/mm² in many test areas. The average number of plaques as determined in the entire cerebral cortex was more than 14/mm² in three cases 19, 20 and 21 years of age. However, in the three oldest dogs (21.5, 24

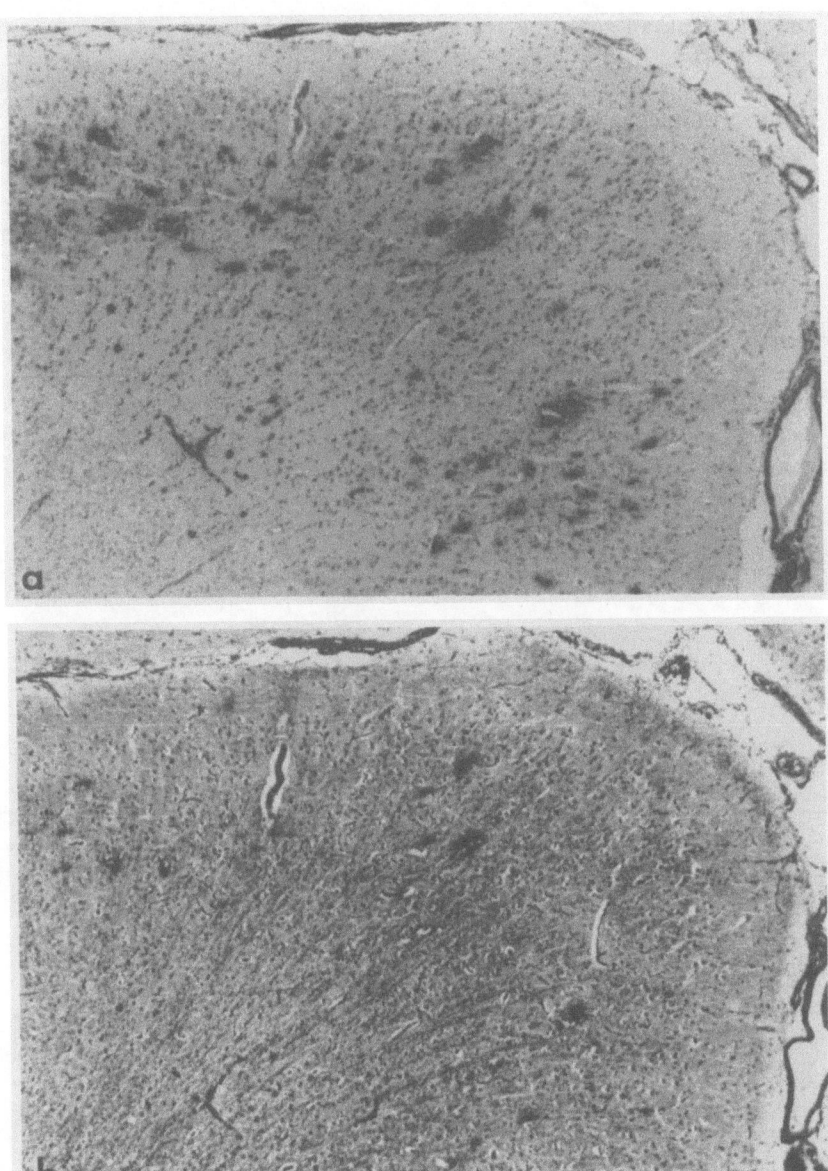

Fig. 1. Gapless serial sections from the cingulate gyrus stained for beta-protein with mAb 4G8 (a) and Bielschowsky's silver method (b). The number of plaques detected with the silver method is significantly lower (× 90)

and 26.5 years of age), the average number of plaques was below 3.5/ mm^2.

When adjacent sections were stained with Bielschowsky's silver method, no plaques could be detected in most cases. In only two dogs, plaques were identified, and their number was significantly lower than that seen on immunohistochemically stained sections (Fig. 1).

The number and morphological features of the plaques varied in different cortical layers. In contrast to the human brain, the molecular layer (layer I) of the cortex was almost free of amyloid deposits (Figs. 1, 2). Layers II and III usually contained numerous relatively small plaques. On beta-protein immunostaining, they appeared dense and well demarcated from the surrounding tissue. The deeper cortical layers (layers IV–VI) were also heavily affected. Often, neighboring plaques seemed to fuse and to form large aggregates. The density of the im-munostained material in these aggregates was usually lower than in the plaques in the superficial layer (Fig. 3).

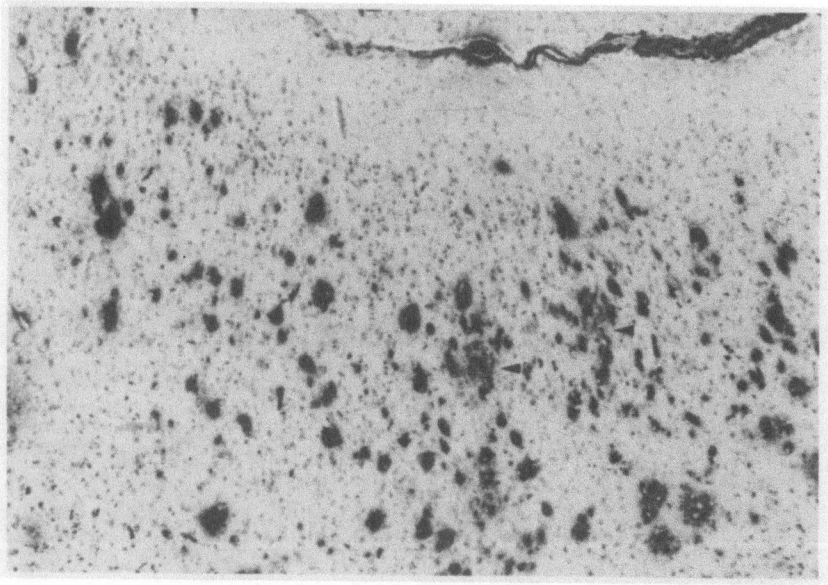

Fig. 2. Parietal medial gyrus of 19-year-old dog. Beta-protein immunostain. Amyloid deposits are seen in the form of relatively small primitive plaques in layer II–III and large plaques in layers IV–VI of the cortex. Confluence of the primitive plaques into large amyloid deposits (arrows). Darkly stained material within neurons corresponds to lipofus-cin. Amyloid is also seen in the walls of meningeal vessels (× 90)

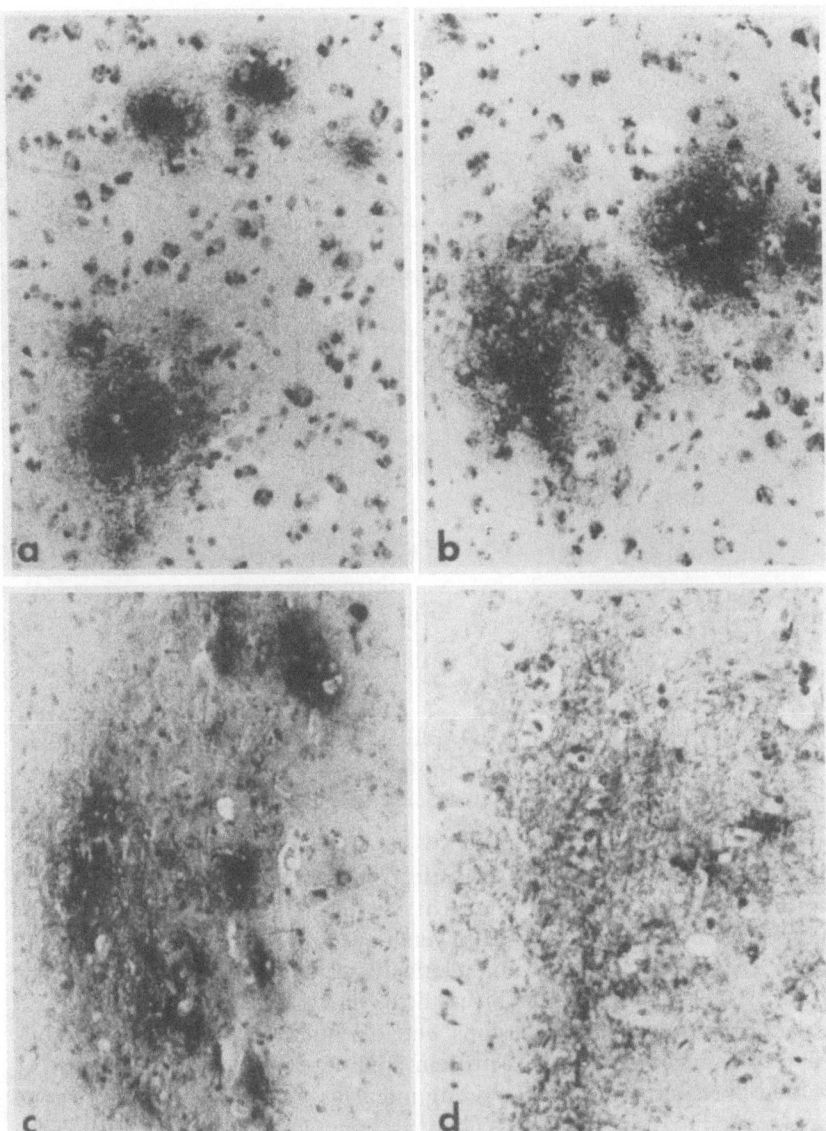

Fig. 3. Scattered and confluent primitive plaques (**a** and **b**) in the putamen of 19-year-old dog. Granular staining in neurons represents lipofuscin (× 300). Fusion of the plaques into large aggregat of amyloid in layers V – VI of the orbital gyrus (**c**) (× 150). Large, diffuse amyloid deposit in layer VI of the suprasylvian gyrus (**d**) (× 300)

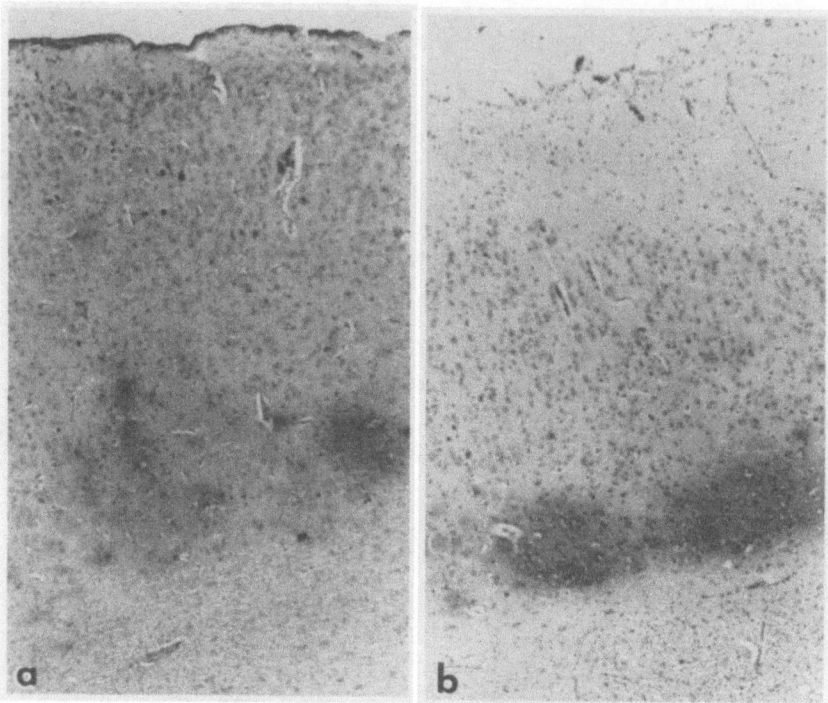

Fig. 4. Diffuse amyloid deposits in deeper layers of the cingulate gyrus (**a**) and piriform cortex (**b**) of a 24-year-old dog (× 80)

The second form of amyloid deposits in the cerebral cortex were diffuse infiltrations of the brain tissue (Fig. 4). These lesions usually appeared in the deeper cortical layers (layers IV–VI). At this location, they coexisted with large but relatively scanty focal amyloid deposits (plaques). The shape of the diffuse deposits was very variable and irregular. The borders of these lesions were often ill-defined. Generally, they appeared pale on anti-beta-protein immunostaining, giving them a cloud-like appearance. The staining intensity with one individual lesion was relatively homogeneous. These areas contained neurons, glial cells and blood vessels. The size of the diffuse amyloid deposits varied within an extremely wide range. Their diameter could reach up to several hundred µm. Some of these lesions appeared to spread into neighboring cortical layers. In the deep cortical layers, the diffuse amyloid deposits occurred together with large focal deposits (plaques). Often, these plaques seemed to fuse and to form aggregates. The staining density of these aggregates was decreased. Therefore, it was not always possible to

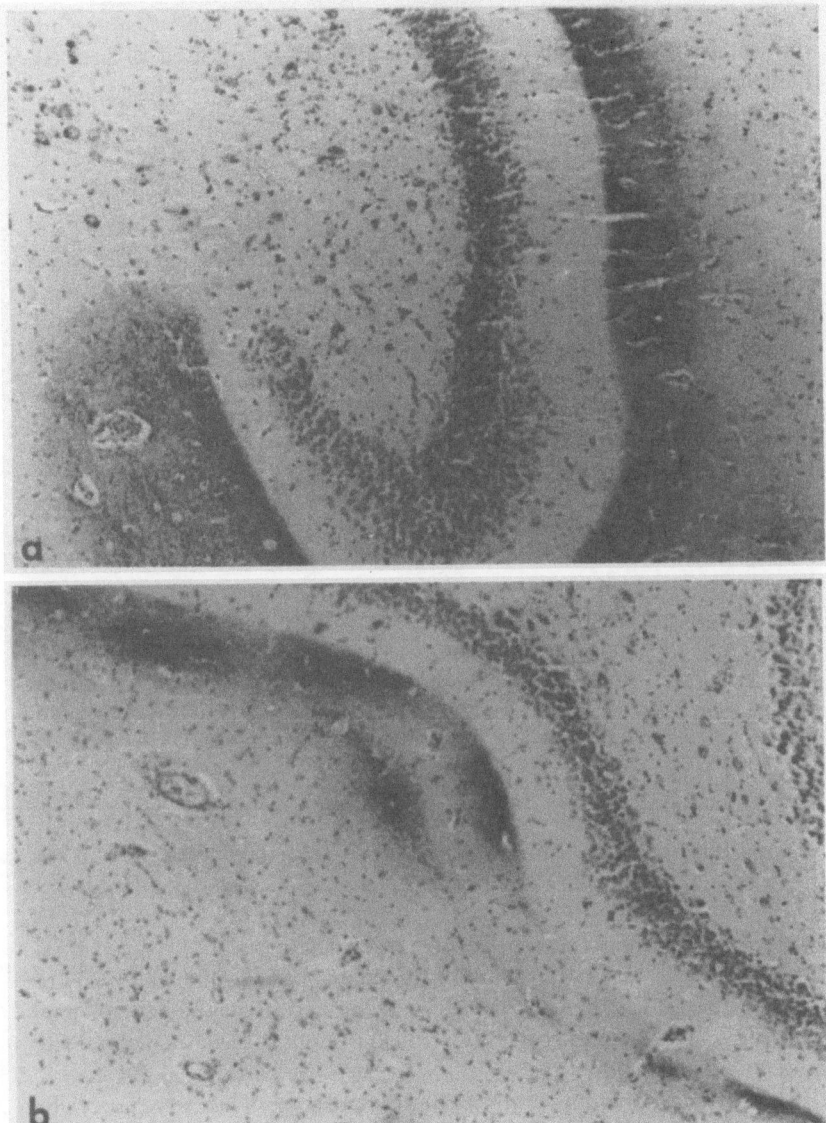

Fig. 5. Homogeneous, diffuse (a) and non-homogeneous, diffuse (b) amyloid deposits in the molecular layer of the hippocampus of an 18.5-year-old dog. Sharp border of the amyloid deposits between molecular layer and stratum radiale (× 100)

make a clear distinction between aggregates of several plaques and diffuse amyloid deposits. Morphologically, there appeared to be intermediate forms between these two typical lesions.

Hippocampus

In all brains studied, amyloid deposits were observed in the hippocampus. The predominant form was the diffuse amyloid deposit (Fig. 5). These deposits were usually homogeneous in texture, had a ribbon-like shape and filled most of the molecular layer of the hippocampal formation. The border with the stratum radiale (which was always unaffected) was very sharp (Fig. 6). Within the molecular layer, the diffuse amyloid deposits were in continuity with regions containing plaques and aggregates of plaques which seemed to be undergoing fusion into homogeneous amyloid masses (Fig. 5).

White matter

In the dog brain, plaques in the white matter developed only sporadically. Massive amyloid deposits in the deep layers of the cerebral cortex appeared to infiltrate the adjacent subcortical white matter. Occasionally, few plaques were observed in the anterior part of the internal capsule.

Basal ganglia

Amyloid deposits in the basal ganglia are usually presented in the form of plaques. Their number, size and shape were very variable between different basal ganglia. Great differences could even be observed within specific nuclei or regions of these structures.

In the caudate nucleus, amyloid deposits were found in six of the seven brains investigated. Plaques were numerous in the ventrolateral part (Fig. 7 a). Sometimes, they were also present in the dorsolateral and periventricular parts.

In all brains, plaques were seen in the claustrum. They were especially numerous in the ventral region of the anterior and central parts (Fig. 7 b).

Amyloid deposits were found in the putamen of five dogs (Fig. 7 b), in the amygdala (usually in the basolateral and basomedial nucleus) of four and in the globus pallidus of two of the seven dogs examined.

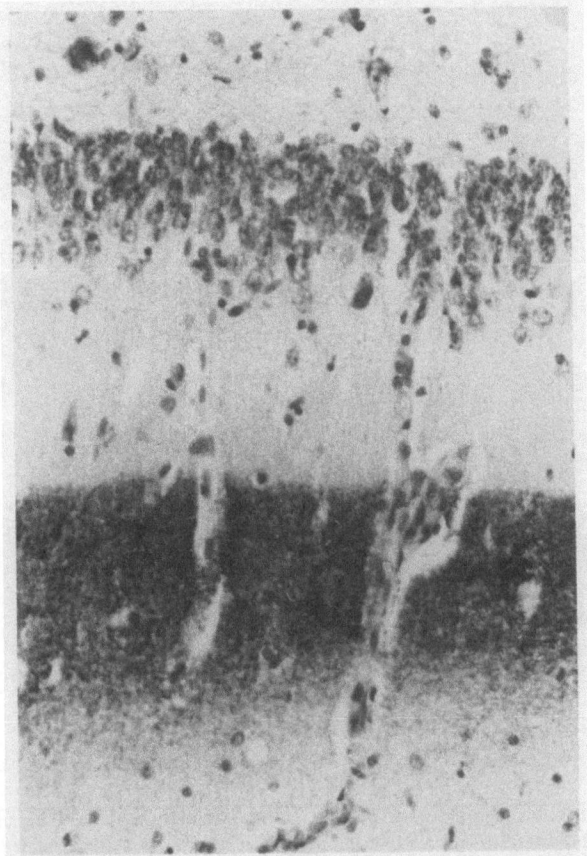

Fig. 6. Homogeneous diffuse amyloid deposit in molecular layer of the hippocampus. Note the sharp border with stratum radiale (× 350)

Diencephalon

In three cases, plaques were observed in the thalamus. They were especially numerous in the ventrolateral nucleus but could also be found in the mediodorsal and reticular nuclei and the lateroposterior part of the thalamus. Only sporadically did they occur in the lateral geniculate body. Plaques in the hypothalamus were very rare.

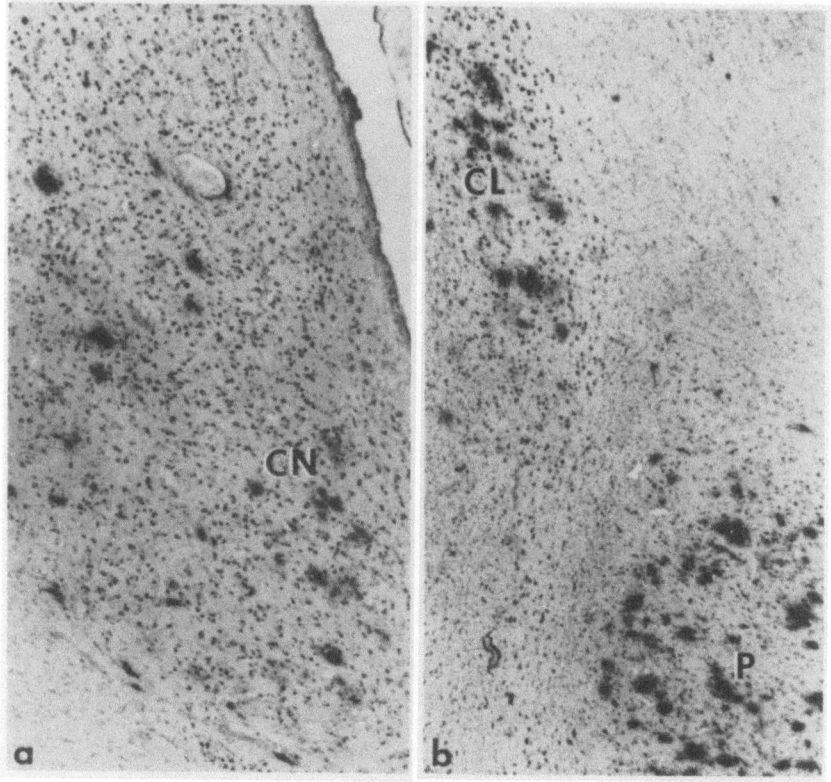

Fig. 7. Primitive plaques in caudate nucleus (CN) (a), claustrum (CL) and putamen (P) (b) of 19-year-old dog. Many nerve cells filled with lipofuscin are also stained (× 80)

Cerebellum

In the cerebellum, plaques were found in one case only. Most of them were located in the granular cell layer, but a few were also seen in the dentate nucleus.

Amyloid angiopathy

Amyloid deposits in the walls of blood vessels were very common in all investigated brains in both meningeal and parenchymal vessels (Fig. 8). In the cerebral cortex of the younger dogs, vessels with amyloid deposits were arranged in clusters. Vessels affected by amyloid angiopathy were seen mainly in areas free of amyloid infiltrates. In the oldest

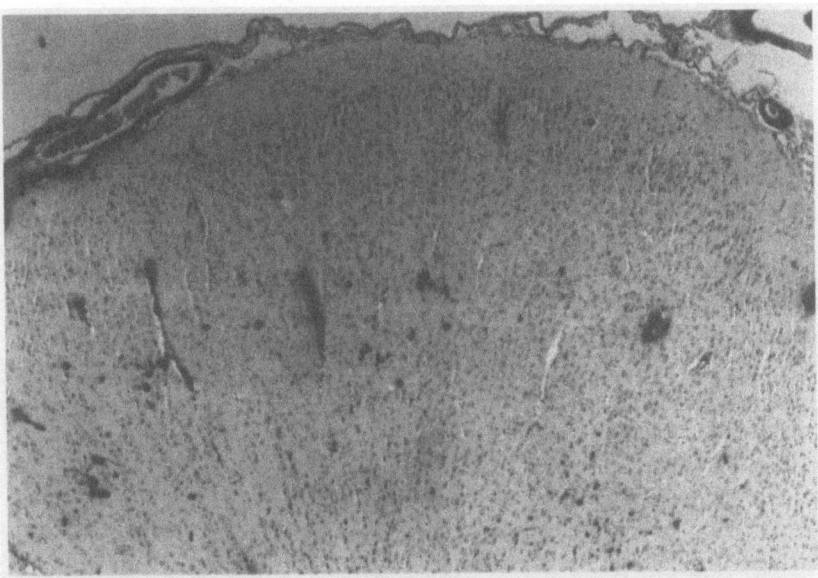

Fig. 8. Amyloid angiopathy in parietal medial gyrus of 19-year-old dog. Amyloid deposits are seen in the walls of meningeal and cortical vessels (arrows) (× 80)

dogs (24 and 26.5 years of age), the number of affected vessels was highest. In these cases, they were more diffusely distributed and were present mostly in areas with focal and diffuse amyloid deposits. With age, amyloid angiopathy increased not only in the cerebral cortex but also in the basal ganglia and the thalamus. In the withe matter, amyloid angiopathy developed only sporadically, generally in the larger vessels.

Neurofibrillary changes

No immunoreactivity with the mAb to tau could be observed in any of the seven dog brains investigated. The absence of neurofibrillary tangles was confirmed by the examination of sections stained with Bielschowsky's silver method.

Discussion

Using two mAbs (4G8 and 6E10) to different regions of the beta-protein molecule, we have shown large numbers of lesions displaying the morphology of primitive and classical plaques and diffuse amyloid deposits in the brains of aged dogs. The immunocytochemical characteristics of these lesions, combined with their morphological appearance, suggest a strong similarity to the human type of lesions and make it likely that, as in the human, the beta-protein (Glenner and Wong, 1984) or A4 (Masters et al., 1985) is the main proteinaceous constituent of the plaques and diffuse amyloid deposits found in aged dogs.

Large numbers of NP have been proposed as a diagnostic criterion for the neuropathological diagnosis of AD (Khachaturian, 1985; Wisniewski and Merz, 1985). Application of this criterion to the dog brains examined in this study would yield a positive diagnosis in four cases. Two points, however, have to be kept in mind: (1) no neurofibrillary tangles were detected in the brains investigated and (2) NIH criteria for diagnosis of AD are based on sections stained with silver. In our experience, the Bielschowsky stain demonstrates a number of plaques comparable to that seen with immunohistochemical methods in human material (Wisniewski et al., 1989a). However, in the dog brain, the Bielschowsky stain showed only few if any plaques. Whether this fact represents a difference in sensitivity between these two methods or other differences between the amyloid deposited remains to be determined. The reported absence of plaques in 25 dogs 1 to 16 years of age studied by Ball et al. (1983) and the low incidence of plaques in 20 dogs between 14 and 20 years of age studied by von Braunmuhl (1956) are probably the result of these investigators using silver methods for the detection of plaques.

The spectrum of morphological appearance of focal (plaques) and diffuse amyloid deposits in the dogs showed a high degree of similarities with the lesions found in humans (Wisniewski et al., 1989b). However, there were also important differences such as presence of few, if any, classical plaques with a clear central core surrounded by an unstained zone and peripheral wisps of amyloid fibers. Also, the dog plaques were negative when the tau antibody was used, indicating that the neurites of the primitive plaques in the dogs do not contain neurofibrillary aggregates made of paired helical filaments. In this respect, these plaques resemble the type of lesions found in young cases of Down syndrome (Mann and Esiri, 1989) and some aged non-demented people (Katzman et al., 1988; Crystal et al., 1988; Barcikowska et al., 1989) rather than the type of plaque typically associated with AD (Dickson et al., 1988; Shin et al., 1989):

The deeper layers of the cortex were more extensively affected in dogs than in humans. Unlike in humans, the molecular layer in dogs was almost completely free of amyloid deposits. Plaques in layers IV to VI were larger and more numerous than in superficial layers. They showed a tendency to form larger conglomerates mixed with diffuse amyloid deposits. In the human cortex, the preferential location of NPs in the second, third and fifth layers has been noted by several authors (Tomlinson et al., 1970; Pearson et al., 1985; Duyckaerts et al., 1986; Rafalowska et al., 1988).

In dog brains, we found individual differences in the number of plaques, diffuse amyloid deposits and amyloid angiopathy, e.g., one 21.5-year old dog had many lesions, whereas the other 21-year-old dog had very few. Of interest was the fact that the three oldest dogs had fewer plaques and less diffuse amyloid infiltration than the dogs 18.5 to 21 years of age. If confirmed on a larger series of dogs, this observation may indicate that there are individual or breed differences in susceptibility to brain amyloidosis or that with age, the rate of parenchymal amyloid formation is reduced. On the other hand, with age, there seemed to be a trend for progressive amyloid angiopathy.

The analysis of the distribution of amyloid deposits in the entire cortical ribbon on 11–12 serial slabs showed differences in the density of lesions in cortical lobes and gyri. Marked differences in density of lesions between adjacent gyri suggest that cytoarchitectonics may play a role in determining the distribution of plaques. Some authors observed that the distribution of NPs changes at the border of cytoarchitectonic fields (De Lacoste et al., 1989) and varies in cortical lobes (Struble et al., 1985). The variability observed in the dog cortex probably not only reflects local anatomical and functional specificity but is also related to the age of the animals.

In dogs, as in humans, some plaques appeared to be associated with vessels and others not. There was no correlation between amyloid angiopathy and diffuse amyloid deposits.

In summary, the presence of beta amyloid deposits in the form of numerous plaques, extensive diffuse amyloid deposits and amyloid angiopathy in all aged dogs studied, together with the fact that the dog is more available than aged monkeys, makes this animal an excellent model for studying beta protein amyloidogenesis.

Acknowledgements

The authors wish to thank Maureen Stoddard for editing the manuscript, Dr. Judy Shek, Mary Lee, Cathy Wang and Jadwiga Wegiel for expert technical assistance. Monoclonal antibody Tau-1 was the generous gift of Dr. L. I. Binder. Supported in part by NIH grants PO-1-AGO-4220-06 and PO-1-HD-22634-03, and funds from the New York State Office of Mental Retardation and Developmental Disabilities.

References

Ball MJ, MacGregor J, Fyfe IM, Rapoport SI, London ED (1983) Paucity of morphological changes in the brains of ageing beagle dogs: further evidence that Alzheimer lesions are unique for primate central nervous system. Neurobiol Aging 4: 127–131

Bancher C, Grundke-Iqbal I, Iqbal K, Kim KS, Wisniewski HM (1989) Immunoreactivity of neuronal lipofuscin with monoclonal antibodies to the amyloid beta-protein. Neurobiol Aging 10: 125–132

Barcikowska M, Wisniewski HM, Bancher C, Grundke-Iqbal I (1989) About the presence of paired helical filaments in dystrophic neurites participating in the plaque formation. Acta Neuropathol 78: 225–231

Binder LI, Frankfurter A, Rebhun LI (1985) The distribution of tau in the mammalian central nervous system. J Cell Biol 101: 1371–1378

Cork LC, Powers RE, Selkoe DJ, Davies P, Geyer JJ, Price DL (1988) Neurofibrillary tangles and senile plaques in aged bears. J Neuropathol Exp Neurol 47: 629–641

Crystal H, Dickson D, Fuld P, Masur D, Scott R, Mehler M, Masdeu J, Kawas C, Aronson M, Wolfson L (1988) Clinico-pathologic studies in dementia: nondemented subjects with pathologically confirmed Alzheimer's disease. Neurology 38: 1682–1687

Dayan AD (1971) Comparative neuropathology of ageing. Brain 94: 31–42

De Lacoste M-C, Sparkman DR, Pollan KS, Hirstein N, Mihailoff GA, White CL (1989) Laminar and regional distribution of plaques and tangles in the cerebral cortex of Alzheimer patients. J Neuropathol Exp Neurol 48: 351 (Abstr 152)

Dickson DW, Farlo J, Davies P, Crystall H, Fuld P, Yen SH (1988) Alzheimer's disease. A double-labeling immunohistochemical study of senile plaque. Am J Pathol 132: 86–101

Duyckaerts C, Hauw JJ, Bastenaire F, Piette F, Poulain C, Rainsard V, Javoy-Agid F, Berthaux P (1986) Laminar distribution of neocortical senile plaques in senile dementia of the Alzheimer type. Acta Neuropathol (Berl) 70: 249–256

Glenner GG, Wong CW (1984) Alzheimer's disease: initial report of the purification and characterization of a novel cerebrovascular amyloid protein. Biochem Biophys Res Commun 120: 885–890

Grundke-Iqbal I, Iqbal K, Tung YC, Quinlan M, Wisniewski HM, Binder LI (1986) Abnormal phosphorylation of the microtubule-associated protein T (tau) in Alzheimer cytoskeletal pathology. Proc Natl Acad Sci USA 83: 4913–4917

Grundke-Iqbal I, Iqbal K, George L, Tung YC, Kim KS, Wisniewski HM (1989) Amyloid protein and neurofibrillary tangles coexist in the same neuron in Alzheimer disease. Proc Natl Acad Sci USA 86: 2853–2857

Katzman R, Terry R, DeTeresa R, Brown T, Davies P, Fuld P, Renbing X, Peck A (1988) Clinical, pathological, and neurochemical changes in dementia: a subgroup with preserved mental status and numerous neocortical plaques. Ann Neurol 23: 138–144

Khachaturian ZS (1985) Diagnosis of Alzheimer's disease. Arch Neurol 42: 1097–1105

Kim KS, Miller DL, Sapienza VJ, Chen CMJ, Bai C, Grundke-Iqbal I, Currie JR, Wisniewski HM (1988) Production and characterization of monoclonal antibodies reactive to synthetic cerebrovascular amyloid peptide. Neurosci Res Commun 2: 121–130

Kitamoto T, Ogomori K, Tateishi J, Prusiner SB (1987) Formic acid pretreatment enhances immunostaining of cerebral and systemic amyloids. Lab Invest 57: 230–236

Kitt CA, Price DL, Struble RG, Cork LC, Wainer BH, Becher MW, Mobley WC (1984) Evidence for cholinergic neurites in senile plaques. Science 226: 1443–1445

Kitt CA, Struble RG, Cork LC, Mobley WC, Walker LC, Joh TH, Price DL (1985) Catecholaminergic neurites in senile plaques in prefrontal cortex of aged nonhuman primates. Neuroscience 16: 105–115

Kosik KS, Orecchio LD, Binder L, Trojanowski JQ, Lee VMY, Lee G (1988) Epitopes that span the tau molecule are shared with paired helical filaments. Neuron 1: 817–825

Mann DMA, Esiri MM (1989) The pattern of acquisition of plaques and tangles in the brains of patients under 50 years of age with Down's syndrome. J Neurol Sci 89: 169–179

Masters CL, Simms G, Weinman NA, Multhaup G, McDonald BL, Beyruther K (1985) Amyloid plaque core protein in Alzheimer disease and Down's syndrome. Proc Natl Acad Sci USA 82: 4245–4249

Osetowska E (1966) Etude anatomopathologique sur le cerveau de chiens senile. In: Luthy F, Bischoff A (eds) Proceedings of the Fifth International Congress of Neuropathology. Excerpta Medica, Amsterdam, pp 497–502

Pearson RCA, Esiri MM, Hiorns RW, Wilcock GK, Powell TPS (1985) Anatomical correlates of the distribution of the pathological changes in the neocortex in Alzheimer disease. Proc Natl Acad Sci USA 82: 4531–4534

Price DL, Cork LC, Struble RG, Kitt CA, Price DL Jr, Lehmann J, Hedreen JC (1985) Neuropathological, neurochemical, and behavioral studies of the aging nonhuman primate. In: Davis RT, Leathers CW (eds) Behavior and pathology of aging in Rhesus monkeys. Monographs in primatology, vol 8. Alan R Liss, New York, pp 113–135

Rafalowska J, Barcikowska M, Wen GY, Wisniewski HM (1988) Laminar distribution of neuritic plaques in normal aging, Alzheimer's disease and Down's syndrome. Acta Neuropathol 77: 21–25

Selkoe DJ, Bell DS, Podlisny MB, Price DL, Cork LC (1987) Conservation of brain amyloid proteins in aged mammals and humans with Alzheimer's disease. Science 235: 873–877

Shin R-W, Ogomori K, Kitamoto T, Tateishi J (1989) Increased tau accumulation in senile plaques as a hallmark in Alzheimer's disease. Am J Pathol 134: 1365–1371

Struble RG, Price DL Jr, Cork LC, Price DL (1985) Senile plaques in cortex of aged normal monkeys. Brain Res 361: 267–275

Tomlinson BE, Blessed G, Roth M (1970) Observations on the brains of demented old people. J Neurol Sci 11:205–242

Vaughan DW, Peters A (1981) The structure of neuritic plaques in the cerebral cortex of aged rats. J Neuropathol Exp Neurol 40:472–487

Von Braunmuhl A (1956) Kongophile Angiopathie und senile Plaques bei greisen Hunden. Arch Psychiat Nervenkr 194:396–414

Walker LC, Kitt CA, Schwam E, Buckwald B, Garcia F, Sepinwall J, Price DL (1987) Senile plaques in aged squirrel monkeys. Neurobiol Aging 8:291–296

Walker LC, Kitt CA, Struble RG, Cork LC, Price DL (1986) Heterogeneity of neurites in senile plaques of aged rhesus monkeys (Abstract). 16th Annual meeting, Washington, DC, November 9–14, 1986 (Soc Neurosci Abstr 12:272)

Wisniewski HM, Bancher C, Barcikowska M, Wen GY, Currie J (1989 b) Spectrum of morphological appearance of amyloid deposits in Alzheimer's disease. Acta Neuropathol 78:337–347

Wisniewski HM, Ghetti B, Terry RD (1973) Neuritic (senile) plaques and filamentous changes in aged rhesus monkeys. J Neuropathol Exp Neurol 32:566–584

Wisniewski HM, Johnson AB, Raine CS, Kay WK, Terry RD (1970) Senile plaques and cerebral amyloidosis in aged dogs. Lab Invest 23:287–296

Wisniewski HM, Merz GS (1985) Neuropathology of the aging brain and dementia of the Alzheimer type. In: Gaitz CM, Samorajski T (eds) Aging 2000: our health care destiny, vol 1. Biomedical issues. Springer, Berlin Heidelberg New York Tokyo, pp 231–243

Wisniewski HM, Terry RD (1973) Morphology of the aging brain, human and animals. In: Ford DH (ed) Progress in brain research, vol 40. Proceedings of a symposium sponsored by the International Society of Psychoneuroendocrinology, and held at Downstate Medical Center, State University of New York, Brooklyn, NY, USA, June 26–29, 1972. Elsevier, Amsterdam, pp 167–186

Wisniewski HM, Wen GY, Kim KS (1989 a) Comparison of four staining methods on the detection of neuritic plaques. Acta Neuropathol 78:22–27

Correspondence: H. M. Wiesniewski, M.D., Ph.D., NYS Institute for Basic Research in Developmental Disabilities, 1050 Forest Hill Road, Staten Island, NY 10314, U.S.A.

Immuncytochemistry

Immunocytochemical and ultrastructural pathology of nerve cells in Alzheimer's disease and related disorders *

H. Lassmann [1,2], P. Fischer [1,2], C. Bancher [1,4], and K. Jellinger [3]

[1] Neurological Institute, University of Vienna,
[2] Institute of Brain Research, Austrian Academy of Science, and [3] Ludwig Boltzmann Institute for Clinical Neurobiology, Lainz Geriatric Hospital, Vienna, Austria
[4] New York State Institute for Basic Research in Developmental Disabilities, New York, U.S.A.

Summary

Neurofibrillary tangles in Alzheimer's disease were analysed by light and electron microscopic immunocytochemistry using antibodies against cytoskeletal antigens and ubiquitin. Our studies indicate that the antigenicity of paired helical filaments is profoundly changed during the stages of tangle formation and maturation. In addition, tangle formation may lead to changes in other neuronal cytoskeletal elements. A comparison of Alzheimer neurofibrillary tangles with cytoplasmic fibrillary inclusions in other CNS diseases indicates that ubiquitin binding to abnormal proteins may be a common denominator of such cytopathological conditions.

Introduction

Neurofibrillary tangles are one of the main pathological alterations in Alzheimer's disease (Alzheimer, 1907). They consist of bundles of pathological argyrophilic filaments, located in the cytoplasm of affected neurons. Recent immunocytochemical and biochemical studies have identified microtubule associated tau protein and ubiquitin as essential components of the paired helical filaments (PHF) or straight tubules (ST), which ultrastructurally build up the neurofibrillary tangles (Brion et al., 1985; Wischik et al., 1988; Mori et al., 1987; Perry et al., 1987). In PHF,

* The study was partly funded by a grant from the Ministry for Science and Research, Austria.

tau proteins appear to be pathologically phosphorylated (Grundke Iqbal et al., 1986 b). In addition to tau and ubiquitin, epitopes of phosphorylated neurofilaments (Anderton et al., 1982; Sternberger et al., 1985) and high molecular weight microtubule associated protein 2 (MAP 2, Kosik et al., 1984) have been identified. Besides these protein components sulfated glycosaminoglycanes or glycoconjugates were found (Snow et al., 1987; Szumanska et al., 1987).

It is at present controversial as to what exent other components than those listed above are constituents of PHF. Several "PHF-specific" antibodies have been raised (Ihara et al., 1983; Wang et al., 1984; Wolozin et al., 1986). However many of them, in the course of a more detailed immunological characterization, turned out to recognize epitopes of tau (Nukina et al., 1988) or ubiquitin (Perry et al., 1987, 1989). Furthermore, the incorporation of amyloid β-peptide into neurofibrillary tangle material (Masters et al., 1985) remains controversial.

In the present study we have analysed the antigenic profile of PHF in the course of tangle formation and maturation by immunocytochemistry, focusing on the following specific questions:

1. Are there differences in the antigenic profile between early, mature and late (extracellular, "ghost") tangles?
2. Are there changes in the normal neuronal cytoskeletal elements during tangle formation and maturation?
3. How do different stages of neurofibrillary tangles quantitatively correlate with the clinical expression of the disease?
4. What is the relationship to other central nervous system diseases with cytoskeletal pathology?

Material and methods

Our studies on Alzheimer's disease were based on autopsy material obtained from a prospective study on clinico-pathological correlations in ageing and dementia (Fischer et al., 1988). Material from other central nervous system diseases with cytoskeletal pathology (Parkinson's disease, diffuse Lewy body disease, progressive supranuclear palsy) were obtained from the autopsy collection of the Neurological Institute and the Neurological Department of the City Hospital Lainz (Bancher et al., 1987, 1989 a, b). Paraffin sections were stained with hematoxylin/eosin, Klüver myelin stain, Bodian and Bielschowski silver impregnation, Congo red and thioflavin S. Glycosaminoglycanes were visualized by Alcian blue with different concentrations of magnesium chloride (Snow et al., 1987).

Immunocytochemical studies were performed on paraffin-embedded material, fixed in buffered 4% formaldehyde. Polyclonal antibodies against MAP 1, MAP 2, tau, tubulin and PHF and monoclonal antibodies against tau, PHF, phosphorylated and non-phosphorylated neurofilament epitopes and glial fibrillary acidic protein were used (Bancher et al., 1989 a). The methods of immunocytochemistry, double staining and immune electron microscopy have been described in detail earlier (Vass et al., 1986; Bancher et al., 1989 a, b; Brunner et al., 1989).

Quantitative evaluation was performed in standardized regions of the left cerebral hemispheres, including frontal, parietal and temporal cortex and hippocampus. The number of plaques and tangles in different maturation stages was counted on sections stained with Bielschowsky silver impregnation and with immunocytochemistry using tau-1 and 3-39 (ubiquitin) monoclonal antibodies. In total, 15 microscopic fields, covering a total area of 6 mm^2 were evaluated per patient and region.

Statistical analysis was performed at the computer center of the University of Vienna, using Statistical Package for the Social Sciences, procedure NPAR CORR, which calculated Spearman rank correlation coefficients.

Results

Changes in the antigenic profile during formation and maturation of Alzheimer neurofibrillary tangles

According to Alzheimer (1911) three stages of tangle maturation can be distinguished. Early tangles (stage 1) are small granular or rod-like cytoplasmic inclusions in the pericarya of affected neurons. Mature tangles (stage 2) appear as the classical flame-shaped neurofibrillary masses, which fill in large parts of the neuronal cytoplasm and extend into the apical dendrite. Late, extracellular "ghost" tangles (stage 3) are loose masses of fibrillar material, which by their triangular outline reflect the shape and location of the degenerated neuron. In addition to these tangle stages, immunocytochemistry with certain antibodies against tau (Bancher et al., 1989 a) and AD brain homogenates (Alz 50; Hyman et al., 1987) made it possible to identify very early changes in neurons, reflected by diffuse accumulation of immunoreactive material in the cytoplasm (stage 0; Bancher et al., 1989 a).

The antigenic profile of tangles at different maturation stages is summarized in Table 1. Early tangle stages are mainly recognized by anti-tau antibodies, but show little or no reactivity with the other markers used. In addition we found characteristic differences in the antigenic profile between classical intracellular tangles and extracellular "ghost" tangles. Intracellular tangles are especially well recognized by both, the polyclonal tau and the monoclonal (Tau-1) antibodies. Reactivity with the monoclonal PHF/ubiquitin antibody 3-39 was variable. Furthermore, a significant percentage of intracellular tangles reacted with the

Table 1. Antigenic profile of Alzheimer neurofibrillary tangles at different stages of maturation

	Stage 0	Stage 1	Stage 2	Stage 3
Cytoskeletal AG:				
tau (polyclonal)	(+)	+	+	(+)
Tau-1	+	+	+	−
SMI 31	−	−/(+)	−/(+)	−
SMI 34	−	−/(+)	−/(+)	−
SMI 33	−	−	−	−
MAP 2	−	−/(+)	−/(+)	−
MAP 1	−	−	−	−
tubulin	−	−	−	−
GFAP	−	−	−	+
PHF/Ubiquitin				
3–39	−	−/+	−/+	+
5–25*	−	−	−/(+)	−/(+)
Amyloid				
Congo red	−	−	(+)	+
thioflavin S	−	−	(+)	+
β-peptide	−	−	−	−/+ *
Glycosaminoglycanes				
Alcian blue (low Mg)	−	−	−	+
Alcian blue (high Mg)	−	−	−	(+)

−/+ = only some tangles stained
(+) = weak reactivity
−/+ * = exceptional tangels stained

neurofilament antibodies SMI 31 and 34 and with a polyclonal anti-MAP 2 serum. On the contrary, extracellular "ghost" tangles mainly reacted with the monoclonal anti-ubiquitin antibody 3-39 and with anti-GFAP. There was only weak reactivity of "ghost" tangles with polyclonal tau and no reaction with monoclonal Tau-1 or neurofilament antibodies nor with anti-MAP 2 serum.

A similar inverse pattern of reactivity between classical intracellular and extracellular tangles was also found with silver impregnations and with reactions for glycosaminoglycanes or amyloid. Interestingly, exceptional extracellular "ghost" tangles, but no intracellular tangles reacted with monoclonal antibodies against the amyloid β-peptide.

Changes of normal cytoskeleton in tangle-bearing neurons

To answer these questions we performed double immunostaining with anti-tau and anti-MAP 1, anti-MAP 2, anti-tubulin or anti-NF respectively. Although tangle-bearing neurons frequently showed irregular, distorted dendrites, the pericaryal and dendritic reactivity of these neurons for MAP's or tubulin was not different from that of normal neurons. However, neurons with early (especially stage 1) tangles frequently revealed cytoplasmic reactivity in the pericaryon and dendrites with a monoclonal antibody against a phosphorylated epitope of neurofilaments (SMI 31).

Correlation of neurofibrillary tangles with clinical dementia

A detailed quantitative analysis of tangles and plaques was performed on 10 cases with variable dementia scores, monitored in the prospective clinico-pathological study (Table 2). The majority of tangles represented classical intracellular inclusions (stage 2), whereas only a minority were classified as stage 0. "Ghost" tangles were found in 5/10 cases in the subiculum but only in 3/10 cases in cortex. The number of "ghost" tangles in cortex was low. We found a significant correlation between the severity of clinical dementia and the number of classical stage 1 and stage 2 tangles in the respective brain areas, which was better than the correlation with senile plaques (Table 2). The correlation between the number of stage 3 ("ghost") tangles and the severity of the clinical disease, although statistically significant in the frontal and parietal cortex, was difficult to interpret, since only cases with exceptionally severe dementia showed this alteration.

Table 2. Spearman rank correlation coefficients of senile plaques (core type) and tangle stages with mini-mental state score

	Cortex			Hippocampus	
	Frontal	Parietal	Temporal	CA1	CA3
Plaques:	−.510	−.630	−.300	−.661	−.462
NFT 0	−.709 [a]	−.801 [a]	−.505	−.393	−.093
	(0− 4)	(0− 0.2)	(0− 3)	(0− 7)	(0−3)
NFT 1	−.784 [b]	−.704 [a]	−.527	−.667	−.863 [b]
	(0− 6)	(0− 1)	(0− 8)	(0−13)	(0−4)
NFT 2	−.784 [b]	−.804 [a]	−.586 [a]	−.667	−.842 [b]
	(0−62)	(0−46)	(0−32)	(0−29)	(0−5)
NFT 3	−.621 [a]	−.868 [b]	−.344	−.647	−.647
	(0− 5)	(0−13)	(0−21)	(0− 8)	(0−3)

[a] $p \, 0.05$; [b] $p \, 0.01$.
The values in brackets represent minimal and maximal counts/mm^2.

Relation to cytoskeletal changes in other diseases

As described by others before, reactivity of intracytoplasmic fibrillary inclusions with antibodies against ubiquitin is not a feature specific for Alzheimer's disease. Similar reactivity is found in Pick bodies, Lewy bodies, Rosenthal fibers and Marinesco bodies. In our material no reactivity of PSP tangles was found with the monoclonal PHF/ubiquitin antibody 3-39 (Bancher et al., 1987). Although in another study (Manetto et al., 1988) ubiquitin was found in PSP tangles, it could only be demonstrated with one out of the 5 monoclonal antibodies used.

However, besides ubiquitin, several different cytoskeletal elements appear to be associated with these various inclusions. They include tau (AD-tangles, PSP tangles and Pick bodies), neurofilaments (Lewy bodies in Parkinson's disease and diffuse Lewy body disease) and glial fibrillary acidic protein (Rosenthal fibers).

Discussion

In our present study we confirmed earlier studies by showing that diffuse cytoplasmic reactivity with anti-tau may be a very early cellular change in the pathogenesis of NFT (Joachim et al., 1987; Bancher et al., 1989 a). Similar results have also been described for the Alz-50 antigen (Wolozin et al., 1986), which in fact may be antigenically related to tau (Nukina et al., 1988; Ksiesak Reding et al., 1988 a, b).

Of major interest appear to be the changes in the antigenic profile of neurofibrillary tangles when they are released into the extracellular space after cell death and become "ghost" tangles. One change is the loss of reactivity with the Tau-1 monoclonal antibody, which takes place although reactivity with polyclonal tau antibodies is partially preserved. Studies on chemically isolated, pronase-treated PHF indicate that only parts of the carboxy terminal of the tau molecule are present in the undegradable backbone of the fibrils (Wischik et al., 1988). Although also ubiquitin epitopes have been reported to disappear from extracellular "ghost" tangles (Joachim et al., 1987) the epitope recognized by the antibody 3-39 is well preserved in ghost tangles and appears to be better accessible compared to intracellular tangles. These results, taken together, could indicate that tau/ubiquitin complexes in neurofibrillary tangles can be partially degraded by extracellular enzymes, leaving the backbone of PHF, which then contains only fragments of tau and ubiquitin.

Another interesting feature of "ghost" tangles is that staining reactions for amyloid (Congo red and thioflavin S; Defossez and Delacourte, 1987) and for glycosaminoglycanes (Alcian blue) are especially prominent in comparison to that in intracellular PHF. Based on the antigenic shift NFT undergo during their evolution process and on their different staining properties, we propose the following sequence of events, which may take place during the formation and maturation of each of these lesions: An as yet unknown stimulus leads to an imbalance of the neuronal cytoskeletal phosphorylation/dephosphorylation system. The diffuse intracytoplasmic accumulation of abnormally phosphorylated tau and the Alz-50 antigen, respectively, precedes the formation of NFT. These abnormal proteins polymerize into filamentous structures and associate with ubiquitin, which represents a fruitless attempt of proteolytic degradation. Degeneration of the neurons exposes the NFT to the enzymatic environment of the extracellular space, leading to conformational changes or, possibly, to partial degradation. Ingrowth of glial cells is reflected by staining with anti-GFAP (Probst et al., 1982). Some ghost tangles may form a nidus for the polymerization of the β-peptide into amyloid fibers or become dressed by glycosaminoglycanes. However, the very low numbers of "ghost" tangles in areas of severe cortical atrophy indicates that the CNS tissue has the capacity to completely degrade the extracellular debris of NFT.

As described by others (Delaere et al., 1989) we found a good correlation between clinical severity of dementia and the number of classical tangles in the central nervous system. The number of extracellular "ghost" tangles, however, was low, even in cases with most severe clinical dementia. Thus, functional impairment of tangle-bearing nerve cells may, at a certain point of tangle maturation, be so severe that there is little functional difference between degenerated or severely compromised, but still live neurons. In addition, nerve cell death with formation of "ghost" tangles may be a very late event in AD pathology and may be precipitated by additional, for instance ischemic, factors due to amyloid angiopathy in the affected regions.

Finally it is interesting to note that the formation of complexes between cytoskeletal elements and ubiquitin is not a unique feature of AD tangles but appears to be a mechanism common to many different neuronal and glial injuries, which leads to accumulation of fibrillar material in the cytoplasm.

References

Alzheimer A (1907) Über eine eigenartige Erkrankung der Hirnrinde. Allg Z Psychiatrie 64:107

Alzheimer A (1911) Über eigenartige Krankheitsfälle des späten Alters. Z Ges Neurol Psychiatr 4:356–385

Anderton BH, Breinburg D, Downes MJ, Green PJ, Tomlinson BE, Ulrich J, Wood JN, Kahn J (1982) Monoclonal antibodies show that neurofibrillary tangles and neurofilaments share antigenic determinants. Nature 298:84–86

Bancher C, Lassmann H, Budka H, Grundke Iqbal I, Iqbal K, Wiche G, Seitelberger F, Wisniewski HM (1987) Neurofibrillary tangles in Alzheimer's disease and progressive supranuclear palsy: antigenic similarities and differences. Acta Neuropathol (Berl) 74:39–46

Bancher C, Brunner C, Lassmann H, Budka H, Jellinger K, Wiche G, Seitelberger F, Grundke Iqbal I, Iqbal K, Wisniewski HM (1989a) Accumulation of phosphorylated tau precedes the formation of neurofibrillary tangles in Alzheimer's disease. Brain Res 477:90–99

Bancher C, Lassmann H, Budka H, Jellinger K, Grundke Iqbal I, Iqbal K, Wiche G, Seitelberger F, Wisniewski HM (1989b) An antigenic profile of Lewy bodies: immunocytochemical indication for protein phosphorylation and ubiquitination. J Neuropathol Exp Neurol 48:81–93

Brion JP, Passarier H, Nunez J, Flament-Durand J (1985) Immunologic determinants of tau protein are present in neurofibrillary tangles. Arch Biol 95:229–235

Brunner C, Lassmann H, Waehnelt TV, Matthieu JM, Linington C (1989) Differential ultrastructural localization of myelin basic protein, myelin/oligodendroglial glycoprotein and 2,3-cyclic nucleoside 3-phosphodiesterase in the CNS of adult rats. J Neurochem 52:296–304

Defossez A, Delacourte A (1987) Transformation of degenerating neurofibrils into amyloid substance in Alzheimer's disease: histochemical and immunohistochemical studies. J Neurol Sci 81:1–10

Delaere P, Duyckaerts C, Brion JP, Poulain V, Hauw JJ (1989) Tau, paired helical filaments and amyloid in the neocortex: a morphometric study of 15 cases with graded intellectual status in ageing and senile dementia of Alzheimer type. Acta Neuropathol (Berl) 77:645–653

Fischer P, Gatterer G, Marterer A, Danielczyk W (1988) Non specificity of semantic impairment in dementia of Alzheimer's type. Arch Neurol 45:1341–1343

Grundke Iqbal I, Iqbal K, Tung YC, Quinlan M, Wisniewski HM, Binder LI (1986) Abnormal phosphorylation of the microtubule associated protein tau in Alzheimer cytoskeletal pathology. Proc Natl Acad Sci USA 83:4913–4917

Hyman BT, Van Hoesen GW, Woloczin B, Davis P, Kromer LJ, Damasio AR (1987) Alz 50 antibody recognizes Alzheimer-related neuronal changes. Ann Neurol 23:371–379

Ihara Y, Abraham C, Selkoe DJ (1983) Antibodies to paired helical filaments do not recognize normal brain proteins. Nature 304:727–730

Joachim CL, Morris JH, Selkoe DJ, Kosik KS (1987) Tau epitopes are incorporated into a range of lesions in Alzheimer's disease. J Neuropathol Exp Neurol 46:611–622

Kosik KS, Duffy LK, Dowling MM, Abraham C, McCluskey A, Selkoe DJ (1984) Microtubule-associated protein 2: monoclonal antibodies demonstrate the selective incorporation of certain epitopes in Alzheimer neurofibrillary tangles. Proc Natl Acad Sci USA 81:7941–7945

Ksiezak-Reding H, Binder LI, Yen SH (1988a) Immunochemical and biochemical characterization of tau-proteins in normal and Alzheimer's disease brains with Alz 50 and tau-1. J Biol Chem 263:7948–7953

Ksiezak-Reding H, Davis P, Yen SH (1988 b) Alz 50, a monoclonal antibody to Alzheimer's disease antigen, cross-reacts with tau proteins from bovine and normal human brain. J Biol Chem 263:7943–7947

Manetto V, Perry G, Tabaton M, Mulvihill P, Fried V, Smith H, Gambetti P, Autilio-Gambetti L (1988) Ubiquitin is associated with abnormal cytoplasmic filaments characteristic of neurodegenerative disease. Proc Natl Acad Sci USA 85:4501–4505

Masters CL, Multhaup G, Simms G, Pottglesser J, Martins RN, Beyreuther K (1985) Neuronal origin of cerebral amyloid: neurofibrillary tangles of Alzheimer's disease contain the same protein as the amyloid of plaque cores and blood vessels. EMBO J 4:2757–2763

Mori H, Kondo J, Ihara Y (1987) Ubiquitin is a component of paired helical filaments. Science 235:1641–1644

Nukina N, Kosik KS, Selkoe DJ (1988) The monoclonal antibody, Alz 50, recognizes tau proteins in Alzheimer's disease brain. Neurosci Lett 87:240–246

Perry G, Friedmann R, Shaw G, Chau V (1987) Ubiquitin is detected in neurofibrillary tangles and senile plaque neurites of Alzheimer disease brains. Proc Natl Acad Sci USA 84:3033–3036

Perry G, Mulvihil P, Fried VA, Smith HT, Grundke Iqbal I, Iqbal K (1989) Immunochemical properties of ubiquitin conjugates in the paired helical filaments of Alzheimer's disease. J Neurochem 52:1523–1528

Probst A, Ulrich J, Heitz PU (1982) Senile dementia of Alzheimer type: astroglial reaction to extracellular neurofibrillary tangles in the hippocampus. An immunohistochemical and electron-microscopic study. Acta Neuropathol (Berl) 57:75–79

Snow AD, Willmer JP, Kisilewski R (1987) Sulphated glycosaminoglycans in Alzheimer's disease. Hum Pathol 18:506–510

Sternberger NH, Sternberger LA, Ulrich J (1985) Aberrant neurofilament phosphorylation in Alzheimer disease. Proc Natl Acad Sci 82:4274–4276

Szumanska G, Vorbrodt AW, Mandybur TI, Wisniewski HM (1987) Lectin histochemistry of plaques and tangles in Alzheimer's disease. Acta Neuropathol 73:1–11

Vass K, Lassmann H, Wekerle H, Wisniewski HM (1986) The distribution of Ia-antigen in the lesions of rat acute experimental allergic encephalomyelitis. Acta Neuropathol 70:149–160

Wang GP, Grundke Iqbal I, Kascak RJ, Iqbal K, Wisniewski HM (1984) Alzheimer neurofibrillary tangles: monoclonal antibodies to inherent antigen(s). Acta Neuropathol 62:268–275

Wischik CM, Novak M, Thogersen HL, Edwards PC, Runswick MJ, Jakes R, Walker JE, Milstein C, Roth M, Klug A (1988) Isolation of a fragment of tau derived from the core of paired helical filament of Alzheimer's disease. Proc Natl Acad Sci USA 85:4506–4510

Wolozin BL, Bruchnicki A, Dickson DW, Davis P (1986) A neuronal antigen in the brains of Alzheimer patients. Science 232:648–650

Correspondence: Doz. Dr. H. Lassmann, Neurological Institute, University of Vienna, Schwarzspanierstrasse 17, A-1090 Wien, Austria.

Choline-acetyltransferase immunoreactivity in the hippocampal formation of control subjects and patients with Alzheimer's disease

G. Ransmayr [1*], P. Cervera [1], E. Hirsch [1], M. Ruberg [1], L. B. Hersh [2], C. Duyckaerts [3], J.-J. Hauw [3], and Y. Agid [1]

[1] INSERM U. 289, Hôpital de la Salpêtrière, Paris, France
[2] Department of Biochemistry, University of Texas Health Science Center, Dallas, TX, U.S.A.
[3] Laboratoire de Neuropathologie R. Escourolle, Hôpital de la Salpêtrière, Paris, France

Summary

Cholinergic fibers and terminals were visualized, with a polyclonal antiserum to human choline-acetyltransferase (ChAT), in hippocampi of 6 patients with Alzheimer's disease (AD) and 4 control brains. The densities of senile plaques (SP) and neurofibrillary tangles (NFT) differed significantly between subregions in the AD brains. In contrast, the significant loss of terminals in the AD brains was homogenous in CA4, CA1 and the subiculum and presumably not related to the densities of SP and NFT in these subregions.

Introduction

Biochemical and histochemical studies have shown that the activities of the cholinergic marker enzymes ChAT and acetylcholinesterase (AChE) decrease in the hippocampus of AD patients (Davies, 1979; Perry et al., 1980). Hippocampal cholinergic structures have been visualized until now with AChE histochemistry (Perry et al., 1980). This method stains, however, also non-cholinergic structures. The present immunohistochemical study was undertaken to describe with a specific

* *Permanent address:* University Clinic of Neurology, Anichstraße 35, A-6020 Innsbruck, Austria.

antiserum to human ChAT (Bruce et al., 1985) the cholinergic system in the hippocampal formation (hippocampus proper, subiculum, gyrus dentatus) from patients with AD and controls.

Patients and methods

Four normal control brains and 6 brains of AD patients, exempt from other diseases, were studied. The clinical diagnosis was based on the criteria of DSM-III and confirmed post-mortem by the presence of numerous senile plaques (SP) and neurofibrillary tangles (NFT) in the frontal (area 10) and temporal neocortex (area 21) (Khatchaturian, 1985). Age (75–92 years in the controls, 67–87 years in the AD-patients) and post-mortem delay (5–19 hours in the controls, 4–28 hours in the AD brains) were not statistically different in the 2 groups. The hippocampi were fixed and cryoprotected (Graybiel et al., 1987), deep-frozen in powdered dry ice, stored at −80° and cut on a sliding microtome into frontal 40 μm sections. ChAT-immunohistochemistry was performed every 2 mm, with a polyclonal antiserum against human ChAT (Bruce et al., 1985) diluted 1:200 [double-bridge peroxidase-antiperoxidase method (Graybiel et al., 1987)]. Adjacent sections were stained with cresyl violet, hematoxylin-eosine, and Bodian silver impregnation. Qualitative description was performed on all sections, quantitative evaluation, on 5 equidistant sections selected from equivalent levels of each hippocampus, counterstained with thioflavin-S to visualize SP and NFT.

In part one of the study, two independent observers estimated the densities of ChAT-positive unbeaded fibers in the distal and proximal part of the fimbria and the alveus at the level of CA2 and CA1 as well as the densities of ChAT-positive terminals in the stratum pyramidale of subiculum, CA1, CA2, CA3, CA4 and stratum polymorphe and moleculare of the dentate gyrus. The distribution of these structures was found to be homogenous within the subregions and along the rostrocaudal extension of the hippocampus (Ransmayr et al., 1989). Therefore, random microscopic fields (0.1 × 0.14 mm) were chosen to quantify the densities of unbeaded fibers and terminals in these subregions: In 2 randomly chosen fields (each 0.1 × 0.14 mm) per subregion and section, all unbeaded fibers were counted and related to the surface area; punctate immunopositive structures (= fiber varicosities and axon segments) were counted in 2 random fields per region and section within 20 elements of a superimposed grid (square openings of 81 μm each). The densities of ChAT-positive fibers and punctate structures were homogeneous in the subregions throughout the rostrocaudal extent of the hippocampus (Kruskal-Wallis ANOVA, H-values < 4.6) (Ransmayr et al., 1989). Therefore, mean regional densities per brain were calculated and the controls and Alzheimer brains compared with the Student's t test.

In the second part of the study, thioflavin-S positive SPs and NFTs were quantified on five sections from 4 control and 5 AD-brains in the whole stratum pyramidale of CA4, CA1 and the subiculum and related to the surface areas of these regions. In one control and one AD brain section, the densities of punctate immunoprecipitates were determined throughout the entire surface of the regions, field by field (each 0.1 × 0.14 mm), in 4 homogenously distributed square elements (each 81 μm^2) of a grid. The densities of punctate ChAT-positive elements were homogenous along the radial extent of CA4, CA1 and the subiculum (one factor ANOVA, F < 2.0), and were, thus, determined on the remaining sections only along the axial (circular) extent of these subregions. The densities of ChAT-positive terminals, SP and NFT in the subregions varied between and within brains, but were not statistically different along the rostrocaudal extent of the hippocampus (one-factor ANOVA, F < 0.6). Thus, mean regional densities for each brain were calculated and the AD and control brains compared (Student's t-test).

Results

No ChAT-positive cell bodies were found in the hippocampus of AD and control brains. ChAT-positive unbeaded fibers projected to the hippocampus via the fimbria and the white matter of the temporal lobe and entered the neuronal layers, where networks of fine beaded fibers and punctate immunoprecipitates were found. In all subregions of the AD brains, ChAT-positive unbeaded fibers and terminals were found to be significantly (by 28–46%) reduced. Means and single standard deviations of the mean regional densities of NFT, SP and terminals in CA$_4$, CA$_1$ and the subiculum (stratum pyramidale) of the AD and the control brains are visualized in the figure. There were significantly higher densities of SPs and NFTs in the AD brains as compared to the controls; the loss of ChAT-positive terminals was homogenous in CA4, CA1 and the subiculum (31 to 33%), whereas the densities of SP and NFT differed considerably in these regions.

Discussion

The hippocampi examined did not contain cholinergic cell bodies, which contrasts to what was suspected from AChE-histochemistry (Perry et al., 1980). The perikarya of the cholinergic afferences visualized in the fimbria and the white matter of the temporal lobe are presumably located in the medial septal nucleus, the diagonal band of Broca and the nucleus basalis of Meynert (Mesulam et al., 1983). In contrast to an inhomogenous distribution of NFT and SP (e. g. 4-fold densities of SP in the subiculum and CA1 as compared to CA4), the cholinergic terminals in the different subregions of the stratum pyramidale were homogenously reduced (by one third) in the AD group as compared to the controls (Fig. 1). This finding suggests that SP and NFT located in the hippocampal pyramidal layer of CA4, CA1 and the subiculum might not be the major factors for the loss of cholinergic terminals in these subregions in AD.

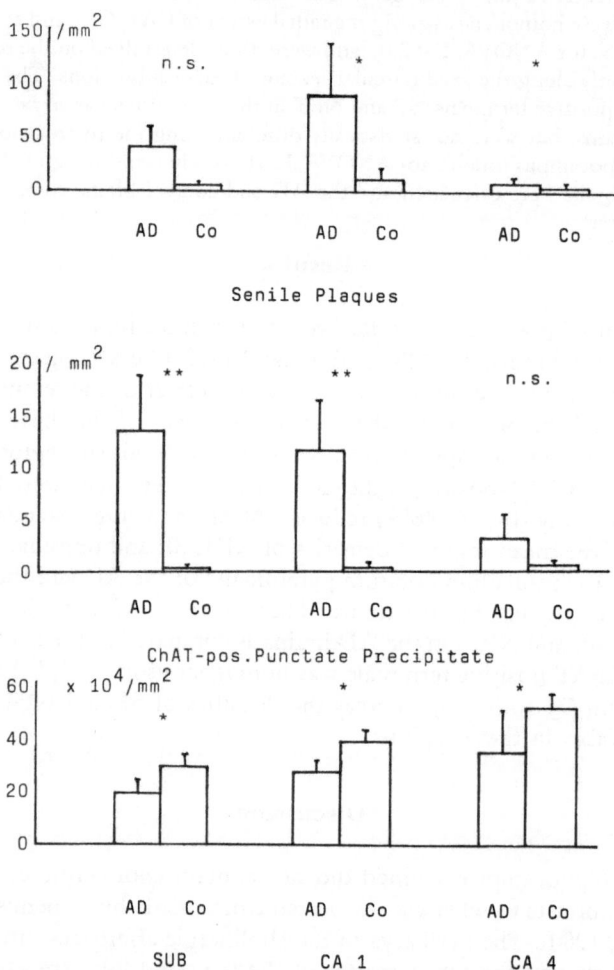

Fig. 1. Mean densities and single standard deviations of SP, NFT, and ChAT-positive punctate structures (terminals) in the stratum pyramidale of the Subiculum (Sub), CA1, and CA4 in the AD (N=5) and the control (Co) group (N=4), as determined in part 2 of the study. Comparison of the 2 groups with the Student's two-sided t-test (* p<0.05; ** p<0.01; *** p<0.001)

References

American Psychiatric Association (1980) Diagnostic and statistical manual of mental disorders, 3rd edn. APA, Washington

Bruce G, Wainer BH, Hersh LB (1985) Immunoaffinity purification of human choline acetyltransferase: comparison of the brain and placental enzymes. J Neurochem 45:611–620

Davies P (1979) Neurotransmitter-related enzymes in senile dementia of the Alzheimer type. Brain Res 171:319–327

Graybiel AM, Hirsch EC, Agid YA (1987) Differences in tyrosine-hydroxylase-like immunoreactivity characterize the mesostriatal innervation of striosomes and extrastriosomal matrix at maturity. Proc Natl Acad Sci 84:303–307

Khatchaturian ZS (1985) Diagnosis of Alzheimer's disease. Conference report. Arch Neurol 42:1097–1105

Mesulam M-M, Mufson EJ, Wainer BH, Levey AI (1983) Central cholinergic pathways in the rat: an overview based on an alternative nomenclature (Ch1-Ch6). Neuroscience 10:1185–1201

Perry RH, Blessed G, Perry EK, Tomlinson BE (1980) Histochemical observations on cholinesterase activity in the brain of elderly normal and demented (Alzheimer-type) patients. Age Ageing 9:9–16

Ransmayr G, Cervera P, Hirsch E, Ruberg M, Hersh LB, Duyckaerts C, Hauw J-J, Delumeau C, Agid Y (1989) Choline-acetyltransferase-like immunoreactivity in the hippocampal formation of control subjects and patients with Alzheimer's disease. Neuroscience 32:701–714

Correspondence: Prof. Y. Agid, MD, PhD, INSERM U. 289, Hôpital de la Salpêtrière, 47, Boulevard de l'Hôpital, F-75651 Paris, Cedex 13, France.

Calbindin immunoreactive neurones
in Alzheimer-type dementia

**Y. Ichimiya[1], H. Arai[1], R. Iizuka[1], P. C. Emson[2],
and C. Q. Mountjoy[3]**

[1] Department of Psychiatry, Juntendo University School of Medicine, Tokyo, Japan
[2] MRC Group, Institute of Animal Physiology and Genetics Research, Babraham,
United Kingdom
[3] Department of Psychiatry, University of Cambridge, Cambridge, United Kingdom

Summary

An antibody to the calcium binding protein, calbindin D28k (CaBP), was
used to study the number and size of CaBP immunoreactive (CaBP-Ir) neurones
in the cerebral cortex and the nucleus basalis of Meynert (nbM) of postmortem
human brains from controls and from patients with Alzheimer-type dementia
(ATD).

Compared to the controls, the number and size of CaBP-Ir neurones in the
cerebral cortex and the nbM were reduced in the ATD.

These findings suggest that CaBP containing neurones are damaged in ATD.

Introduction

It is well known that elevated internal Ca^{++} leads to cell injury
(Campbell, 1985), and that the excessive elevation of intracellular Ca^{++}
results in cell death (Cotman et al., 1987).

In mammalian brains, a number of neurones contain the calcium
binding proteins such as parvalbumin (PVA) (Heizmann et al., 1987) and
calbindin D28k (CaBP) (Parmentier et al., 1987). These proteins are
believed to be involved in the buffering or transport of internal Ca^{++}
(Heizmann et al., 1987), and possibly in reducing neuronal vulnerability
to excitotoxic effects of excitatory amino acids (Rothman et al., 1987).

In Alzheimer-type dementia (ATD), we have been investigating the
expression of calcium binding proteins to determine whether neurones

containing these proteins are damaged in ATD. We recently reported a specific loss of PVA containing neurones in ATD (Arai et al., 1987). We have also observed that the number and size of CaBP immunoreactive neurones in the cerebral cortex (Ichimiya et al., 1988) and the nucleus basalis of Meynert (nbM) (Ichimiya et al., 1989) are reduced in ATD.

In this paper, we mentioned CaBP immunoreactive neurones in ATD brain.

Material and methods

Postmortem brains from control patients and neurologically diagnosed ATD patients were used for our serial immunohistochemical studies. The detail of these patients was reported previously (Ichimiya et al., 1988, 1989). Blocks of tissues, including the frontal, temporal and parietal cortices and the nbM, were taken from formalin fixed brains as reported previously (Ichimiya et al., 1986, 1988). Prior to sectioning, the blocks were transferred to 30% sucrose (w/v) in 0.1 M phosphate buffer saline (PBS) and stored at 4°C until required. 20 µm sections were cut from the blocks on a freezing microtome.

Calbindin D28k (CaBP) antibodies were raised in rabbit against purified chick intestinal CaBP. The CaBP antibody was used at a final dilution of 1:500. In order to visualize CaBP immunoreactivity (CaBP-Ir), peroxidase-antiperoxidase (PAP) and/or immunofluorescence techniques were used. For studies of possible co-existence of CaBP and cholin acetyltransferase (ChAT), CaBP-Ir was detected using the rabbit anti-CaBP primary antibody and a FITC labelled sheep anti-rabbit Ig-G (1:100). After photography the coverslips were removed, sections washed in PBS and the sections then processed to visualize ChAT immunoreactivity using a mouse monoclonal anti-ChAT (Boehringer) (2 µg/ml) and the PAP technique.

Counts and size of CaBP-Ir neurones were made with the aid of a Quantimet 720 image analyser as reported previously (Mountjoy et al., 1983).

Results

In the controls, CaBP-Ir neurones were observed all layers of the cerebral cortex with exception of layer I, and almost all the large neurones in the nbM were also CaBP-Ir. Some CaBP-Ir neurones in the nbM were also ChAT immunoreactive (Fig. 1).

Compared to the controls, the number and the mean size of CaBP-Ir neurones in the cerebral cortex and the nbM were reduced in the ATDs (Table 1).

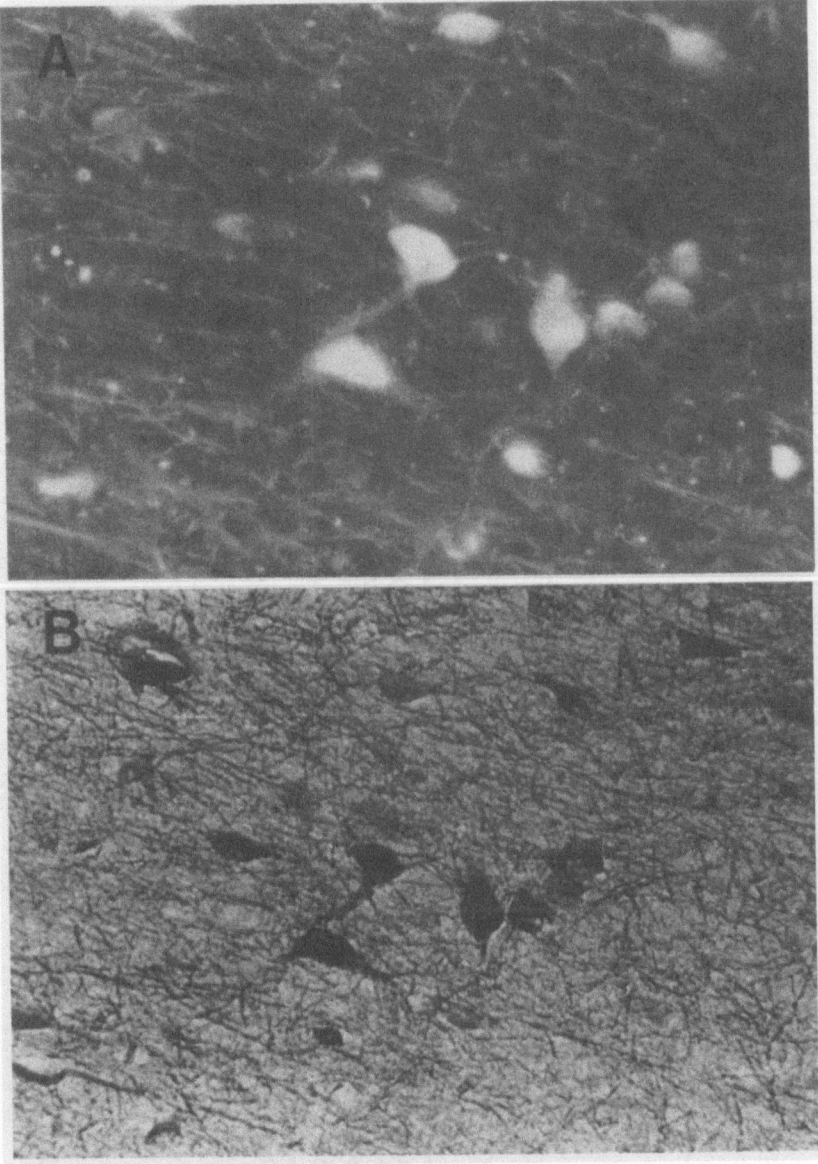

Fig. 1. The co-expression of CaBP and ChAT immunoreactivity in human nbM neurones. CaBP immunoreactie neurones revealed by immunofluorescence (FITC) (**A**), are also positive for ChAT immunoreactivity (**B**). × 200

Table 1. Total number and mean size of CaBP immunoreactive neurones

	Number		Size (μm^2)	
	C	ATD	C	ATD
Cerebral cortex { frontal	43.3± 7.2 (5)	16.1± 2.7 (5)	80.2± 8.4 (5)	59.6± 9.7 (5)
temporal	41.6± 3.7 (5)	15.7± 3.1 (5)	88.1±11.5 (5)	67.8± 8.1 (5)
parietal	43.5± 3.6 (4)	21.9± 2.8 (5)	84.8± 8.6 (4)	50.5± 4.6 (5)
Nucleus basalis of Meynert	870.0±23.3 (3)	222.0±45.3 (5)	778.6±35.5 (3)	647.3±69.6 (5)

Values: mean ± S.D., ( ): number of subjects, *C* controls, *ATD* Alzheimer-type dementia

Discussion

CaBP immunoreactivity was well preserved in the cerebral cortex and the nbM neurones on the controls. Postmortem stability of CaBP has been previously demonstrated using Western blot analysis (Pochet et al., 1985).

In ATD brain, a reduction of CaBP content in the cerebral cortex has been reported by radioimmunoassay (McLachlan et al., 1987). As expected those observation, the number and size of CaBP-Ir cortical neurones were decreased in the ATD brains.

Our findings in normal control human nbM are similar to the earlier observation of CaBP-Ir nbM neurones in monkey brains (Celio et al., 1985). Neuropathologically, neuronal loss in the nbM in ATD is well established (Ichimiya et al., 1986). As expected from the observation on the ChAT staining of ATD brains (Nagai et al., 1983), the number of CaBP-Ir nbM neurones was reduced dramatically. The co-expression of CaBP and ChAT immunoreactivities in some control nbM neurones suggest that CaBP might co-exist, at least partially, in the nbM cholinergic neurones.

These observations suggest that CaBP containing neurones are widely affected in ATD brain, and that a failure of calcium homeostasis may be associated with the development of the pathology of ATD.

References

Arai H, Emson PC, Mountjoy CQ, Carassco LH, Heizmann CW (1987) Loss of parvalbumin-immunoreactive neurones from cortex in Alzheimer-type dementia. Brain Res 418:164–169

Campbell AK (1985) Intracellular calcium. John Wiley & Sons Ltd, Chichester New York Brisbane Toronto Singapore

Celio MR, Norman AW (1985) Nucleus basalis Meynert neurons contain the vitamin D-induced calcium-binding protein (calbindin-D 28k). Anat Embryol 173:143–148

Cotman CW, Iversen LL (1987) Excitatory amino acids in the brain-focus on NMDA receptors. TINS 10:263–272

Heizmann CW, Berchtold MW (1987) Expression of parvalbumin and other Ca^{++}-binding proteins in normal and tumor cells: a topical review. Cell Calcium 8:1–41

Ichimiya Y, Arai H, Kosaka K, Iizuka R (1986) Morphological and biochemical changes in the cholinergic and monoanimergic systems in Alzheimer-type dementia. Acta Neuropathol (Berl) 70:112–116

Ichimiya Y, Emson PC, Mountjoy CQ, Lawson DEM, Heizmann CW (1988) Loss of calbindin-28K immunoreactive neurones from the cortex in Alzheimer-type dementia. Brain Res 475:156–159

Ichimiya Y, Emson PC, Mountjoy CQ, Lawson DEM, Iizuka R (1989) Calbindin immunoreactive cholinergic neurones in the nucleus basalis of Meynert in Alzheimer-type dementia. Brain Res 499:402–406

McLachlan DRC, Wong L, Bergeron C, Baimbridge KG (1987) Calmodulin and calbindin D28k in Alzheimer disease. Alzheimer Disease and Associated Disorders 1:171–179

Mountjoy CQ, Roth M, Evans NJR, Evans HM (1983) Cortical neuronal counts in normal elderly controls and demented patients. Neurobiol Aging 4:1–11

Nagai R, McGeer PL, Peng JH, McGeer EG, Dolman CE (1983) Choline acetyltransferase immunohistochemistry in brains of Alzheimer's disease patients and controls. Neurosci Lett 36:195–199

Parmentier M, Ghysels M, Rypens F, Lawson DEM, Pasteels JL, Pochet R (1987) Calbindin in vertebrate classes: immunohistochemical localization and Western blot analysis. Gen Comp Endocrinol 65:399–407

Pochet R, Parmentier M, Lawson DEM, Pasteels JL (1985) Rat brain synthesizes two 'vitamin D-dependent' calcium-binding proteins. Brain Res 345:251–256

Rothman SM, Olney JW (1987) Excitotoxicity and NMDA receptor. TINS 10:299–302

Correspondence: Dr. Y. Ichimiya, Department of Psychiatry, Juntendo University School of Medicine, 2-1-1 Hongo, Bunkyo-ku, Tokyo, 113, Japan.

Neurochemistry

Lactate production and glycolytic enzymes in sporadic and familial Alzheimer's disease

S. Sorbi, M. Mortilla, S. Piacentini, G. Tesco,
S. Latorraca, B. Nacmias, S. Tonini, and L. Amaducci

Department of Neurology and Psychiatry,
University of Florence, Florence, Italy

Summary

In this paper we report the results of an investigation of lactate production and of the activity of phosphofructokinase (PFK), lactate dehydrogenase (LDH) and hexokinase (HK) in skin cultured fibroblasts from familial and sporadic Alzheimer's disease patients. We found that the activity of HK was deficient in some patients with the familial dominant form of Alzheimer's disease. These results indicate that some alterations on the energy metabolism are present in some subjects with the familial Alzheimer's disease but not in sporadic cases.

Introduction

Alterations in cerebral metabolic rate, cerebral oxygen consumption, cerebral blood flow and cerebral glucose utilization have been among the earliest and best documented alterations in dementias (Quastel, 1932; Sokoloff, 1966). The changes in vivo metabolic rate and the decrease of neurotransmitter system activities and of energy metabolism markers in post-mortem samples are well documented in Alzheimer's disease (AD) but how and why these changes occur in Alzheimer's neuronal cells is still unknown.

Besides these results reduction in the activity of glycolytic enzymes has been documented in the brain of patients with AD and more recently an increase of lactate production has been observed in cultured skin fibroblasts from AD patients. Perry et al. (1980) and Sorbi et al. (1983) have reported a decreased activity of the pyruvate dehydrogenase complex (PDHC) in affected areas of AD and Huntington brain. Bowen et al.

(1979) observed that 19 of 37 neuroconstituents examined were reduced in AD post-mortem brain. Among them the principal decrease was in the activity of phosphofructokinase (PFK), a key enzyme in glycolysis that may be the rate-limiting enzyme in cerebral glucose utilization (Lowrey and Passanneau, 1964). Iwangoff et al. (1980) have also reported a reduction in PFK activity with age and a marked decrease (90% of age-matched control values) in Alzheimer cerebral cortex. The abnormality in PFK observed in post-mortem brain seems to persist, in at least some patients, in cultured skin fibroblasts even after several passages (Sorbi and Blass, 1983). Studies on cultured skin fibroblasts have also shown abnormalities in PDHC and PFK activities in C-21 trisomy Down patients (Sorbi et al., 1983; Sorbi and Blass, 1983) who are at very high risk to develop a dementia if they live long enough. Recently Sims et al. (1985, 1987) have observed an increased lactate production and decreased CO_2 production in cultured skin fibroblasts from AD patients. Contrasting results have been reported, however, by Peterson and Goldman (1986) who failed to observe changes in lactate production in similar cell cultures from AD patients. Moreover, the study of Peterson and Goldman (1986) indicate that in cultured skin fibroblasts there is an overall alteration of cell metabolism. They reported, in fact, changes in glucose oxidation, calcium content, protein and DNA synthesis in aged and AD fibroblasts.

To further investigate the involvement and glycolysis in cultured cells from AD patients we have studied lactate production, PFK, hexokinase (HK) and lactate dehydrogenase (LDH) activities in cultured skin fibroblasts from patients affected by familial Alzheimer's disease (FAD), their unaffected relatives, sporadic AD, apparently normal controls and pathological controls.

Materials and methods

Skin fibroblasts have been studied from 3 FAD patients (2 brothers and a cousin from the same family), 3 non-blood relatives (spouses) of this family and 6 age-matched unrelated apparently normal controls. Leukocytes were also used to study PFK and HK in 3 other FAD patients, 5 sporadic AD and 5 apparently normal age-matched controls.

Fibroblast cell cultures were obtained from forearm skin biopsy. Cells were grown in Dulbecco's modified Eagle's medium, supplemented with 20% FBS and harvested at confluence, 5 to 7 days after a 3:1 subculture, by mild trypsinization. Periodic testing revealed no contamination by PPLO. Protein content and cells numbers were similar in all lines studied as well as passage numbers. All cell lines were coded and studied in at least 3 different experiments, at least in triplicate in each experiment. Leukocytes were obtained by 30 ml of blood collected from over-night fasted patients using a plastic syringe containing heparin.

PFK activity was measured using the method of Racker (1947) as modified by Sorbi and Blass (1983).

LDH activity was measured according to Clark and Nicklas (1970). HK activity was measured according to Racker (1947). Protein concentration was measured according to Lowry et al. (1951).

Results

Lactate production was comparabe in cultured skin fibroblasts from FAD affected members of the family studied, their unaffected relatives and 3 controls (Table 1).

The activities of fibroblast PFK and LDH were comparable in all groups studied (Table 2).

Fibroblast HK activity was significantly decreased in the FAD patients compared to the unaffected relatives. All affected members of this family had HK activity below the lowest of control values.

Leukocyte HK activity was also significantly decreased in patients with FAD but not in sporadic AD and PFK activity was normal in all AD patients (Table 3).

Table 1. Lactate production in Alzheimer's disease

FAD (3)	858 ± 401
Unaffected relatives (5)	756 ± 126
Controls (3)	676 ± 242

Lactate production (Mean $\pm$ SD) is expressed as nmole/mg prot/hour

Table 2. Fibroblasts PFK, LDH and HK in Alzheimer's disease

	PFK	LDH	HK
Controls (6)	34.3 ± 11	997 ± 27	25.3 ± 6
FAD (3)	39.6 ± 11	621 ± 12	19.3 ± 2 [a]
Unaffected relatives (3)	28.8 ± 12	486 ± 72	29.0 ± 5

[a] $p < 0.05$
Enzyme activities are expressed as nmol/min/mg prot. All experiments were done in triplicate in at least three different experiments.

Table 3. Leukocyte PFK and HK in Alzheimer's disease

	PFK	HK
Controls (5)	27.11 ± 9	55.9 ± 12
FAD (3)	20.9 ± 9	36.1 ± 5 [a]
AD (5)	40.5 ± 19	56.3 ± 24

[a] $p < 0.05$

Enzyme activities are expressed as nmol/min/mg prot. All experiments were done in triplicate in at least three different experiments.

Discussion

In vivo studies of cerebral metabolic rate for oxygen and glucose, and of cerebral blood flow, have consistently disclosed significant reductions in Alzheimer's brain. Moreover, there is now evidence that metabolic changes associated with AD are not confined to the brain.

In this study we have reported evidence of an alteration of the catalytic activity of the glycolytic rate-limiting enzyme HK in affected patients with the familial form of AD. Enzyme values were decreased either versus unaffected relatives or apparently normal unrelated controls. The decrease in HK activity was present in cultured skin fibroblasts and in leukocytes from these patients. However, HK activity was normal in cells from patients with sporadic AD. In FAD patients lactate concentration was normal contrasting with results reported by Sims et al. (1985).

The significance of the finding of abnormalities in peripheral tissues in AD is not yet clear. It is difficult to assess the role of metabolic changes in peripheral cells in the pathogenesis of AD. The finding of any abnormality in cultured cells even after several cell passages should indicate that such abnormality is independent of exogenous factors and linked, at least in a broad sense, to the cellular genoma. Moreover, the finding of a consistent abnormality in peripheral cells may lead to a biological marker to be used as a test for diagnosis.

Our results suggest a possible heterogeneity within AD patients, providing evidence that at least a subgroup with FAD have a modification in the activity of hexokinase. Further studies are in progress to evaluate the frequency of this finding among FAD patients and to characterize that abnormality at a molecular level.

Acknowledgement

This work was supported by CNR (Consiglio Nazionale delle Ricerche), grants 85.00772.56 and 86.01951.56.

References

Bowen DM, White P, Spillane JA, Goodhart MJ, Curzon G, Iwangoff P, Mayer-Rouge W, Davison AN (1979) Accelerated aging or selective neuronal loss as an important cause of dementia? Lancet i: 11–12

Clark JB, Nicklas WJ (1970) The metabolism of rat brain mitochondria. J Biol Chem 245: 4724–4731

Iwangoff P, Armbruster R, Enz A, Mayer-Rouge W (1980) Glycolytic enzymes from human autopsy brain cortex: normal aged and demented cases. Mech Aging Dev 14: 203–209

Lowrey OH, Passaneau J (1964) The relation between substrates and enzymes of glycolysis in brain. J Biol Chem 139: 31–38

Lowry OH, Rosenbrough NJ, Farr AL, Randall RJ (1951) Protein measurement with the Folin phenol reagent. J Biol Chem 193: 265–275

Perry EK, Perry RH, Tomlinson BE, Blessed G, Gibson PH (1980) Coenzyme A acetylating enzymes in Alzheimer's disease: possible cholinergic "compartments" of pyruvate dehydrogenase. Neurosci Lett 18: 105–109

Peterson C, Goldman JE (1986) Alterations in calcium content and biochemical processes in cultured skin fibroblasts from aged and Alzheimer donors. Proc Natl Acad Sci USA 83: 2758–2762

Quastel JH (1932) Biochemistry and mental disorder. Lancet ii: 1417–1425

Racker R (1947) Spectrophotometric measurement of hexokinase and phosphohexokinase activity. J Biol Chem 167: 843–854

Sims N, Finegan JM, Blass JP (1985) Altered glucose metabolism in fibroblasts from patients with Alzheimer's disease. N Engl J Med 313: 638–639

Sims N, Finegan JM, Blass JP (1987) Altered metabolic properties of cultured skin fibroblasts in Alzheimer's disease. Ann Neurol 21: 451–457

Sokoloff L (1966) Cerebral circulatory and metabolic changes associated with aging. Res Publ Assoc Res Nerv Ment Dis 41: 237–241

Sorbi S, Bird ED, Blass JP (1983) Decreased pyruvate dehydrogenase complex activity in Huntington and Alzheimer brain. Ann Neurol 13: 72–73

Sorbi S, Blass JP (1983) Fibroblast phosphofructokinase in Alzheimer's disease and Down's syndrome. Banbury Rep 15: 297–307

Correspondence: Dr. S. Sorbi, Department of Neurology, University of Florence, Viale Morgagni, 85, I-50134 Florence, Italy.

Impairment of cerebral glucose metabolism parallels learning and memory dysfunctions after intracerebral streptozotocin

R. Nitsch, G. Mayer, R. Galmbacher, G. Galmbacher, V. Apell, and S. Hoyer

Department of Pathochemistry and General Neurochemistry, University of Heidelberg, Heidelberg, Federal Republic of Germany

Summary

In early-onset Alzheimer's disease, the cerebral glucose metabolism is disturbed in a characteristic manner. Here, we attempted to mimic these alterations by intracerebral injection of the pancreatic islet cell toxin streptozotocin (STZ) into the rat brain ventricles. This treatment resulted in a reduced arteriovenous difference (AVD) of glucose, an increased AVD of lactate, whereas the AVDs of oxygen and carbon dioxide remained unchanged. In addition, learning and memory functions were impaired. These alterations may be related to a disturbance of the local action of insulin on the brain. Furthermore, the approach reported here may provide a model for the study of the early pathogenetic events of Alzheimer's disease.

Introduction

In early-onset Alzheimer's disease, a number of metabolic alterations of the brain have been observed soon after the onset of initial clinical symptoms. The cerebral metabolic rate of glucose is reduced in combination with an increased cerebral release of lactate (Hoyer et al., 1988). In addition, massive amounts of ammonia, aspartate and glycine are released from the brain (Hoyer and Nitsch, 1989). This pattern of alterations shares similarities with a number of changes that occur during an insulin-deficient metabolic state of classical insulin-dependent tissues like muscle or adipose tissue (cf. Rosen, 1987). The precise role of insulin in the brain is unclear as yet. However, neuromodulatory effects have been reported (Palovcik et al., 1984; Sakaguchi and Bray, 1987) as well as metabolic actions (Phillips and Coxon, 1976; Pellegrino et al., 1987). Furthermore, insulin stimulates choline acetyltransferase activity in retina neurons (Kyriakis et al., 1987), inhibits neuronal norepinephrine uptake (Boyd et al., 1985) and stimulates the release of norepinephrine and dopamine (Sauter et al., 1983). In addition, the neuronal uptake of the amino acid neurotransmitter glycine is stimulated by insulin (Yorek et al., 1987).

From a behavioural point of view, insulin in the brain has been shown to induce a sleep-promoting effect (Danguir and Nicolaidis, 1984) and to be involved in the central regulation of food intake (Woods et al., 1979).

Insulin receptors are present throughout the brain in a highly characteristic distribution (Werther et al., 1987; Hill et al., 1986; Corp et al., 1986; Havrankova et al., 1978). The site of synthesis of cerebral insulin may be localized within the brain as shown by in situ hybridization of insulin mRNA in periventricular hypothalamic cells (Young, 1986). Furthermore, neuronal cell cultures have been shown to synthesize (Schechter et al., 1988) and to release insulin (Clark et al., 1986). Moreover, molluscan neurons were shown to produce the precursor of an insulin-related peptide (Smit et al., 1988) and transgene experiments implied that pancreatic endocrine cells are ontogenetically related to neurons (Alpert et al., 1988). On the other hand, it can not be excluded as yet, that circumventricular nerve endings take up insulin from the blood and transport it into deeper regions of the brain (cf. Baskin et al., 1987).

In experimental animals, the insulin secretion of the pancreas is inhibited by streptozotocin (STZ) (Zucker and Archer, 1988). STZ is a DNA-methylating agent (Fram et al., 1988) which destroys pancreatic islet-cells (Kolb-Backofen et al., 1988) and thus causes diabetes mellitus. In addition, the post-binding autophosphorylation of the insulin receptor kinase is decreased in STZ-diabetes (Kadowaki et al., 1984) and hepatocytes from streptozotocin diabetic animals are insulin-resistant (Hussin and Skett, 1988). Thus, the metabolic action of insulin may be suppressed by STZ via an additional defect of the insulin-receptor. In this study, we injected STZ in a subdiabetogenic dose directly into the ventricular system of the rat brain in order to inhibit the cerebral insulin-mediated effects. We show, that the cerebral glucose metabolism is reduced and that deficits in learning and memory behaviour occur after intracerebral administration of STZ.

Materials and methods

Intraventricular injection

One year old male Wistar rats (486 ± 55 g) were anaesthesized with chloral hydrate (240 mg/kg bw i.p., in a 4% solution). STZ (Sigma, München, FRG), in the subdiabetogenic dose of 1–1.5 mg/kg bw was dissolved immediately before injection in 9 µl of artificial CSF, and injected into the left lateral ventricle under stereotactical guidance. Control animals received 9 µl of artificial CSF alone. Artificial CSF consisted of 120 mM NaCl, 3 mM KCl, 1.15 mM $CaCl_2$, 0.8 mM $MgCl_2$, 27 mM $NaHCO_3$ and 0.33 mM NaH_2PO_4, pH adjusted to 7.2 by 5% CO_2-insufflation, T = 37 °C. The injection coordinates were: 2 mm lateral, 0.7 mm caudal and 4 mm ventral, from bregma at the piamatral level.

Cerebral arterio-venous differences

For determination of the cerebral arterio-venous differences (AVDs) of glucose, lactate, O_2 and CO_2, the animals (control: n = 13, STZ day 21: n = 13, STZ day 7: n = 8) were anaesthesized (Halothan 0.5%, N_2/O_2: 70/30) and artificially ventilated under controlled conditions (Rhema animal respirator 4600). Arterial blood gases, pH, HCO_3, base excess, blood pressure, heart rate, body temperature as well as the temperature of the head, were followed up closely and maintained strictly within the normal range. The superior sagittal venous sinus was exposed with a straight milling-cutter. A 24G micro-catheter (Braun, Melsungen, FRG), attached to a heparinized gas-tight syringe (Hamilton), was stereotactically placed into the superior sagittal venous sinus. Cerebro-venous blood (2.5 ml) from the sinus and the same amount of blood from the catheterized femoral artery were drawn simultaneously at a constant rate (0.625 ml/min). O_2- and CO_2-contents in the obtained samples were measured immediately thereafter (Korning 168 auto-analyzer). Serum was separated and processed for determination of glucose and lactate using commercial test kits (Boehringer, Mannheim, FRG). In order to minimize interassay variation of these tests, both parameters were determined by means of quadruple measurements. Arteriovenous differences were calculated for each animal individually.

Psychometrical testing

The following tests were carried out on day six after intraventricular injection:

1. Heidelberg attention test (Mayer et al., 1988; Nemeth et al., 1989): This procedure tests the attentional behaviour to low extracutaneous electrical stimuli which cause neither pain nor escape behaviour. The first reaction is related to sensory attention (SA), the stimulus is located in the sensory modalities itself. The second reaction is related to cognitive attention (CA), which adresses higher brain centers and the functioning of the brain as a

whole. The cognitive attention reaction leads to goal-directed behavioural patterns which require controlled cognitive processing.

2. Single-trial two-compartment passive avoidance test: In the learning trial, (one hour after the initial trial) the rat was put into the black compartment of the test chamber. Twenty seconds later one footshock (50 Hz, 1 mA, 1 sec) was applied. After 3 minutes the rat was placed into the light compartment for the retention trial.

Statistical analysis

Statistical analysis of the obtained data was performed by two-tailed Mann-Whitney U-test for unpaired samples. Differences between groups were accepted as significant at $p < 0.05$. Homogeneity of variances was estimated by the F-test.

Results

Arterio-venous differences

On day 21 after STZ-injection, AVD-glucose was significantly decreased by 36% (Fig. 1 A), whereas the AVD-lactate was increased by factor two (Fig. 1 B). On day 7, these alterations were less pronounced and thus failed to achieve statistical significance. However, the variations of AVD-glucose and AVD-lactate within the STZ-treated group were significantly higher (F-test).

AVD-O_2 as well as AVD-CO_2 were unchanged on day 7 and on day 21, respectively (Figs. 1 C and D).

Peripheral blood levels

The levels of glucose, lactate, ketone bodies, O_2 and CO_2 in the peripheral arterial blood were indifferent between the three groups (data not shown).

Psychometrical testing

On day 6 after STZ-injection, the values of sensory attention remained unchanged as compared to the control group (Fig. 2 A). The cognitive attention values, however, were significantly elevated in the STZ-treated group (Fig. 2 B). In comparison to controls, a 150% increase of stimulus intensity was necessary to elicit comparable attentive

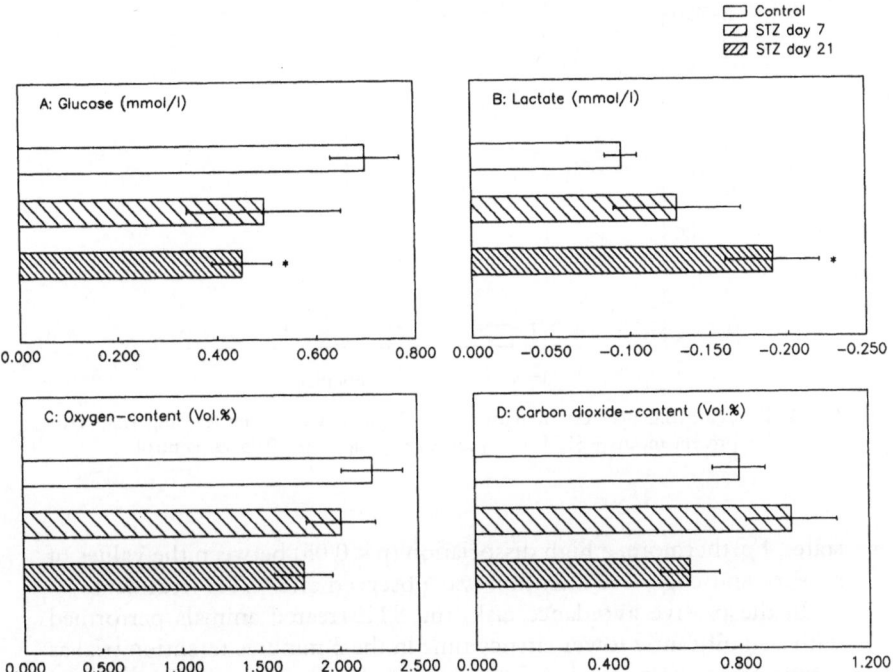

Fig. 1. Cerebral arterio-venous differences of (**A**) glucose (mmol/l); (**B**) lactate (mmol/l); (**C**) oxygen-content (vol%); (**D**) carbon dioxide-content (vol%) 7 and 21 days after intraventricular injection of STZ (1–1.5 mg/kg bw). Bars represent means ± SEM; n = 13 for both, control and STZ days 21 group; n = 8 for the STZ day 7 group; * p < 0.05 vs. control

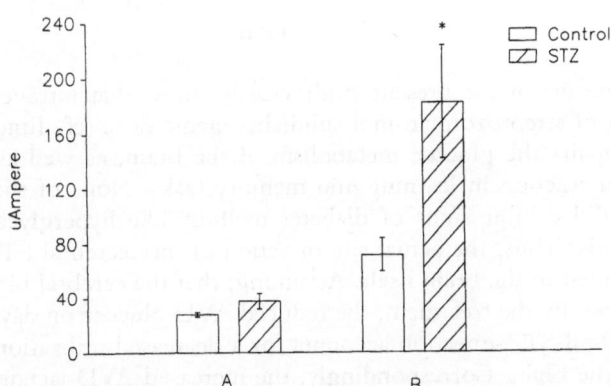

Fig. 2. Attention test on day 6 after injection. (**A**) sensory attention, (**B**) cognitive attention. Bars represent means ± SEM, n = 8 for each group; * p < 0.05 vs. control

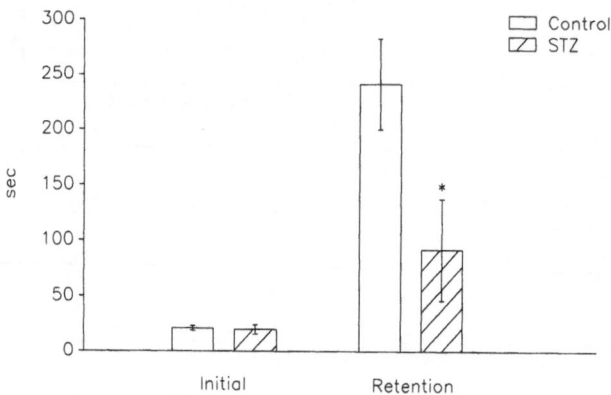

Fig. 3. Passive avoidance test. Retention trial 3 minutes after the learning trial. Bars represent means ± SEM; n = 8 for each group; * p < 0.05 vs. control

states. Furthermore, a high dissociation (p < 0.05) between the values of sensory and cognitive attention was observed after STZ-treatment.

In the passive avoidance task, the STZ-treated animals performed with a significantly lower latency time in the 3 minutes retention trial as compared to the control group (Fig. 3), whereas in the initial trial, the groups were indifferent. It was interesting to observe, that the STZ-treated animals, although showing anxiety-free behaviour in the light compartment, entered very soon the dark compartment where they subsequently showed the immobilizing autonomic response pattern of state anxiety.

Discussion

The results of the present study clearly show, that intraventricular injection of streptozotocin in a subdiabetogenic dose (cf. Junod et al., 1969) impairs the glucose metabolism of the brain, as well as the behavioural reaction in learning and memory tasks. None of the treated animals did exhibit signs of diabetes mellitus like hyperglycaemia or ketonaemia. Thus, the actual site of action of intracerebral STZ seems to be limited to the brain itself. Assuming, that the cerebral blood flow is unaltered by the treatment, the reduced AVD-glucose on day 21 after intracerebral STZ-injection accounts for a decreased utilization of glucose by the brain. Correspondingly, the increased AVD-lactate reflects an increased cerebral release of lactate. The onset of the metabolic action of intracerebral STZ is located around day 7 after injection as shown by a significantly greater variation of the AVD-glucose and AVD-lactate values in the group STZ day 7 (Figs. 1 A and B). The reduced utilization of glucose and the high release of lactate may be caused by an inhibition of pyruvate dehydrogenase, an enzyme complex, which is stimulated by insulin (Rinaudo et al., 1985). A reduction of pyruvate dehydrogenase activity may result in a reduced glycolytic flux and an accumulation of unused glucose in brain tissue (unpublished results) with the consequence of a feedback-mediated inhibition of further glucose uptake. It can not be excluded, however, that the glucose transporter system may be affected directly by the treatment (cf. Steinfelder and Joost, 1988). Although the cerebral glucose metabolism is impaired after intraventricular STZ, the overall oxidation and production of CO_2 by the brain is maintained normal as shown by unchanged AVDs of O_2 and CO_2. This implicates, however, that other substrates than glucose are oxidized. Since keton-bodies were not present in the blood of either controls or STZ-treated animals, amino acids or phospholipids are likely to be oxidized. These substitutional substrates may be derived from intrinsic cerebral sources and thus cause functional and structural damage of neuronal and glial cells, respectively.

The behavioural findings of this study show that intraventricular injection is followed by an impairment of cognitive attentional processing. Interestingly, after STZ-injection, the first and second reaction valued diverged considerably. This is in consistence with previous findings in aged animals (Mayer et al., 1988; Nemeth et al., 1989). Furthermore, the performance of the 3 minutes short-term memory for inhibitory avoidance is decreased after intraventricular STZ-treatment.

Although the results reported here, clearly show that intraventricular STZ exhibits impairments of biochemical as well as behavioural parameters, the actual site of action remains to be speculative. First, intrinsic cerebral synthesis and release of insulin (c.f. Young, 1986; Schechter et al., 1988; Clarke et al., 1986) may be inhibited to argue from analogy to the highly specific action on pancreatic islet cells were STZ inhibits the insulin-secretion by 90% (Bolaffi et al., 1987). Second, cerebral insulin receptor signalling may be impaired, and thus, the action of insulin be reduced (c.f. Kadowaki et al., 1984). However, in the present state of investigation, unspecific toxic effects of STZ can not be excluded with certainty. Thus, further investigations concerning the synthesis of insulin peptide and mRNA, as well as the insulin receptor signalling after intracerebral STZ are needed.

Taken together, the results obtained in the present study very well match the alterations of cerebral metabolism observed in early onset Alzheimer's disease (Hoyer et al., 1988). Thus, the approach reported here may provide a model not only for the investigation of the early pathogenetic events of Alzheimer's disease but also for the development of new strategies of treatment for Alzheimer's disease.

References

Alpert S, Hanahan D, Teitelman G (1988) Hybrid insulin genes reveal a developmental lineage for pancreatic endocrine cells and imply a relationship with neurons. Cell 53:295–308

Baskin DG, Figlewicz DP, Woods SC, Porte D, Dorsa DM (1987) Insulin in the brain. Ann Rev Physiol 49:335–347

Bolaffi JL, Nagamatsu S, Harris J, Grodsky GM (1987) Protection by thymidine, an inhibitor of polyadenosine diphosphate ribosylation, of streptozotocin inhibition of insulin secretion. Endocrinology 120:2117–2122

Boyd FT, Clarke DW, Muther TF, Raizada MK (1985) Insulin receptors and insulin modulation of norepinephrine uptake in neuronal cultures from rat brain. J Biol Chem 260:15880–15884

Clarke DW, Mudd L, Boyd FT, Fields M, Raizada MK (1986) Insulin is released from rat brain neuronal cells in culture. J Neurochem 47:831–836

Corp ES, Woods SC, Porte D, Dorsa DM, Figlewicz DP, Baskin DG (1986) Localization of ^{125}I-insulin binding sites in the rat hypothalamus by quantitative autoradiography. Neurosci Lett 70:17–22

Danguir J, Nicolaidis S (1984) Chronic cerebroventricular infusion of insulin causes selective increase of slow wave sleep in rats. Brain Res 306:97–103

Fram RJ, Marinus MG, Volkert MR (1988) Gene expression in E. coli after treatment with streptozotocin. Mutat Res 198:45–51

Havrankova J, Roth J, Brownstein M (1978) Insulin receptors are widely distributed in the central nervous system of the rat. Nature 272:827–829

Hill JM, Lesniak MA, Pert CB, Roth J (1986) Autoradiographic localization of insulin receptors in rat brain: prominence in olfactory and limbic areas. Neuroscience 17:1127–1138

Hoyer S, Nitsch R (1989) Cerebral excess release of neurotransmitter amino acids subsequent to reduced cerebral glucose metabolism in early-onset dementia of Alzheimer type. J Neural Transm 75:227–232

Hoyer S, Oesterreich K, Wagner O (1988) Glucose metabolism as the site of the primary abnormality in early-onset dementia of Alzheimer type? J Neurol 235:143–148

Hussin AH, Skett P (1988) Lack of effect of insulin in hepatocytes isolated from streptozotocin-diabetic male rats. Biochem Pharmacol 37:1683–1689

Junod A, Lambert AE, Stauffacher W, Renold AE (1969) Diabetogenic action of streptozotocin: relationship of dose to metabolic response. J Clin Invest 48:2119–2139

Kadowaki T, Kasuga M, Akanuma Y, Ezaki O, Takaku F (1984) Decreased autophosphorylation of the insulin receptor-kinase in streptozotocin diabetic rats. J Biol Chem 259:14208–14216

Kolb-Bachofen V, Epstein S, Kiesel U, Kolb H (1988) Low-dose streptozotocin-induced diabetes in mice. Diabetes 37:21–27

Kyriakis JM, Hausman RE, Peterson SW (1987) Insulin stimulates choline acetyltransferase activity in cultured embryonic chicken retina neurons. Proc Natl Acad Sci USA 84:7463–7467

Mayer G, Nemeth G, Hoyer S (1988) Psychometrie altersabhängiger Aufmerksamkeitsveränderungen – Darstellung an der Ratte. Z Gerontol 21:87–92

Nemeth G, Mayer G, Hoyer S (1989) A new psychometric test of attention-related behaviour in rats; its validity in the aging process. Arch Gerontol Geriatr 8:29–36

Palovcik RA, Phillips MI, Kappy MS, Raizada MK (1984) Insulin inhibits pyramidal neurons in hippocampal slices. Brain Res 309:187–191

Pellegrino DA, Miletich DJ, Albrecht RF (1987) Effects of superfused insulin on cerebral cortical glucose utilization in awake goats. Am J Physiol 253 (Endocrinol Metab 16): E418–E427

Phillips ME, Coxon RV (1976) Effect of insulin and phenobarbital on uptake of 2-deoxyglucose by brain slices and hemidiaphragms. J Neurochem 27:643–645

Rinaudo MT, Curto M, Bruno R (1985) Effect of insulin on the pyruvate dehydrogenase complex in the rat brain. Ital J Biochem 34:229–238

Rosen OM (1987) After insulin binds. Science 237:1452–1458

Sakaguchi T, Bray GA (1987) Intrahypothalamic injection of insulin decreases firing rate of sympathetic nerves. Proc Natl Acad Sci USA 84:2012–2014

Sauter A, Goldstein M, Engel J, Keta K (1983) Effect of insulin on central catecholamines. Brain Res 260:330–333

Schechter R, Holtzclaw L, Sadiq F, Kahn A, Devaskar S (1988) Insulin synthesis by isolated rabbit neurons. Endocrinology 123:505–513

Smit AB, Vreugdenhil E, Ebberink RHM, Geraerts WPH, Klootwijk J, Joosse J (1988) Growth-controlling molluscan neurons produce the precursor of an insulin-related peptide. Nature 331:535–538

Steinfelder HJ, Joost HG (1988) Inhibition of insulin-stimulated glucose transport in rat adipocytes by nucleoside transport inhibitors. Febs Lett 227:215–219

Werther GA, Hogg A, Oldfield BJ, McKinley MJ, Fidor R, Allen AM, Mendelsohn FOK (1987) Localization and characterization of insulin receptors in rat brain and pituitary gland using in vitro autoradiography and computerized densitometry. Endocrinology 121:1562–1570

Woods SC, Lotter EC, McKay LD, Ponte D Jr (1979) Chronic intracerebroventricular infusion of insulin reduces food intake and body weight of baboons. Nature 282:503–505

Yorek MA, Dunlap JA, Ginsberg BH (1987) Amino acid and putative neurotransmitter transport in human Y79 retinoblastoma cells. J Biol Chem 262:10968–10993

Young WS (1986) Periventricular hypothalamic cells in the rat brain contain insulin mRNA. Neuropeptides 9:93–97

Zucker PF, Archer MC (1988) Streptozotocin toxicity to cultured pancreatic islets of the syrian hamster. Cell Biol Toxicol 4:349–357

Correspondence: R. Nitsch, M. D., Department of Pathochemistry and General Neurochemistry, University of Heidelberg, Im Neuenheimer Feld 220–221, D-6900 Heidelberg, Federal Republic of Germany.

Choline levels, the regulation of acetylcholine and phosphatidylcholine synthesis, and Alzheimer's disease

R. J. Wurtman[1], I. H. Ulus[2], J. K. Blusztajn[3], I. Lopez G.-Coviella[1], M. Logue[4], C. Mauron[1], and J. H. Growdon[4]

[1] Department of Brain and Cognitive Sciences, Massachusetts Institute of Technology, Cambridge, MA, U.S.A.
[2] Department of Pharmacology, Uludag University Medical School, Bursa, Turkey
[3] Department of Pathology, Boston University School of Medicine, Boston, MA
[4] Department of Neurology, Massachusetts General Hospital, Boston, MA, U.S.A.

Summary

Circulating choline is converted to membrane phosphatidylcholine (PC) in all cells, and, additionally, to acetylcholine (ACh) within cholinergic neurons. Neither of the enzymes that initiate these processes (choline kinase and choline acetyltransferase) is substrate-saturated; hence supplemental choline can accelerate the formation of both compounds, as well as acetylcholine release from physiologically-active neurons. We have found, using a human neuronal cell line (LA-N-2), that the PC in membranes contributes choline for ACh synthesis. Moreover, using superfused slices of rat brain, we have shown that, when free choline is inadequate to sustain ACh synthesis, the use of the PC "reservoir" for this purpose may deplete membrane PC and other membrane constituents. Hence choline availability may be rate-limiting for membrane formation, as well as for ACh synthesis.

The possibility that abnormalities in choline utilization may be involved in Alzheimer's disease is demonstrated by the very high levels (twice control) of a PC breakdown product, glycerophosphocholine (GPC), in samples of Alzheimer's disease brains. This increase could reflect accelerated PC breakdown or an impairment in enzymatic processes that allow free choline to be salvaged from the GPC.

Introduction

Brain slices and minces continue to make and, upon depolarization, to release large amounts of acetylcholine without exhibiting reductions in tissue acetylcholine levels, when perfused with a choline-free medium (Bhatnagar and MacIntosh, 1967; Collier et al., 1972; Maire and Wurtman, 1985; Quastel et al., 1936; Weiler et al., 1979; Weiler et al., 1983). However, addition to the choline-free medium of hemicholinium-3 (HC-3), a drug which blocks choline's high-affinity uptake into cholinergic terminals, stops the release of acetylcholine from nerve terminals, and decrease tissue acetylcholine levels (Maire and Wurtman, 1985). We interpreted these findings as suggesting that some of the choline used by the slices for acetylcholine synthesis originated in the PC of neuronal membranes (Maire and Wurtman, 1985): Depolarization would cause a net hydrolysis of this PC by accelerating its hydrolysis and/or slowing its synthesis, and the choline thus released into extracellular fluid would be taken up into cholinergic terminals for acetylation (Blusztajn and Wurtman, 1983).

If this explanation were correct, it might be predicted that prolonged stimulation of cholinergic neurons superfused without choline would also reduce their PC content; moreover, this reduction might be blocked by adding choline to the medium. Additional studies using superfused electrically-stimulated rat brain slices, tested these predictions (Ulus et al., 1989).

Materials and methods

Slices of striata or cerebella 0.3 mm thick were transferred to a superfusion chamber (volume 1.0 ml) and placed between two Ag/AgCl$_2$ stimulation electrodes. The chamber was maintained at 37 °C, and the slices were superfused (0.6 ml/min) with a physiological solution (mM: NaCl, 120; KCl, 3.5; CaCl$_2$, 1.3; MgSO$_4$, 1.2; NaH$_2$PO$_4$, 1.2; NaHCO$_3$, 25; glucose, 10; eserine salicylate, 0.02) continuously bubbled with 95% O$_2$ and 5% CO$_2$. Following a twenty minute equilibration period, some slices were removed for determination of ACh, choline, phospholipids, protein and DNA contents (referred to as "tissue initial"), and the remaining slices were then superfused for 90 to 340 minutes during which they were alternately maintained at rest for twenty minutes and stimulated (amplitude of 30 V), also for twenty minutes, for 2, 4, 6 or 8 stimulation periods. Perfusates representing each entire 20-minute rest or stimulation period were assayed for ACh and choline (Fonnum, 1969; Gilberstadt and Russell, 1984); slices were taken from the superfusion medium ten minutes after their last stimulation period and assayed for ACh, choline (Torn and Aprison, 1966; Spanner et al., 1976; Goldberg and McCaman, 1973), phospholipids (Folch et al., 1957; Svanborg and Svennerholm, 1961; Touchstone et al., 1980), protein (Lowry et al., 1951), and DNA (Labarca and Paigen, 1980), (referred to as "tissue final").

Results

The striatal slices released ACh and choline into the medium during superfusion with the choline-free medium. The rate of ACh release was 4.2 ± 0.8 pmol/µg DNA/20 min (mean $\pm$ SEM, n = 11) during the first collection period; this remained constant for three hours among slices superfused at rest. When the slices were stimulated, ACh release increased about 4-fold, to 15.4 ± 2.1 pmol/µg DNA/20 min (mean $\pm$ SEM, n = 11); this quickly returned to basal levels when the stimulation was terminated.

Initial studies used slices stimulated for four 20-minute periods. Slices continued to release ACh at the higher rate for all four periods. The rate of choline efflux was 66.0 ± 8.0 pmol/µg DNA/20 min (mean $\pm$ SEM, n = 14) during the first collection period; this gradually fell, during the next three hours, to 40.0 ± 8.0 pmol/µg DNA/20 min (mean $\pm$ SEM, n = 4) in unstimulated slices, or to 36.0 ± 4.0 pmol/µg DNA/20 min (mean $\pm$ SEM, n = 9) in stimulated slices. The rate of choline efflux showed a tendency to increase while the tissue was being stimulated; however, this tendency was not statistically significant.

Cerebellar slices failed to release detectable amounts of ACh into the superfusate at rest or during periods of electrical stimulation. They did, however, release free choline at an initial rate of 22.0 ± 2.0 pmol/µg DNA/20 min (mean $\pm$ SEM, n = 12) and a final rate of 12.0 ± 2.0 pmol/µg DNA/20 min (mean $\pm$ SEM, n = 12). Choline efflux from cerebellar slices was not altered by electrical stimulation.

Stimulation of the striatal slices for four 20-minute periods significantly decreased their contents of phospholipids (by 14%, compared with phospholipid levels in their tissue initials) and of proteins (by 10.5%). In contrast, stimulation failed to affect the phospholipid or protein contents of the cerebellar slices. Superfusion alone did not affect phospholipid levels in either tissue.

When the striatal slices were stimulated for up to eight periods, ACh release continued at unchanged rates, the tissues liberating during the 340 minutes an amount of ACh about three times their initial ACh levels. Addition of exogenous choline (10–40 µM) to the superfusion medium enhanced ACh release, both basally and during stimulation. These increases were concentration-dependent and were statistically significant in the presence of 20 (105%) or 40 µM (255%) of choline. When the slices were superfused with a choline-free solution, their total phospholipid contents (examined initially and after 2, 4, 6 or 8 stimulation periods) declined at the rate of 2.7% per stimulation period. This decrease was highly correlated with the number of stimulation periods ($r = -0.98$; $p < 0.0001$). Addition of exogenous choline (10–40 µM) to the superfusion medium partially or completely protected the slices from phospholipid depletion, for as many as six stimulation periods (Table 1). Moreover, the phospholipid contents of the slices were actually increased after two periods of electrical stimulation in the presence of 40 μM choline (Table 1).

The major phospholipids in brain membranes, besides PC, are phosphatidylserine (PS) and phosphatidylethanolamine (PE); together, these three phosphatides account for 85% of total brain phospholipids (Ansell, 1973). Moreover, the choline in PC represents about 80% of the total membrane-bound choline in the brain (Ansell, 1973) suggesting that PC would be the cellular phospholipid reservoir most likely to provide free choline for ACh synthesis. To determine whether the reduction in *total* brain phospholipids observed when stimulated slices were superfused without choline was entirely due to PC, or also included other brain phospholipids, we fractioned and quantified these compounds in tissues obtained before or after two, four, six and eight stimulation periods.

Stimulation in the choline-free medium was associated with decreases not only in membrane PC but also, proportionately, in PE and PS (Table 1). After 4 (or 8) stimulation-rest periods, the levels of total phospholipid, PC, PE, or PS were 83.2 ± 3.3 (or $77.5 \pm 3.9\%$), 89 ± 5.2 (or $76.6 \pm 2.5\%$), 84.7 ± 7.7 (or $77.3 \pm 2.4\%$) and 89.2 ± 2.3 (or $78.2 \pm 1.0\%$) of initial values, respectively. In seven separate perfusion studies, the ratio of PC to the total phospholipid content [or to the sum of the three main structural phospholipids $(PC + PE + PS)$] was 0.33 ± 0.04 [or 0.41 ± 0.05] initially and 0.34 ± 0.04 [or 0.43 ± 0.08], subsequent to eight stimulation periods. Addition of choline to the superfusion medium protected the slices from the declines in the three phosphatides, i.e., PE and PS as well as PC (Table 1). The protective effect of choline on each phosphatide was apparently concentration-dependent. At 20 μM, none of their levels declined even after as many as six stimulation periods. Protection was partial and occurred for up to as many as two to six stimulation periods in the presence of 10 μM choline (Table 1). Addition of 40 μM choline to the superfusion medium actually *increased* slice levels of PC, PE and PS by 19%, 24% and 23%, respectively (Table 1), after two periods of electrical stimulation (i.e., compared with those in controls obtained at the beginning of superfusion). The small but significant depletion of protein caused by stimulation was also blocked by adding choline to the superfusion medium (Table 1).

Table 1. Effects of repeated electrical stimulations on choline, acetylcholine, phospholipid and protein contents of striatal slices superfused with choline-free or choline-containing media

Compounds (µM) Choline	TI	TF-2	TF-4	TF-6	TF-8
Choline					
0	16± 3	22± 6	16± 5	17± 3	15± 2
10	ND	ND	ND	ND	ND
20	16± 3	43± 8	34± 3**	26± 4	16± 4
40	ND	ND	ND	ND	ND
ACh					
0	49± 3	60±13	42±11	49± 9	22± 3*
10	47± 3	36± 8	37±12	42±12	30±14
20	47± 2	42± 4	45± 2	44± 2	20± 3*
40	49± 3	64±14	54± 8	39± 3	25± 4*
Protein					
0	65± 2	62± 2	61± 1*	58± 1*	54± 1*
10	62± 2	61± 2	63± 2	60± 5	56± 2*
20	63± 2	66± 7	57± 2	61± 3	54± 3*
40	66± 3	61± 5	63± 4	61± 3	52± 2*
Total Phospholipid					
0	735±12	666±41	621±27*	585±22*	570±31*
10	686±21	672±46	650±46	716±54	574±47*
20	739±23	704±37	704±31	730±46	573±20*
40	750±14	828±22**	726±65	700±37	616±21*
Phosphatidylcholine					
0	256±14	225±24*	223±20*	222±27*	196±15*
10	248± 9	228±10	237±26	243±16	206±19*
20	246±27	267±12	247±15	241± 8	210± 8*
40	248±15	295±29**	264±12	292±18**	209±14*
Phosphatidylethanolamine					
0	248± 8	206± 8*	203±12*	201±21*	187±18*
10	243±19	184±30	186±11*	204±24	171±13*
20	251±18	240±20	243±11	245±32	175±34*
40	242±17	304±32**	232± 9	258±24	199±18*
Phosphatidylserine					
0	97± 8	92± 8	83± 8*	78± 6*	77± 8*
10	92± 4	86±14	84±13	85± 6	81±12
20	95±10	98± 4	99± 7	92± 7	82± 5*
40	92± 6	117± 6**	105± 5	104± 6	83± 6*

Significantly lower (* p 0.05) or higher (** p 0.05) than corresponding TI value

TI – tissue initial content

TF – final tissue content after the number of stimulation periods indicated.

Discussion

Hence slices from rat striatum continue to make large amounts of ACh when superfused with a choline-free medium, and under these conditions tissue contents of PC and other structural membrane phospholipids actually decrease significantly (Table 1). Addition of increasing concentrations (10–40 µM) of choline to the superfusion medium increases the release of ACh both at rest and during electrical stimulation, in proportion to the choline concentration. Moreover, the exogenous choline also protects the slices from losses of PC, PE, PS, and protein (Table 1).

These results confirm and extend an earlier report from our laboratory (Maire and Wurtman, 1985) showing that ACh release is enhanced by superfusing striatal slices with choline, both during electrical stimulation and at rest. Increases in ACh release in the presence of exogenous choline have also been described in brain slices incubated with high potassium concentrations, which depolarize the neurons, but not in slices at rest (Millington and Goldberg, 1982; Weiler et al., 1983). This discrepancy may result, in the latter studies, from the artefactual accumulation of exogenous choline, liberated from the slices, in the incubation media; i.e., this endogenous choline may have raised effective tissue choline concentrations well above those at which choline acetyltransferase was capable of responding to additional choline. [Choline concentrations in the media reportedly rise to 20–40 µM, causing slices to accumulate very high free choline concentrations (Bhatnagar and MacIntosh, 1967; Millington and Goldberg, 1982; Weiler et al., 1979; Weiler et al., 1983).] Under our experimental conditions (using relatively rapid superfusion, and a small amount of tissue relative to the medium's volume), the liberation of choline from the slices elevated choline concentrations in the superfusate only 0.5 µM, or even less; and the choline contents of the slices (Table 1) did not exceed those usually seen in vivo.

Brain tissue is known to produce and release free choline in vivo (Bream et al., 1987; Dross and Kewitz, 1972; Spanner et al., 1976) and during incubation in vitro (Bhatnagar and MacIntosh, 1967; Collier et al., 1972; Freeman and Jenden, 1976; Jope and Jenden, 1979; Kosh et al., 1980; Maire and Wurtman, 1985; Wecker, 1985; Wecker, 1986; Weiler et al., 1983; Zeisel, 1985). In both situations, the source of the choline appears to be membrane phospholipids. When we superfused striatal slices in a choline-free medium, the slices lost a fraction of their PC, as well as of their PE and PS, during the course of the experiments (Table 1). During the same period, the amount of ACh plus choline recovered in the superfusate was similar to the amount of choline lost from membrane PC, but several orders of magnitude larger than the intracellular pools of choline and ACh (Table 1).

The release of choline from a lipid-bound form can occur in all areas of the brain (Kosh et al., 1980): in vitro, the process apparently is more active in the striatum, which is rich in cholinergic activity, than in cortex (Jope and Jenden, 1979) or cerebellum. Other studies have suggested, however, that the release of tissue choline from a bound form is not related to the extent of its cholinergic innervation (Freeman and Jenden, 1976; Kosh et al., 1980). Also, as shown in the present study, the reductions in the PC and other membrane phospholipids were not simply related to the production of free choline but apparently also depended on an increased demand for free choline to sustain ACh synthesis. When the striatal slices were superfused with a choline-free medium but not depolarized, they released choline into the medium *without* reducing their phospholipid contents; in contrast, when stimulated electrically, they released slightly more choline and much more ACh, and in that circumstance, levels of PC and other membrane phospholipids fell in proportion to the duration of stimulation (Table 1) – and thus to the total amounts of ACh released. Moreover, when slices were provided with adequate free choline and stimulated electrically, they produced and released even *more* ACh, but without depleting their membrane of PC and other phosphatides (Table 1). Electrical stimulation failed to alter the phospholipid contents of the cerebellar slices, a tissue in which stimulation was not associated with ACh release.

Consistent with the view that neuronal PC is a reservoir of choline for ACh synthesis, and that this reservoir can be depleted when more choline is needed for ACh synthesis, we find that the phospholipid contents of electrically-stimulated striatal slices do not decline if the stimulation-induced release of ACh is blocked by tetrodotoxin (unpublished observation), or if the cholinesterase inhibitor eserine is omitted from the superfusion medium (Ulus and Wurtman, 1988). When cholinesterase is active, the breakdown of intrasynaptic ACh generates free choline that can be re-used for ACh synthesis; this probably reduces the need for membrane PC to provide free choline. The ability of cholinesterase inhibitors to potentiate the "autocannibalism" of neuronal membrane phospholipids may have important clinical implications if patients with Alzheimer's disease are to receive such drugs for prolonged periods. Until it is established for sure *whether* a given cholinesterase inhibitor has this effect, it would seem prudent that it be co-administered with an agent that will provide the brain with supplemental choline.

The possibility that membrane constituents are depleted when neuronal tissues lack adequate free choline is supported by other data showing that PC levels and the number of synaptic vesicles in the cat's

superior cervical ganglion fell significantly with stimulation of the pre-
ganglionic nerve trunk if the uptake of exogenous choline was blocked
by hemicholinium-3 (Parducz et al., 1976), or if the ganglia were perfused with a choline-free Locke solution (Parducz et al., 1986). The
vesicles reappeared immediately if the stimulated ganglia were provided
with choline for two minutes (Parducz et al., 1986).

The protective effect of exogenous choline on the stimulus-induced
depletion of membrane phosphatides depends on its concentration.
Twenty μM choline in the superfusion medium (which is in the physiological range for plasma choline; Blusztajn and Wurtman, 1983) completely protected the slices from phospholipid depletion for as many as
six stimulation periods (Table 1), while 10 μM was less effective
(Table 1). In the presence of 40 μM choline, the phospholipid contents
of the slices sometimes actually increased. After four stimulation-rest
cycles, the levels of total phospholipids and of PC or PS in the striatal
slices were highly ($r = 0.91$, 0.99 or 0.94, respectively) and significantly
($p < 0.001$) correlated with the choline concentration in the medium. It
has recently been shown that the activities of choline kinase (which
catalyzes the phosphorylation of choline to form phosphocholine in the
first step of PC biosynthesis via the CDP-choline pathway) and of
choline acetyltransferase in superior cervical ganglia show reciprocal
fluctuating changes as a function of the extracellular choline concentrations in the incubating medium (Ando et al., 1987). Neuronal depolarization diminishes the activity of choline kinase and increases that of
choline acetyltransferase (Ando et al., 1987), thus possibly shunting the
choline into ACh synthesis. However, when the choline concentrations
in the medium were $40 - 50$ μM, neuronal depolarization diminished
choline acetyltransferase activity and increased that of choline kinase
(Ando et al., 1987). Hence, the main determinant of the choline's actual
fate in cholinergic neurons may be the relative activities of choline kinase
and choline acetyltransferase which, in turn, depend upon extracellular
choline concentrations. It has been shown that giving supplemental
choline to laboratory animals can increase brain levels of phosphocholine (Millington and Wurtman, 1982), total phospholipids (Wecker,
1985; Wecker, 1986), and PC (Pendley et al., 1983; Wecker, 1985; Wecker,
1986) while a choline-deficient diet significantly decreases PC levels
(Pendley et al., 1983). Interestingly, the 19% increase in PC levels observed in slices superfused with 40 μM choline was associated with a
10% increase in total phospholipids (Table 1). Our present results, taken
with these in vivo observations, suggest that PC levels and even, con-
ceivably, the *total amounts of membrane* in cholinergic neurons may be
choline-dependent.

The above data are compatible with the view that PC's degradation provides choline for acetylcholine synthesis, and that this process is accelerated when neuronal activity is increased, thereby decreasing tissue phospholipid contents. However, they are also consistent with a slowing of PC synthesis, which might secondarily suppress the incorporation of such other constituents as PS, PE and protein into membranes. Little information is available about the processes that might mediate the accelerated PC breakdown or slowed PC synthesis. We have recently shown that PC in a human cholinergic cell line (LA-N-2) contributes choline used for acetylcholine synthesis (Blusztajn et al., 1987). However, in striatal slices it is likely that the decrease in PC and other major membrane phospholipids also reflects changes in membranes of non-cholinergic cells, perhaps in synaptic contact with cholinergic neurons. Acetylcholine reportedly can increase the availability of free choline by a mechanism mediated by muscarinic receptors (Brehm et al., 1987; Corradetti et al., 1983), and we observe that such receptors activate both phospholipase C and D in the above cholinergic cell line (Sandmann and Wurtman, in press).

In vivo, circulating and, presumably, extracellular choline levels are about 10 μM in fasting animals and humans (Tucek, 1984) but can rise to 30–40 μM after ingestion of PC-rich foods (Magil et al., 1981; Zeisel et al., 1980). Administration of choline to laboratory animals elevates plasma and brain choline levels, and sometimes (Haubrich and Pflueger, 1979; Millington and Wurtman, 1982) but not always (Wecker and Schmidt, 1980; Weiler et al., 1979) brain ACh; it also enhances ACh release (Koshimura et al., 1990) and can produce postsynaptic changes suggestive of enhanced cholinergic neurotransmission (Haubrich and Pflueger, 1979; Ulus and Wurtman, 1976; Ulus et al., 1977; Ulus et al., 1978; Wecker and Schmidt, 1979). Our present data indicate that elevating choline concentrations from the fasting range (10 μM) to 20–40 μM enhances the release of ACh without changing tissue ACh levels. Steady-state brain ACh concentrations are known to decrease when the activity of cholinergic neurons is increased pharmacologically, e.g., after administration of atropine, pentylenetetrazol, or fluphenazine (Jope, 1982; Schmidt and Wecker, 1981; Wecker, 1985; Wecker, 1986; Wecker and Schmidt, 1979). Choline, administered either as the salt or as PC, prevents these decreases (Jope, 1982; Schmidt and Wecker, 1981; Wecker, 1985; Wecker, 1986; Wecker and Schmidt, 1979). Moreover, choline administration augments cholinergic transmission during increased cholinergic activity (Ulus et al., 1978). The most parsimonious conclusion from these data is that under conditions of stimulated ACh release the normal supply of choline may not keep up with the demand for ACh synthesis, unless it is supplemented by giving choline or PC, and that the increased levels of choline present in the extracellular medium after administration of choline can indeed be utilized to support ACh synthesis. In this way, cholinergic transmission can increase while steady-state ACh concentrations are protected from drastic changes.

In the absence of supplemental choline, extracellular free choline concentrations (10 μM) may not be sufficient to sustain membrane phospholipid levels in frequently-firing cholinergic neurons. This propensity might be exaggerated in old subjects with diminished choline transport across the bloodbrain barrier (Mooradian, 1988) or, as indicated above, among people taking cholinesterase inhibitors (Ulus and Wurtman, 1988).

Cholinergic neurons are unique in that they alone use choline for two purposes, i.e., the syntheses of their neurotransmitter, ACh, and of phospholipid constituents of their membranes. The selective vulnerability of certain cholinergic neurons in neurodegenerative disorders involving the brain (e.g., Alzheimer's disease) or motor neurons may result from this property, i.e., the over-utilization or slower production of membrane PC and the shunting of the choline towards ACh synthesis (Wurtman et al., 1985). Consistent with this conclusion are recent reports describing elevations in the levels of phospholipid breakdown products within brains of untreated patients suffering from Alzheimer's disease (Barany et al., 1985; Miatto et al., 1986). We have measured brain levels of the PC metabolite glycerophosphocholine (GPC) in cortical and cerebellar brain samples from 11 patients with Alzheimer's disease, as well as in control patients without brain disease (n = 7) or Down's syndrome (n = 5). Average control GPC levels in the Alzheimer's disease patients were more than twice those observed in any other group. GPC levels in cerebellum, a brain region lacking significant cholinergic innervation, were also elevated, suggesting that the disease process is not restricted to the brain cells most vulnerable to it.

The elevation in brain GPC could reflect accelerated PC breakdown [and that of glycerophosphoethanolamine (Pettegrew, 1988) accelerated PE breakdown, compatible with the propensity of "autocannibalism" to affect all of the membrane constituents (Table 1)], or perhaps an impairment in the enzymes that allow the choline in GPC to be salvaged for subsequent reutilization. Studies are in progress to determine which process is operating.

Acknowledgements

These studies were supported in part by a grant from the National Institute of Mental Health (MH-28783) and by funds from the Center for Brain Sciences and Metabolism Charitable Trust.

References

Ansell GB (1973) Phospholipids in the nervous system. In: Ansell GB, Hawthorne JN, Dawson RMC (eds) Form and function of phospholipids. Elsevier, New York, pp 377–422

Ando M, Iwata M, Takahama K, Nagata Y (1987) Effects of extracellular choline concentration and K^+ depolarization on choline kinase and choline acetyltransferase activities in superior cervical sympathetic ganglia excised from rats. J Neurochem 48: 1448–1453

Barany M, Chang Y-C, Arus C, Rustan TR, Frey WH (1985) Increased glycerol-3-phos-phorylcholine in post-mortem Alzheimer's brain. Lancet i: 517

Bhatnagar SP, MacIntosh FC (1967) Effects of quaternary bases and inorganic cations on acetylcholine synthesis in nervous tissue. Can J Physiol Pharmacol 45: 249–268

Blusztajn JK, Wurtman RJ (1983) Choline and cholinergic neurons. Science 221: 614–620

Blusztajn JK, Richardson UI, Liscovitch M, Mauron C, Wurtman RJ (1987) Synthesis of acetylcholine from choline derived phosphatidylcholine in a human neuronal cell line. Proc Natl Acad Sci 84: 5474–5477

Blusztajn JK, Richardson UI, Liscovitch M, Mauron C, Wurtman RJ (1987) Phospholipids in cellular survival and growth. In: Hanin I, Ansell GB (eds) Lecithin: technological, biological and therapeutic aspects. Plenum Press, New York, pp 85–94

Brehm R, Lindmar R, Loffelholz K (1987) Muscarinic mobilization of choline in rat brain in vivo as shown by the central arterio-venous difference of choline. J Neurochem 48: 1480–1485

Browning ET (1971) Free choline formation by cerebral cortical slices from rat brain. Biochem Biophys Res Comm 45: 1585–1590

Collier B, Poon P, Salehmoghaddam S (1972) The formation of choline and of acetylcholine by brain in vitro. J Neurochem 19: 51–60

Corradetti R, Lindmar R, Loffelholz K (1983) Mobilization of cellular choline by stimulation of muscarine receptors in isolated chicken heart and rat cortex in vivo. J Pharmacol Exp Ther 226: 826–832

Dross K, Kewitz H (1972) Concentration and origin of choline in rat brain. Naunyn Schmiedebergs Arch Pharmacol 274: 91–106

Folch J, Lees M, Sloane-Stenley GH (1957) A simple method for the isolation and purification of total lipids from animal tissue. J Biol Chem 226: 497–503

Fonnum F (1969) Radiochemical microassays for the determination of choline acetyltransferase and acetylcholinesterase activities. Biochem J 115: 465–472

Freeman JJ, Jenden DJ (1976) The source of choline for acetylcholine synthesis in brain. Life Sci 19: 949–962

Gilberstadt ML, Russell JA (1984) Determination of picomole quantities of acetylcholine and choline in physiological salt solutions. Anal Biochem 138: 78–85

Goldberg AM, McCaman RE (1973) The determination of picomole amounts of acetylcholine in mammalian brain. J Neurochem 20: 1–8

Haubrich DR, Wang PFL, Clody DE, Wedeking PW (1975) Increase in rat brain acetylcholine induced by choline or deanol. Life Sci 17: 975–980

Haubrich DR, Pflueger AB (1979) Choline administration: central effects mediated by stimulation of actetylcholine. Life Sci 24: 1083–1090

Jope RS (1982) Effects of phosphatidylcholine administration to rats on choline in blood and choline and acetylcholine in brain. J Pharmacol Exp Ther 220: 322–328

Jope RS, Jenden DJ (1979) Choline and phospholipid metabolism and the synthesis of acetylcholine in rat brain. J Neurosci Res 4: 69–82

Kosh JW, Dick RM, Freeman JJ (1980) Choline postmortem increase: effect of tissue, agitation, pH and temperature. Life Sci 27: 1953–1959

Koshimura K, Miwa S, Lee K, Hayashi Y, Hasegawa H, Hamahata K, Fujiwara M, Kimura M, Itokawa I (1990) Effects of choline administration on in vivo release and biosynthesis of acetylcholine in the rat striatum as studied by in vivo microdialysis. J Neurochem 54: 533–539

Labarca CA, Paigen K (1980) A simple, rapid, and sensitive DNA assay procedure. Anal Biochem 102: 344–352

Lowry OH, Rosenbrough NJ, Fall AL, Randall RJ (1951) Protein measurement with the Folin phenol reagent. J Biol Chem 193: 265–275

Magil SG, Zeisel SH, Wurtman RJ (1981) Effects of ingesting soy or egg lecithins on serum choline, brain choline and brain acetylcholine. Nutrition 111: 166–170

Maire J-C, Wurtman RJ (1985) Effects of electrical stimulation and choline availability on the release and contents of acetylcholine. J Physiol (Paris) 80: 189–195

Miatto O, Gonzales RG, Buonanno F, Growdon JH (1986) In vitro ^{31}P NMR spectroscopy detects altered phospholipid metabolism in Alzheimer's disease. Can J Neurol Sci 13: 535–539

Millington WR, Goldberg AM (1982) Precursor dependence of acetylcholine release from rat brain in vitro. Brain Res 243: 263–270

Millington WR, Wurtman RJ (1982) Choline administration elevates brain phosphorylcholine concentrations. J Neurochem 38: 1748–1752

Mooradian AD (1988) Blood-brain barrier transport of choline is reduced in the aged rat. Brain Res 440: 328–332

Parducz A, Kiss Z, Joo F (1976) Changes of the phosphatidylcholine content and the number of synaptic vesicles in relation to the neurohumoral transmission in sympathetic ganglia. Experienta 32: 1520–1521

Parducz A, Joo F, Toldi J (1986) Formation of synaptic vesicles in the superior cervical ganglion of cat: choline dependency. Exp Brain Res 63: 221–224

Pendley II CE, Horrocks LA, Mervis RF (1983) The effect of dietary choline on brain phospholipid content. In: Sun GY, Bazan N, Wu J-Y, Porcellatti AY (eds) Neural membranes. Humana Press, Clifton NJ, pp 171–190

Pettegrew JW, Panchalingam K, Moossy J, Martinez J, Rao G, Boller F (1988) Correlation of phosphorus-31 magnetic resonance spectroscopy and morphologic findings in Alzheimer's disease. Arch Neurol 45: 1093–1096

Quastel JH, Tennenbaum M, Wheatley AHM (1936) Choline ester formation in, and cholinesterase of, tissues in vitro. Biochem J 30: 1668–1681

Sandmann J, Wurtman RJ (in press) Phospholipase D and phospholipase C in human cholinergic neuroblastoma (LA-N-2) cells: modulation by muscarinic agonists and protein kinase C. In: Nishizuka Y (ed) Advances in second messenger and phosphoprotein research. Raven Press, New York (in press)

Schmidt DE, Wecker L (1981) CNS effects of administration: evidence for temporal dependence. Neuropharmacology 20: 535–539

Spanner S, Hall RC, Ansell GB (1976) Arteriovenous differences of choline and choline lipids across the brain of rat and rabbit. Biochem J 154: 133–140

Svanborg A, Svennerholm L (1961) Plasma total lipids, cholesterol, triglycerides, phospholipids and free fatty acids in a healthy Scandinavian population. Acta Med Scand 169: 43–49

Torn M, Aprison MH (1966) Brain acetylcholine studies: a new extraction procedure. J Neurochem 13: 1533–1544

Touchstone JC, Chen JC, Beaver KM (1980) Improved separation of phospholipids on thin layer chromatography. Lipids 15: 61–62

Tucek S (1984) Problems in the organization and control of acetylcholine synthesis in brain neurons. Prog Biophys Molec Biol 44: 1–41

Ulus IH, Wurtman RJ (1976) Choline administration: activation of tyrosine hydroxylase in dopaminergic neurons of rat brain. Science 194: 1060–1061

Ulus IH, Wurtman RJ (1988) Prevention by choline of the depletion of membrane phosphatidylcholine by a cholinesterase inhibitor. N Engl J Med 318: 191

Ulus IH, Hirsch MJ, Wurtman RJ (1977) Transsynaptic induction of adrenomedullary tyrosine hydroxylase activity by choline: evidence that choline administration can increase cholinergic transmission. Proc Natl Acad Sci 74: 798–800

Ulus IH, Scally MC, Wurtman RJ (1978) Enhancement by choline of the induction of adrenal tyrosine hydroxylase by phenoxybenzamine, 6-hydroxydopamine, insulin or exposure to cold. J Pharmacol Exp Ther 204: 676–682

Ulus IH, Wurtman RJ, Mauron C, Blusztajn JK (1989) Choline increases acetylcholine release and protects against the stimulation-induced decrease in phosphatide levels within membranes of rat corpus striatum. Brain Res 484: 217–227

Wecker L (1985) The utilization of supplemental choline by brain. In: Hanin I (ed) Dynamics of cholinergic function. Plenum Press, New York, pp 851–857

Wecker L (1986) Neurochemical effects of choline supplementation. Can J Physiol Pharmacol 64: 329–333

Wecker L, Schmidt DE (1979) Central cholinergic function: relationship to choline administration. Life Sci 25: 375–384

Wecker L, Schmidt DL (1980) Neuropharmacological consequences of choline administration. Brain Res 184: 234–238

Weiler MH, Misgeld U, Bak IL, Jenden DJ (1979) Acetylcholine synthesis in rat neostriatal slices. Brain Res 176: 401–406

Weiler MH, Bak IJ, Jenden DJ (1983) Choline and acetylcholine metabolism in rat neostriatal slices. J Neurochem 41: 473–480

Wurtman RJ, Blusztajn JK, Maire J-C (1985) "Autocannibalism" of choline-containing membrane phospholipids in the pathogenesis of Alzheimer's disease. Neurochem Int 7: 369–372

Zeisel SH (1985) Formation of unesterified choline by rat brain. Biochem Biophys Acta 835:331–343

Zeisel SH, Growdon JH, Wurtman RJ, Magil SG, Logue M (1980) Normal plasma choline responses to ingested lecithin. Neurology 30:1226–1229

Correspondence: Dr. R. J. Wurtman, Department of Brain and Cognitive Sciences, Bldg. E25-604, Massachusetts Institute of Technology, Cambridge, MA 02139, U.S.A.

Acetylcholine synthesis and membrane phospholipids

K. J. Martin and L. Widdowson

Department of Pharmacology, University of Cambridge, Cambridge, United Kingdom

Summary

The effect of chronic treatment with pyritinol on the phospholipid composition of synaptosomal membranes has been examined in rats. The treatment does not significantly change phospholipid content of synaptosomal membranes obtained from the cortex of young rats; in old rats an increase in total phospholipid content and phosphatidylcholine was observed. The results suggest that the pyritinol induced increase in cerebral ACH levels does not occur at the expense of membrane phospholipids.

Introduction

Chronic treatment of old rats with pyritinol has been shown to increase ACH levels and ACH release in some areas of the CNS (Martin and Vyas, 1987), suggesting an increase in ACH turnover. Recently Blusztajn et al. (1987) provided evidence that in cultures of human cholinergic cells ACH can be synthesized from choline derived from the degradation of endogenous phosphatidylcholine. The authors point out that enhanced demand for and inadequate supply of choline might result in the depletion of cholinephospholipids within cell membranes that might ultimately compromise membrane viability. We have, therefore, examined the effects of chronic treatment of rats with pyritinol on the phospholipid content and composition of synaptosomal membranes obtained from the cortex of both old and young animals.

Material and methods

Male Sprague Dawley rats were used in all experiments. Young rats were about six months old, old rats were ex-breeding stock between 18 and 20 months old. Pyritinol in powder form (200 mg/kg per day) was added to the normal diet and it was established that all was consumed. After 2–3 weeks of treatment the animals were killed by decapitation, the brain was removed quickly and cerebral cortex, mainly the frontal region, and the striatum were dissected out. A crude synaptosomal fraction was prepared (Gray and Whittaker, 1962), the lipids were extracted, separated using thin layer chromatography and phosphorous was determined following the methods described by Stein and Smith (1982).

Results

The effect of pyritinol treatment on the phospholipid content of synaptosomal membranes is shown in Fig. 1. There is no significant change in the tissue obtained from young animals but in old rats the phospholipid content actually shows a small but significant increase, both in the cortex and in the striatum.

The separation of phospholipids obtained from treated and untreated old animals yielded the results shown in Fig. 2. Only phosphatidyl-choline is significantly increased in the treated animals but the data suggest that the pyritinol induced increase in total phospholipid content involves an increase in all three major groups of phospholipids. It appears that the increase in ACH levels found after pyritinol treatment of old animals is not associated with a decrease in total phospholipids and does not lead to a selective decrease in phosphatidylcholine, the most obvious source for choline.

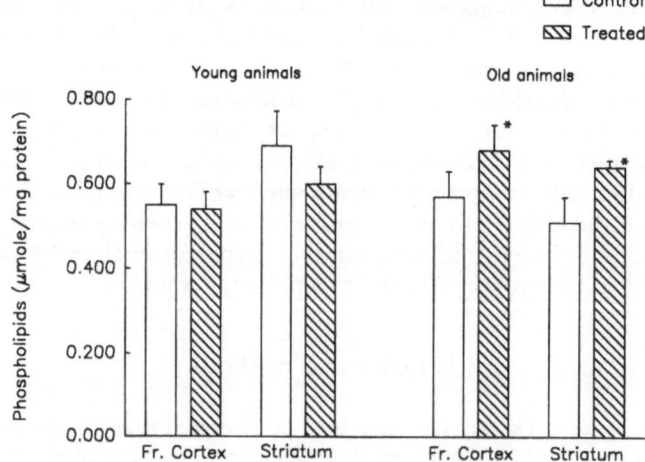

Fig. 1. The effects of treatment with pyritinol on the phospholipid content of crude synaptosomes obtained from the frontal cortex and the striatum of old and young rats. Ten to twelve animals were used in each group. The asterisk indicates a difference from the control at $p < 0.05$

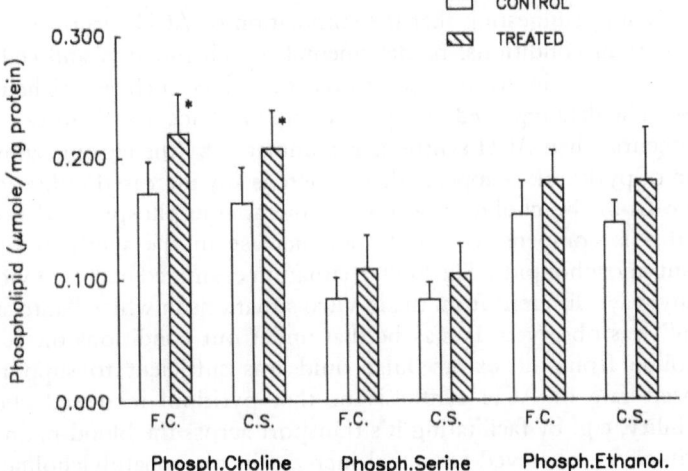

Fig. 2. The effects of pyritinol treatment on the concentrations of phophatidylcholine, phosphatidylethanolamine and phosphatidylserine in crude synaptosomes obtained from the frontal cortex (*F.C.*) and the corpus striatum (*C.S.*) of old rats. Ten rats were used in each group. The asterisk indicates a differnece from the control p < 0.05

Our values for the phospholipids of the crude synaptosomal preparation used (expressed per mg of protein) are lower than those reported in the literature (Breckenridge et al., 1972) for purified synaptosomal membranes. This probably reflects the different distribution of phospholipids and proteins: phospholipids are confined almost entirely to the membranes while proteins are also present in the cytoplasm trapped inside the synaptosomes.

Discussion

The purpose of this investigation was to see whether there is any evidence that chronic treatment of old rats with pyritinol, which increases ACH levels and probably ACH release in the CNS, leads to a decrease in the phospholipid content of synaptosomal membranes. This question is relevant since Blusztajn et al. (1987) provided direct evidence that in cultures of human cholinergic cells increased ACH turnover may be associated with a decrease in membrane bound phosphatidylcholine. Similarly, Ulus et al. (1989) showed that electrically stimulated rat striatal slices will continue to release ACH over long periods, but that, when the slices are incubated in the absence of external choline, the ACH synthesis is associated with a decline in membrane levels of phosphatidylcholine. The term "autocannibalism" was used to describe this phenomenon, suggesting that the stimulation of ACH turnover might, under certain conditions, be detrimental to cell function and cell survival, especially in neurodegenerative disorders such as Alzheimer's disease. The data reported here provide no evidence that "autocannibalism" occurs when ACH synthesis is stimulated by the chronic administration of pyritinol. It appears that if there is any increased utilisation of choline from the choline reservoir in membrane phosphatidylcholine, then this is compensated for by an increase in the synthesis of the relevant phospholipids. The in vivo situation examined here is obviously in many ways different from the in vitro situations in which "autocannibalism" was observed. It may be that under our conditions the supply of choline from the extracellular fluid was sufficient to support an increased rate of ACH synthesis or that pyritinol increased choline availability, e.g. by facilitating it's transport across the blood-brain barrier. Since we observed an actual increase in phosphatidylcholine and total phospholipids it is also conceivable that stimulation of phospholipid-turnover and -level is the primary event, enabling an increase in ACH synthesis.

References

Blusztajn JK, Liscovitch M, Richardson UI (1987) Synthesis of acetylcholine from choline derived from phosphatidylcholine in a human neuronal cell line. Proc Natl Acad Sci 84: 5474–5477

Breckenridge WC, Gombos G, Morgan IG (1972) The lipid composition of adult rat brain synaptosomal plasma membranes. Biochim Biophys Acta 266: 605–707

Gray EG, Whittaker VP (1962) The isolation of nerve endings from brain. An electron-microscopic study of cell fragments divided by homogenisation and centrifugation. J Anat (London) 96: 79–88

Martin KJ, Vyas S (1987) Increase in acetylcholine concentrations in the brain of 'old' rats following treatment with pyrithioxin (Encephabol). Br J Pharmacol 90: 561–565

Stein J, Smith G (1982) Techniques in the life sciences. Biochemistry – B4/1. Elsevier Biomedical

Ulus IH, Wurtman RJ, Mauron C, Blusztajn JK (1989) Choline increases acetylcholine release and protects against the stimulation-induced decrease in phophatide levels within membranes of rat corpus striatum. Brain Res 484: 217–227

Correspondence: Dr. K. J. Martin, Department of Pharmacology, University of Cambridge, Tennis Court Road, Cambridge, CB2 1QJ, United Kingdom.

Hippocampal and cardiovascular effects of muscarinic agents

J. Loudon, F. Brown, M. Clark, and **G. Riley**

Beecham Pharmaceuticals Research Division, Harlow, Essex, United Kingdom

Summary

Muscarinic full and partial agonists were compared in rats for induction of hippocampal EEG effects (RSA) and falls in blood pressure and heart rate. Full agonists induced strong RSA and strong, higher potency, cardiovascular (CVS) effects. Partial agonists induced weak RSA and weaker, lower potency, CVS effects. RSA involves the septo-hippocampal cholinergic system which degenerates in dementia. It is inferred that partial agonists will cause weaker anti-dementia effects but less undesirable side effects than full agonists.

Introduction

Degeneration of the ascending septo-hippocampal and basal fore-brain-cortical cholinergic neuronal pathways is a principal feature of Alzheimer's dementia. The few attempts at replacement therapy, using brain penetrating tertiary-amine muscarinic agonists, to compensate for the loss of the cholinergic neurones, has achieved only very limited success. A wide range of side effects (e.g. salivation, diaphoresis, miosis, nausea, emesis, hypotension, bradycardia, dyspnoea, lassitude, depression, tremor) limits the therapeutic use of muscarinic agonists in dementia. It was possible that some side effects would be less pronounced if partial agonists rather than full agonists were used (e.g. tremor and hypothermia were absent or reduced with partial agonists in mice). If partial agonists were selective for septo-hippocampal or basal forebrain-cortical effects over other CNS and peripheral effects, then they could have better therapeutic potential than full agonists. Some full and partial agonists (Fig. 1) were compared for ability to increase rhythmical slow wave activity (RSA, theta rhythm) in the hippocampal EEG, and to

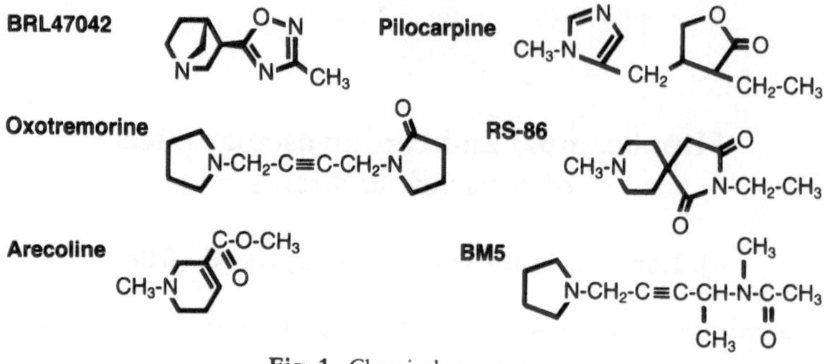

Fig. 1. Chemical structures

depress the cardiovascular system (CVS) in rats anaesthetised with urethane. Muscarinic hippocampal RSA is dependent upon septohippocampal connections in anaesthetised rats whereas conscious recordings are complicated by movement related non-muscarinic RSA (Bland, 1986).

Material and methods

Male hooded lister rats (Olac UK) were anaesthetised with urethane i.p. The femoral vein and carotid artery were cannulated for drug injections and for recording blood pressure (BP) and heart rate (HR). Concentric bipolar EEG electrodes (Rhodes Medical Instruments, model NE-100) were inserted into the dorsal hippocampus at or close to the CA1 pyramidal layer, with fine adjustment to obtain typical standardised RSA responses (Bevan, 1984). The EEG effects of drugs were recorded while peripheral effects were blocked by injection of N-methylatropine (NMA). The EEG was displayed on an oscilloscope and recorded digitally via a Cambridge Electronic Design, model 1401 laboratory computer interface. Effects were analysed quantitatively as power changes in the spectrum. CVS effects were measured without protection of NMA.

Drugs were also assessed by inhibition of the binding of the muscarinic ligands ^{3}H-oxotremorine-M (1.88 nM ^{3}H-OXO-M, agonist) and ^{3}H-quinuclidinyl benzilate (0.27 nM ^{3}H-QNB, antagonist) in rat cerebral cortex membranes in vitro. Muscarinic full agonists inhibit ^{3}H-OXO-M binding with much higher potency than for ^{3}H-QNB binding. Antagonists show approximately equal potency for inhibition of binding of both ligands. Partial agonists show higher potency for inhibition of binding of ^{3}H-OXO-M than ^{3}H-QNB, but the ratio is less marked than that shown by agonists. The ratio of IC50's ^{3}H-QNB/^{3}H-OXO-M is generally > 100 for full agonists, in the range $20-80$ for partial agonists, and near to 1 for antagonists, and thus gives a good indication of the muscarinic potencies and characters (full, partial agonist, antagonist or inactivity) of drugs (Brown et al., 1988).

Results

The activities of compounds on the EEG and CVS and on inhibition of ligand binding are summarised quantitatively in Table 1. BRL 47042, oxotremorine and arecoline, in inhibition of ligand binding, exhibited results typical of muscarinic full agonists whereas pilocarpine, RS-86 and

Table 1. Activities of muscarinic agonists and partial agonists

	RSA [a]	BP [b]	HR [b]	OXO-M [c]	QNB/OXO-M [d]
Full agonists					
BRL 47042	0.00056	0.0002	0.0004	2.8	360
Oxotremorine	0.032	0.009	0.007	17	195
Arecoline	0.1	0.02	0.02	113	227
Partial agonists					
BM5	0.32 W	0.56 I	0.56 I	25	28
Pilocarpine	0.32 W	0.56 I	0.56	334	54
RS-86	0.32	0.32	0.2	430	42

[a] Median effective dose mg/kg; [b] doses for 50% decrease; [c] IC50 (nM) for inhibition of ^{3}H-OXO-M; [d] ratio of IC50's (greater ratios denote greater agonist character); *I* inactive; *W* weak effect

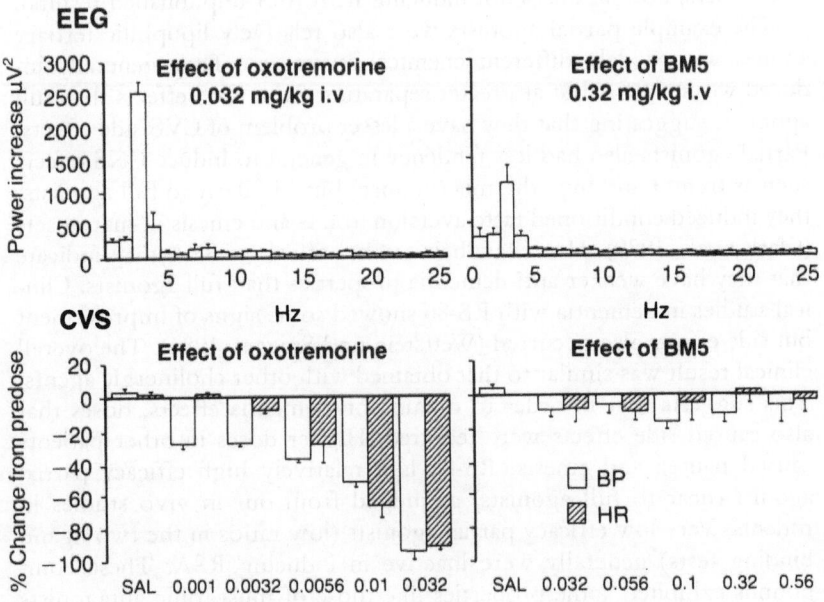

Fig. 2. Example EEG and CVS effects of oxotremorine and BM5

BM5 exhibited results typical of muscarinic partial agonists. These muscarinic full agonists exhibited high potency in increasing RSA; they caused large power increases concentrated in the 3–5 Hz band of the power spectrum. They were even more potent on the CVS where 50% falls in BP and HR occurred at doses about 4-fold less than for induction of RSA. The partial agonists generally exhibited low potency weak RSA effects and even less potent and weaker CVS effects. Typical RSA and CVS effects of oxotremorine (full agonist) and BM5 (partial agonist) for example are shown in Fig. 2.

Discussion

The three selected muscarinic full agonists represented widely different chemical structures (Fig. 1) and all induced strong RSA and CVS effects. The CVS effects occurred at lower doses than the CNS effects. This tendency may have been exaggerated by anaesthesia and i.v. injection, but it would be of interest for potential anti-dementia therapy if compounds were identified with greater selectivity for CNS effects. The three full agonists were relatively lipophilic and showed greater CNS selectivity than many other less lipophilic tertiary and quaternary amine muscarinic agonists. These less lipophilic types often caused CVS failure or fatal dyspnoea at doses not inducing RSA (our unpublished results).

The example partial agonists were also relatively lipophilic tertiary amines with widely different chemical structures. They generally induced weaker RSA but at greater separation from CVS effects than full agonists, suggesting that they have a lesser problem of CVS side effects. Partial agonists also had less tendency in general to induce CNS effects such as tremor and hypothermia (in mice) but, similarly to full agonists, they induced conditioned taste aversion in rats and emesis in marmosets (Clark et al., 1989). However their weaker effect on RSA may indicate that they have weaker anti-dementia properties than full agonists. Clinical studies in dementia with RS-86 showed some signs of improvement but side effects also occurred (Wettstein und Spiegel, 1984). The overall clinical result was similar to that obtained with other cholinergic agents. This suggests that in order to obtain anti-dementia effects, doses that also caused side effects were required. Higher doses in other patients caused nausea and emesis. RS-86 is a relatively high efficacy partial agonist (near to full agonists) as judged from our in vivo studies in rodents. Very low efficacy partial agonists (low ratios in the two ligand binding tests) generally were inactive in inducing RSA. These compounds exhibited some properties like those of muscarinic antagonists and may exaggerate dementia. Thus the differences shown here between the full and partial agonists, as classes, could be crucial to their relative therapeutic potential in dementia. We suggest that a more selective profile of the properties described is required in a muscarinic compound for successful therapeutic use in dementia.

References

Bevan P (1984) Effect of muscarinic agents on the electrical activity recorded from the hippocampus: a quantitative approach. Br J Pharmacol 82:431–440

Bland BH (1986) The physiology and pharmacology of hippocampal formation theta rhythms. Prog Neurobiol 26:1–54

Brown F, Clark M, Graves D, Hadley M, Hatcher J, McArthur R, Riley G, Semple J (1988) Variation of muscarinic activities of oxotremorine analogues. Drug Dev Res 14:343–347

Clark MSG, Hatcher J, Brown F (1989) Relative aversive potency of muscarinic agonists. TIPS [Suppl 4]:100

Wettstein A, Spiegel R (1984) Clinical trials with the cholinergic drug RS 86 in Alzheimer's disease (AD) and senile dementia of the Alzheimer type (SDAT). Psychopharmacology 84:572–573

Correspondence: Dr. F. Brown, Beecham Pharmaceuticals Research Division, Coldharbour Road, The Pinnacles, Harlow, Essex CM19 5AD, United Kingdom.

Cholinergic and monoaminergic neuromediator systems in DAT. Neuropathological and neurochemical findings

G. Moll[1], W. Gsell[1], I. Wichart[2], K. Jellinger[2], and P. Riederer[1]

[1] Department of Psychiatry, Division of Clinical Neurochemistry, University of Würzburg, Federal Republic of Germany
[2] Lainz Hospital, Ludwig Boltzmann Institute for Clinical Neurobiology, Department of Neurology, Vienna, Austria

Summary

The patterns of cholinergic (activity of choline acetyltransferase) and monoaminergic (concentrations of noradrenaline, dopamine, homovanillic acid, serotonin and 5-hydroxyindoleacetic acid) neuromediator systems were studied in postmortem tissue of the cortical lobes in each seven cases of patients with dementia of Alzheimer type (DAT) and controls. The findings show severe degenerations of neurons and deficits of neurotransmitters in all neuromediator systems. These results suggest that a substitutional therapy of only one neuromediator deficiency will not be clinically effective.

Introduction

The rise in life expectancy has led to an increase in the number of elderly people in most industrialized countries. It is reported that 15% of the european population is over 65 years (Grundy, 1983), with the tendency that the percentual part of the elderly in the whole population is increasing. The consequence is an increasing prevalence of age-related disorders, which is about 5% in people over 65 years and 20 to 30% in people over 85 years (Nielsen, 1962; Kay et al., 1970). Dementia of the Alzheimer type and Parkinson's disease are the most common types of age-related progressive degenerative brain disorders. Both are characterized by degeneration of certain neuron systems in vulnerable regions while sparing other cell groups. The neuronal degeneration is accompanied by characteristic neurochemical alterations in affected brain areas.

In dementia of Alzheimer type (DAT), our neuropathological findings and concerning neurochemical alterations, neuromediators (noradrenergic, dopaminergic and serotoninergic systems) and the cholinergic synthezising enzyme [choline acetyltransferase (CAT)], will be outlined and compared to the findings in the literature.

Material and methods

Seven cases with no clinical and neuropathological signs of dementia and seven cases with severe neuropathological changes [age-matched (see Table 1), all cases histologically verified] were selected for investigation.

Number of senile plaques and neurofibrillary tangles were chosen as neuropathological parameters for diagnosis and determination of severity of brain degeneration. Loss of neurons or decline in neuron density was determined microscopically (see Table 1).

The activity of the acetylcholine synthezising enzyme choline acetyltransferase (CAT) was determined by a radioenzymatic method (Fonnum, 1975).

The following neuromediators and related metabolites were determined by HPLC-ECD: noradrenaline, dopamine, homovanillic acid, serotonin and 5-hydroxyindole acetic acid (Sofic, 1986).

For statistic analysis Wilcoxon rank sum test was performed.

Table 1. Composition of groups

	n	Sex	Age (years)	Post-mortem delay (h)
Control	7	5 f, 2 m	75.9 ± 2.7	6.3 ± 2.1
DAT	7	2 f, 5 m	78.6 ± 2.9	3.7 ± 0.8

No neuroleptics were supplied 3 months prior to death, no prolonged agonal states were observed

Results and discussion

The cholinergic neuromediator system

The first neurochemical findings showed deficits in the cholinergic system leading to the "cholinergic hypothesis" of DAT (Bowen et al., 1976; Davies and Maloney, 1976; Perry et al., 1977). This was supported by clinical findings of Bartus et al. (1982), who found that memory functions could be influenced by pharmacologic manipulations of the cholinergic system.

The pericarya of neurons of the nucleus basalis Meynert (NbM), which are embedded in substantia innominata, a basal forebrain structure, are rich in choline acetyltransferase (CAT) and acetylcholine esterase activity. 90% of axones which arise from these pericarya innervate the neocortex (Mesulam et al., 1983). Therefore deficits in cholinergic function of the neocortex might result from a loss of neurons in the NbM.

Compared to an age-matched control group, we found that neuron density decreased in NbM by 65% (p < 0.001) in DAT (see also Table 2). This is in line with data from the literature (*age 60 to 70:* loss of neurons between 56 to 75% compared to a control group; decrease in neuron density between 44 to 66%; Arendt et al., 1983; Tagliavini and Pilleri, 1983; Rogers et al., 1985; Ichimiya et al., 1986; Mann et al., 1986; *age 70 to 80:* loss of neurons between 42 to 58% and decrease in neuron density between 33 to 62% compared to a control group; Perry et al., 1982; Mann et al., 1984, 1985, 1986; Chui et al., 1986). The variation in both the loss of neurons and neuron density between different authors is due to the low number of cases and varying composition of groups. In addition neuropathological data of control groups are often missing and the area of NbM and its great pericarya is not sufficient determined. Furthermore, in the most papers, the DAT-group is not defined by unique clinical and neuropathological criteria.

Summarizing our own data and these from literature, it is evident that the decline is greatest in patients with an early onset of the degenerative process (see also Mann et al., 1986).

Most of the neurons of NbM project into the neocortex. There, the activity of CAT decreases by 86% (p < 0.001), with a maximum decrease in the temporal lobe by 91%. As expected, there is no unique distribution of CAT-activity over the whole cortex. Moreover, Perry et al. (1984) showed that CAT-activity differs even within one area, with maximum activity in cell layers II and IV.

Table 2. Neuropathological findings

	SP, NFT/mm^2	NbM[a]	LC[b]	SN[a]	NRD[a]
Control	0−4	80.6±4.4	77.8±7.2	87.8±6.8	42.3±7.8
DAT	>15	27.5±4.0	37.3±4.9	67.0±7.4	18.7±3.1
% of control		35 ***	48 ***	76 **	44 ***

Mean ± standard deviation; *SP* senile plaques; *NFT* neurofibrillary tangles (Khatchaturian, 1985); *LC* locus coeruleus; *SN* substantia nigra; *NRD* nucleus raphe dorsalis; statistic: Wilcoxon rank sum test; ** p < 0.01; *** p < 0.001; [a] neuron density; [b] cell counts

The integrity and interconnections of basal forebrain (NbM) and temporal lobe (regio entorhinalis and hippocampus) play an important role in memory functions (Damasio et al., 1985), and show severe deficits in DAT patients.

The noradrenergic neuromediator system

The cholinergic system is not the only one involved in the degenerative process.

Locus coeruleus (LC) is the most important noradrenaline synthezising nucleus, containing about half of all noradrenergic pericarya.

Our own data show a decrease in maximum neuron number in LC by 52% (p < 0.001) according to the (see also Table 2) data from literature (*age 60 to 70:* 47 to 86% reduction in maximum neuron number compared to a control group; Vijayashankar and Brody, 1979; Mann et al., 1984; Ichimiya et al., 1986; *age 70 to 80:* 48 to 71% reduction in maximum neuron number; Chui et al., 1986; Mann et al., 1982, 1984, 1986). As observed in NbM, the loss of neurons is dependent on age with severiest degeneration in early onset cases (see above) and in patients with familiar Alzheimer's disease (Etienne et al., 1986; Jellinger, 1986; Mountjoy, 1986).

LC is projecting to various areas of the neocortex, gyrus cinguli, hippocampus, corpus amygdaloideum, nucleus accumbens, thalamus, hypothalamus, NbM and Nucleus raphe dorsalis (NRD). Therefore deficits in the noradrenergic function of these areas should result from loss of neurons in the LC.

Our data show a significant reduction in noradrenaline of 68% in the parietal lobe (p = 0.170) and of 65% in the temporal lobe (p = 0.017) and also in corpus amygdaloideum by 48% (p = 0.026). Within the other projecting areas mentioned above there is no statistically significant alteration in noradrenaline content. Nearly all investigations reported in literature demonstrated a decline in noradrenaline in various brain regions. With statistical significance (student's t-test) a decrease in noradrenaline was reported by Arai et al. (1984) and Rossor et al. (1984) in gyrus cinguli, by Yates et al. (1983) in corpus mamillare, by Arai et al. (1984) in thalamus, by Yates et al. (1981, 1983), Mann et al. (1982, 1987) and Arai et al. (1984) in hypothalamus, by Nyberg et al. (1985) in globus pallidus, by Mann et al. (1982, 1987) and Nyberg et al. (1985) in caudate nucleus and by Adolfsson et al. (1979), Arai et al. (1984), Nyberg et al. (1985) and Reinikainen et al. (1988) in putamen.

As in the cholinergic system a severe reduction in neuron number in the synthezising area (LC) and a decline in neuromediator concentration in the iso- and allocortex is observed. In the non-cortical projection areas mentioned above, there was no evident decrease in noradrenaline concentration. This might refer to a primary lesion in the parietal and temporal isocortex and the amygdaloid allocortex.

The dopaminergic neuromediator system

For a long time the dopaminergic neuromediator system was suspected to be not involved into the degenerative process.

The most important dopamine synthesizing areas are the cell group A 9, the zona compacta of substantia nigra (SN), and nearby cell group A 10 (area of the ventral tegmentum). They mainly contribute in the dopaminergic innervation of the investigated brain regions, with A 9 projecting into the striatum (caudate nucleus and putamen) and into the amygdala and A 10 projecting into the frontal cortex, gyrus cinguli, nucleus accumbens and the entorhinal cortex (Nieuwenhuys et al., 1981). In SN reduction in neuron density is 24% compared to the control group. This is the lowest one of all neuromediator synthesizing regions investigated, according to data from literature (*age 60 to 70:* 12 to 27% decline in neuron number as compared to a control group; Mann et al., 1984, 1987; Tabaton et al., 1985; *age 70 to 80:* 6 to 17% decline in neuron number; Mann et al., 1984, 1985, 1987; Chui et al., 1986). Mann et al. (1987) found that neuron number of ventral tegmentum decreases by about the same degree as observed in other neuromediator systems (43 to 61%, depending on age).

The dopamine concentrations are not significantly reduced in any brain region investigated, but decrease in hippocampus [DA: 53% ($p = 0.117$)] and corpus amygdaloideum [DA: 51% ($p > 0.5$)] is remarkable. In contrast, the concentration of the dopamine metabolite homovanillic acid was significantly reduced in frontal (61%; $p = 0.018$), parietal (63%; $p = 0.035$) and temporal (60%; $p = 0.011$) lobes and in hippocampus by 55% ($p = 0.017$). The nigrostriatal dopaminergic system, which is mainly affected in the pathogenesis of Morbus Parkinson, is not involved at all in DAT (caudate nucleus: DA 130%; HVA 107%; putamen: DA 99%, HVA 94%; each $p > 0.5$).

The serotoninergic neuromediator system

The most important serotonin synthesizing nuclei are nucleus raphe dorsalis (NRD), nucleus tegmentalis dorsalis (NTD) and nucleus centralis superior (NCS). Compared to NTD (decrease in neuron number by

12%, Mann et al., 1984, 1985) and NCS (37%; Tabaton et al., 1985), NRD is mainly affected by reduction of neuron density (56%, p < 0.001).

NRD projects to LC, SN, caudate nucleus, putamen, nucleus accumbens, globus pallidus, thalamus, corpus mamillare, corpus amygdaloideum, hippocampus and neocortex. We could not find reported decreases in serotonin and 5-HIAA contents in the striatum, globus pallidus, thalamus, corpus mamillare and gyrus cinguli (Adolfsson et al., 1979; Arai et al., 1984; Reinikainen et al., 1988). According to Gottfries et al. (1986) and Reinikainen et al. (1988), we obtained an affection of the serotoninergic system in the hippocampus (significant reduction in 5-HIAA by 66%; p = 0.033) as compared to controls.

In the frontal (5-HT: 70%, p = 0.030; 5-HIAA: 54%, p = 0.073), parietal (5-HT: 77%, p = 0.002; 5-HIA: 54%, p = 0.001) and temporal (5-HT: 66%, p = 0.07; 5-HIAA: 55%, p = 0.001) lobes we found a reduction in serotonin and in 5-HIAA content. The serotoninergic system, which plays an important role in memory and learning processes (Monadori, 1981; Vachon and Roberge, 1981; Kandel and Schwartz, 1982) is therefore damaged mostly in the cortical and hippocampal regions (5-HT: 63%, p = 0.383; 5-HIAA: 66%, p = 0.033).

Conclusions

The cholinergic and all three monoaminergic neuromediator systems were affected in DAT, predominantly in iso- and allocortical brain regions.

These results suggest that a substitutional therapy of only one neuromediator deficiency will not be clinically effective. Authors who report a monosystemic therapy especially for the substitution of the cholinergic system may be right when substituting shortly after onset of degeneration. Our data indicate that they might not succeed in progredient cases of degeneration. We suggest a combined therapy to maintain and enforce the function of the involved neuromediator systems, e.g. lecithin/choline, phosphatidylserin, pyritinol, piracetam for the cholinergic system (Wurtman, 1988), selective MAO inhibitors for the dopaminergic (1-deprenyl) (Tariot et al., 1987a, b) and noradrenergic systems (moclobemide, brofaromine) and serotoninergic drugs, like precursor amino acids with or without inhibitors of the peripherally acting aromatic amino acid decarboxylase or MAO-inhibitors, in order to substitute the serotoninergic deficiency. Furthermore, agonists should be developed in order to stimulate postsynaptic receptors independently from presynaptic degeneration.

Acknowledgement

This work was supported by Hirnliga e.V., Liga zur Erforschung, Erkennung und Behandlung der Hirnleistungsstörungen (Morbus Alzheimer), Heidelberg, F.R.G.

References

Adolfsson R, Gottfries C-G, Roos BE, Winblad B (1979) Changes in the brain catecholamines in patients with dementia of Alzheimer type. Br J Psychiatry 135:216–223

Arai H, Kosaka K, Iizuka R (1984) Changes of biogenic amines and their metabolites in postmortem brains from patients with Alzheimer-type dementia. J Neurochem 17:388–393

Arendt T, Bigl V, Arendt A, Tennstedt A (1983) Loss of neurons in nucleus basalis of Meynert in Alzheimer's disease, paralysis agitans and Korsakoff's disease. Acta Neuropathol 61:101–110

Bartus RT, Dean RL, Beer B, Lippa AS (1982) The cholinergic hypothesis of geriatric memory dysfunction. Science 217:408–417

Bowen DM, Smith CP, White P, Davison AN (1976) Neurotransmitter-related enzymes and indices of hypoxia in senile dementia and other abiotrophies. Brain 99:459–496

Chui HC, Mortimer JA, Slager U, Zarrow C, Bondareff W, Webster DD (1986) Pathologic correlates of dementia in Parkinson's disease. Arch Neurol 43:991–995

Damasio AR, Graff-Radford NR, Eslinger PJ, Damasio H, Kassell N (1985) Amnesia following basal forebrain lesions. Arch Neurol 42:263–271

Davies P, Maloney AJ (1976) Selective loss of cholinergic neurons in Alzheimer's disease. Lancet ii:1403

Etienne P, Robitaille Y, Wood P, et al (1986) Nucleus basalis neuron loss, neuritic plaques and choline acetyltransferase activity in advanced Alzheimer disease. Neuroscience 19:1279–1291

Fonnum F (1975) A rapid radiochemical method for the determination of choline acetyltransferase. J Neurochem 24:407–409

Gottfries C-G, Bartfai T, Carlsson A, Eckernaes SA, Svennerholm L (1986) Prog Neuropsychopharmacol Biol Psychiatry 10:405–413

Grundy E (1983) Demography and old age. J Am Geriatr Soc 31:325–336

Ichimiya Y, Arai H, Iizuka R (1986) Morphological and biochemical changes in the cholinergic and monoaminergic system in Alzheimer-type dementia. Acta Neuropathol 70:12–116

Jellinger K (1986) Overview of morphological changes in Parkinson's disease. In: Yahr MD, Bergmann KJ (eds) Advances in neurology, vol 45. Raven Press, New York, pp 1–18

Kay DWK, Bergmann K, Foster EM, McKechnie AG, Roth M (1970) Mental illness and hospital usage in the elderly: a random sample follow-up. Compr Psychiatry II:26–35

Khatchaturian ZS (1985) Diagnosis of Alzheimer's disease. Arch Neurol 42:1097–1105

Mann DMA, Yates PO, Hawkes J (1982) The noradrenergic system in Alzheimer's and multi-infarct dementias. J Neurol Neurosurg Psychiatry 45:113–119

Mann DMA, Yates PO, Marcyniuk B (1984) Alzheimer's presenile dementia, senile dementia of Alzheimer type and Down's syndrome in middle age form an age related continuum of pathological changes. Neuropathol Appl Neurobiol 10:185–207

Mann DMA, Yates PO, Marcyniuk B (1985) Some morphometric observations on the cerebral cortex and hippocampus in presenile Alzheimer's disease, senile dementia of the Alzheimer type and Down's syndrome in middle age. J Neurol Sci 69:139–159

Mann DMA, Yates PO (1986) Neurotransmitter deficits in Alzheimer's disease and in other dementing disorders. Human Neurobiol 5:147–158

Mann DMA, Yates PO, Marcyniuk B (1987) Dopaminergic neurotransmitter systems in Alzheimer's disease and in Down's syndrome at middle age. J Neurol Neurosurg Psychiatry 50:341–344

Mesulam MM, Mufson EJ, Levey AL, Wainer BHJ (1983) Cholinergic innervation of cortex by the basal forebrain. J Comp Neurol 214:170–197

Mountjoy CQ (1986) Correlations between neuropathological and neurochemical changes. Br Med Bull 42:81–85

Nielsen J (1962) Gerontopsychiatric period-prevalence investigation in a geographically delimited population. Acta Psychiatr Scand 38:307–330

Nieuwenhuys R, Voogd J, van Huijzen C (1981) The human central nervous system. A synopsis and atlas. Springer, Berlin Heidelberg New York

Nyberg P, Adolfsson R, Hardy JA, Nordberg P, Wester P, Winblad B (1985) Catecholamine topochemistry in human basal ganglia. Comparison between normal and Alzheimer brains. Brain Res 333:139–142

Perry EK, Gibson PH, Blessed G, Perry RH, Tomlinson BE (1977) Neurotransmitter enzyme abnormalities in senile dementia. J Neurol Sci 34:247–265

Perry RH, Candy JM, Perry EK, Irving D, Blessed G, Fairburn AF, Tomlinson BE (1982) Extensive loss of choline acetyltransferase activity is not reflected by neuronal loss in the nucleus basalis of Meynert in Alzheimer's disease. Neurosci Lett 33:311–315

Perry EK, Atack JR, Perry RH, Hardy JA, Dodd PR, Edwardson JA, Blessed G, Tomlinson BE, Fairbairn AF (1984) Intralaminar neurochemical distribution in human midtemporal cortex: comparison between Alzheimer's disease and the normal. J Neurochem 42:1402–1410

Reinikainen KJ, Paljaervi L, Huuskonen M, Soininen H, Laakso M, Riekkinen PJ (1988) A post-mortem study of noradrenergic, serotoninergic and GABAergic neurons in Alzheimer's disease. J Neurol Sci 84:101–116

Rogers JD, Brogan D, Mirra SS (1985) The nucleus basalis of Meynert in neurological disease: a quantitative morphological study. Ann Neurol 17:163–170

Rossor MN, Iversen LL, Reynolds GP, Mountjoy CQ, Roth M (1984) Neurochemical characteristics of early and late onset types of Alzheimer's disease. Br Med J 288:961–964

Sofic E (1986) Dissertation. Technical University of Vienna, Austria

Tabaton M, Schenone A, Romagnoli P, Mancardi GL (1985) A quantitative and ultrastructural study of substantia nigra and nucleus centralis superior in Alzheimer's disease. Acta Neuropathol 68:218–223

Tagliavini F, Pilleri G (1983) Basal nucleus of Meynert: a neuropathological study in Alzheimer's disease, simple senile dementia, Pick's disease and Huntington's chorea. J Neurol Sci 62:243–260

Tariot PN, Sunderland T, Weingartner H, Murphy DL, Welkowitz JA, Thompson K, Cohen RM (1987a) Cognitive effects of L-deprenyl in Alzheimer's disease. Psychopharmacology 91:489–495

Tariot PN, Cohen RM, Sunderland T, Newhouse PA, Yount D, Mellow AM, Weingartner H, Mueller EA, Murphy DL (1987b) L-deprenyl in Alzheimer's disease. Arch Gen Psychiatry 44:427–433

Vijayashankar N, Brody H (1979) A quantitative study of the pigmented neurons in the nuclei coeruleus and subcoeruleus in man as related to aging. J Neuropathol Exp Neurol 38:490–497

Wurtman RJ (1988) Cholinergic brain neurons and the dementias associated with old age: toward the development of effective drugs. In: Maurer K, Wurtman RJ (eds) Organic brain disorders. Recent neurobiochemical findings, diagnostic procedures and consequences for treatment. Vieweg, Braunschweig, pp 11–17

Yates CM, Ritchie IM, Simpson J, Maloney AFJ, Gordon SA (1981) Noradrenaline in Alzheimer-type dementia and Down's syndrome. Lancet ii:39–40

Yates CM, Simpson J, Gordon SA, Maloney AFJ, Allison Y, Ritchie IM, Urquart A (1983) Catecholamines and cholinergic enzymes in pre-senile and senile Alzheimer-type dementia and Down's syndrome. Brain Res 280:119–126

Correspondence: Dr. G. Moll, Department of Psychiatry, Division of Clinical Neurochemistry, University of Würzburg, Füchsleinstrasse 15, D-8700 Würzburg, Federal Republic of Germany.

Alterations in catecholamine neurons in the locus coeruleus in dementias of Alzheimer's and Parkinson's disease

V. Chan-Palay and E. Asan

Neurology Clinic, University Hospital, Zürich, Switzerland

Summary

A differentiation can be made between the LC in normal brain, in SDAT and PD for diagnostic purposes, based on the findings concerning the morphological alterations of the TH-immunoreactive neurons, the topographical distribution of neuron loss within the length of the LC, and, to some extent, the total reduction in cell number.

Introduction

Senile dementia of the Alzheimer's type (SDAT) is neuropathologically characterized by severe cortical atrophy and cell loss as well as a high index of dementia as measured by numbers of neurofibrillary tangles (NFT) and neuritic plaques (NP) in neocortex and hippocampus. In addition, several subcortical afferent projection systems are disturbed in the disease, namely those based on acetylcholine, norepinephrine (NE) and serotonin. The occurrence of extrapyramidal signs in SDAT suggests involvement of dopaminergic pathways in some cases. Investigations of the functional role of the locus coeruleus-NE system in SDAT have previously focused on the study of the locus coeruleus (LC) cellular neuropathology and measurements of NE content in the various cortical projection areas of the LC. Quantitative investigations using the neuromelanin pigment as a marker for NE neurons have demonstrated a reduction in neuron numbers in the LC in most cases of SDAT with a high incidence of neuropathologic markers like NFT, NP and, occasionally, Lewy bodies in the remaining neurons. Recent studies using catecholamine biosynthetic enzymes have also demonstrated a loss of LC-

NE neurons in SDAT, though with different results as to the total neuron numbers counted in control and SDAT cases. This cell loss from the LC was reported to be topographically arranged. NE-level, dopamine-beta-hydroxylase (DBH)-activity and the levels of several other NE markers have been shown to be decreased in LC projection areas both in ante-mortem biopsies and in post-mortem brain tissue indicating a deficiency of the NE-transmission in SDAT. Correlation between cortical plaque formation and cortical NE-levels and LC neuron loss in the anterior and central regions of the LC known to project to these areas in animals have been reported. Also, the severity of cortical plaque incidence and the degree of reduction of LC neuron number has been observed to be correlated, though no direct correlation between the severity of dementia and the extent of LC damage has been demonstrated. A recent study, however, has shown a positive correlation of the occurrence of depression in SDAT and the decrease in LC neuron number.

Lesions of brainstem nuclei including the LC, with neuropathologic changes such as Lewy bodies and NFT in PD have been described many years ago and Parkinsonian state and LC-lesions similar to those found in the disease are caused by the administration of the neurotoxin 1-methyl-4-phenyl-1,2,3,6-tetrahydropyridine in the macaque monkey. Even though PD is principally a disorder of locomotion, it is now generally accepted that in a number of patients progressive mental impairment occurs in the course of the disease. Some authors have reported dementia in more than 50% of cases of PD. According to the responsiveness to L-dopa treatment it has been postulated that "two separate disorders can be distinguished in PD: an exclusive motor disorder occurring in a younger population with a longer and more "benign" course and a better response to L-dopa; and another, a motor followed by a cognitive disorder occurring in an older population with a more fulminant course and a poorer response to L-dopa". There has been a controversy over the distinction of a "cortical" dementia found in SDAT and a "subcortical" dementia present in PD patients. Several authors have claimed that in neuropsychological tests the dementia of SDAT, characterized mainly by aphasia, amnesia, agnosia and apraxia, can be distinguished from that found in PD patients, where the dementia is characterized by slowness of mental processing, forgetfulness, impaired cognition, apathy and depression, while no psychopathological difference between demented PD and SDAT patients was found by others. Some investigators have suggested that the dementia in PD displays a pattern of impairment typical for a lesion of the frontal lobe and a laterality of the disease has been suggested based on the finding that patients with greater disease involvement on the left side of the body showed greater neuropsychological impairment than those more affected on the right body side. The question whether the incidence of cortical plaques and tangles is correlated to the severity of dementia in PD is still also somewhat controversial. Early reports have shown more frequent occurrence of NP and NFT in demented PD patients than in non-demented, suggesting coincidental SDAT in these patients. Other authors could not demonstrate a positive correlation between NFT and NP formation and dementia, but reported a severe cell loss in the LC more frequently in demented PD patients than in those without symptoms of dementia. A correlation between the coeruleo-cortical NE-system and dementia has also been suggested based on modifications in the number of adrenergic receptors in demented PD patients.

Material and methods

Normal controls included the brains of 11 patients, 4 male and 7 female, ranging in age from 43 to 89 years, with no clinical history of neurological or psychiatric disease as confirmed by postmortem gross and microscopic neuropathological examination. Vital data and selection criteria for all control cases were described in detail (Chan-Palay and Asan, 1989 a). Data assembled in three paradigm control cases in the age group of the patients in the Alzheimer's and Parkinson's groups served as control values for quantitative analyses. Appropriate levels of normal mental function in patients was shown by results of between 22 and 26 from a possible 30 points in the last available mini-mental status tests. In the group of senile dementia of the Alzheimer's type (SDAT) cases the brains of 8 patients that had been clinically diagnosed were studied, 2 male and 6 female cases with ages ranging from 71 to 85 years. Cases of dementia due to other neurological disorders, such as ischemia, multiple infarcts, Pick's disease etc. were excluded. Postmortem delays ranged from 3 to 16 hours, with postmortem delays of 5 hours and less in 5 of the cases. Counts of neuritic plaques and neurofibrillary tangles were made on Bodian-silver-stained preparations. For all cases in this group, the counts yielded moderate to high indices of neurofibrillary plaques and tangles in the examined areas, which is indicative of SDAT. Seriously impaired mental function in these patients was indicated by a score of 0 to 5 points in the last available mini-mental status tests.

In the study of cases with Parkinson's disease 7 diagnosed patients were included, ranging in age from 76 to 90 years, three male and four female cases. Clinically, two of the patients had PD responsive to L-dopa treatment without symptoms of dementia (P−D), five patients had histories of rapidly progressive dementia, with onset 2−3 years before death. Of these five demented patients, three were responsive to treatment with L-dopa (P + D); two were atypical and their Parkinsonian symptoms did not respond to L-dopa treatment (P + D/L-dopa nonresponsive). The postmortem delays ranged from 3.5 to 21 hours, and was less than 7 hours in four of the cases. The clinical diagnosis of Parkinson's

disease was confirmed at autopsy by both gross and microscopic neuropathological examination. The substantia nigra showed considerable cell loss, loss of
pigmentation, numerous Lewy bodies, and gliosis pathognomic of Parkinson's
disease in every case. Counts of neurofibrillary tangles and neuritic plaques were
performed as described for the SDAT cases and yielded indices slightly higher
than in the age-matched control and non-demented Parkinson's cases for the
demented Parkinson's patients. The last available mini-mental status test scores
were 22 to 26 for the non-demented Parkinson's disease group and 11 to 16 for
the demented patients.

Fixation and immunocytochemistry

The protocols used for fixation of the studied brainstems and immunocytochemistry were described in detail in preceeding papers (Chan-Palay and Asan,
1989a, b; Chan-Palay et al., 1990).

Computer-assisted quantitative morphological analyses

The computer system and the recording procedures used have been described
in detail (Chan-Palay and Asan, 1989a, b). Briefly, the immunocytochemically
stained serial brainstem sections were reassembled in the correct anatomical
order, and the LC outline in the coronal plane, its rostral and caudal borders were
determined, and its rostrocaudal length on both sides of the brainstem was
calculated. For the computer-assisted measurements of morphological parameters of TH-immunoreactive LC neurons, and for the mapping and counting of
neurons and the three dimensional reconstruction procedure for the analysis of
neuron distribution in the LC an IBM-AT-mouse based user-interactive image
analysis system with the Cellmate program (Bioquant, Tenn.) was used. Outlines
of individual cell somata and dendritic arbors were recorded for calculations of
soma areas and dendritic arbor length. Plots of these recordings served to
illustrate alterations in individual neuron morphology. To ensure comparability
all quantitative measurements of neuronal parameters were carried out on immunoreactive neurons stained with the PAP-method. Cell counts were performed on all reassembled sections of one TH-immunoreacted series of sections
by cursor-marking the localization of whole cell bodies using different symbols
signifying different morphological classes of cells (see below) in all focus levels
throughout the entire extent of the LCs of both sides of the brainstem. TH-immunoreactive cells of the locus subcoeruleus and the pars cerebellaris loci
coerulei were recorded but not counted. Neuron numbers on partially damaged
sections were approximated either from the numbers counted on immediately
adjacent TH-immunoreacted sections or from those counted in the contralateral
LC of the same section and the ipsilateral LCs of the preceeding and following
sections of the same series. Total neuron numbers were calculated by the computer for the entire LCs and, by their differing recording symbols, differentiated
according to the four neuron classes: Large multipolar (LM), large "bipolar"
(LB), small multipolar (SM) and small "bipolar" (SB) neurons. The recordings

of the reassembled sections were then aligned to match as closely as possible the situation in the intact brains, and an image of the three dimensional distribution of the neurons was created by the computer.

Results

Identification of NE-producing neurons is done by immunocyto-chemical demonstration of two NE biosynthetic enzymes, tyrosine hy-droxylase (TH) and DBH, and immunoreactions are visualized by the peroxidase-antiperoxidase (PAP) and immunogold-silver-staining (IGSS) methods. It is demonstrated that the reactions with antisera against TH and DBH yield equivalent results and that both immunocytochemical visualization methods allow detailed analysis of neuronal morphology. The neurons of the human LC fall into four distinct classes: large multipolar neurons with round or multiangular somata (LM), large elliptical "bipolar" neurons (LB), small multipolar neurons with round or multiangular somata (SM) and small ovoid "bipolar" (SB) neurons. Though most of the neurons contain neuromelanin pigment, some of the neurons of the larger type lack pigmentation. Dendritic arborization in all neuron types is extensive and computer-assisted quantitative mea-surements of the neuronal structure parameters soma size, dendritic arbor length, surface area and volume are given. Comparison of neu-ronal morphology in different age groups shows that even though the soma areas of LC neurons of all four classes are decreased in older normal adult brain, the dendritic arborization is equally extensive. De-tailed mapping of the immunoreactive neurons and computer-assisted three dimensional reconstruction of the LCs are used to analyze the morphology of the nucleus as a whole. According to cellular distribution patterns, the LC is divided into rostral, middle and caudal parts with neurons scattered over a large area rostrally, tightly clustered in the middle and very densely packed in the caudal part. Small neurons pre-dominate in all parts, but the relative contribution of larger cells decreas-es in a rostro-caudal direction. Small bipolar neurons are the most frequent cells of the caudal part and display distinct dorsomedial-ventro-lateral orientation. These general morphological characteristics are the same in all age groups, but cell density in rostral and middle parts is decreased in old age, while the relative frequency of large cells is in-creased especially in the rostral LC. No age-dependent decrease in nucle-ar length is observed. Assessment of neuron numbers documents a cell loss of 27% to 55% in older adult brains. Cell loss is topographically arranged, being highest in the rostral part, lower in the middle and virtually absent in the caudal part. Quantitative assessment of the distri-bution of the different morphological neuron classes confirms the obser-

Table 1.

Case	Age	Sex	Neuron number $\times 10^3$
Control	79	m	47.5
Control	78	f	40.9
SDAT (mild)	78	m	34.0
SDAT (severe)	74	f	18.8
SDAT (severe)	77	f	5.7
P	76	f	31.1
P+D	83	f	23.3
P+D/L-dopa non-responsive	79	m	2.5

vations mentioned above, suggesting that especially in the rostral part of the LCs of older adult brains loss of smaller cells is comparatively higher than of larger cells. The computer-generated three dimensional reconstruction provides the possibility of visualizing LC shape and cell distribution closely approximating the situation in the intact brain and facilitates the detection of morphological differences of the LCs in individual brains (see Fig. 1 a–f). After the studies of the controls, a detailed qualitative and quantitative investigation of the morphology and distribution of the NE neurons in the human locus coeruleus in two classes of neurodegenerative disorders involving dementia, the senile dementia of the Alzheimer's type (SDAT) and Parkinson's disease (PD) is undertaken. In SDAT, the four basic LC neuron classes found in the normal human brain are recognizable in the remaining cells, but the cell somata are generally larger, the cell bodies are swollen and misshapen, and the dendrites are forshortened and thick and less branched than in neurons of control LCs. Quantitative analysis confirms the qualitative observations. The reduction of absolute numbers of LC–NE neurons in paradigm cases of SDAT and PD as compared to controls are shown in Table 1.

Fig. 1. Three dimensional computer reconstruction of the LC of a younger control case, 55 yr, **(a)**; an older control case, 78 yr, case **(b)**; a case of mild SDAT with comparatively little cell loss, 78 yr, case **(c)**; a severe SDAT case with extreme cell loss, 77 yr, case **(d)**; a PD case without dementia, 76 yr, case **(e)**; and a PD case with dementia, L-dopa non-responsive, 82 yr, case **(f)**. The reconstruction is viewed from dorsal, shifted in a 25° angle from the plane of the figure. R = right, L = left LC. The outline of the fourth ventricle is drawn on every fourth section, and each TH-immunoreactive neuron on all the recorded sections is marked by a dot. Note the cell loss which occurs mainly in the rostral part of the older control case (b) as compared to the younger control case (a). Note also the high neuronal loss present predominantly in the rostral and middle parts in both SDAT cases (c) and (d). In the PD case with dementia/L-dopa non-responsive (f) cell loss is present throughout the nucleus

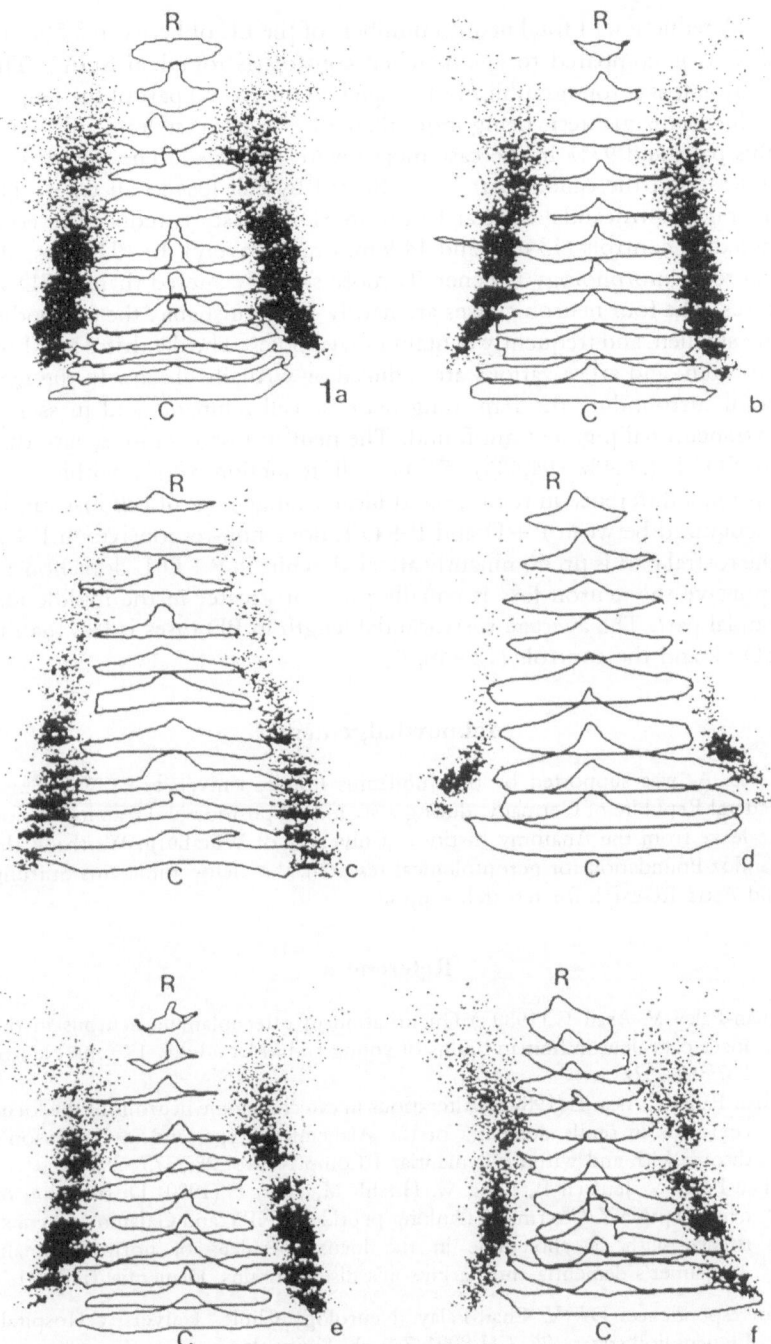

A reduction of total neuron numbers of the LC of between 3.5% and 87.5% as compared to age-matched controls is found in SDAT. This neuron loss is topographically arranged: in the rostral part of the LC, the reduction is greatest, being more than 28% in the case least affected in this part, and 97% in the case most severely affected. The middle part is less, and the caudal part least affected by cell loss in all cases. The average rostrocaudal nuclear length in SDAT cases is reduced as compared to controls (13 mm and 14.9 mm respectively). In PD cases, the neuronal morphology is generally more severely altered than in SDAT cases. The four neuron classes are hardly distinguishable, the cell bodies are swollen, and frequently contain Lewy bodies. The dendrites are short and thin, and arborizations are reduced or virtually absent. In the neuropil surrounding the remaining neurons cell remnants and masses of extraneuronal pigment are found. The neuron loss is more severe than in SDAT (26.4% – 94.4%). While cell reduction varies within each group, a difference in the topographical arrangement of cell loss can be recognized between P + D and P + D/L-dopa non-responsive: in P + D the rostral part is predominantly affected, while in P + D/L-dopa non-responsive the neuron loss is equally great or greater in the middle and caudal part. The average rostrocaudal length in PD cases is less than in SDAT and the controls (12.4 mm).

Acknowledgements

E. A. was supported by the Jubiläums Fonds, University of Würzburg, Federal Republic of Germany, during a six month postdoctoral training period on leave from the Anatomy Institute, University of Würzburg. We thank the Sandoz Foundation for gerontological research, the Geigy Jubiläums Stiftung and Astra Research for research support.

References

Chan-Palay V, Asan E (1989a) Quantitation of catecholamine neurons in the locus coeruleus in human brains of young and older adults. J Comp Neurol 287:357–372
Chan-Palay V, Asan E (1989b) Alterations in catecholamine neurons of the locus coeruleus in senile dementia of the Alzheimer's type, and in Parkinson's disease with and without dementia. J Comp Neurol 287:373–392
Chan-Palay V, Jentsch B, Lang W, Höchli M, Asan E (1990) Distribution of neuropeptide Y, C-terminal flanking peptide of NPY and Galanin and coexistence with catecholamine in the locus coeruleus of normal human, Alzheimer's dementia and Parkinson's disease brains. Dementia 1:18–31

Correspondence: Dr. V. Chan-Palay, Neurology Clinic, University Hospital, Frauenklinikstrasse 26, CH-8091 Zürich, Switzerland.

Tyrosine hydroxylase, tryptophan hydroxylase, biopterin and neopterin in the brains and biopterin and neopterin in sera from patients with Alzheimer's disease

T. Nagatsu[1], M. Sawada[2], M. Hagihara[1], N. Iwata[1], H. Arai[3],
and R. Iizuka[3]

[1] Department of Biochemistry, Nagoya University School of Medicine, Nagoya
[2] Institute of Comprehensive Medical Science, Fujita Health University, Toyoake
[3] Department of Psychiatry, School of Medicine, Juntendo University, Tokyo, Japan

Summary

The activities of tyrosine hydroxylase (TH), tryptophan hydroxylase (TPH), and the concentrations of biopterin (BP) and neopterin (NP) were examined in postmortem brains from histologically verified patients of senile dementia of Alzheimer type (SDAT). The results suggest that the reductions in TH, TPH, and BP may be related to the reduction in monoamine neurotransmitters, due to destruction of monoaminergic neurons in SDAT, and may be an event independent of the cholinergic dysfunction. Serum BP levels were also significantly reduced in patients with SDAT.

Introduction

A marked loss of presynaptic cholinergic indices in the cerebral cortex and limbic structures is known to be the main neurochemical event in senile dementia of Alzheimer type (SDAT) (Bowen et al., 1976; Davies and Maloney, 1976; Perry et al., 1977; Rossor et al., 1982b). In parallel with these chemical changes, cholinergic neurons situated in basal forebrain and projecting to cerebral cortex and limbic structures were found to be degenerated in SDAT (Whitehouse et al., 1981; Rossor et al., 1982a; Candy et al., 1983; Henke and Lang, 1983).

On the other hand abnormalities of the metabolism of noradrenaline, dopamine, and serotonin (Adolfsson et al., 1979; Cross et al., 1981, 1983; Arai et al., 1984) as well as neuropathological abnormalities in the brainstem monoamine-containing nuclei (Hirano and Zimmerman, 1962; Ishii, 1966; Forno, 1978; Bondareff et al., 1981; Mann and Yates, 1983) had also been demonstrated.

The cholinergic hypofunction in SDAT could cause reduction in monoamines. However, the reductions in monoamines in SDAT are not necessarily linked to the well-established cholinergic hypofunction, but instead might be an independent event that is related to reductions in monoamine-synthesizing enzymes associated with monoamines, due to destruction of monoaminergic neurons. However, there have been only few reports on the monoamine-synthesizing enzymes in SDAT except for dopamine β-hydroxylase (Cross et al., 1981).

We have examined tyrosine hydroxylase (TH), tryptophan hydroxylase (TPH), and their biopterin (BP) cofactor in the brains of normal controls and patients with SDAT in comparison with monoamines and their metabolites (Sawada et al., 1987). TH (Nagatsu et al., 1964) and TPH (Lovenberg et al., 1976; Ichiyama et al., 1970; Friedman et al., 1972) are the rate-limiting monooxygenases for the biosynthesis of catecholamines and serotonin and require a tetrahydropterin as a cofactor. L-erythro-tetrahydrobiopterin (BPH$_4$) is the naturally occurring cofactor for TH (Brenneman and Kaufman, 1964) and TPH (Lovenberg et al., 1967). BPH$_4$ is synthesized from GTP via D-erythro-7,8-dihydroneopterin triphosphate. BPH$_4$ can be measured as total L-erythro-biopterin (BP) after oxidation and D-erythro-7,8-dihydro-neopterin triphosphate as total D-erythro-neopterin (NP) after oxidation and subsequent phosphatase treatment (Nagatsu et al., 1981, 1984).

Materials and methods

Cases

Human brains were obtained postmortem from patients of SDAT, who are diagnosed clinically (Arai et al., 1983) and histologically, and from eight age-matched control subjects without any history of neurological and psychiatric diseases. Mean values of age of death (years), and postmortem intervals (hours) of SDAT patients and controls were: 73.5 and 67.8, and 7.3 and 9.2, respectively. In the postmortem brains of SDAT patients, neuronal loss, Alzheimer's neurofibrillary tangles, and senile plaques were observed in the cerebral cortex, subcortical nuclei, and lower brainstem, and therefore the diseases were histologically diagnosed as advanced stages.

No SDAT patients had been given neuroleptics for at least the last 8 months before death with the exception of one patient who had been given sulpiride (100 mg/day) up to 3 days prior to death. Another SDAT patient had been given benzodiazepine (estazolam) (2 mg/day). None of the patients had received antidepressants or opiates.

Fresh brains were divided midsagittally; one half of each brain was stored at −70 °C in a sealed bag for biochemical assays and the other half was fixed in formalin for neuropathological examination. Brain dissection was performed using a micro-punching technique (Arai et al., 1984).

TH activity was assayed by the method of Nagatsu et al. (1979) using HPLC with electrochemical detection: TPH activity was assayed by the method of Sawada et al. (1985) using HPLC-fluorometric detection. Total L-erythro-BP derived from L-erythro-BPH$_4$ (Nagatsu et al., 1981) and total D-erythro-NP from D-erythro-7,8-dihydro-NP triphosphate (Nagatsu et al., 1984) were determined by specific radio immunoassays. Protein was assayed by the method of Bradford (1976).

Total BP and total NP concentrations were also examined in sera from 30 controls and 9 SDAT patients, shown in Table 2. TH and TPH are not present in serum, but BP and NP are present in serum.

Table 1. Correlation coefficients between tyrosine hydroxylase (TH), tryptophan hydroxylase (TPH), total biopterin (BP), biogenic amines, and their metabolites

	TH	TPH	BP
DA	0.85 [c] (Pallidum ext. segm.)	−	−
HVA	0.84 [c] (Pallidum ext. segm.)	−	0.77 [a] (Subs. nigra) 0.69 [a] (Hippocampus)
5HIAA	−	0.72 [a] (Subs. nigra)	0.62 [a] (Pallidum ext. segm.) 0.76 [b] (Hippocampus)

Significance: [a] $p < 0.05$, [b] $p < 0.01$, [c] $p < 0.001$
DA dopamine; *HVA* homovanillic acid; *5HIAA* 5-hydroxyindoleacetic acid.

Table 2. Total biopterin (BP) and total neopterin (NP) in sera from controls and SDAT patients

	Age	N	BP	NP	NP/BP ratio
			(pmol/ml serum)		
Controls	73± 8 (62−84)	30	12.7±3.8	43.6±14.4	3.41±0.80
SDAT	73±15 (62−95)	9	8.4±3.2 [a]	39.1±14.9	5.15±2.32 [a]

Mean ± SD. Significantly different from controls, [a] $p < 0.05$.

Noradrenaline (NA), dopamine (DA), serotonin (5HT), homovanillic acid (HVA) and 5-hydroxyindoleacetic acid (5HIAA) were determined by HPLC with electrochemical detection (Arai et al., 1984).

Results and discussion

In various brain regions, significant correlation was observed between TH and BP, and between TPH and BP. However no significant correlation was observed between BP and NP. The results suggest some unknown roles of NP in the human brain besides the role as a precursor of BP. Table 1 shows the brain regions with high correlation coefficients between TH and dopamine or homovanillic acid, between TPH and 5-hydroxyindoleacetic acid, and between BP and homovanillic acid or 5-hydroxyindoleacetic acid.

Distribution of TH, TPH, BP, and NP in control brains, and the mean values (percent of control) in brains of SDAT patients are shown in Figs. 1–4.

TH activity (Fig. 1) was high in the substantia nigra, caudate nucleus, putamen, substantia innominata, and nucleus accumbens. High TPH acitvity (Fig. 2) was found in the raphe nucleus. The BP concentration (Fig. 3) was high in the caudate nucleus, whereas the NP concentration (Fig. 4) was high in the locus coeruleus, thalamus, and putamen.

TH activities (Fig. 1) in 12 regions of the brain from SDAT patients were lower than those of the control brains, but the difference was

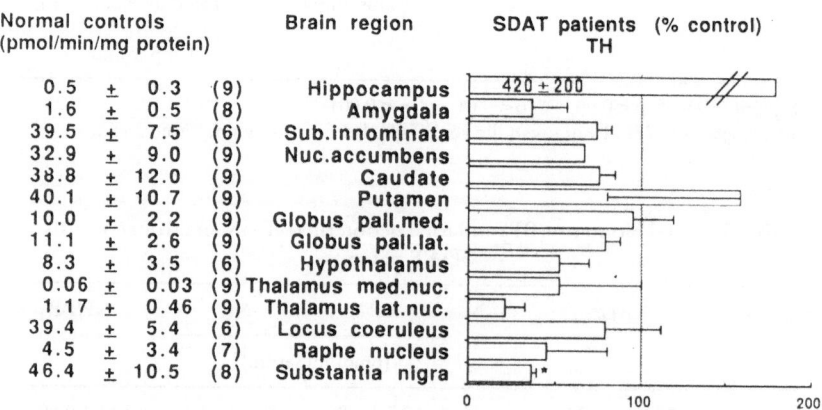

Fig. 1. Distribution of tyrosine hydroxylase (TH) in the brains of normal controls and the mean values of TH in the brains from patients with SDAT. Columns and bars represent mean ± SEM as percent of the mean control values for each region. * p < 0.05

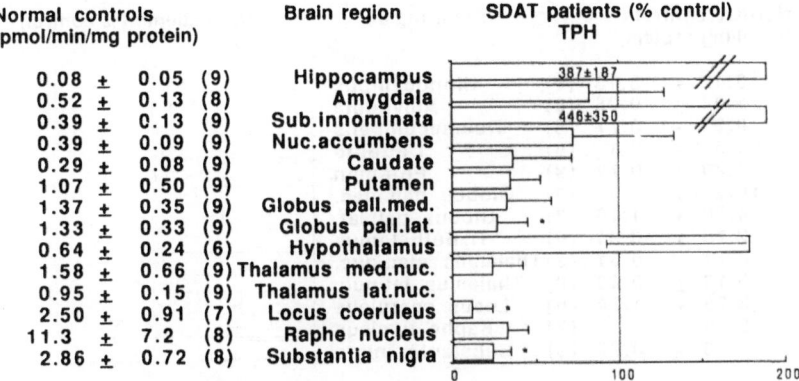

Normal controls (pmol/min/mg protein)	Brain region	SDAT patients (% control) TPH
0.08 ± 0.05 (9)	Hippocampus	387±187
0.52 ± 0.13 (8)	Amygdala	
0.39 ± 0.13 (9)	Sub.innominata	446±350
0.39 ± 0.09 (9)	Nuc.accumbens	
0.29 ± 0.08 (9)	Caudate	
1.07 ± 0.50 (9)	Putamen	
1.37 ± 0.35 (9)	Globus pall.med.	
1.33 ± 0.33 (9)	Globus pall.lat.	*
0.64 ± 0.24 (6)	Hypothalamus	
1.58 ± 0.66 (9)	Thalamus med.nuc.	
0.95 ± 0.15 (9)	Thalamus lat.nuc.	
2.50 ± 0.91 (7)	Locus coeruleus	*
11.3 ± 7.2 (8)	Raphe nucleus	
2.86 ± 0.72 (8)	Substantia nigra	*

Fig. 2. Distribution of tryptophan hydroxylase (TPH) in the brains of normal controls and the mean values of TPH in the brains from patients with SDAT. Columns and bars represent mean ± SEM as percent of the mean control values for each region. * $p < 0.05$

significant only in the substantia nigra. The SDAT values were higher in the hippocampus and putamen than those of controls, but the difference was not significant.

TPH activities (Fig. 2) in 11 regions of the brains from SDAT patients were lower than those of controls, and the values were significantly low in the globus pallidus lateral segment, locus coeruleus, and substantia nigra, and in the hippocampus, substantia innominata, and hypothalamus, the SDAT values were higher than those of controls, but the differences were not significant.

BP levels (Fig. 3) were reduced significantly in the substantia nigra and putamen from SDAT patients, but NP contents (Fig. 4) in various brain regions from SDAT patients did not change significantly.

More significant reduction of TPH than TH supports the previous reports which showed the reduction of serotonin and 5-hydroxyindoleacetic acid (Cross et al., 1983; Arai et al., 1984), and suggests that the serotonergic system might be more widely affected than catecholaminergic systems in SDAT.

In the locus coeruleus BP content decreased significantly, and this may be related to the noradrenergic deficits in the brains of SDAT (Arai et al., 1984).

It is noteworthy that the decreases of TH, TPH, and BP were observed in the substantia nigra.

TH and TPH are not present in serum, but BP and NP are present in serum. Thus, serum BP and NP concentrations were measured in normal controls and patients with SDAT. As shown in Table 2, a significant reduction was observed only in serum BP, and therefore, the

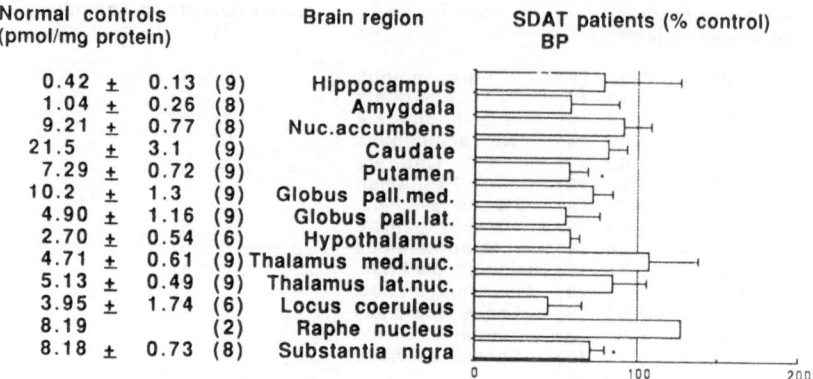

Fig. 3. Distribution of biopterin (BP) in the brains of normal controls and the mean values of BP in the brains from patients with SDAT. Columns and bars represent mean ± SEM as percent of the mean control values for each region. * p < 0.05

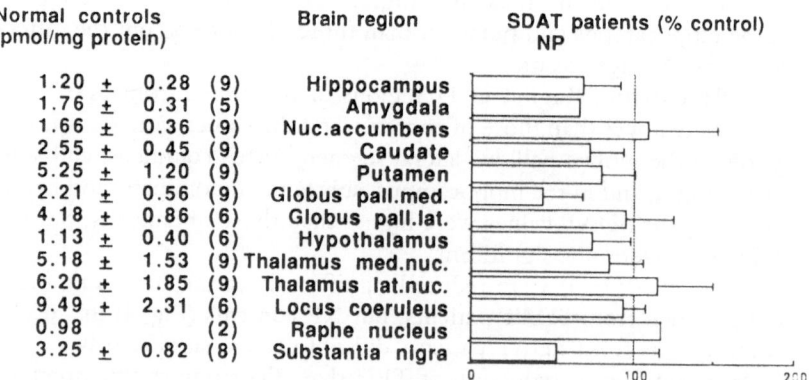

Fig. 4. Distribution of neopterin (NP) in the brains of normal controls and the mean values of NP in the brains from patients with SDAT. Columns and bars represent mean ± SEM as percent of the mean control values for each region

NP/BP ratio was significantly increased in patients with SDAT. The results suggest that BP synthesis from GTP may also be impaired not only in the brain but also in peripheral tissues in patients with SDAT.

The results on the reduction of TH, TPH and BP in SDAT patients suggest that the reduction in monoamines in SDAT may be an event independent of the cholinergic dysfunction and may be related to reduction in monoamine-synthesizing enzymes, due to destruction of monoaminergic neurons.

References

Adolfsson R, Gottfries CG, Roose BE, Winblad B (1979) Changes in brain catecholamines in patients with dementia of Alzheimer-type. Br J Psychiatry 135:216–223

Arai H, Kobayashi K, Ikeda K, Nagao Y, Ogihara R, Kosaka K (1983) A computed tomography study of Alzheimer's disease. J Neurol 229:60–77

Arai H, Kosaka K, Iizuka R (1984) Changes of biogenic amines and their metabolites in postmortem brains from patients with Alzheimer-type dementia. J Neurochem 43:388–393

Bondareff W, Mountjoy CQ, Roth M (1981) Selective loss of neurons of adrenergic projection to cerebral cortex (nucleus locus coeruleus) in senile dementia. Lancet i:783–784

Bowen DM, Smith CB, White P, Davison AN (1976) Neurotransmitter-related enzymes and indices of hypoxia in senile dementia and the abiotrophies. Brain 99:457–496

Bradford MM (1976) A rapid and sensitive method for the quantitation of microgram quantities of protein utilizing the principle of protein-dye binding. Anal Biochem 72:248–254

Brenneman AR, Kaufman S (1964) The role of tetrahydropteridines in the enzymatic conversion of tyrosine to 3,4-dihydroxyphenylalanine. Biochem Biophys Res Commun 17:177–183

Candy JM, Perry RH, Perry EK, Irving D, Blessed G, Fairbairn AF, Tomlinson BE (1983) Pathological changes in the nucleus of Meynert in Alzheimer's and Parkinson's disease. J Neurol Sci 59:227–289

Cross AJ, Crow TJ, Perry RH, Blessed G, Tomlinson BE (1981) Reduced dopamine beta-hydroxylase activity in Alzheimer's disease. Br Med J 282:93–94

Cross AJ, Crow TJ, Johnson JA, Joseph MH, Perry EK, Perry RH, Blessed G, Tomlinson BE (1983) Monoamine metabolism in senile dementia of Alzheimer type. J Neurol Sci 60:383–392

Davies P, Maloney AJ (1976) Selective loss of control cholinergic neurons in Alzheimer's disease. Lancet ii:1430

Forno LS (1978) The locus coeruleus in Alzheimer's disease. J Neuropathol Exp Neurol 37:614

Friedman PA, Kappelman AH, Kaufman S (1972) Partial purification and characterization of tryptophan hydroxylase from rabbit bind brain. J Biol Chem 247:4165–4173

Henke H, Lang W (1983) Cholinergic enzymes in neocortex, hippocampus and basal forebrain of non-neurological and senile dementia of Alzheimer-type patients. Brain Res 267:281–291

Hirano A, Zimmerman HM (1962) Alzheimer's neurofibrillary changes: a topographic study. Arch Neurol 7:227–242

Ichiyama A, Nakamura S, Nishizuka Y, Hayaishi O (1970) Enzymatic studies on the biosynthesis of serotonin in mammalian brain. J Biol Chem 245:1699–1709

Ishii T (1966) Distribution of Alzheimer's neurofibrillary changes in the brain stem and hypothalamus of senile dementia. Acta Neuropathol 6: 181–187

Lovenbeg W, Jequire E, Sjoerdsma A (1967) Tryptophan hydroxylase: measurement in pineal gland, brain stem and carcinoid tumors. Science 155: 217–219

Mann DMA, Yates PO (1983) Serotonin nerve cells in Alzheimer's disease. J Neurol Neurosurg Psychiatry 46: 96–98

Nagatsu T, Levitt M, Udenfriend S (1964) Tyrosine hydroxylase. The initial step in norepinephrine biosynthesis. J Biol Chem 239: 2910–2917

Nagatsu T, Oka K, Kato T (1979) Highly sensitive assay for tyrosine hydroxylase activity by high-performance liquid chromatography. J Chromatogr 163: 247–252

Nagatsu T, Yamaguchi T, Kato T, Sugimoto T, Matsuura S, Akino M, Tsushima S, Nakazawa N, Ogawa H (1981) Radioimmunoassay for biopterin in body fluids and tissues. Anal Biochem 110: 182–189

Nagatsu T, Sawada M, Yamaguchi T, Sugimaoto T, Matsuura S, Akino M, Nakazawa N, Ogawa H (1984) Radioimmunoassay for neopterin in body fluids and tissues. Anal Biochem 141: 472–480

Perry EK, Gibson PH, Blessed G, Perry RH, Tomlinson BE (1977) Neurotransmitter enzyme abnormalities in senile dementia. J Neurol Sci 34: 247–265

Rossor MN, Svendsen C, Hunt SP, Mountjoy CQ, Roth M, Iversen LL (1982a) The substantia innominata in Alzheimer's disease: a histochemical and biochemical study of cholinergic marker enzymes. Neurosci Lett 28: 217–222

Rossor MN, Garrett NJ, Johnson AL, Mountjoy CQ, Roth M, Iversen LL (1982b) A post mortem study of the cholinergic and GABA systems in senile dementia. Brain 105: 313–330

Sawada M, Nagatsu T, Nagatsu I, Ito K, Iizuka R, Kondo T, Narabayashi H (1985) Tryptophan hydroxylase activity in the brains of controls and parkinsonian patients. J Neural Transm 62: 107–115

Sawada M, Hirata Y, Arai H, Iizuka R, Nagatsu T (1987) Tyrosine hydroxylase, tryptophan hydroxylase, biopterin, and neopterin in the brains of normal controls and patients with senile dementia of Alzheimer type. J Neurochem 48: 760–764

Whitehouse PJ, Price DL, Clark AW, Coyle TT, DeLong M (1981) Alzheimer disease: evidence for a selective loss of cholinergic neurons in the nucleus basalis. Ann Neurol 10: 122–126

Correspondence: Dr. T. Nagatsu, Department of Biochemistry, Nagoya University School of Medicine, 65 Tsurumai, Showa-ku, Nagoya 466, Japan.

Postreceptorial enhancement of neurotransmission for the treatment of cognitive disorders

H. Wachtel, R. Horowski, and **P. A. Löschmann**

Research Laboratories of Schering AG, Berlin and Bergkamen, Federal Republic of Germany

Summary

Based on the assumption that the impairment of cognitive functions in normal ageing, depression in the elderly, Morbus Parkinson or Morbus Alzheimer reflects the gradually different injury of the signal propagation in circumscribed neuronal systems due to degenerative and functional deficiencies, it is proposed to strengthen the signalling function of the remaining intact neurons in neurodegenerative disorders via manipulation of the intraneuronal availability of second messengers. This is illustrated for the noradrenergic system where the therapeutic usefulness of selective neuronal cyclic AMP phosphodiesterase inhibitors like rolipram or denbufylline appears particularly promising.

Introduction

The main functional changes of the central nervous system in advanced age are affective disorders and impaired intellectual performance. Thus, the incidence of cognitive impairment in connection with mood disturbances is most frequently observed in normal ageing, depression in the elderly as well as neurodegenerative disorders like Morbus Alzheimer and Parkinson's disease. A constant finding in these conditions is, beside the well known decline in cholinergic neurons in Alzheimer's or dopaminergic cells in Parkinson's disease, the degeneration of noradrenaline (NA) neurons in the locus coeruleus (LC) (Brody, 1976; Mann and Yates, 1979; Yijayashankar and Brody, 1979; Wree et al., 1980; Tomlinson et al., 1981; Mann et al., 1984; Bondareff and Mountjoy, 1986; Chan-Palay, 1989). In the present paper the central NA neurotransmission is taken as an example to demonstrate the pre- and postsynaptic deficits of signal propagation in age-associated and neurodegenerative impairment of cognitive function and to show up the rationale for treatment strategies aimed at the manipulation of the transmission process *beyond* the first messengers.

Physiological aspects

In the mammalian brain the NA innervation of many subcortical and cortical structure involved in the regulation of affect and cognition has its origin in the LC of the brain stem (Lindvall and Björklund, 1983; Levitt et al., 1984; Moore and Card, 1984). The functional significance of the NA system of the LC has been described in numerous reviews; according to Olpe et al. (1985) the functions of the LC can be subdivided into a vegetative-emotional and a cognitive sphere. The former comprises the regulation of affects, the latter comprises the regulation of arousal, vigilance and attention which are important for information processing and memory.

Pathophysiological aspects

Morphological findings

A progressive loss of NA neurons of the LC takes place in the human brain in the course of normal ageing (Brody, 1976; Mann and Yates, 1979; Yijayashankar and Brody, 1979; Wree et al., 1980; Tomlinson et al., 1981; Bondareff and Mountjoy, 1986) and, more marked, in Alzheimer's disease (Tomlinson et al., 1981; Mann et al., 1984) and Morbus Parkinson (Chan-Palay, 1989).

Biochemical findings

In accordance with the loss of LC neurons, reduced NA levels (Robinson, 1975; Adolfsson et al., 1979; Gottfries, 1980; Pradhan, 1980) and reduced activity of catecholamine synthesing enzymes (McGeer, 1978) are found in human cerebral tissue in the course of normal aging, in Alzheimer's disease (Adolfsson et al., 1979; Perry et al., 1981; Iversen et al., 1983) and in Parkinson's disease (Riederer et al., 1977). Studies on peripheral human tissue suggest that there is a reduction in β-receptor mediated adenylate cyclase (AC) activation with reduced cAMP accumulation in old age and in patients with Alzheimer's disease (Ebstein et al.,

1986). Recently a marked decrease of AC activity has been shown in the hippocampus of patients with Morbus Alzheimer (Lemmer et al., 1989).

Electrophysiological findings

In senile animals a dramatic decrease was found in the spontaneous discharge frequency of LC neurons, with a shift towards slow frequencies as compared to juvenile animals (Olpe and Steinmann, 1982). Since the reduction in the average discharge rate of these neurons is presumably accompanied by diminished release of NA, these findings suggest a presynaptic reduction in NA turnover in old age. In addition, in senile animals a reduced responsiveness to NA administered by iontophoresis was found postsynaptically at neurons of the cortex (Jones and Olpe, 1983) and at pyramidal cells of the hippocampus (Bickford-Wimer et al., 1988). It was demonstrated both for the pyramidal cells of the hippocampus and for the NA-sensitive neurons of the cortex that the NA affect is mediated via cAMP as second messenger (Madison and Nicoll, 1986).

Pharmacological aspects

On the basis of the findings described so far, the pathophysiological changes in noradrenergic neurotransmission in age-associated and neurodegenerative disorders are characterised by (a) a *quantitative* aspect (loss of NA neurons of the LC) and (b) a *qualitative* aspect (functional alterations of NA signal propagation). The functional changes affect both the *presynaptic* component of noradrenergic transmission (decrease in the discharge frequence of the remaining LC neurons with reduction of NA turnover) and also the *postsynaptic* component (decreased responsiveness of the effector cell to the first messenger NA and, in addition, diminished formation of the second messenger cAMP due to reduced AC activity).

This combination of morphological and functional pre- and postsynaptic deficits is thought to result in a marked impairment of the function of cortical and limbic projection areas of the LC which are involved in the regulation of cognition and mood. In view of the diminished responsiveness of the effector cells to NA a therapeutic effect of NA agonists seems unlikely. Therefore, pharmacological manipulations should aim at improving the deficient response of the effector cell via mechanisms located *beyond* the first messenger receptors to increase the availability of the second messenger cAMP for an appropriate physio-

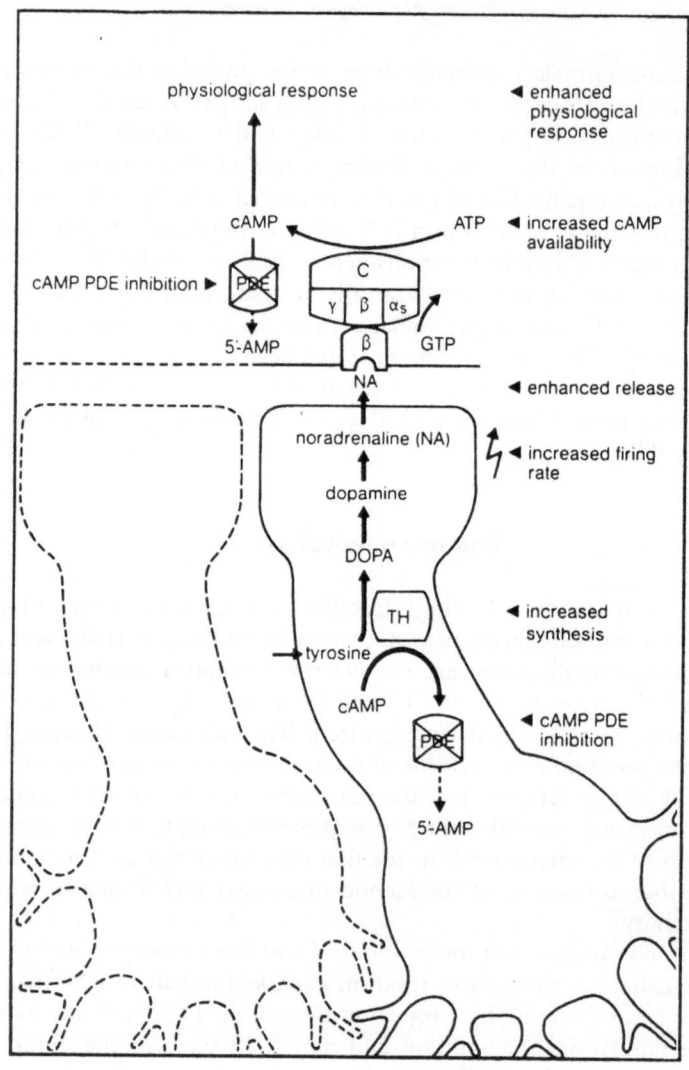

logical response (Fig. 1). Conceivable approaches are direct activation of the catalytic subunit of AC, e.g. by forskolin, to increase the formation of cAMP, or selective inhibition of neuronal cAMP phosphodiesterase (PDE), e.g. by rolipram, to prevent cAMP degradation. In addition to prevention of cAMP catabolism (Wachtel, 1982), rolipram increases the discharge frequency of LC neurons (Scuvée-Moreau et al., 1987), increases NA turnover in the CNS (Kehr et al., 1985) and stereoselectively binds with high affinity to a cAMP PDE isoenzyme predominantly located in hippocampal and cortical areas (Schneider et al., 1986). Rolipram and a stable analogue of cAMP have previously been shown to exert antiamnestic effects in animals (Chute et al., 1981; Randt et al., 1982), indicating the usefulness of this approach for the treatment of age-associated and neurodegenerative impairment of cognitive function. Finally, the lack of anticholinergic effects (Ross et al., 1988; Wachtel et al., 1988) and the clinically established antidepressant activity of rolipram (Horowski and Sastre-y-Hernandez, 1985; Bobon et al., 1988) might be useful additional properties for the management of affective disturbances accompanying cognitive impairment.

Fig. 1. *Effect of rolipram on neurotransmission in neurodegenerative disorders.* Schematic representation of the defects in noradrenergic synaptic neurotransmission in ageing and neurodegenerative disorders and the influence thereon of selective inhibitors of neuronal cAMP phosphodiesterase (PDE), e.g. rolipram. Under *normal* conditions tyrosine is taken up by neurones of the locus coeruleus (LC) which synthesise noradrenaline (NA) via DOPA and dopamine; hydroxylation of tyrosine via tyrosine hydroxylase (TH), an enzyme activated by cAMP-dependent protein kinase, is the rate-limiting step. Depolarisation of the LC neurone releases NA into the synaptic cleft. Activation by NA of postsynaptic β-receptors (β) is transduced via regulatory subunits (α_s, β, γ) to stimulate the catalytic subunit (C) of adenylate cyclase (AC), which forms cAMP from ATP. Part of the cAMP activates cAMP-dependent protein kinases which regulate the cellular response of the effector cell. Conversion of cAMP to $5'$-AMP by PDE is the only known enzymatic mechanism for terminating the action of cAMP; thus PDE plays a crucial role in determining the intensity and duration of the physiological response to the intracellular signal. Defects in signal transmission in *ageing and neurodegenerative disorders* comprise a *presynaptic* component with (a) loss of LC neurons, (b) decreased firing rate of the remaining neurons, (c) decreased NA release and a *postsynaptic* component with (a) reduced dendritic arborisation, (b) decreased NA receptor responsiveness, (c) decreased AC activity, (d) decreased cAMP formation and (e) diminished effector cell response. Defective signal transmission is restored by enhancement of intraneuronal cAMP availability by selective inhibition of cAMP PDE, e.g. by rolipram: *postsynaptically* via prolongation of the second messenger stimulus in the effector cell and *presynaptically* via (a) increased NA synthesis by cAMP-dependent activation of TH, (b) increased firing rate of LC neurons and (c) increased NA release

References

Adolfsson R, Gottfries CG, Roos BE, Windblad B (1979) Changes in the brain catecholamines in patients with dementia of the Alzheimer type. Br J Psychiatry 135:216–223

Bobon D, Breulet M, Gerard-Vandenhove MA, Guiot-Goffioul F, Plomteux G, Sastre-y-Hernandez M, Schratzer M, Troisfontaines B, von Frenckell R, Wachtel H (1988) Is phosphodiesterase inhibition a new mechanism of antidepressant action? A double-blind double dummy study between rolipram and desipramine in hospitalized major and/or endogenous depression. Eur Arch Psychiatry Neurol Sci 238:2–6

Bickford-Wimer PC, Miller JA, Freedman R, Rose GM (1988) Age-related reduction in responses of rat hippocampal neurons to locally applied monoamines. Neurobiol Aging 9:173–179

Bondareff W, Mountjoy CQ (1986) Number of neurons in nucleus coeruleus in demented and non-demented patients: rapid estimation and correlated parameters. Neurobiol Aging 7:297–300

Brody H (1976) An examination of cerebral cortex and brainstem aging. In: Terry RD, Gershon S (eds) Neurobiology of aging. Raven Press, New York, pp 177–181

Chan-Palay V (1989) Alterations in catecholamine neurons in the locus coeruleus in dementias of Alzheimer's and Parkinson's diseases. J Neural Transm [P-D Sect] 1:19–20

Chute DL, Villinger JW, Kirton NF (1981) Testing cyclic AMP mediation of memory: reversal of α-methyl-p-tyrosine-induced amnesia. Psychopharmacology 74:129–131

Ebstein RP, Oppenheim G, Ebstein BS, Amiri Z, Stessman J (1986) The cyclic AMP second messenger system in man: the effects of heredity, hormones, drugs, aluminum, age and disease on signal amplification. Prog Neuropsychopharmacol Biol Psychiatry 10:323–353

Gottfries CG (1980) Biochemistry of dementia and normal aging. Trends Neurosci 3:55–57

Horowski R, Sastre-y-Hernandez M (1985) Clinical effects of the neurotropic selective cAMP phosphodiesterase inhibitor rolipram in depressed patients: global evaluation of the preliminary reports. Current Ther Res 38:23–29

Iversen L, Rossor M, Reynolds G, Hills R, Roth M, Mountjoy C, Foote S, Morrison J, Bloom F (1983) Loss of pigmented dopamine-beta-hydroxylase positive cells from locus coeruleus in senile dementia of Alzheimer's type. Neurosci Lett 39:95–100

Jones RSG, Olpe HR (1983) Altered sensitivity of forebrain neurones. Neurobiol Aging 4:97–99

Kehr W, Debus G, Neumeister R (1985) Effects of rolipram, a novel antidepressant, on monoamine metabolism in rat brain. J Neural Transm 63:1–12

Lemmer B, Ohm T, Bohl J (1989) Reduced basal and stimulated adenylate cyclase activity in postmortem hippocampus of Alzheimer patients. Naunyn-Schmiedebergs Arch Pharmacol 339 [Suppl]:R 106

Levitt P, Rakic R, Goldman-Rakic PS (1984) Comparative assessment of monoamine afferents in mammalian cerebral cortex. In: Descarries L, Reader TR, Jasper HH (eds) Monoamine innervation of cerebral cortex. Alan R Liss, New York, pp 41–60

Lindvall O, Björklund A (1983) Dopamine and noradrenaline-containing neuron systems: their anatomy in rat brain. In: Emson PC (ed) Chemical neuroanatomy. Raven Press, New York, pp 229–255

Madison DV, Nicoll RA (1986) Cyclic adenosine 3',5'-monophosphate mediates β-receptor actions of noradrenaline in rat hippocampal pyramidal cells. J Physiol 372:245–259

Mann DMA, Yates PO (1979) The effects of ageing on the pigmented cells of the human locus coeruleus and substantia nigra. Acta Neuropathol 47:93–98

Mann DMA, Yates PO, Marcyniuk B (1984) A comparison of changes in the nucleus basalis and locus coeruleus in Alzheimer's disease. J Neurol Neurosurg Psychiatry 47:201–203

McGeer EG (1978) Aging and neurotransmitter metabolism in the human brain. In: Katzman R, Terry RD, Bick KL (eds) Alzheimer's disease, senile dementia and related disorders. Raven Press, New York, pp 427–440

Moore RY, Card JP (1984) Noradrenaline-containing neuron systems. In: Björklund A, Hökfelt T (eds) Classical neurotransmitters in the CNS, part I. Handbook of chemical neuroanatomy, vol 2. Elsevier, Amsterdam, pp 123–156

Olpe HR, Steinmann MW (1982) Age-related decline in the activity of noradrenergic neurons of the rat locus coeruleus. Brain Res 251:174–176

Olpe HR, Steinmann MW, Jones RSG (1985) Electrophysiological perspectives on locus coeruleus: its role in cognitive versus vegetative functions. Physiol Psychol 13:179–187

Perry EK, Tomlinson BE, Blessed G, Perry RH, Cross AJ, Crow TJ (1981) Neuropathological and biochemical observations on the noradrenergic system in Alzheimer's disease. J Neurol Sci 51:279–287

Pradhan SN (1980) Central neurotransmitters and aging. Life Sci 26:1643–1656

Randt CT, Judge ME, Bonnet KA, Quartermain D (1982) Brain cyclic AMP and memory in mice. Pharmacol Biochem Behav 17:677–680

Riederer P, Birkmayer W, Seemann D, Wuketich S (1977) Brain noradrenaline and 3-methoxy-4-hydroxyphenylglycol in Parkinson's syndrome. J Neural Transm 41:241–251

Robinson DS (1975) Changes in monoamine oxidase and monoamines with human development and aging. Fed Proc 34:103–107

Ross CE, Toon S, Rowland M, Murray GH, Meya U (1988) A study to assess the anticholinergic activity of rolipram in healthy elderly volunteers. Pharmacopsychiatry 21:222–225

Schneider HH, Schmiechen R, Brezinski M, Seidler J (1986) Stereospecific binding of the antidepressant rolipram to brain protein structures. Eur J Pharmacol 127:105–115

Scuvée-Moreau J, Giesbers J, Dresse A (1987) Effect of rolipram, a phosphodiesterase inhibitor and potential antidepressant, on the firing rate of central monoaminergic neurons in the rat. Arch Int Pharmacodyn Ther 288:43–49

Tomlinson BE, Irving D, Blessed G (1981) Cell loss in the locus coeruleus in senile dementia of Alzheimer type. J Neurol Sci 49:419–428

Vijayashankar N, Brody H (1979) Quantitative study of the pigmented neurons in the nuclei locus coeruleus and subcoeruleus in man as related to aging. J Neuropathol Exp Neurol 38:490–497

Wachtel H (1982) Characteristic behavioral alterations in rats induced by rolipram and other selective adenosine cyclic 3',5'-monophosphate phosphodiesterase inhibitors. Psychopharmacology 77:309–316

Wachtel H, Löschmann PA, Pietzuch P (1988) Absence of anticholinergic activity of rolipram, an antidepressant with a novel mechanism of action, in three different animal models in vivo. Pharmacopsychiatry 21:218–221

Wree A, Braak H, Schleicher A, Zilles K (1980) Biomathematical analysis of the neuronal loss in the aging human brain of both sexes, demonstrated in pigment preparations of the pars cerebellaris loci coerulei. Anat Embryol 160:105–119

Correspondence: Dr. H. Wachtel, Research Laboratories, Schering AG, Müllerstrasse 170–178, D-1000 Berlin 65.

Excitatory dicarboxylic amino acid and pyramidal neurone neurotransmission of the cerebral cortex in Alzheimer's disease

D. M. Bowen and S. L. Lowe

Institute of Neurology, Department of Neurochemistry, London, United Kingdom

Summary

Drug development for Alzheimer's disease has so far had little success, possibly due to failure to correct abnormalities of neurones other than the corticopetal. Most enquiries indicate that cortical inhibitory neurotransmitters are neither selectively nor critically affected. It is highly probable that shrinkage or loss of corticocortical association fibres occurs. This change appears to be circumscribed, clinically relevant and to involve neurotransmitter glutamate. Modulatory domains of the N-methyl-D-aspartate receptor complex are identified as sites at which potential therapeutic agents might be safely directed.

Introduction

Cortical pyramidal neurones were implicated in Alzheimer's original report and an apparently early change was reported by the late Saul Korey as "a decrease in the glutamic acid group" (Korey et al., 1961) in a brain *biopsy* study of Alzheimer's disease (AD). Another pioneer (Corsellis, 1962), on the basis of the widespread histopathology in *autopsy* samples, concluded: "Obviously, therefore, if treatment is to be effective it must be started before the rot sets in" (Corsellis, 1979). While AD is a progressive disorder, this is rarely acknowledged in the interpretation of postmortem biochemical studies, leading some to the pessimistic conclusion that a successful neurotransmitter therapy will not evolve. The disease is often approached optimistically as a genetic trait that is potentially identifiable, treatable and preventable yet recent research fails to establish linkage to chromosome 21 for late onset families. It is possible that an understanding of the early onset familial AD defect at

the molecular level could eventually help identify the biochemical seque-
lae involved in production of common (sporadic) AD, hence leading to
treatment. The discovery of the corticopetal cholinergic abnormality
suggested the basis for a rational pharmacological treatment, although
development has so far had little success. This may be due to failure to
correct abnormalities of neurones other than the corticopetal rather than
the lack of a suitable centrally-acting cholinomimetic drug. It is argued
herein that restoration of function of a second critical neurotransmitter
system may be an achieveable goal. The early biochemical finding of
reduced cerebrospinal fluid homovanillic acid content may now be interpreted in terms of loss or dysfunction of corticofugal projections (as
reviewed Procter et al., 1988). The discovery of the cholinergic nature
of the nucleus basalis of Meynert suggested that cortical perikaryal
pathology in AD involves glutamatergic neurones (Bowen et al., 1983).

Degeneration of corticocortical association fibres

This may be clinically relevant, from results of a number of histological and positron emission topographical (PET) enquiries. Firstly, degeneration of these fibres in AD is indicated by the distribution of
neurofibrillary tangles and senile plaques and loss of columns of pyramidal neurones, based on a shorter cortical ribbon (as reviewed, Procter et
al., 1988, 1989a). Secondly, loss of neurones giving rise to association
fibres appears to occur relatively early in the disease as pyramidal cell
counts (corrected for any cortical thinning) are reduced in cortical layer
III of biopsy tissue from the temporal lobe (Neary et al., 1986). Thirdly,
shrinkage and loss of pyramidal neurones from the temporal and parietal
lobes may be the major cause of cerebral atropy in AD. These regions
show apparent glucose hypometabolism by PET; measurements by this
technique are sensitive to atrophy and hypometabolism is not seen in
vitro; the scanning results may therefore actually index, in part, the early
shrinkage or loss of pyramidal cells giving rise to corticocortical association fibres. This change seems to be clinically relevant because neuropsychological test scores significantly correlate with both scanning
data (as reviewed, Procter et al., 1989) and pyramidal cell counts in layer
III (Neary et al., 1986). It is highly probable, therefore, that in AD
shrinkage or loss of association fibres occurs from circumscribed areas,
mainly the temporal and parietal lobes. Moreover, this change appears
to be clinically relevant.

Review of methods and results

Excitatory dicarboxylic amino acids (EDAAs) are probably neurotransmitters of pyramidal neurones forming corticocortical association fibres

Although quantitative electron microscopic data indicate that pyramidal cells constitute the most abundant neuronal type in the mammalian neocortex, the neurotransmitter(s) associated with this cell type has not been established unequivocally. Major candidates are EDAAs, principally L-aspartate and L-glutamate. There is a variety of evidence for this, largely from studies of the corticostriatal pathway. Firstly, selective retrograde labelling of cortical pyramidal neurones has been demonstrated following injection of D-[³H]-aspartate (a non-metabolizable 'false' neurotransmitter) into rat neostriatum. Secondly, the neostriatum responds to electrophysiological stimulation of the cortex and this 'signal' can be mimicked by application of EDAAs to the neostriatum and blocked by EDAA antagonists. Thirdly, a large number of both in vitro and in vivo studies have shown reduced EDAA neurotransmitter function in the neostriatum following ablation or undercutting of the frontal cortex. Finally, although assessment of EDAA neurotransmitter function in response to stimulation of the cortex has been problematical, increased extracellular concentrations of radiolabelled glutamate and endogenous glutamate and aspartate in the neostriatum has been shown to be evoked by both electrical and chemical stimulation of the cortex (as reviewed, Palmer et al., 1989). This is supported by immunochemical localization to pyramidal neurones in layer V of phosphate activated glutaminase, albeit one of several enzymes which catalyze glutamate synthesis, and of glutamate itself (see Akiyama et al., 1989). The possibility of corticocortical glutamatergic pathways is particularly difficult to establish, in part as a technique has not been developed for destroying pyramidal neurones selectively in the cerebral cortex of experimental animals. Human as well as rat cerebral cortex has a high-affinity uptake system for D-[³H]-aspartate and shows Ca^{2+}-dependent, K^+-stimulated release of aspartate and glutamate (see Peinado and Mora, 1986; Procter et al., 1988). These glutamatergic terminals could be thalamocortical projections (Peinado and Mora, 1986), the other major source of excitatory cortical fibres. Thalamic pathology in AD is less prominent than cortical changes (e.g., Najlerahim and Bowen, 1988). D-[³H]-aspartate uptake is reduced by 50 percent in temporal cortex from promptly performed autopsies of AD patients, using a technique designed to minimise influences of artifacts and epiphenomena (Procter et al., 1988). It is highly probable, therefore, that a large proportion of the glutamatergic terminals in the cerebral cortex are of corticocortical fibres.

Glutamate content of the grey matter from biopsy samples of AD patients is also reduced (Procter et al., 1988) and the value for individual subjects relates to pyramidal neurone density in layer III (Lowe and Bowen, 1990). The most straightforward explanation is that loss of corticocortical association fibres has had a major influence on the glutamate content of these samples. These results provide new evidence that an EDAA, probably L-glutamate, is the principal neurotransmitter of corticocortical association fibres.

EDAA neurotransmission and magnitude of dementia

No study has so far been reported that seeks to directly relate an index of glutamatergic neurotransmission to the magnitude of dementia. There is, however, evidence that EDAA neurotransmitter dysfunction relates to memory and behavioural disturbance. Firstly, such dysfunction may be prominent in critical structures, the hippocampus (Hyman et al., 1987) and amygdala (Francis et al., 1987). Secondly, the histological and PET enquiries, loss of tissue glutamate and reduced D-aspartate uptake in AD, suggest that in AD degeneration occurs of a proportion of glutamatergic corticocortical neurones and that this is clinically relevant. Thirdly, the presence in the cortex of pyramidal cells with synaptic targets of other pyramidal neurones and the abundance of both cortical L-glutamate binding sites and pyramidal cells suggests that EDAA neurotransmission is critical for sustaining the activity of corticocortical neurones (as reviewed Peinado and Mora, 1986; Procter et al., 1988). Finally, the primary agonist site of the N-methyl-D-aspartate (NMDA) receptor complex is the best characterized binding site for L-glutamate and animal studies implicate the complex in memory and behaviour (i.e., experience-dependent changes in kittens visual cortex, initiation of long-term potentiation, acquisition of place learning in rodents and storage of information during learning in chicks; as reviewed, McCabe and Horn, 1988).

Preserved inhibitory cortical neurotransmitters

The majority of cortical interneurones stain with antisera against gamma-aminobutyric acid (GABA) or its biosynthetic enzyme glutamic acid decarboxylase (GAD). Thus, GABA constitutes a major inhibitory transmitter system in the cortex, accounting for as many as 30 percent of all neurones. Assessment postmortem has been complicated by *artefact and epiphenomena*. Large reductions in cortical GAD activity determined postmortem in AD were originally published alongside some of the first reports of diminished choline acetyltransferase activity. However, detailed studies carried out then, of brain tissue obtained from diagnostic craniotomies and experimental animals, indicated that loss of GAD activity alone was attributable to the terminal hypoxia associated with protracted death (as reviewed, Lowe et al., 1988). Thus, no change in GAD activity was found in a recent study where AD and control subjects were carefully matched for agonal state (Reinikainen et al., 1988). Tissue atrophy is another factor which may confound interpretation of data. This is because the practice of reporting results relative to unit mass does not make allowance for any reduction in volume of brain structure. Shrinkage or loss of some structures, but not others, may lead to the reporting of an increase in the markers of unaffected structures, such as increased GABA content of frontal cortex from AD biopsy tissue (Lowe et al., 1988).

The reductions found in GABA content of most postmortem series are less substantial and widespread than those reported for a group of subjects that included only pathologically severe examples of the disease. Large and widespread reductions in uptake sites of GABAergic neurones have also been reported postmortem. However, since this was based upon active uptake determinations in tissue that had been frozen, thawed and subfractionated it is difficult to exclude the possibility that inappropriate preparations were produced in disease-affected tissue. Indeed, others find preservation of this uptake site (assayed using a ligand binding technique) in all regions examined, except the temporal cortex. Moreover, GAD activity as well as GABA content of cortical biopsy tissue are not reduced (as reviewed, Lowe et al., 1988) and an attempt at treating AD with a GABA agonist was unsuccessful (Mohr et al., 1986). Glycine and taurine are also thought to function as inhibitory neurotransmitters and content is unchanged in AD, based on both postmortem (Ellison et al., 1986; Perry et al., 1987; Lowe SL, Bowen DM, unpublished) and biopsy tissue (Lowe and Bowen, 1990). Most enquires indicate therefore, that in AD cortical inhibitory neurones are neither selectively nor critically affected.

Discussion

AD is a fatal disorder with widespread social and economic implications so a therapy is urgently required. Tacrine (1,2,3,4-tetrahydro-9-aminoacridine) might be a modestly effective treatment. Various pharmacological paradigms related to EDAA neurotransmission have been shown in rodents to be modified both in vivo and in vitro by the drug. These effects are only observed with high concentrations so it is unlikely that tacrine acts through EDAA neurotransmission in humans; if tacrine is efficacious in AD patients it is likely to be through brain cholinergic systems (Steele et al., 1989 a).

Treatment of corticocortical neurone dysfunction

A proportion of glutamatergic corticocortical neurones in the temporal and parietal lobes, that probably receive an extensive EDAA input, seem to degenerate in AD. Is this due to excitotoxicity? Does this cause changes in memory and behaviour? Can the activity of the remaining neurones be safely restored and protected with a drug active towards the NMDA-receptor complex? Will this provide clinical benefit? These questions cannot be answered without further research but two modulatory domains of the NMDA receptor complex have already been identified at which potential therapeutic agents might be directed. The extent of occupancy of the modulatory domains by glycine, spermidine and any other endogenous compounds, as well as the magnitude of neuronal loss, are important factors that will determine the success of this therapeutic approach. However, a number of properties of the complex remain unchanged suggesting that pharmacological manipulation of either the polyamine (Carter et al., 1989) or glycine (Procter et al., 1989 a) domain with a partial agonist may provide an approach for restoring and protecting the activity of the remaining pyramidal neurones. Firstly, the rank order of binding of different amino acids to the glycine domain is probably not different between AD and control. Secondly, the rank order of potency of $(+)$MK-801, $(-)$MK-801, TCP, SKF10047 and dextrorphan to displace [3H]MK-801 is also not different (Procter et al., 1989 a), as are the IC_{50} values for inhibition by Zn^{2+} of [3H]MK-801 binding under various conditions (Steele et al., 1989 b). Thirdly, the primary agonist (i.e., glutamate) recognition site may also not be different between AD and control (Procter et al., 1989 a). There is, however, an apparent disease-related reduction in efficacy of glycine at promoting binding of NMDA-receptor-ionophore ligands (Procter et al., 1989 b).

Acknowledgements

This work would not have been possible without those that helped in the collection and classification of samples, in particular, Professor J. A. N. Corsellis and Drs. M. M. Esiri, A. W. Procter and D. Neary. Supported in part by Astra Alab and the Brain Research Trust ("Miriam Marks Department").

References

Akiyama H, McGeer PL, Itagaki S, McGeer EG, Kaneko T (1989) Loss of glutaminase-positive cortical neurones in Alzheimer's disease. Neurochem Res 14:353–358

Bowen DM, Allen SJ, Benton JS, Goodhardt MJ, Haan EA, Palmer AM, Sims NR, Smith CCT, Spillane JA, Esiri MM, Snowden JS, Wilcock GK, Davison AN (1983) Biochemical assessment of serotonergic and cholinergic dysfunction and cerebral atrophy in Alzheimer's disease. J Neurochem 41:266–272

Carter C, Rivy JP, Scatton B (1989) Ifenprodil and SL 82.0715 are antagonists at the polyamine site of the N-methyl-D-aspartate (NMDA) receptor. Eur J Pharmacol 164:611–612

Corsellis JAN (1962) Mental illness and the ageing brain, Maudsley Monograph 9. Oxford University Press, London

Corsellis JAN (1979) Alzheimer's disease: early recognition of potentially reversible deficits. In: Glen AIM, Whalley LD (eds) Alzheimer's disease. Early recognition of potentially reversible deficits. Churchill Livingstone, Edinburgh, pp 6

Ellison DW, Beal MF, Mazurek MF, Bird ED, Martin JB (1986) A postmortem study of amino acid neurotransmitters in Alzheimer's disease. Ann Neurol 20:616–621

Francis PT, Pearson RCA, Lowe SL, Neal JN, Stephens PH, Powell TPS, Bowen DM (1987) The dementia of Alzheimer's: an update. J Neurol Neurosurg Psychiatry 50:242–243

Hyman BT, Van Hoesen GW, Damansio AR (1987) Alzheimer's disease: glutamate depletion in the hippocampal perforant pathway zone. Ann Neurol 22:37–40

Korey SR, Scheinberg L, Terry R, Stein A (1961) Studies in presenile dementia. Trans Am Neurol Assoc 86:99–102

Lowe SL, Francis PT, Procter AW, Palmer AM, Davison AN, Bowen DM (1988) Gamma-aminobutyric acid concentration in brain tissue at two stages of Alzheimer's disease. Brain 111:785–799

Lowe SL, Bowen DM (1990) Glutamic acid concentration in brains of patients with Alzheimer's disease. Biochem Soc Trans (in press)

McCabe BJ, Horn G (1988) Learning and memory: regional changes in N-methyl-D-aspartate receptors in the chick brain. Proc Natl Acad Sci (USA) 85:2849–2855

Mohr E, Bruno G, Foster N, Gillespie M, Cox C, Hare TA, Tamminga C, Fedio P, Chase TN (1986) GABA-agonist therapy for Alzheimer's disease. Clin Neuropharmacol 9:257–263

Najlerahim A, Bowen DM (1988) Regional weight loss of the cerebral cortex and some subcortical nuclei in senile dementia of the Alzheimer type. Acta Neuropathol (Berl) 75:509–512

Neary D, Snowden JS, Mann DMA, Northern B, Bowen DM, Sims NR, Yates PO, Davison AN (1986a) Alzheimer's disease: a correlative study. J Neurol Neurosurg Psychiatry 49:229–237

Perry TL, Young VW, Bergeron C, Hansen S, Jones K (1987) Amino acids, glutathione and glutathione transferase activity in the brains of patients with Alzheimer's disease. Ann Neurol 21:331–336

Procter AW, Palmer AM, Francis PT, Lowe SL, Neary D, Murphy E, Doshi R, Bowen DM (1988) Evidence of glutamatergic denervation and possible abnormal metabolism in Alzheimer's disease. J Neurochem 50:790–802

Procter AW, Wong EHF, Stratmann GC, Lowe SL, Bowen DM (1989a) Reduced glycine stimulation of [^{3}H]MK801 binding in Alzheimer's disease. J Neurochem 53:698–704

Procter AW, Stirling JM, Stratmann GC, Cross AJ, Bowen DM (1989b) Loss of glycine-dependent radioligand binding to the N-methyl-D-aspartate-phencyclidine receptor complex in patients with Alzheimer's disease. Neurosci Lett 101:62–66

Palmer AM, Hutson PH, Lowe SL, Bowen DM (1989) Extra-cellular concentrations of aspartate and glutamate in rat neostriatum following chemical stimulation of frontal cortex. Exp Brain Res 75:659–663

Peinado JM, Mora F (1986) Glutamic acid as a putative transmitter of the interhemispheric corticocortical connections in the rat. J Neurochem 47:1598–1603

Reinikainen KJ, Paljarvi L, Huuskonen M, Soininen H, Laakso M, Riekkinen PJ (1988) A postmortem study of noradrenergic, serotonergic and GABAergic neurones in Alzheimer's disease. J Neurol Sci 84:101–116

Steel JE, Palmer AM, Lowe SL, Bowen DM (1989a) The influence of tetrahydro-9-aminoacridine in excitatory amino acid neurotransmission in vivo and in vitro. Br J Pharmacol 96:353

Steel JE, Palmer HM, Stratmann GC, Bowen DM (1989b) The N-methyl-D-aspartate receptor complex in Alzheimer's disease: reduced regulation by glycine but not zinc. Brain Res 500:369–373

Correspondence: Dr. D. M. Bowen, Institute of Neurology, Department of Neurochemistry, 1, Wakefield Street, London, WC1N 1PJ, United Kingdom.

The sequence within the two polyadenylation sites of the A4 amyloid peptide precursor stimulates the translation

F. de Sauvage[1], V. Kruys[2], and J. N. Octave[1]

[1] Laboratoire de Neurochimie, Université Catholique de Louvain
[2] Université Libre de Bruxelles, Belgium

Summary

cDNA probes specific for the A4 amyloid peptide precursor hybridize with a 3.2–3.4 kb mRNA doublet which can be attributed to the use of two polyadenylation sites. Different chimeric mRNAs were synthesized by in vitro transcription of the coding region of the chicken lysozyme or the chloramphenicol acetyl transferase followed by two 3' untranslated regions of the A4 amyloid peptide precursor mRNA using the two possible polyadenylation sites. In vivo translation of these mRNA constructs in Xenopus oocytes indicates that the long mRNAs using the second polyadenylation site produce higher amounts of proteins as compared to the short mRNAs. This effect on the translation is not related to a higher stability of the long mRNA in oocytes.

Introduction

Senile plaques, found in the brain of patients with Alzheimer's disease (AD) contain an amyloid core from which a 4.2 kd A4 peptide was isolated as a major constituent (Masters et al., 1985). Isolation and sequence of cDNA clones indicated that the A4 peptide is obtained from a larger A4 amyloid peptide precursor (A4 APP) (Kang et al., 1987). cDNA probes corresponding to the A4 APP hybridize to a 3.2–3.4 kb mRNA doublet expressed in normal brain and other tissues. Our results show that this doublet corresponds to the use of two polyadenylation sites. Furthermore, the sequence contained between them increases the translation of reporter genes like those of chicken lysozyme or chloramphenicol acetyl transferase (CAT).

Material and methods

Northern blot analysis

Total RNA was prepared by the guanidinium/CsCl method (Chirgwin et al., 1979). Ten µg of glyoxal denaturated RNA were fractionated on 1% agarose gel, in 10 mM sodium phosphate buffer (pH 7), as described by Thomas (1980), and transfered on Hybond N membranes (Amersham). Prehybridization (4 h) and hybridization with the probes (16 h) were carried out at 45 °C or 65 °C in 5 × SSC, 50% formamide, 10 × Denhaerdt, 0.1% SDS.

Injection into oocytes

Oocytes were injected with 50 nl of mRNA dissolved in water and adjusted to a concentration of 0.1 mg/ml and incubated for 6 hours at 18 °C in Barth's medium (0.01 ml per oocyte) containing [^{35}S] methionine (9 µci per oocyte), 10% bovin serum albumin, 1% trasylol. Oocytes were lysed and immunoprecipited with an anti-chicken lysozyme antibody according to the method descibed by Huez et al. (1983). Analysis of immunoprecipited proteins was performed by 20% polyacrylamide/NaDodSO$_4$ gel electrophoresis.

CAT assay

The CAT activity in oocytes injected with the chimeric CAT constructs was measured by the usual procedure (Gorman et al., 1982).

Analysis of mRNA stability

Total RNA was extracted from a batch of 10 oocytes at different incubation times after injection. The amount corresponding to 2 oocytes was submitted to Northern blot analysis using a probe specific for the lysozyme mRNA.

Results

Northern blot analysis

An anti-sense riboprobe encompassing 50 nucleotides between the two polyadenylation sites of the 3' untranslated region (UTR) of the A4 APP was used in Northern blot analysis. The results presented in Fig. 1 indicate that it recognizes only the 3.4 kb band instead of the two bands recognized by a 1 kb riboprobe corresponding to the EcoR1 fragment encoding part of the A4 APP precursor (Kang et al., 1987).

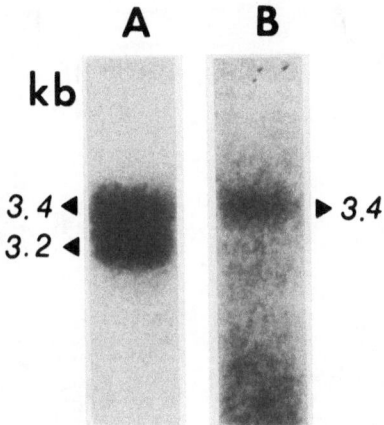

Fig. 1. Northern blot analysis of human brain RNA. The anti-sense riboprobes used for the hybridization were the 1 kb A4 APP riboprobe (A) or the 50 nucleotides riboprobe encompassing the sequence between the two polyadenylation sites (B)

In vivo translation of chimeric mRNAs

Since antibodies able to immunoprecipitate the A4 APP were not available, two 3' UTR of the A4 APP mRNA using the two possible polyadenylation sites were subcloned downstream the coding region of the chicken lysozyme into a SP64 plasmid. When the chimeric messengers obtained by in vitro transcription were translated for 6 h in Xenopus oocytes, the mRNA containing the long 3'UTR had synthesized more protein than the mRNA containing the short 3'UTR (Fig. 2 A). The scanning of the autoradiogram indicates that the long mRNA produces about three times more protein than the short one.

In order to exclude a possible role of the coding region on the stimulation of the translation, other mRNA constructs were obtained by inserting the two different 3' UTR of the A4 APP downstream the sequence coding for the CAT. After 6 h of in vivo translation in Xenopus oocytes, the CAT activity (Fig. 2 B) was two times more abundant in cells microinjected with the mRNA containing the long 3' UTR of the A4 APP.

Effect of the 3' UTR on the stability of the chimeric RNAs

The difference in translational efficiency described above could have two explanations. Either the 3' UTR directly affects the translation, or

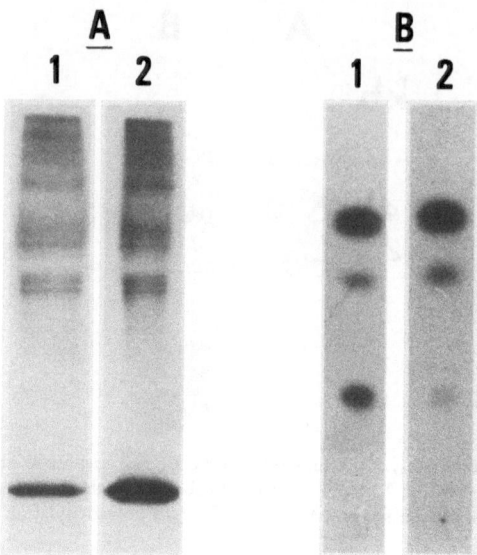

Fig. 2. Autoradiograms obtained after 6 h of in vivo translation of the lysozyme mRNA construct (A) or the CAT mRNA construct (B). In (A), the labelled proteins obtained after translation of the mRNA construct containing the short (1) or the long (2) 3′ UTR of the A4 APP mRNA were immunoprecipited by an anti-lysozyme antibody and analyzed by SDS PAGE. In (B), the CAT activity was measured after translation of the CAT mRNA construct with the short (1) or the long (2) 3′ UTR of the A4 APP mRNA

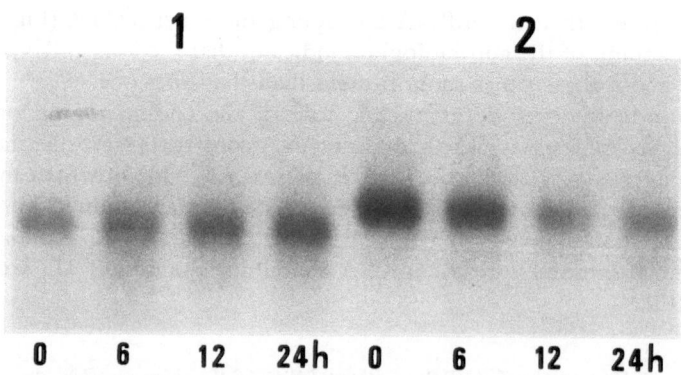

Fig. 3. Analysis of the stability of the chimeric RNAs. Northern blot analysis using a lysozyme specific cDNA probe was performed on RNA isolated from oocytes injected for 0, 6, 12 or 24 h with the lysozyme mRNA constructs containing the short (1) or the long (2) 3′ UTR of the A4 APP

the short mRNA is degraded more rapidly in oocytes. In order to test the latter hypothesis, the stability of the lysozyme chimeric RNAs in oocytes was studied by Northern blot analysis at different time intervals after microinjection. Figure 3 shows that mRNA with the short 3' APP UTR was stable over a period of at least 24 h, whereas partial degradation of mRNA with the long 3' UTR was already observed 6 h after the microinjection of the cells.

Discussion

We report herein that the 3.2–3.4 kb mRNA doublet expressed in normal brain and other tissues is due to alternative use of two polyadenylation sites at the 3' untranslated region of the A4 APP mRNA. Furthermore, the sequence contained between the 2 polyadenylation sites increases the translation of the lysozyme and CAT reporter genes. By measuring the stability of the chimeric RNAs injected into oocytes, we have observed that the more efficiently translated mRNAs are not more stable, indicating that the increased production of protein is not related to the mRNA stability but to their increased translation.

An increase of some forms of the A4 APP mRNA has been reported in small AD brain areas but it remains difficult to show significant increase of A4 APP mRNA in brain areas where the senile plaques are found (Bahmanyar et al., 1987; Cohen et al., 1988; Lewis et al., 1988; Goedert, 1988; Palmert et al., 1988). Whether the proportion of the 3.4 kb A4 APP mRNA is greater in brain areas containing plaques awaits Northern blots analysis and in situ hybridization experiments.

References

Bahmanyar S, Higgins GA, Godgaber D, Lewis DA, Morrison JH, Wilson MC, Shankar SK, Gajdusek DC (1987) Localization of amyloid β protein messenger RNA in brains from patients with Alzheimer's disease. Science 237:77–80

Chirgwin JM, Przybyla AE, MacDonald RJ, Rutter WJ (1979) Isolation of biological active ribonucleic acid from sources enriched in ribonuclease. Biochemistry 18:5294–5299

Cohen ML, Golde TE, Usiak MF, Younkin LH, Younkin SG (1988) In situ hybridization of nucleus basalis neurons shows increased β-amyloid mRNA in Alzheimer disease. Proc Natl Acad Sci USA 85:1227–1231

Goedert M (1988) Neuronal localization of amyloid beta protein precursor mRNA in normal human brain and in Alzheimer's disease. EMBO J 6:3627–3632

Gorman CM, Moffat LF, Howards BH (1982) Recombinant genomes which express chloramphenicol acetyl transferase in mammalian cells. Mol Cell Biol 2:1044–1051

Huez G, Cleuter Y, Bruck C, Van Vloten-Doting L, Goldbach R, Verdiun B (1983) Translational stability of plant viral RNAs microinjected into living cells; influence of a 3'-poly(A) segment. Eur J Biochem 130:205–209

Kang J, Lemaire HG, Unterbeck A, Salbaum MJ, Masters CL, Grzeschik KH, Multhaup G, Beyreuther K, Muller-Hill B (1987) The precursor of Alzheimer's disease amyloid A4 protein resembles a cell surface receptor. Nature 325:733–736

Lewis DA, Higgins GA, Young WG, Goldgaber D, Gajdusek DC, Wilson MC, Morrison JH (1988) Distribution of precursor amyloid-β-protein messenger RNA in human cerebral cortex: relationship to neurofibrillary tangles and neuritic plaques. Proc Natl Acad Sci USA 85:1691–1695

Masters CL, Simms G, Weinman NA, Multhaup G, McDonald BL, Beyreuther K (1985) Amyloid plaque core protein in Alzheimer's disease and Down syndrome. Proc Natl Acad Sci USA 82:4242–4249

Palmert MR, Golde TE, Cohen ML, Kovacs DM, Tanzi RE, Gusella JF, Usiak MF, Younkin LH, Younkin SG (1988) Amyloid protein precursor messenger RNAs: differential expression in Alzheimer's disease. Science 241:1080–1084

Strickland S, Huarte J, Belin D, Vassalli A, Rickles RJ, Vassalli JD (1988) Antisense RNA directed against the 3' noncoding region prevents dormant mRNA activation in mouse oocytes. Science 241:680–684

Thomas PS (1980) Hybridization of denatured RNA and small DNA fragments transferred to nitrocellulose. Proc Natl Acad Sci USA 77:5201–5205

Correspondence: Dr. J. N. Octave, Laboratoire de Neurochimie, UCL 1352, Avenue Hippocrate 10, B-1200 Bruxelles, Belgium.

Alzheimer-like changes of cortical amino acid transmitters in elderly Down's syndrome

G. P. Reynolds[1], K. G. Mercer[1], C. E. J. Warner[1], J. S. Lowe[1], and D. M. A. Mann[2]

[1] Department of Pathology, University of Nottingham Medical School, Queen's Medical Centre, Nottingham
[2] Department of Pathology, University of Manchester, Manchester, United Kingdom

Summary

The concentrations of glutamate and GABA were determined in samples of frontal and temporal cortex and hippocampus taken post mortem from nine patients with Down's syndrome and nine control subjects. These results were correlated with neuropathological findings and compared to changes found in patients with Alzheimer's disease. GABA demonstrated an Alzheimer-like reduction in the temporal cortex and hippocampus, although this was not found in the frontal cortex. These deficits correlated with cell density but not plaque or tangle counts. Glutamate was significantly reduced only in the hippocampus, but found not to be changed in Alzheimer's disease.

Introduction

An invariable feature of the brain in Down's syndrome (DS) is the eventual development of the neuropathology of Alzheimer's disease (AD) in patients living beyond middle age. These histopathological abnormalities, neurofibrillary tangles and cortical neuritic plaques, are found in parallel with the neurotransmitter losses and other biochemical indicators of neuronal degeneration which also occur in AD. Thus cortical, and some subcortical, deficits in choline acetyltransferase occur in adult DS (Yates et al., 1980; Godridge et al., 1987) along with losses of noradrenaline and 5-hydroxytryptamine (Reynolds and Godridge, 1985; Godridge et al., 1987). Neuropathological studies (e.g. Mann et al., 1985 b) have indicated that these neurotransmitter changes reflect

cell losses in the subcortical nuclei containing the corresponding neuronal cell bodies. In AD, the amino acid transmitters GABA and glutamate have also been reported to be diminished (Rossor et al., 1984;
Sasaki et al., 1986), as have their uptake sites (Simpson et al., 1988;
Hardy et al., 1987), which presumably reflects losses of other neuronal
groups, notably in the cortex. In a preliminary study (Reynolds and
Warner, 1988) of five DS cases we have shown deficits of hippocampal
glutamate and indications of GABA losses in temporal cortex and
hippocampus. We report here how these amino acids are affected in a
larger series of patients with DS, how they compare with changes in AD
and how they relate to indicators of neurodegeneration.

Methods

We have measured the concentrations of GABA and glutamate in postmortem tissue from nine cases of DS (age 34–65, mean 53 y) and nine control
subjects (age 35–63, mean 55 y) matched for age and post-mortem delay. In
addition, tissue samples from a series of patients with a neuropathological
diagnosis of AD (mean age 76 y, n = 8) and an appropriate control group (mean
age 68 y, n = 10) were similarly studied. The amino acids were measured after
separation of o-phthalaldehyde derivatives by HPLC with electrochemical detection as previously described (Reynolds and Warner, 1988). A further series of
nine DS cases (age 38–72, mean 58 y), containing four from the above series,
underwent neuropathological investigation by one of us (DMAM). This was
performed by formalin fixation of one cerebral hemisphere and measurement of
neuritic plaque, neurofibrillary tangle and neuronal densities using standard
histological techniques (Mann et al., 1985 a). Neurochemical measurements were
performed on the other hemisphere to permit correlation of transmitter concentrations with these indicators of Alzheimer-like neurodegeneration. Non-parametric statistics were used throughout.

Results

Figure 1 shows that there is a substantial and highly significant deficit
in hippocampal glutamate in DS (mean 962 vs. 1397 µg/g for controls),
which is not found in the other cortical regions studied. We observed an
apparent decrease in glutamate concentrations with age in both control
subjects and DS patients in the hippocampus, although the results indicated a glutamate deficit in even the younger DS patients who did not
exhibit plaques or tangles. In this study glutamate exhibited no reduction in the tissue from AD patients although in the hippocampus and
temporal cortex the AD series shows significantly greater variances
($F = 3.48$, 4.29 respectively, $p < 0.05$ in each case).

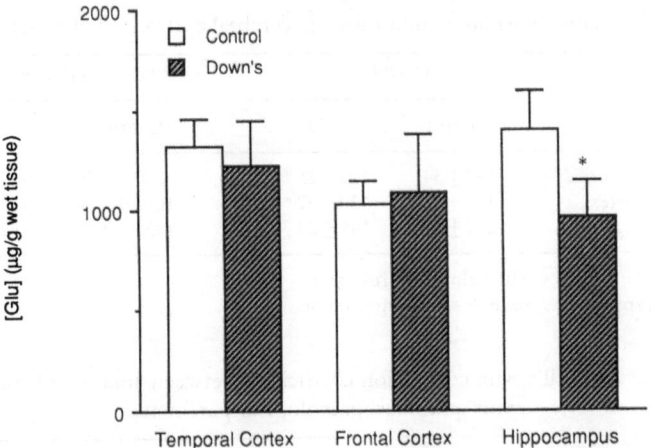

Fig. 1. Glutamate in brain tissue in Down's syndrome; * p < 0.05

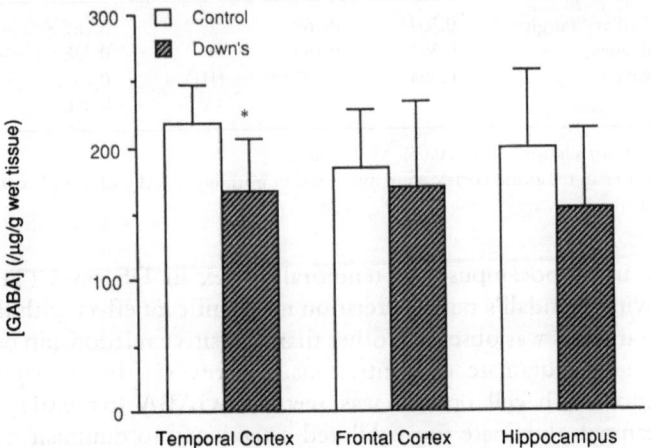

Fig. 2. GABA in brain tissue in Down's syndrome; * p < 0.05

GABA (Fig. 2) showed a significant decrease only in the temporal cortex (mean 168 vs. 218 µg/g for controls) although its concentration was diminished by a similar amount (23%) in the hippocampus, while in AD (Table 1), GABA is reduced significantly in both neocortical areas, although the hippocampus exhibits a similar (if non-significant) deficit.

In order to understand the possible relationship between these results and the neuropathological changes in DS we have compared amino acid concentrations with senile plaques, neurofibrillary tangles and neuronal

Table 1. Neurotransmitter amino acids in cerebral cortex in Alzheimer's disease

	GABA		Glutamate	
	Control	AD	Control	AD
Temporal cortex	187±34	122±25**	1247±216	1512±447
Frontal cortex	179±52	130±22*	1092±185	1273±282
Hippocampus	208±52	149±23	1455±301	1479±562

* $p < 0.05$;   ** $p < 0.01$ below control values.
Values expressed as mean ± s.d. in µg/g tissue

Table 2. Kendall's rank correlation coefficients between amino acid concentrations and neuropathological parameters

	Hippocampus			Temporal cortex	
	Glutamate	GABA		Glutamate	GABA
Neurofibrillary tangles	0.507*	0.169		0.167	−0.222
Senile plaques	0.394	0.169		0.338	−0.338
Cell density	0.394	0.620**	(III)	0.167	0.444*
			(V)	−0.056	0.222

Significant correlations: * $p < 0.05$; ** $p < 0.02$.
Cell densities in temporal cortex were measured in both layers III and V and correlated as indicated

density in hippocampus and temporal cortex in DS cases (Table 2). Employing Kendall's rank correlation no significant effect with the first two parameters was observed, other than a positive relationship between tangles and glutamate concentrations. However a highly significant correlation with cell density was seen for GABA ($p = 0.011$), while hippocampal glutamate too exhibited a tendency to diminish with decreasing cell density ($p = 0.07$).

Discussion

It would appear from these preliminary results that hippocampal glutamate is more affected in DS than in most cases of AD, although the difference in variance shows that at least some of the AD cases have abnormal glutamate levels. The reports mentioned above relating to glutamatergic loss in AD have not always been consistently reproduced; certainly there have been studies reporting no loss of glutamate in AD cortex (Gramsbergen et al., 1987). This may relate to age or other differences between the groups studied or that glutamate concentrations provide a poorer assessment of glutamatergic integrity than, say, measurement of glutamate uptake sites. Nevertheless it is likely that the decrease in glutamate concentrations in DS does reflect a more profound loss of glutamate containing neurons, particularly since it does not reflect the pattern of GABA deficit (as discussed in the case of Huntington's disease by Reynolds and Pearson, 1987). Thus the glutamatergic deficit in DS hippocampus indicated here is all the more robust. As we have previously observed (Reynolds and Warner, 1988), the reduction of glutamate concentrations also found in younger patients suggests that in DS the deficit is present prior to the full development of the neuropathological features.

Since cortical GABA has been previously reported as reduced only in the younger patients in AD (Rossor et al., 1985), the present results appear to agree with our interpretation (Godridge et al., 1987) that DS resembles this more severely affected sub-group, at least in respect of some neurochemical deficits. However, the GABA deficit in DS appears to be restricted to the temporal lobe, since no significant decrease compared with control values could be seen in the frontal cortex.

The comparison with neuropathological features indicates that these deficits of GABA in DS are more related to the loss of cortical neurones than to the other indicators of AD. Mountjoy et al. (1984) made a similar observation in a series of AD cases, although they also found a negative relationship of GABA with tangle counts in the temporal cortex. The strong correlation of GABA with cell density in the hippocampus, along with the lack of any negative correlation of GABA with the major pathological indicators of AD, senile plaques and neurofibrillary tangles, is certainly notable in this area. It is tempting to suggest that, in these DS cases at least, two neurodegenerative processes might be occurring, one related to the classical indicators of Alzheimer's disease and one in which a further GABAergic loss occurs in the hippocampus. This latter effect might relate to the accelerated aging process apparent in DS.

The unexpected positive correlations between hippocampal glutamate concentrations and plaque and tangle counts (reaching significance for the latter) cannot be easily understood. However hippocampal atrophy is substantial in DS and AD and this factor might tend to increase the relative density of glutamatergic terminals. Thus if the atrophy were correlated with tangle formation, an apparent relationship with glutamate concentration could emerge. This effect would be unrelated to the glutamatergic deficit in this area which occurs, to some extent at least, prior to the formation of plaques and tangles. We can only speculate whether this deficit is a result of a neurodegenerative process or might reflect a developmental abnormality.

References

Gramsbergen JBP, Mountjoy C, Rossor MN, Reynolds GP, Roth M, Korf J (1987) A correlative study of hippocampal cation shifts and amino acids and clinico-pathological data in Alzheimer's disease. Neurobiol Aging 8:487–494

Godridge H, Reynolds GP, Czudek C, Calcutt NA, Benton M (1987) Alzheimer-like neurotransmitter deficits in adult Down's syndrome brain tissue. J Neurol Neurosurg Psychiatry 50:775–778

Hardy J, Cowburn R, Barton A, Reynolds GP, Lofdahl E, O'Carroll A-M, Wester P, Winblad B (1987) Region-specific loss of glutamate innervation in Alzheimer's disease. Neurosci Lett 73:77–80

Mann DMA, Yates PO, Marcyniuk B (1985a) Correlation between senile plaque and neurofibrillary tangle counts in cortex and subcortical structures in Alzheimer's disease. Neurosci Lett 56:51–55

Mann DMA, Yates PO, Marcyniuk B, Ravindra CR (1985b) Pathological evidence for neurotransmitter deficits in Down's syndrome of middle age. J Ment Defic Res 29:125–135

Mountjoy CQ, Rossor MN, Iversen LL, Roth M (1984) Correlation of cortical cholinergic and GABA deficits with quantitative neuropathological findings in senile dementia. Brain 107:507–518

Reynolds GP, Godridge H (1985) Alzheimer-like monoamine deficits in adults with Down's syndrome. Lancet ii:1368–1369

Reynolds GP, Pearson SJ (1987) Decreased glutamic acid and increased 5-hydroxytryptamine in Huntington's disease brain. Neurosci Lett 78:233–238

Rossor MN, Iversen LL, Reynolds GP, Mountjoy CQ, Roth M (1984) Neurochemical characteristics of early and late onset types of Alzheimer's disease. Br Med J 288:961–964

Sasaki H, Muramoto O, Kanazawa I, Arai H, Kosaka K, Iizuka R (1986) Regional distribution of amino acid transmitters in postmortem brains of presenile and senile dementia of Alzheimer type. Ann Neurol 19:263–269

Simpson MDC, Cross AJ, Slater P, Deakin JFW (1988) Loss of cortical GABA uptake sites in Alzheimer's disease. J Neural Transm 71:219–226

Yates CM, Simpson J, Maloney AFJ, Gordon A, Reid AH (1980) Alzheimer-like cholinergic deficiency in Down's syndrome. Lancet ii:979

Correspondence: Dr. G. P. Reynolds, Department of Biomedical Science, The University of Sheffield, Sheffield S10 2TN, United Kingdom.

Characteristics of learning deficit induced by ibotenic acid lesion of the frontal cortex in rats

C. Hara[1], N. Ogawa[1], A. Mitani[2], S. Masuda[2], and K. Kataoka[2]

[1] Department of Pharmacology, and [2] First Department of Physiology,
Ehime University School of Medicine, Ehime-ken, Japan

Summary

The present study examined influences of ibotenic acid lesions of the dorso-lateral frontal cortex (DFC) and medial prefrontal cortex (MPC) having cholinergic input from the nucleus basalis of Meynert (NBM) on retention of discrimination avoidance learning in rats. The DFC and MPC lesions impaired the retention similar to the NBM lesion. In contrast to the NBM lesion, the DFC and MPC lesions reduced GAD activity, but not ChAT, in the cortex. Therefore, memory dysfunction based on the nerve cell loss of the frontal cortex may be involved in the GABAergic neuronal system.

Introduction

Alzheimer's disease (AD) and senile dementia of Alzheimer type (SDAT) will be considered as the same disease process, a view broadly accepted at this time. The final goal of this study is to develop an animal model of SDAT for preclinically evaluating the therapeutic drugs. AD has classically been defined as a progressive dementia accompanied by the characteristic neuropathological changes such as senile plaques and neurofibrillary tangles which are the hallmark of AD. However, there are no known animals which show the characteristic neuropathological changes. On the other hand, AD is also characterized by biochemical abnormalities, including many neurotransmitter defects. The memory dysfunction of AD has been associated with a cortical cholinergic deficiency (DeFeudis, 1988). The cholinergic component of AD can be modeled in the rat by ibotenic acid lesion of the nucleus basalis of Meynert (NBM). Ibotenic acid lesion of NBM of the rat reduces choline acetyltransferase (ChAT) activity in the dorsolateral frontal cortex (DFC), medial prefrontal cortex (MPC) and parieto-temporal cortex (Mayo et al., 1984). On the other hand, there are many papers reporting that the NBM lesion induces learning impairment in animals. However, relationship between memory function and these cortical regions is not clear yet. Therefore, the present study examined influence of nerve cell loss induced by ibotenic acid in these frontal cortex regions on retention of discrimination avoidance learning (DAL) in rats in connection with an animal model of memory dysfunction of AD. In addition, influence of ibotenic acid lesion in the frontal cortex regions on ChAT and glutamic acid decarboxylase (GAD) activities was also assessed.

Materials and methods

Male Wistar strain rats (8 weeks old) were used as subjects. The animals were daily trained DAL using the two compartment shuttle box to discriminate the positive conditioned stimulus (PCS) from the negative conditioned stimulus (NCS) to avoid foot shock. The CS were used two pure tones with different frequencies. The foot shock was applied after PCS for 6 sec. The avoidance response that the animals moved into the opposite compartment during the PCS for 6 sec prior to foot shock was recorded as correct response (CR). The transfer to the opposite compartment during the NCS for 6 sec was regared to be faulse response (FR). The NCS was not followed by foot shock. Transfer between compartments in the shuttle box during the intertrial period was recorded as a spontaneous response. Following the daily training of DAL consisting of the 20 trials of PCS and the 20 trials of NCS, only the animals which showed both over 80% CR and under 20% FR were subjected to the brain lesion by ibotenic acid. Ibotenic acid (7.5 μg/0.75 μl for 3 min) was injected bilaterally into DFC or MPC, or unilaterally into NBM according to the brain atlas of Paxinos and Watson (1986). Sham-operated group was injected saline. On 1, 2, 3 and 4 weeks after the surgery, the DAL of 40 trials was loaded. After terminating the experiment, the lesioned site was verified histologically. The ChAT and GAD activities were evaluated on 2 weeks after the surgery by the methods of Fonnum (1969), and Albers and Brady (1959), respectively. The statistical evaluation was performed by ANOVA followed by Scheffe's test and Student-t test.

Results

Body weight grain in each ibotenic acid lesioned group was not different from that of the sham-operated group except on 1 week after the surgery (data not shown). Figure 1 shows influence of ibotenic acid lesion in DFC, MPC and NBM on retention of DAL. The sham-operated group showed marked retention of DAL. The DFC and MPC lesioned

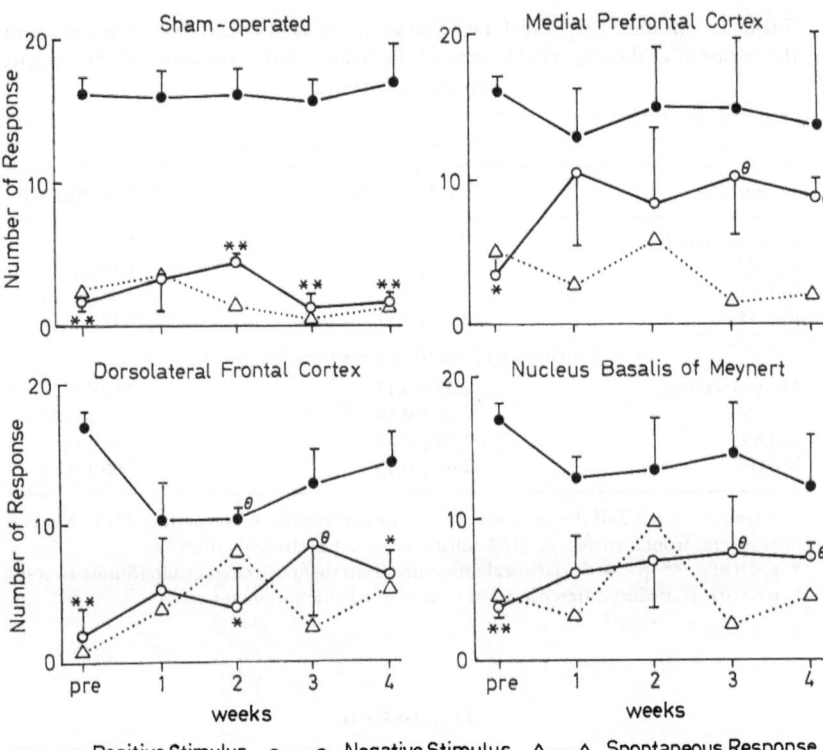

Fig. 1. Influence of ibotenic acid lesion in the dorsolateral frontal cortex, medial prefrontal cortex and nucleus basalis of Meynert on retention of discrimination avoidance learning in rats. Each group consists of 4 animals. * $p < 0.05$, ** $p < 0.01$: statistical difference from positive response (Scheffe's test). θ $p < 0.05$: statistical difference from sham-operated group (Scheffe's test)

groups revealed impairment of the retention from 1 week after the surgery similar to the NBM lesioned group. The impairment was characterized by attenuation of CR and augmentation of FR. Table 1 indicates ChAT and GAD activities in the cortex on 2 weeks after the surgery when the marked impairment of DAL was observed. The NBM lesioned group showed reduction of ChAT activity, but not GAD, only in the lesion side of the cortex. The DFC and MPC lesioned groups showed reduction of GAD activity, but not ChAT, in the cortex.

Table 1. Choline acetyltransferase and glutamic acid decarboxylase activities in the cortex after ibotenic acid lesions of the frontal cortex regions and the nucleus basalis of Meynert

Choline acetyltransferase (n mole/mg/hr)		
Lesion site	Right hemisphere	Left hemisphere
Sham-operated	2.23 ± 0.19	2.15 ± 0.21
r-NBM	1.40 ± 0.16 **, [a]	1.19 ± 0.15
bi-DFC	2.33 ± 0.22	2.33 ± 0.13
bi-MPC	2.23 ± 0.21	2.10 ± 0.14
Glutamic acid decarboxylase (nmole/mg/hr)		
Sham-operated	2.23 ± 0.17	2.03 ± 0.28
r-NBM	2.15 ± 0.30	2.20 ± 0.22
bi-DFC	1.98 ± 0.21 *	1.85 ± 0.34
bi-MPC	2.05 ± 0.30	1.60 ± 0.54

Abbreviations: *r-NBM* the right side of the nucleus basalis of Meynert; *bi-DFC* bilateral dorsolateral frontal cortex; *bi-MPC* bilateral medial prefrontal cortex.
* $p < 0.05$,   ** $p < 0.01$: statistical difference from sham-operated group (Student-t test).
[a] $p < 0.01$: statistical difference from contralateral side (Student-t test)

Discussion

The present study examined whether memory dysfunction could be induced by nerve cell loss of the frontal cortex regions (DFC and MPC) having cholinergic input from NBM. The results showed that the nerve cell loss of DFC and MPC induced by ibotenic acid impaired retention of DAL. The impairment seems to be selective on memory, since unconditioned response was not impaired and there was no co-relation between DAL response and spontaneous response. On the other hand, the result of the biochemical study showed that the NBM lesion induced only reduction of ChAT activity in the lesion side of the cortex, whereas DFC and MPC lesions reduced only GAD activity in the cortex. The results are in accordance with those of the kainic acid study of Lehmann et al. (1980). Our pharmacological study showed that DM9384, a new pyrrolidone derivative which activates ChAT (Kawajiri et al., 1988) and improves learning deficit induced by GABA antagonists (Nabeshima et al., 1988), and tetrahydroaminoacridine, a centrally acting cholinesterase inhibitor (Heilbronn, 1961), improved the impairment of the retention observed in DFC, MPC and NBM (data not shown). Therefore, the memory dysfunction based on nerve cell loss of the frontal cortex may be involved in the GABAergic neural system. Thus, the model of memory dysfunction induced by the frontal cortex lesion may be applicable for an animal model of AD, since the characteristic neuropathological changes of AD are conspicuous in the neocortex (Price, 1986).

Acknowledgement

This work was partly supported by the Mitsui Life Social Welfare Foundation.

References

Albers WR, Brody RO (1959) The distribution of glutamic decarboxylase in the nervous system of the rhesus monkey. J Biol Chem 234:926–928

DeFeudis FV (1988) Cholinergic system and Alzheimer's disease. Drug Dev Res 14:95–109

Fonnum F (1969) Radiochemical micro assays for the determination of choline acetyltransferase and acetylcholinesterase activities. Biochem J 115:465–472

Heilbronn E (1961) Inhibition of cholinesterase by tetrahydroaminoacridine. Acta Chem Scand 15:1386–1390

Kawajiri S, Sakurai T, Ojima H, Akashi A (1988) Effect of DM9384, a new pyrrolidone derivative, on learning behavior and cerebral choline acetyltransferase activity in rats. Psychopharmacology 96 [Suppl]:306

Lehmann J, Nagy JI, Atmadja S, Fibiger HC (1980) The nucleus basalis magnocellularis: the origin of a cholinergic projection to the neocortex of the rat. Neuroscience 5:1161–1174

Mayo W, Dubois B, Ploska A, Javoy-Agid F, Agid Y, Le Moal M, Simon H (1984) Cortical cholinergic projections from the basalis forebrain of the rat, with the special reference to the prefrontal cortex innervation. Neurosci Lett 47:149–154

Nabeshima T, Noda Y, Tohyama K, Harrer S, Kameyama T (1988) Antiamnesic effects of DM9384, a pyrrolidone derivative, on drug-induced amnesia animal models. Psychopharmacology 96 [Suppl]:305

Paxinos G, Watson C (1986) The rat brain in stereotaxic coordinates, 2nd edn. Academic Press, Sydney

Price DL (1986) New perspectives on Alzheimer's disease. Ann Rev Neurosci 9:489–512

Correspondence: C. Hara, Ph. D., Department of Pharmacology, Ehime University School of Medicine, Shigenobu-cho, Onsen-gun, Ehime 791-02, Japan.

Memory loss by glutamate antagonists: an animal model of Alzheimer's disease?

W. J. Schmidt

Zoologisches Institut, Abteilung Neuropharmakologie, University of Tübingen, Federal Republic of Germany

Summary

Spatial learning of rats was tested in T- and 8-arm radial mazes. MK-801 (0.16 and 0.33 mg/kg i.p.) that blocks glutamatergic transmission at the NMDA receptor, caused deficits in short-term memory and induced a strong tendency for perseveration. These symptoms are reminiscent of some clinical features of Alzheimer's disease. In connection with anatomical, neurochemical and electrophysiological data, a glutamatergic deficit may tentatively be assumed to be implicated in this disease.

Introduction

The mechanisms inducing cellular neuropathology underlying Alzheimer's disease (AD) are still unknown and seem not to be amenable to therapeutic treatment in the near future. Although there can be no doubt about the need for a causal therapy preventing cell death, we must also try to substitute pharmacologically for the transmitter deficits caused by the cell loss. Neurochemical studies have revealed decreases in markers for a variety of brain neurotransmitters. A deficit in cholinergic projections to the cortex have received most attention and present treatment of AD focusses on a substitution of this deficit. However, the clinical efficacy of this therapy is not satisfying. During the last years some evidence has accumulated indicating that – among other transmitter deficits – a loss of the excitatory transmitter glutamate (GLU) plays a crucial role in learning and memory impairments seen in patients suffering from AD.

Anatomical evidence

GLU is an excitatory transmitter of intra-cortical association fibers and of corticofugal efferents. GLU is localized in pyramidal cells or in Layer III of the cortex respectively. In AD, the neurofibrillary tangles, mainly found at the beginning of pathways, were found in pyramidal cell bodies and in cortical association fibers. Additionally tangles were found to be arranged in clusters coinciding with the origin of cortical efferents (for review see Pearson et al., 1985).

The hippocampus is thought to be involved in learning and memory. It receives its major cortical input via the perforant pathway that originates in the entorhinal cortex and uses GLU as its transmitter (Hyman et al., 1986). Degeneration of the perforant pathway would disconnect the cortex from the hippocampus and is thought to occur in AD (for review see Deutsch and Morihisa, 1988).

Neurochemical evidence

In AD a decreased GLU-level in spinal fluid and brain tissue has been reported as well as a decrease in ^{3}H-D-aspartate (presynaptic) uptake indicating a decreased number of glutamatergic neurons (Procter et al., 1988).

Electrophysiological evidence

Long-term potentiation (LTP) is a form of synaptic plasticity and is considered to represent the cellular substrate for learning and memory (LTP can last for 37 days in young rats, but lasts only 17 days in old rats). GLU is a prerequisite for LTP-induction and a deficient glutamatergic transmission or a pharmacological blockade of the NMDA subtype of GLU receptors, block LTP-induction (for review see Gustafsson and Wigström, 1988).

Pharmacological evidence

According to their preferred agonists GLU receptors are classified into kainate-, quisqualate- and N-methyl-D-aspartate (NMDA) receptors. A blockade of the NMDA receptor using amino-5-phosphonovaleric acid (AP-5) i.c.v. has been reported to inhibit LTP in the hippocampus (see above) and spatial learning in rats. Visual discrimination learning remained unaffected (Gustafsson and Wigström, 1988). This experiment has succeded in showing the connection between electrical phenomena on the single cell level with alterations at the behavioural level.

The present study addresses the question whether a selective NMDA antagonist [(+)-5-methyl-10,11-dihydroxy-5H-dibenzo (a, d) cyclohepten-5,10 imine] (MK-801) given systemically affects spatial learning and memory in rats. MK-801 is a non competitive antagonist that binds to a side within a ion channel (when in its open state) gated by the NMDA receptor.

Material and methods

Delayed alternation in the T-maze

The rats (N = 10 for drug and N = 10 for saline) were rewarded with a food pellet for entering that arm which was not visited on the previous run. Thus, the rat was only rewarded when alternating between the two arms. After having collected the food, the rat was manually put back into the start box for a delay of 30 s. Each rat was given 10 runs per day for 5 days.

Spatial reversal

The rats were only rewarded when visiting the right arm of a T-maze. 10 runs per day were given for 7 days. On day 8 food was presented within the left arm of the T-maze; 10 runs were given.

8-arm radial maze tasks

Rats (N = 5/5) were tested in a radial maze in the following tasks: 1. Spontaneous alternation (no food reward). 2. Reinforced alternation (all arms were baited with food pellets). 3. Delayed alternation (all arms baited, a delay of 1 min was interposed between the fourth and the fifth run); reentries into previously visited arms were scored as errors. 4. To differentiate between short-term and long-term memory, only four arms of the maze were baited: short-term (working-) memory errors were scored when a rat entered an arm from which food has already been collected. Long-term (reference-) memory errors were scored when an arm was visited that was never baited. 5. Allocentric place navigation: a rat starting in different randomly selected arms had to navigate to only one baited arm remaining in a fixed location in the room.

Results

T-maze

In the delayed alternation test MK-801 (0.16 mg/kg i.p.) treated rats made more errors than controls (p ≤ 0.002 U-test) (Bischoff et al., 1988).

Spatial reversal: During the first 7 days, when rats had to learn to find food in only one arm of the T-maze, MK-801 treated rats learned as quick as controls. But on day 8 when food was presented in the opposite arm, the MK-801 treated animals exhibited a strong tendency to perseverate: 80% of the runs were directed towards the empty arm in MK-801 treated rats; as compared to 40% in control rats ($p \leq 0.05$).

8-arm radial maze

In either of the alternation tasks MK-801 (0.16 and 0.33 mg/kg i.p.) treated rats made much more errors as compared to controls. In the delayed alternation task, the post delay errors were enhanced by MK-801 (Bischoff et al., 1988). However only short-term (working-) memory was affected. Long-term memory remained unchanged (for all results $p \leq 0.05$). Allocentric place navigation was not impaired by MK-801 (0.16 mg/kg i.p.).

Discussion

The results show that MK-801 worsened short-term memory but only in tests when the rats had to alternate. In the place navigation task in T- and 8-arm maze, MK-801 did not worsen learning. This and deficits in spatial reversal indicate a strong tendency to perseverate.

Perseveration and spatial memory loss (especially in a novel environment) are some of the clinical features of AD. Thus, in connection with anatomical, neurochemical and electrophysiological data, a glutamatergic deficit may tentatively be assumed to be implicated in AD.

NMDA receptor blockade in animals might be a model to screen memory enhancing drugs.

References

Bischoff C, Tiedtke PI, Schmidt WJ (1988) Learning in an 8-arm-radial maze: effects of dopamine- and NMDA-receptor-antagonists. In: Elsner N, Barth FG (eds) Sense organs – Interfaces between environment and behaviour. Proceedings of the 16th Göttingen Neurobiology Conference. G Thieme, Stuttgart New York, p 358

Deutsch SI, Morihisa JM (1988) Glutamatergic abnormalities in Alzheimer's disease and a rationale for clinical trials with L-glutamate. Clin Neuropharmacol 11:18–35

Gustafsson B, Wigström H (1988) Physiological mechanisms underlying long-term potentiation. TINS 11:156–162

Hyman BT, VanHoesen GW, Kromer LJ, Damasio AR (1986) Perforant pathway changes and the memory impairment of Alzheimer's disease. Ann Rev Neurol 20:472–481

Pearson RCA, Esiri MM, Hiorns RW, Wilcock GK, Powell TPS (1985) Anatomical correlates of the distribution of the pathological changes in the neocortex in Alzheimer disease. Proc Natl Acad Sci USA 82:4531–4534

Procter AW, Palmer AM, Stratmann GC, Bowen DM (1986) Glutamate/aspartate-releasing neurons in Alzheimer's disease [Letter]. N Engl J Med 314:1711–1712

Correspondence: Prof. Dr. W. J. Schmidt, Zoologisches Institut der Universität Tübingen, Abteilung Neuropharmakologie, Mohlstrasse 54/1, D-7400 Tübingen 1, Federal Republic of Germany.

Convulsant properties of methylxanthines, potential cognitive enhancers in dementia syndromes

J. Deckert[1,2], P. F. Morgan[2], K. A. Jacobson[3], J. W. Daly[3], and P. J. Marangos[2]

[1] Universitäts-Nervenklinik, Würzburg, Federal Republic of Germany
[2] Unit on Neurochemistry, NIMH, Biological Psychiatry Branch, and
[3] Laboratory of Biological Chemistry, NIDDK, Bethesda, MD, U.S.A.

Summary

One of the major side effects of the methylxanthines and adenosine receptor antagonists caffeine and theophylline is the provocation of seizures. With the availability of highly selective adenosine receptor antagonists it was of interest to investigate if they would also induce seizures. In experiments with mice it was found that also highly selective adenosine receptor antagonists like Xanthine Amine congener are able to induce seizures. The therapeutic usefulness of highly selective adenosine receptor antagonists as e.g. cognitive enhancers in dementia syndromes is thus limited by this serious side effect.

Introduction

The leading syndrome in dementia syndromes is cognitive impairment. Cognitive enhancers like cholinergic drugs and benzodiazepine receptor inverse agonists have therefore been suggested and are currently investigated as potential therapeutic agents in dementia syndromes (Sarter et al., 1988).

Two cognitive enhancers in use for centuries are the methylxanthines caffeine and theophylline. They are consumed all over the world in coffee, tea and soft drinks (Rall, 1985). Their predominant mechanism of action in nontoxic concentrations is to antagonize the neurodepressant actions of the endogenous neuromodulator adenosine at its receptors although at higher concentrations interaction with central-type benzodiazepine receptors and phosphodiesterases may be relevant (Daly et al., 1983).

Recently highly specific, high affinity methylxanthine adenosine receptor antagonists have been developed (Jacobson et al., 1986; Deckert et al., 1988) and they have been discussed as therapeutic agents in dementia syndromes (Williams and Jarvis, 1988). However, caffeine and theophylline in higher concentrations are known to induce seizures (Rall, 1985), an effect that severely limits their therapeutic usefulness.

We therefore decided to investigate whether a highly specific, high affinity adenosine receptor antagonist like Xanthine Amine Congener (XAC) would also be able to induce seizures.

Material and methods

Naive male Swiss albino mice (25–30 g) were maintained on a 12 h light/12 h dark cycle with ad lib access to food and water.

They were infused with convulsants through a lateral tail vein at a constant flow rate of 313 μl/min using a 25 gauge butterfly needle fed from an infusion pump (Harvard 22) as described before (Morgan and Stone, 1982). The latency between starting infusions and the onset of convulsions was measured, and from this measurement, the infusion rate, the weight of the animals and the concentration of the convulsant the amount of convulsant per kg of animal weight required to elicit generalized tonic-clonic convulsions (convulsion threshold) was calculated.

XAC was synthesized as described (Jacobson et al., 1986). Caffeine and theophylline were obtained from Sigma, St. Louis, Mo. and several other xanthines were obtained from RBI, Natick, Ma.

Results

XAC induces generalized tonic-clonic convulsions like caffeine and theophylline (Table 1).

Table 1. Seizure threshold of XAC and other xanthines

Xanthine	Seizure threshold (mg/kg)
XAC	39.8± 2.0 (10)
Caffeine	161.3±20.4 (10)
Theophylline	503.1±31.3 (10)
8-Sulphophenyltheophylline	>1000 (5)
1,3-dipropyl-8-p-sulphophenylxanthine	>1000 (5)
3-proxylxanthine (enprofylline)	>1000 (5)

Values represent the mean±SEM with the number of animals in parentheses.

Methylxanthines that do not cross the blood-brain-barrier like 8-sulphophenyltheophylline and 1,3-dipropyl-8-p-sulphophenylxanthine as well as the bronchodilatant 3-propylxanthine (enprofylline) do not induce seizures (Table 1).

If saline is used as a vehicle XAC is only fourfold more potent than caffeine. If however, dimethylsulphoxide is used as a vehicle XAC is fiftyfold more potent than caffeine (data not shown). Accordingly, this effect of XAC could only be seen when it was administered intravenously. No convulsant effect was seen when it was given intraperitoneally (data not shown).

Discussion

These data show that highly selective, high affinity adenosine receptor antagonists like XAC are able to induce generalized tonic-clonic convulsions. This is in good agreement with in vitro data demonstrating proconvulsant properties of methylxanthines which are highly specific adenosine receptor antagonists (Dragunow and Robertson, 1987).

The convulsant potency of XAC compared to caffeine (fifty-fold more potent) appears low given its much higher affinity for adenosine receptors (onethousandfold). However, this may be well explained by its poor penetration into the CNS (Fredholm et al., 1987).

In case of the utilization of specific adenosine receptor antagonists like XAC as cognitive enhancers in dementia syndromes – at least under intravenous administration conditions – seizures are a side effect that has to be taken into account.

References

Daly JW, Butts-Lamb P, Padgett W (1983) Subclasses of adenosine receptors in the central nervous system. Cell Mol Neurobiol 3:69–80

Deckert J, Morgan PF, Bisserbe JC, Jacobson KA, Kirk KL, Daly JW, Marangos PJ (1988) Autoradiographic localization of mouse brain adenosine receptors with an antagonist ((^{3}H)xanthine amine congener) ligand probe. Neurosci Lett 86:121–126

Dragunow M, Robertson HA (1987) 8-cyclopentyl-1,3-dimethylxanthine prolongs epileptic seizures in rats. Brain Res 417:377–379

Fredholm BB, Jacobson KA, Jonzon B, Kirk KL, Li Yo, Daly JW (1987) Evidence that a novel 8-phenyl-substituted xanthine derivative is a cardioselective adenosine receptor antagonist in vivo. J Cardiovasc Pharmacol 9:936–400

Jacobson KA, Ukena D, Kirk KL, Daly JW (1986) (^{3}H)Xanthine amine congener of 1,3-dipropyl-8-phenylxanthine: an antagonist radioligand for
adenosine receptors. PNAS USA 83:4089–4093

Morgan PF, Stone TW (1982) Anticonvulsant actions of the putative γ-aminobu-
tyric acid (GABA)-mimetic, ethylendiamine. Br J Pharmacol 77:525–530

Rall TW (1985) Central nervous system stimulants: the xanthines. In: Gilman
AG, Goodman LS, Rall TW, Murand F (eds) The pharmacological basis of
therapeutics. Mac Millan, New York, pp 589–602

Sarter M, Schneider MM, Stephens DN (1988) Treatment strategies for senile
dementia: antagonist β-carbolines. Trends Neurosci 11:13–17

Williams M, Jarvis MF (1988) Adenosine antagonists as potential therapeutic
agents. Pharmacol Biochem Behav 29:433–441

Correspondence: Dr. J. Deckert, Universitäts-Nervenklinik, Füchsleinstrasse 15,
D-8700 Würzburg, Federal Republic of Germany.

Neurodegenerative diseases:
CSF amines, lactate and clinical findings

E. Sofic[1], J. Fritze[2], G. Schnaberth[3], J. Bruck[3], and P. Riederer[1]

[1] Section on Clinical Neurochemistry, and
[2] Department of Psychiatry, University of Würzburg, Federal Republic of Germany
[3] Ludwig-Boltzmann-Institut zur Erforschung kindlicher Hirnschäden,
Neurologisches Krankenhaus Rosenhügel, Vienna, Austria

Summary

Biogenic amines and their metabolites in lumbar CSF and serum from 43 patients did not significantly differ between dementia of Alzheimer's type (DAT), Binswanger's disease (BD), multiinfarct dementia (MID) and Parkinson's disease (PD). The only significant finding was elevated CSF 5-hydroxyindoleacetic acid (5HIAA) in DAT patients treated with neuroleptics relative to the other groups. The values found in a small number of patients with Pick's, Huntington's (CH) and Creutzfeldt-Jacob's (CJ) disease were within the range of the other groups. CSF lactate, measured in 28 of the patients, tended to be elevated in vascular dementia (BD, MID). A crude estimate of cognitive impairment showed correlations to inner and outer brain atrophy in CT-scans. There were, however, no definite correlations between clinical impairments and CSF or serum findings.

Introduction

Various neurotransmitter deficits have been established in post-mortem brain tissue in neurodegenerative diseases like Alzheimer's disease (DAT) (Mann and Yates, 1986), Parkinson's disease (PD) (Gibb, 1989; Riederer et al., 1988), Huntington's disease (CH) (Sanberg and Coyle, 1984). Less is known about the neurochemical basis of Pick's and Creutzfeldt-Jacob's (CJ) disease (Rossor, 1986) as well as multiinfarct dementia (MID) and Binswanger's disease (BD) (Miller Fisher, 1989). PD is due to a degeneration predominantly of dopaminergic neurons, while mainly cholinergic neurons degenerate in DAT and GABAergic neurons in CH. The possibility to differentiate these disorders in vivo on the basis of the patterns of CSF neurotransmitter concentrations and their metabolites is discussed controversially (Beal and Growdon, 1986; Cummings, 1986; Gibson et al., 1985; Gottfries, 1985; Hollander et al., 1986). Therefore, the present study again addressed this problem, additionally considering the degree of clinical (neurologic and psychiatric) impairment as well as estimates of EEG abnormalities and brain atrophy.

Materials and methods

Lumbar CSF and serum were obtained from 43 patients suffering from DAT (n = 13; f = 11, m = 2; 69.8 ± 7.8 yrs [mean ± SD]), BD (n = 7; f = 6, m = 1; 68.1 ± 6.2 yrs), MID (n = 5; f = 1, m = 4; 68.2 ± 12.2 yrs), PD (n = 11; f = 2; m = 9; 75.9 ± 6.5 yrs), PD with dementia (n = 3; f = 2, m = 1; 75.3 ± 9 yrs), Pick's disease (n = m = 1; 44 yrs), CH (n = 2; f = m = 1; 66/44 yrs), and CJ (n = m = 1; 69 yrs). Rating scales ranging from '0' (absent) to '3' (severe) were used by experienced clinicians (G. S. & J. B.) to estimate the degree of neurologic and cognitive impairment, psychotic phenomena, and bradyphrenia, generalized abnormalities of the EEG as well as inner and outer brain atrophy in cranial CT-scans. CSF cell counts and protein determinations were performed routinely. CSF lactate was determined enzymatically in 28 CSF samples. Norepinephrine (NE), epinephrine (E), dopamine (DA), 3,4-dihydroxyphenylacetic acid (DOPAC), 5-hydroxyindoleacetic acid (5HIAA) and homovanillic acid (HVA) were measured in serum and CSF by HPLC-ECD using dihydroxybenzylamine as internal standard (Sofic, 1986). In serum, serotonin (5HT) could additionally be determined. For statistical analysis, patient groups were split into subgroups depending on the presence or absence of neurotropic/psychotropic drug treatment during the week before lumbar puncture and blood drawing. Thus, most parkinsonians had been treated with L-dopa at any time, DAT patients by preferably low potency neuroleptics. Other centrally active drugs (nootropics like piracetam and ergot alcaloids, one patients on valproate) as well as gender could not be considered in this subgrouping. The statistical tests included one-way ANOVA, Bartlett-test, linear contrasts of Scheffé and U-tests. Differences were considered significant if $p < 0.05$. Moreover, linear and non-parametric correlations (Spearman's rho) were calculated without α-adjustment.

Results

The diagnostic subgroups did not significantly differ in age or duration of illness. However, the genders were unevenly distributed. This could not be considered in the further analysis. DAT patients were significantly more cognitively impaired (df = 6,28; F = 2.84; $p < 0.05$) and prone to psychotic symptoms (F = 2.69; $p < 0.05$). Although insignificant, they expectedly showed less neurologic impairment and more brain atrophy, especially involving an enlargement of the extra-cerebral spaces. CSF protein and cell counts and serum amines and their metabolites did not discriminate between the diagnostic groups. The only significant finding in CSF amines and metabolites (Fig. 1) was an increase of 5HIAA (ANOVA: df=6,28; F=2.73, p<0.05) in DAT patients on neuroleptics relative to drug free DAT (U-test: p<0.01), BD (p<0.05) and MID (p<0.01) patients, and parkinsonians on antiparkin-son drugs (p<0.01). The Scheffé-test revealed only trends. The ratios 5HT/5HIAA and DA/HVA indicating turnover did not discriminate. In correlative calculations, only patients free from antiparkinsonian and neuroleptics drugs were considered (Table 1). There were no correla-tions to age and duration of illness except that these two intercorrelated (r=0.47) and that the latter was related to CSF HVA (r=0.57). Gener-alized EEG abnormalities were unrelated to any of the other parameters. The amount of significant correlations doubled that expected. Half of these correlations, however, were not reproduced by Spearman's rho (Table 1). CSF lactate was significantly elevated in MID and BD in comparison to the other groups (ANOVA: df=5,19; F=6.001; p<0.01;

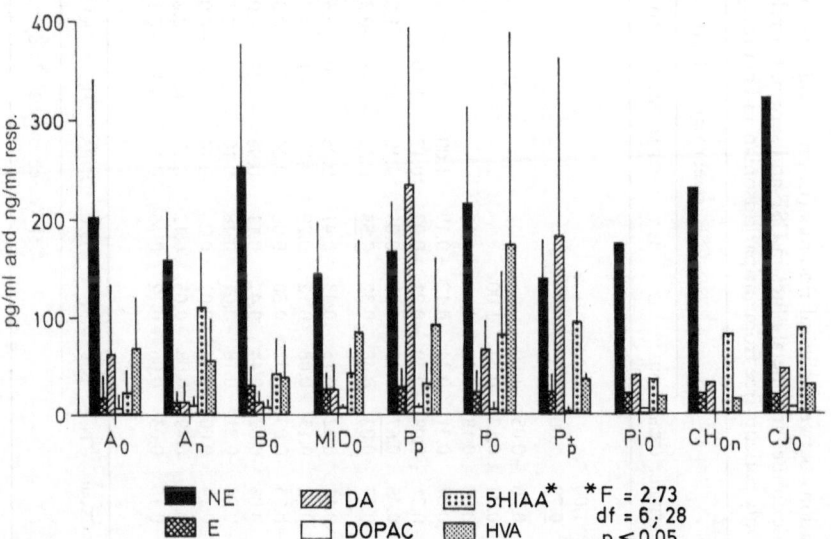

Fig. 1. Biogenic amines (pg/ml) and metabolites (ng/ml) in CSF (mean ±SD). A Alzhei-mer's disease; A_0 off psychotropic drugs (n=4); A_n on neuroleptics (n=9); B_0 Binswan-ger's disease (n=5); MID_0 multiinfarct dementia (n=4), both off medication; P_p Parkin-son's disease (n=6) on L-dopa; P_0 off medication; $P+$ Parkinson's disease with dementia (n=2); Pi Pick's disease (n=1); CH Huntington's disease (n=2); CJ Creutzfeldt-Jacob's disease (n=1)

Table 1. Correlations between clinical findings (Cogn = cognitive impairment), outer (CTout) and inner (CTin) brain atrophy, as well as biogenic amines and metabolites in CSF and serum. Correlations significant both parametrically and non-parametrically are underlined. Only patients free from antiparkinsonian and neuroleptic drugs are considered (n = 19; p < 0.05; 2-sided: r ≥ 0.456)

	Cogn	CTout	CTin	NE	E	DA	DOPAC	5HIAA	HVA	DA/HVA	NE-S	E-S	DA-S	5HIAA	HVA-S	5HT-S	DA/HVA
							CSF (cerebrospinal fluid)							Serum			
Cogn	1.00																
CTout	**0.79**	1.00															
CTin	**0.68**	**0.79**	1.00														
NE-CSF	−0.21	−0.26	−0.05	1.00													
E-CSF	0.03	−0.04	0.12	**0.67**	1.00												
DA-CSF	−0.08	−0.03	0.18	**0.49**	0.40	1.00											
DOPAC	0.39	0.27	0.31	−0.34	0.12	−0.19	1.00										
5HIAA	−0.25	−0.21	−0.08	−0.39	−0.21	−0.02	0.17	1.00									
HVA	0.32	**0.45**	**0.57**	−0.33	−0.28	−0.05	0.19	0.19	1.00								
DA/HVA	−0.19	−0.17	−0.02	**0.45**	0.32	**0.55**	−0.39	−0.41	−0.31	1.00							
NE-S	−0.04	0.11	0.12	−0.15	−0.43	−0.41	0.05	0.00	**0.57**	−0.40	1.00						
E-S	−0.22	−0.02	0.15	0.08	0.12	0.29	0.03	−0.01	−0.04	−0.03	0.08	1.00					
DA-S	0.06	−0.15	−0.03	0.42	**0.50**	0.00	0.24	0.25	0.04	−0.22	0.15	−0.22	1.00				
5HIAA	0.16	0.08	−0.02	0.44	**0.47**	0.11	−0.06	−0.22	0.08	−0.09	−0.05	−0.27	**0.52**	1.00			
HVA-S	0.08	−0.24	−0.21	0.18	−0.01	0.18	−0.19	0.09	−0.17	−0.02	−0.30	−0.13	0.16	0.11	1.00		
5HT-S	−0.10	−0.14	0.00	0.24	0.03	0.23	−0.14	−0.24	−0.09	**0.67**	−0.26	−0.30	−0.12	−0.13	0.29	1.00	
DA/HVA	0.27	0.33	**0.49**	−0.08	−0.05	−0.12	0.17	0.11	**0.84**	−0.30	**0.62**	−0.10	0.37	0.27	−0.35	−0.22	1.00
5HT/5HIAA	−0.03	−0.15	0.03	0.06	−0.12	0.16	−0.13	−0.12	−0.11	**0.56**	−0.27	−0.27	−0.23	−0.38	0.41	**0.92**	−0.31
	Cogn	CTout	CTin	NE	E	DA	DOPAC	5HIAA	HVA	DA/HVA	NE-S	E-S	DA-S	5HIAA	HVA-S	5HT-S	DA/HVA
							CSF (cerebrospinal fluid)							Serum			

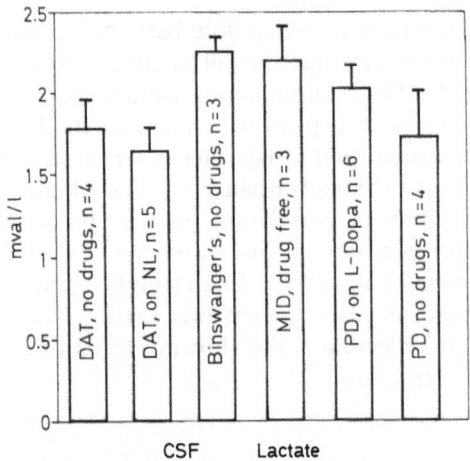

Fig. 2. CSF lactate (mean ± SD) in neurodegenerative diseases. *DAT* dementia Alzheimer type; *MID* multiinfarct dementia; *PD* Parkinson's disease; *NL* neuroleptics. ANOVA: $p < 0.01$; U-tests: $* = p < 0.05$ versus the other groups (except MID versus parkinsonians on L-Dopa)

Fig. 2). While the Scheffé-test revealed significant differences ($p < 0.05$) only between BD as well as MID patients and DAT patients on neuroleptics, U-tests additionally revealed significance ($p < 0.05$) between the former two groups and drug free DAT and PD patients, but not between MID and PD on drugs. ANOVA without subgrouping according to medication, however, revealed no significant differences ($df = 3,24$; $F = 2.684$; $p < 0.1$).

Discussion

As the present study did not include a control group, it cannot contribute to previous findings of reduced CSF HVA and 5HIAA in DAT (Beal and Growdon, 1986; Gottfries, 1985; Hollander et al., 1986; Palmer et al., 1984) and PD (Gibson et al., 1985). The status of CSF amines in CH is controversial (Kurlan et al., 1988; Sanberg and Coyle, 1984). The findings of no differences in CSF HVA (in the absence of probenecid loading) correspond to those of Gibson et al. (1985). The present study was confounded by drug effects where a wash-out of 1 week may have been inadequate. The relevance of drug treatment is underlined by elevated CSF 5HIAA in DAT patients on neuroleptics. Nevertheless, the patterns of amines in CSF are far from diagnostically specific. This is true also for the patterns in serum. Serum NE is especially relevant in parkinsonians where low basal NE and an impaired rise after standing up characterizes the autonomous, orthostatic dysregulation (Turkka et al., 1986). Autonomous dysfunction was not prominent in the PD patients studied possibly explaining the lack of differences between groups. Amines and metabolites in serum and CSF were mostly unrelated (Table 1). Although brain HVA contributes considerably to serum HVA, the lack of correlation (in the absence of a peripheral inhibitor of monoamine oxidase like debrisoquin) conforms to previous findings (Davidson et al., 1987). Interestingly, increased CSF lactate tended to distinguish vascular dementia (MID, BD) from the other disorders possibly reflecting regional malnutrition of brain tissue. This finding deserves replication.

References

Beal MF, Growdon JH (1986) CSF neurotransmitter markers in Alzheimer's disease. Prog Neuropsychopharmacol Biol Psychiatry 10:259–270

Cummings JL (1986) Subcortical dementia. Br J Psychiatry 149:682–697

Davidson M, Giordani AB, Mohs RC, Mykytyn VV, Platt S, Aryan ZS, Davis KL (1987) Control of exogenous factors affecting plasma homovanillic acid concentration. Psychiatry Res 20:307–312

Gibb WRG (1989) Dementia and Parkinson's disease. Br J Psychiatry 154:596–614

Gibson CJ, Logue M, Growdon MD (1985) CSF monoamine metabolite levels in Alzheimer's and Parkinson's disease. Arch Neurol 42:489–492

Gottfries CG (1985) Alzheimer's disease and senile dementia: biochemical characteristics and aspects of treatment. Psychopharmacology 86:245–252

Hollander E, Mohs RC, Davis KL (1986) Antemortem markers of Alzheimer's disease. Neurobiol Aging 7:367–387

Kurlan R, Caine E, Rubin A, Nemeroff CB, Bissette G, Zaczek R, Coyle J, Spielman FJ, Irvine C, Shoulson I (1988) Cerebrospinal fluid correlates of depression in Huntington's disease. Arch Neurol 45:881–883

Miller Fisher C (1989) Binswanger's encephalopathy: a review. J Neurol 256:65–79

Mann MA, Yates PO (1986) Neurotransmitter deficits in Alzheimer's disease and in other dementing disorders. Human Neurobiol 5:147–158

Palmer AM, Sims NR, Bowen DM, Neary D, Palo J, Wikstrom J, Davison AN (1984) Monoamine metabolite concentrations in lumbar cerebrospinal fluid of patients with histologically verified Alzheimer's dementia. J Neurol Neurosurg Psychiatry 47:481–484

Riederer P, Rausch WD, Schmidt B, Kruzik P, Konradi C, Sofic E, Danielczyk W (1988) Biochemical fundamentals of Parkinson's disease. Mt Sinai J Med (NY) 55:21–28

Rossor M (1988) Neurochemical studies in dementia. In: Iversen LL, Iversen SD, Snyder SH (eds) Handbook of psychopharmacology, vol 20. Psychopharmacology of the aging nervous system. Plenum Press, New York London, pp 107–130

Sanberg PR, Coyle JT (1984) Scientific approaches to Huntington's disease. Crit Rev Clin Neurobiol 1:1–44

Sofic E (1986) Untersuchungen von biogenen Aminen, Metaboliten, Ascorbinsäure und Glutathion mittels HPLC-ECD und deren Verhalten in ausgewählten Lebensmitteln und im Organismus von Mensch und Tier. Thesis, TU Vienna

Turkka JV, Juujärvi KK, Lapinlampi TO, Myllylä VV (1986) Serum noradrenaline response to standing up in patients with Parkinson's disease. Eur Neurol 25:355–361

Correspondence: Prof. Dr. E. Sofic, Clinical Neurochemistry, Department of Psychiatry, University of Würzburg, Füchsleinstrasse 15, D-8700 Würzburg, Federal Republic of Germany.

Somatostatin-like immunoreactivity and neurotransmitter metabolites in the cerebrospinal fluid of patients with senile dementia of Alzheimer type and Parkinson's disease

M. Strittmatter[1], H. Cramer[1], C. Reuner[1], D. Strubel[2], and F. Kuntzmann[2]

[1] Department of Neurology and Clinical Neurophysiology, University of Freiburg, Federal Republic of Germany
[2] Department of Geriatrics, University of Strasbourg, France

Summary

The concentration of somatostatin-like immunoreactivity (SLI), the molecular forms of SLI, serotonin and dopamine metabolites in cerebrospinal fluid of patients with senile dementia of Alzheimer type (SDAT), Parkinson's disease (PD) and age-matched control patients were determined by reverse phase HPLC and by specific radioimmunoassay. The mean SLI level in the control group was 29.5 ± 8.9 fmol/ml. In SDAT and PD the mean SLI level was significantly lower (18.6 ± 7.9 fmol/ml, 21.4 ± 8.1 fmol/ml). HPLC separation of SLI yielded four peaks with a preponderance of Somatostatin-14 (SST-14) and Somatostatin-28 (SST-28). In SDAT and PD changes in the molecular pattern of somatostatin indicate an altered biosynthesis and/or processing of somatostatin.

Introduction

Our knowledge of the neurochemical alterations in degenerative brain disease, such as senile dementia of Alzheimer type (SDAT) and Parkinson's disease (PD) is increasing at a fast and sometimes confusing pace. Changes in several transmitter systems have been described (Terry and Davies, 1980; Cross et al., 1981; Whitehouse and Unnerstall, 1988) and loss of cholinergic cells in the basal forebrain appears to be a prominent feature of SDAT (McGeer et al., 1984; Whitehouse et al., 1982). A deficiency of somatostatin attributable to intrinsic cortical neurons has been described in patients with Alzheimer's disease and SDAT (Davies et al., 1981). Marked decreases of SLI were also reported in the CSF of patients with SDAT (Oram et al., 1980).

Less is known about the molecular forms of SLI in SDAT and other types of dementia. Early reports using gel permeation and thin layer chromatography suggested that only somatostatin-14 was present in the human CSF (Kronheim et al., 1977) but HPLC studies have provided evidence for 3–4 peaks of SLI in human CSF co-eluting with high-molecular weight precursor forms (HMW-SST), somatostatin-14 (SST-14) and somatostatin-28 (SST-28) (Beal et al., 1985; Rissler et al., 1986). A reduction of HMW-SST in SDAT brains is reported suggesting a reduction in the rate of biosynthesis and/or an increase in the rate of proteolytic processing of the HMW-SST (Pierotti et al., 1985). The CSF concentrations of neuropeptides, such as somatostatin, seem to reflect partially their passage into extracellular space and hence their synthesis and release from brain cells (Soerensen et al., 1981).

We investigated the molecular forms of SLI, 5-hydroxyindoleacetic acid (5-HIAA) and homovanillic acid (HVA) in the CSF of patients with SDAT and PD because of the clinical interest of evaluating the CSF neurotransmitter and neuropeptide status and possible pathophysiological significance of changes, especially in the molecular pattern of somatostatin.

Materials and methods

104 hospitalized geriatric patients (86 female, 18 male) were studied. A total of 53 patients (50 female, 3 male; mean age 85.6 ± 5.8 years) had SDAT, 38 patients (24 female, 14 male; mean age 79.1 ± 7.4 years) exhibited PD. The high-age control group consisted of 12 inmates of the geriatric department (11 female, 1 male, mean age 82.3 ± 6.7 years) without central nervous disease, intellectual deterioration or major psychiatric disturbances. Diagnoses were established by clinical criteria, including personal history and neurological examination. In 36 patients parkinsonism was of late onset (> 70 years). Severity of parkinsonism was rated according to Hoehn and Yahr (1967). Medication consisted in L-DOPA at various doses in 25 patients. Cognitive impairment was evaluated in all groups by use of global deterioration scale (GDS) of Reisberg et al. (1982).

The lumbar punctures were performed in the morning. First 9 ml of CSF were withdrawn without fractionating, immediately frozen on dry ice and kept congelated at $-30\,°C$ for subsequent studies.

SLI was determined using a specific antibody (K-18) raised in rabbits which recognizes the somatostatin molecule at the ring structure. It shows 100% crossreaction with cyclic synthetic SST-14 (SRIF), synthetic SST-28, somatostatin-25, 65% with cyclic des-Ala^1-Gly^2-desamino-Cys^3-somatostatin (CGP 10 941)

(Des-Ala-SST) and no crossreaction with a number of central nervous system and gastrointestinal peptides. The final antibody solution was 1:165000. As a tracer N-tyrosyl-somatostatin was iodinated with Na^{125}I by application of Chloramin-T and immediately purified by column chromatography. After a total incubation period of 92 h in a phosphate buffer system mimicking the conditions of somatostatin in CSF (pH 7.4) free SLI was separated from the antibody-bound species by use of the charcoal separating technique. The sensitivity of the assay was about 6 fmol/ml. The intra-assay and inter-assay coefficients of variations were 5% and 12% respectively. SLI was fractionated by reverse phase HPLC using a C18 column (LiChrosorb RP 18, particle size 5 μm, column size 4.6 × 250 mm). The mobile phase consisted of 0.1 M triethylammonium formate, pH 3.5 (TEAF) with 17% n-propanol as an organic modifier. Aliquots of up to 500 μl were injected onto the column which was perfused isocratically at a flow rate of 1 ml/min. Thirty six fractions each containing 1 ml of eluate were collected. The fractions were lyophilized and reconstituted in assay buffer for subsequent radioimmunoassay. Recovery of total SLI was $83.7 \pm 11.4\%$ (n = 23). 5-HIAA and HVA were determined by electrochemical detection.

For statistical analysis Student's t-test was employed to determine significance indicate by p values. Correlations were established by Pearson's correlation coefficient. All data is expressed as the mean ± standard deviation.

Results

Total SLI was significantly decreased to 18.6 ± 8.7 fmol/ml in SDAT ($p < 0.01$) and to 21.4 ± 8.1 fmol/ml ($p < 0.05$) in PD, compared to controls (29.5 ± 8.9 fmol/ml). In all groups we could exclude a positive correlation between age and SLI, 5-HIAA and HVA. We observed positive significant correlations between HVA and 5-HIAA in controls ($r = 0.68$), in PD ($r = 0.42$) and SDAT ($r = 0.59$). Moreover HVA and 5-HIAA were significantly correlated with SLI in controls ($r = -0.68$, $r = -0.56$) but not in PD and SDAT. The decrease of SLI in SDAT patients was correlated with dementia scores ($r = 0.41$, $p < = 0.01$). When classified according to the stage of deterioration, patients with GDS 5 showed no decrease of SLI, compared with the age-matched controls, while significant decrease was observed in GDS 6 ($p < 0.05$) and GDS 7 ($p < 0.01$) (Table 1).

HVA was lower in advanced SDAT but only in severe cases the decrease of HVA was significant ($p < 0.05$) (Table 1).

In patients with PD HVA and SLI levels depended on L-DOPA treatment (Table 1). Within the parkinsonian group there was no significant correlation between SLI concentration and GDS ratings. After classification of PD patients according to Hoehn and Yahr (H&Y) we observed in cases of slight disability (H&Y 1−3) an insignificant reduc-

Table 1. Clinical data and CSF levels of 5-HIAA, HVA and SLI in controls, patients with SDAT according to GDS and patients with PD classified by L-DOPA treatment

Patient group		Patients n	Sex F/M	Mean age	5-HIAA nmol/ml	HVA nmol/ml	SLI fmol/ml
Control		12	11/1	82.6 ± 6.7	184.6 ± 64.8	364.9 ± 103.8	29.5 ± 8.9
SDAT	GDS 5	6	5/1	84.7 ± 3.9	231.8 ± 70.4	414.5 ± 238.6	28.3 ± 7.7
	GDS 6	16	15/1	83.4 ± 5.6	198.0 ± 72.9	317.8 ± 172.0	$18.3 \pm 9.4*$
	GDS 7	31	30/1	87.0 ± 5.9	178.3 ± 63.7	$273.5 \pm 138.7*$	$16.5 \pm 7.7**$
	total	53	50/3	85.6 ± 5.8	190.3 ± 68.3	302.8 ± 164.7	$18.6 \pm 7.9**$
PD	with L-DOPA	25	17/8	79.6 ± 5.7	162.5 ± 96.5	$583.8 \pm 354.3**$	$23.2 \pm 7.4*$
	without L-DOPA	13	7/6	78.4 ± 5.1	153.0 ± 84.4	$193.4 \pm 133.6**$	$18.0 \pm 7.2**$
	total	38	24/14	79.1 ± 7.4	160.3 ± 92.1	450.2 ± 345.2	$21.4 \pm 8.1**$

Values are mean $\pm$ SD. * $p < 0.05$, ** $p < 0.01$ vs. control by Student's t-test

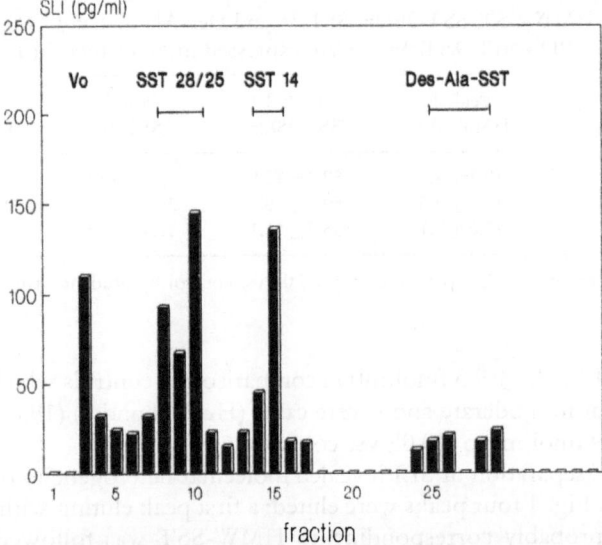

Fig. 1. Reversed phase HPLC elution profile of SLI from a control CSF. Horizontal lines indicate the elution positions of standards

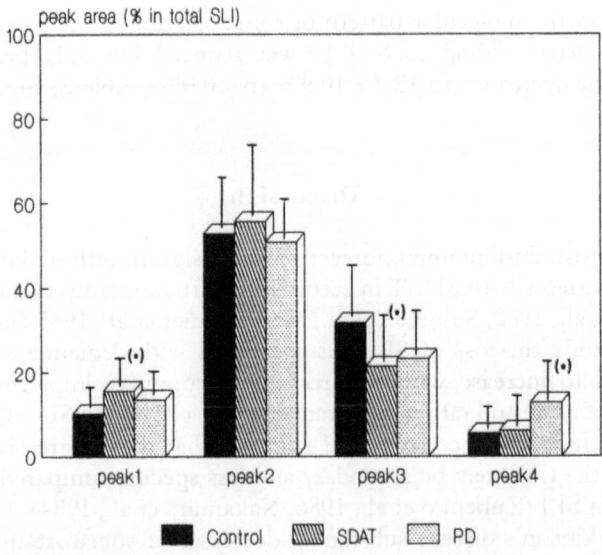

Fig. 2. Histogram showing the molecular pattern of somatostatin-like immunoreactivity in controls and patients with senile dementia of Alzheimer type and Parkinson's disease

Table 2. HMW-SST, SST-28/25, SST-14 and Des-Ala-SST in CSF of controls, PD and SDAT. Values are expressed in % of total SLI

Patient group	Peak 1 HMV-SST	Peak 2 SST-28/25	Peak 3 SST-14	Peak 4 Des-Ala-SST
Control	10.1±4.9	52.7±12.9	31.2±13.6	5.4± 6.1
PD	13.5±6.5	50.8± 8.7	23.2±12.6	12.7±10.9*
SDAT	15.6±7.1*	55.7±11.1	21.4±13.5*	7.4± 8.2

Values are mean ± SD. * p<0.05, ** p<0.01 vs. control by Student's t-test

tion of SLI (23.3±9.3 fmol/ml) in comparison to controls which became significant in moderate and severe cases (H&Y 4 and 5) (19.6±5.9 and 19.2±6.9 fmol/ml, p<0.05 vs. controls).

HPLC separation of SLI revealed molecular heterogeneity of SLI. As shown in Fig. 1 four peaks were eluted: a first peak eluting with the void volume probably corresponding to HMW-SST was followed by two pronounced peaks eluting at retention times of synthetic SST-28/25 and SST-14 and by a small and variable peak co-eluting with Des-Ala-SST. In patients with SDAT we found the following changes in the molecular forms: peak 3 was decreased to 21.4±13.5% (p<0.05) and peak 1 was increased to 15.6±7.1% (p<0.05). PD patients showed less marked changes in the molecular pattern of somatostatin. As well as in SDAT, peak 3 corresponding to SST-14 was reduced but only peak 4 was significant decreased to 12.7±10.9% (p<0.05) (Table 2, Fig. 2).

Discussion

Somatostatin-like immunoreactivity was significantly reduced in the CSF of patients with SDAT in accordance with numerous earlier reports (Wood et al., 1982; Soininen et al., 1984; Cramer et al., 1985 and others). In our study the loss of SLI was correlated with dementia scores and appeared to increase with progression of cognitive impairment. Our finding of a late and rather moderate decrease of HVA in SDAT, together with the independence from SLI suggests that the progressive loss of HVA in the CSF may be secondary and less specific compared with the change in SLI (Zubenko et al., 1986; Nakamura et al., 1984). In patients with Parkinson's disease substantial decreases of somatostatin-like immunoreactivity (SLI) were observed in the CSF by several research groups (Dupont et al., 1982; Cramer et al., 1989), while no significant alteration was found by another (Beal et al., 1986). It was suggested that

SLI levels in CSF and cerebral cortical neurons are only reduced in parkinsonian patients with dementia (Beal et al., 1986; Agid et al., 1985). In our study the deficiency of SLI shows a similar extent compared to SDAT. Although no correlation between the presence or absence of dementia and the SLI concentration was noted our results suggest that a turnover decrease or loss of somatostatin may be closely related to the progression of the disease process of parkinsonism. A pathophysiological link is represented by the functional relationship between monoaminergic and somatostatinergic neurons in extrapyramidal system (Garcia-Sevilla et al., 1978). The presence of multiple forms of SLI in the CSF is in agreement with recent studies (Beal et al., 1985; Rissler et al., 1986) but their quantitive relation is up to now discussed controversely. We have shown a preponderance of SST-14 and SST-28/25 in human CSF. SST-14 probably representing the principal form of SLI, stored in, and released from cortical nerve terminals, is reduced in PD and significantly decreased in SDAT. The also biologically active molecular forms SST-28 and SST-25, which are likely to be intermediates in the biosynthesis of SST-14 show no changes. The increased Des-Ala-SST peak in PD suggests an increase in the rate of proteolytic processing of SST-14 and SST-28. Our finding of qualitative and quantitative changes in the pattern of the molecular forms is in summary compatible with dysregulated synthesis and/or processing of somatostatin in SDAT and PD. However, most of the molecular forms recovered from SLI still require further classification.

In SDAT exists a close association of SLI with senile plaques and neurofibrillary tangles containing abnormal proteinaceous structures (Morrison et al., 1985; Roberts et al., 1985). It remains to be established whether these changes in the molecular pattern of SLI are specific for the neuropathological processes involved in plaque and tangle formation or reflect a generalized alteration of protein biosynthesis and posttranslational processing in SDAT.

References

Agid Y, Javoy-Agid F (1985) Peptides and Parkinson's disease. Trends Neurosci 8:30–35

Beal MF, Mazurek MF, Mc Black PL, Martin JB (1985) Human cerebrospinal fluid somatostatin in neurological disease. J Neurol Sci 71:91–104

Beal MF, Growdon JH, Mazurek MJ, Martin JB (1986) CSF somatostatin-like immunoreactivity in dementia. Neurology 36:294–297

Cramer H, Schaudt D, Rissler K, Strubel D, Warter JM, Kuntzmann F (1985) Somatostatin-like immunoreactivity and substance P-like immunoreactivity in the CSF of patients with senile dementia of Alzheimer type, multiinfarct syndrome and communicating hydrocephalus. J Neurol 232:346–351

Cramer H, Rissler K, Rösler N, Strubel D, Schaudt D, Kuntzmann F (1989) Immunoreactive substance P and somatostatin in the cerebrospinal fluid of senile parkinsonian patients. Eur Neurol 29:1–5

Cross AJ, Crow TJ, Perry EK, Blessed G, Tomlinson BE (1981) Reduced dopamine-beta-hydroxylase activity in Alzheimer's disease. Br Med J 282:93–94

Davies P, Terry RD (1980) Cortical somatostatin-like immunoreactivity in cases of Alzheimer's disease and senile dementia of the Alzheimer type. Brain Res 171:319–327

Dupont E, Christensen SE, Hansen AP, Olivarius BF, Orskow H (1982) Low cerebrospinal fluid somatostatin in Parkinson's disease: an irreversible abnormality. Neurology 32:312–314

Garcia-Sevilla JA, Athee L, Magnusson T, Carlsson A (1978) Opiate-receptor mediated changes in monoamine synthesis in rat brain. Pharm Pharmacol 30:613–621

Hoehn M, Yahr MD (1967) Parkinsonism. Onset, progression, and mortality. Neurology 17:427–444

Kronheim S, Berelowitz M, Pimstone BL (1977) The presence of immunoreactive growth hormone release-inhibiting hormone in normal cerebrospinal fluid. Clin Endocrinol 6:411–415

McGeer PL, McGeer EG, Suzuki J, Dolman CE, Nagai T (1984) Aging, Alzheimer's disease and the cholinergic system of the basal forebrain. Neurology 34:741–745

Morrison JH, Rogers J, Scherr S, Benoit R, Bloom FE (1985) Somatostatin immunoreactivity in neuritic plaques of Alzheimer patients. Nature 314:90–92

Nakamura S, Kshimura K, Kato T, Yamao S, Iijima S, Nagata H, Miyata S, Fujiyoshi K (1984) Neurotransmitters in dementia. Clin Ther 7:18–34

Oram JJ, Edwardson J, Millard PH (1981) Investigation of cerebrospinal fluid neuropeptides in idiopathic senile dementia. Gerontology 27:216–223

Pierotti AR, Harmar AJ, Simpson J, Yates CM (1985) High-molecular weight forms of somatostatin are reduced in Alzheimer's disease and Down's syndrome. Neurosci Lett 63:141–146

Reisberg B, Ferris SH, Mony D, De Leon J, Crook T (1982) The global deterioration scale for assessment of primary degenerative dementia. Am J Psychiatry 139:1136–1139

Rissler K, Cramer H, Schaudt D, Strubel D, Gattaz WF (1986) Molecular size distribution of somatostatin-like immunoreactivity in the cerebrospinal fluid of patients with degenerative brain disease. Neurosci Res 3:213–225

Roberts GW, Crow TJ, Polak JM (1985) Location of neuronal tangles in somatostatin neurones in Alzheimer's disease. Nature 314:92–94

Soininen H, Jolkkonen JT, Keinikainen KJ, Halonen TO, Riekkinen PJ (1984) Reduced cholinesterase activity and somatostatin-like immunoreactivity in the cerebrospinal fluid of patients with dementia of the Alzheimer type. J Neurol Sci 63:167–172

Soerensen KV, Christensen SE, Hansen AP, Ingerslev J, Pedersen E, Orskov H (1981) The origin of cerebrospinal fluid somatostatin: hypothalamic or disperse central nervous system secretion. Neuroendocrinology 32:335–338

Terry RD, Davies P (1980) Dementia of the Alzheimer type. Ann Rev Neurosci 3:77–85

Whitehouse PJ, Unnerstall JR (1988) Neurochemistry of dementia. Eur Neurol 28 (1):36–41

Whitehouse PJ, Price DL, Struble RG, Clark AW, Coyle JT, DeLong MR (1982) Alzheimer's disease and senile dementia: loss of neurons in the basal forebrain. Science 215:1237–1239

Wood PL, Etienne P, Lal S, Gauthier S, Cajal S, Nair NP (1982) Reduced lumbar CSF somatostatin in Alzheimer's disease. Life Sci 31:2073–2079

Zubenko GS, Marquis JK, Volicier L, Direnfeld LK, Langlais PJ, Noxin RA (1986) Cerebrospinal fluid levels of angiotensin-converting enzyme, acetylcholinesterase, and dopamine metabolites in dementia associated with Alzheimer's disease and Parkinson's disease: a correlative study. Biol Psychiatry 21:1365–1381

Correspondence: Dr. H. Cramer, Department of Neurology and Clinical Neurophysiology, University of Freiburg, Hansastrasse 9, D-7800 Freiburg, Federal Republic of Germany.

Nerve growth factor in serum of patients with dementia (Alzheimer type)

U. Stephani[1] and K. Maurer[2]

[1] Kinderklinik, University of Göttingen, and
[2] Psychiatrische Klinik, University of Würzburg, Federal Republic of Germany

Summary

Lack of nerve growth factor (NGF) has been suggested to play a causal role in Alzheimer's disease. Employing sensitive biological and immunological assays for NGF we determined the levels of NGF in sera of controls and patients with Alzheimer's disease. Absence of NGF immunoreactivity and elevated levels of NGF-like bioactivity were found in sera of the patient group. This might be relevant since NGF regulates and increases the transcription rate of many genes, including those for the β-amyloid precursor and the prion protein.

Introduction

In the central nervous system (CNS) of patients with dementia (Alzheimer type) acetylcholine synthesis declines early and cholinergic neurons of the basal forebrain nuclei which project neurites to the cortex cerebri become atrophic. This process results in cortical cholinergic deafferentiation (Perry et al., 1978). Diminished availability of neurotrophic factors has been suggested to play an important role in the disease process (Appel, 1981). Nerve Growth Factor (NGF) is the molecularly best characterised neurotrophic polypeptide. There is strong evidence that NGF may function as a trophic factor for magnocellular basal forebrain cholinergic neurons, i.e. neurons of the diagonal band of Broca, of the medial septal nucleus and of the nucleus basalis of Meynert (for review see: Thoenen et al., 1987). Intrathecal administration of NGF reduces both the basal forebrain cholinergic neuron atrophy and the spatial memory impairment that are associated with normal ageing in rats (Will and Hefti, 1985; Kromer, 1986; Fischer et al., 1987).

The study of the NGF and the NGF-receptor gene expression disclosed no difference of the production of NGF mRNA (Goedert et al., 1986) and NGF-receptor mRNA (Goedert et al., 1989) in the brains of patients with Alzheimer's disease compared to controls. These findings indicate that the NGF gene and the NGF receptor gene expression are not causally involved in the degeneration of basal forebrain cholinergic neurons. However, these findings do not exclude that the biosynthesis and metabolic turn over of the NGF protein and of the NGF receptor protein are altered in Alzheimer's disease.

In order to investigate possible alterations of the NGF system in Alzheimer's disease we analysed whether the presence of NGF in extraneural compartments, i.e. in the blood circulation (S-NGF) was affected by the disease process.

Material and methods

Sera were taken from 7 female and 3 male patients with the classical symptoms of Alzheimer type dementia (diagnosed according to the NINCDS/ADRDA-criteria; mean age 69 years). Sera of 10 age and sex matched patients without neurological diseases served as controls.

Serum NGF levels were evaluated with a sensitive *biological assay* employing embryonic sensory neurons of the chicken as in vitro indicators of NGF bioactivity (Stephani et al., 1987). In this assay the neurite promoting activity of serum (Fig. 1) was titrated. The serum NGF level was calculated from comparison with the dose response curve of purified NGF. Sensitivity of this assay system is 8 pg/ml NGF.

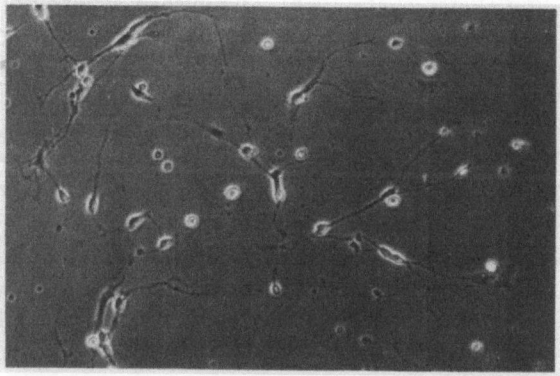

Fig. 1. Neurite promoting activity of the serum of a patient with Alzheimer's dementia
(serum dilution 1:10³, 1 cm = 50 μm)

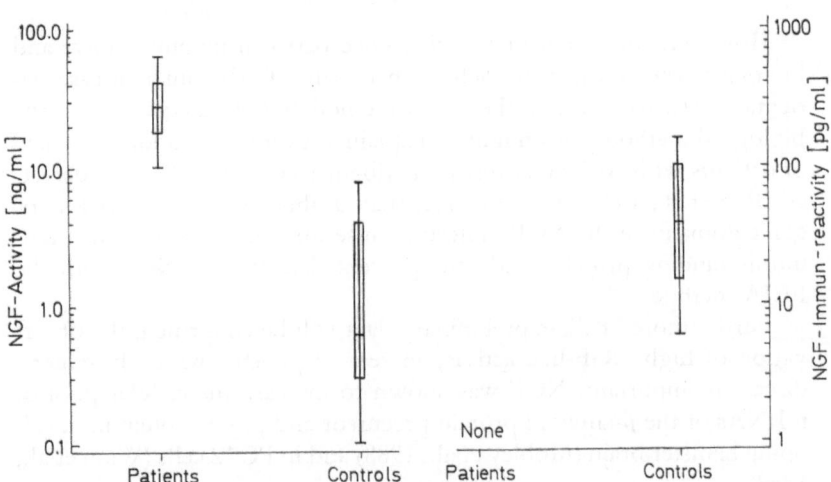

Fig. 2. Senile dementia of Alzheimer type: serum-NGF-levels in 10 patients and controls as determined employing biological and immunological assays. The ends of the vertical lines represent minimal and maximal values, the horizontal lines of the open boxes 25%, 50% and 75% values

In addition, serum NGF levels were measured with a commercially available two site *fluorimetric enzyme immunoassay* (FEIA). Detection limit of this assay system is 4 pg/ml NGF.

Results

In the sera of 10 control patients NGF (S-NGF) levels were 6–150 pg/ml immunoreactive NGF-equivalents and 0.1–8 ng/ml biological NGF-equivalents. In the 10 sera of patients with Alzheimer disease no NGF was detected by the FEIA method whereas by bioassay S-NGF levels equivalent to 9–90 ng/ml NGF were evaluated (Fig. 2).

Discussion

The differences between the results of biological and immunological methods in serum are still a matter of methodological investigations. Possibly, NGF-binding proteins in serum are able to hide antigenic sites more than biologically active sites of the NGF molecule. With the FEIA method only free NGF-molecules are measured in serum, what certainly is an underestimate of the total NGF concentration present in serum.

However, the extent of the difference between immunological and biological values was extremely high in sera of Alzheimer disease patients with zero values in the FEIA method and elevated values in the biological method. This might be explained by a recent finding: In sera of patients with Alzheimer disease antibodies against NGF were detected (Roy et al., 1988). It is possible, that antibodies to NGF mask antigenic domains of the NGF molecules in serum additionally to the ubiquitous binding proteins and thus prevent detection of NGF with the FEIA method.

Furthermore, if these preliminary data will be confirmed, the observation of high NGF-like activity in sera of patients with Alzheimer's disease is important: NGF was shown to increase the cellular pool of mRNAs of the β-amyloid protein precursor and prion protein in developing hamster brain (Mobley et al., 1988) and in PC 12 cells (Wion et al., 1988).

Amyloid deposits and the formation of plaques and tangles are the most striking pathological findings in brains affected by Alzheimer's disease. Further studies are needed to elucidate the possible pathogenetic role of the NGF-system in Alzheimer's disease.

Acknowledgement

Supported by Deutsche Forschungsgemeinschaft (Grant Ste 421/1-1).

References

Appel SH (1981) A unifying hypothesis for the cause of amyotrophic lateral sclerosis, parkinsonism, and Alzheimer disease. Ann Neurol 10:499–505

Fischer W, Wictorin K, Björklund A, Williams LR, Varon S, Gage FH (1987) Amelioration of cholinergic neuron atrophy and spatial memory impairment in aged rats by nerve growth factor. Nature 329:65–68

Goedert M, Fine A, Hunt SP, Ullrich A (1986) Nerve growth factor mRNA in peripheral and central rat tissues and in the human central nervous system: lesion effects in the rat brain and levels in Alzheimer's disease. Mol Brain Res 1:85–92

Goedert M, Fine A, Dawbarn D, Wicock GK, Chao MV (1989) Nerve growth factor receptor mRNA distribution in human brain: normal levels in basal forebrain in Alzheimer's disease. Mol Brain Res 5:1–7

Kromer LF (1986) Nerve growth factor treatment after brain injury prevents neuronal death. Science 235:214–216

Mobley WC, Neve RL, Prusiner SB, McKinley MP (1988) Nerve growth factor increases mRNA levels for the prion protein and the β-amyloid protein precursor in developing hamster brain. Proc Natl Acad Sci USA 85:9811–9815

Perry EK, Tomlinson BE, Blessed G, Bergmann K, Gibson PH, Perryk RH (1978) Correlation of cholinergic abnormalities with senile plaques and mental test scores in senile dementia. Br Med J 2:1457–1459

Roy BF, Sunderland T, Murphy DL, Marchisa JM (1988) Antibody for nerve growth factor detected in patients with Alzheimer's disease. Ann NY Acad Sci 540:398–400

Stephani U, Sutter A, Zimmermann A (1987) Nerve growth factor (NGF) in serum: evaluation of serum NGF levels with a sensitive bioassay employing embryonic sensory neurons. J Neurosci Res 17:25–35

Thoenen H, Bandtlow C, Heumann R (1987) The physiological function of nerve growth factor in the central nervous system: comparison with the periphery. Rev Physiol Biochem Pharmacol 109:145–178

Will B, Hefti F (1985) Behavioural and neurochemical effects of chronic intraventricular injections of nerve growth factor in adult rats with fimbria lesions. Behav Brain Res 17:17–24

Wion D, Le Bert M, Brachet P (1988) Messenger RNAs of β-amyloid precursor protein and prion protein are regulated by nerve growth factor in PC 12 cells. Int J Dev Neurosci 6:387–393

Correspondence: Priv. Doz. Dr. U. Stephani, Kinderklinik der Universität, Robert-Koch-Strasse 40, D-3400 Göttingen, Federal Republic of Germany.

Neuroendocrine dysfunction in early-onset Alzheimer's disease

K. P. Lesch[1], R. Ihl[1], L. Fröhlich[1], R. Rupprecht[1], U. Müller[1], H. M. Schulte[2], and K. Maurer[1]

[1] Department of Psychiatry, University of Würzburg, Würzburg
[2] Department of Medicine I, University of Kiel, Kiel, Federal Republic of Germany

Summary

To explore hypothalamic-pituitary-somatotropic (HPS) and -adrenal (HPA) system integrity in early-onset Alzheimer's disease (AD) 10 drug-naive patients and matched controls received 50 µg GHRH at 9:00 and 100 µg CRH at 18:00 as an i.v. bolus dose. Compared with controls, patients with AD showed attenuated GHRH-induced GH responses and decreased ACTH but normal cortisol secretion following CRH. GH responses to GHRH were negatively correlated with the plasma insulin-like growth factor (IGF-I) concentrations and the severity of dementia. A positive correlation was found between GHRH-evoked GH release and ACTH responses to CRH. The results suggest a pathologic process at the level of the pituitary and/or hypothalamus possibly involving a cholinergic, monoaminergic, and/or peptidergic imbalance in AD and support the view that altered HPS and HPA secretory dynamics in AD are related to the underlying brain dysfunction.

Introduction

Postmortem examination of brain specimen from patients with AD reveal marked cholinergic, monoaminergic and peptidergic deficiences not only in the cerebral cortex but also in subcortical structures, including the hypothalamus (Greenwald et al., 1983; Perry, 1987). In particular, irreversibly decreased neuronal tissue and cerebrospinal fluid somatostatin (SRIH) concentrations in AD are widely replicated findings (Davies and Terry, 1981; Wood et al., 1982), with the significance of these observations further suggested by the identification of the SRIH neurons as a major locus of development of neurofibrillary tangles and plaques. Hypothalamic regions receiving cholinergic and monoaminergic input are also the major sources of peptide-releasing/inhibiting hormones. Since growth hormone (GH) secretion is principally regulated through altered hypothalamic SRIH or GH-releasing hormone (GHRH) output, distorted GH secretory dynamics associated with AD, as detected by reduced sleep-related secretion of GH (Davis et al., 1982), diurnal GH hypersecretion (Christie et al., 1987) and blunted GH responses to the cholinesterase inhibitor edrophonium (Thienhaus et al., 1986), are believed to reflect deficient release of SRIH resulting in HPS system hyperactivity. In addition, a variety of neuroendocrine derangements involving the hypothalamic-pituitary-adrenal (HPA) axis have been described in AD. These abnormalities include decreased corticotropin-releasing hormone (CRH) concentrations and reciprocal changes in CRH receptors in the cortex (Bissette et al., 1985; DeSouza et al., 1986), increased nocturnal cortisol secretion (Davis et al., 1986) and nonsuppression of plasma cortisol concentrations after dexamethasone administration (Raskind et al., 1982; Balldin et al., 1983). In order to explore HPS and HPA system integrity in AD, GHRH and CRH stimulation tests were conducted in patients with early-onset AD and matched controls.

Methods

Ten drug-naive and non-depressed patients (7 women, 3 men), age 49–66 years (mean ± SD, 58.6 ± 6.3) and body mass index (BMI) 23.1 ± 2.6 kg/m² met DSM-III-R criteria for primary degenerative dementia and NINCDS-ADRDA criteria (McKhann et al., 1984) for probable AD. All investigations were performed after the patients' admission to the Department of Psychiatry, University of Würzburg. Patients were evaluated by physical examination, routine laboratory screening and various imaging techniques including event related potentials (ERP), brain electrical activity mapping (BEAM), computed tomography (CT) or magnetic resonance imaging (MRI), single photon emission computed tomography (SPECT) and positron emission tomography (PET). The severity of dementia was assessed with Folstein's Mini-Mental State (FMMS) test (Folstein et al., 1975) and the Brief Cognitive Rating Scale (BCRS) (Reisberg et al., 1983). The mean BCRS and FMMS score was 4.1 ± 1.5 and 16.3 ± 10.0, respectively. Written informed consent was obtained from all patients and/or their spouses. Ten controls were matched for age (55.2 ± 4.6 years), gender (7 women, 3 men) and BMI (23.6 ± 2.9 kg/m²).

All subjects received 50 µg synthetic human GHRH-44 amide at 9:00 and 100 µg synthetic human CRH-41 at 18:00 as an i.v. bolus dose at rest in bed. GH was measured at −15, 0, 15, 30, 45, 60, 90, 120 min and IGF-I was determined at −15 min. Corticotropin (ACTH) and cortisol was measured at −15, 0, 5, 15, 30, 60, 90 and 120 min. GH, IGF-I, ACTH, and cortisol analyses employed standard RIA techniques as previously described (Lesch et al., 1989). Hormone responses were calculated as net areas under the curve (AUC), using trapezoidal integration. The data were analyzed using Mann-Whitney's U test and Spearman's rank order correlation.

Results

Compared with controls, the patients with AD showed attenuated GH responses to GHRH (851 ± 734 vs. 361 ± 666 ng · min/ml; $p < 0.05$) (Fig. 1, Table 1) and decreased ACTH (3136 ± 2798 vs. 1431 ± 825 pg · min/ml; $p < 0.05$) but normal cortisol release (12.9 ± 4.4 vs. $13.9 \pm 7.3 \times 10^3$ ng · min/ml; ns) following CRH (Fig. 1, Table 2). No difference was found in basal GH and IGF-I concentrations. The

Table 1. Baseline growth hormone (GH), GH reponses to GH-releasing hormone and insulin-like growth factor I (IGF-I) in patients with Alzheimer's disease (n = 10) and matched controls

	Basal GH (ng/ml)	AUC GH (ng · min/ml)	IGF-I (U/ml)
Alzheimer's disease	1.8 ± 2.2	361 ± 666*	0.74 ± 0.31
Controls	1.7 ± 1.6	851 ± 734	0.73 ± 0.36

* $p < 0.05$; Mann-Whitney U-test, two-tailed

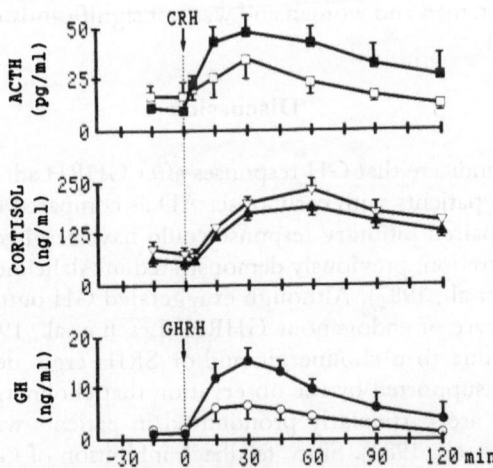

Fig. 1. Time course curves of GH responses to GHRH and CRH-induced ACTH/cortisol release in 10 patients with Alzheimer's disease (o, □, ∆) and 10 normal controls (●, ■, ▲) matched for age, gender and body mass index

Table 2. Baseline cortisol, ACTH and cortisol responses to corticotropin-releasing hormone (CRH) in patients with Alzheimer's disease ($n = 10$) and matched controls

	Basal cortisol (ng/ml)	AUC ACTH (pg · min/ml)	AUC cortisol (ng · min/ml × 10^3)
Alzheimer's disease	78.6 ± 50.0	1431 ± 825*	13.9 ± 7.3
Controls	50.6 ± 23.7	3136 ± 2798	12.9 ± 4.4

* $p < 0.05$; Mann-Whitney U-test, two-tailed

GHRH-induced GH responses and IGF-I concentrations were negatively correlated in the AD patients ($r_s = -0.58$, $p < 0.05$) and across the entire sample ($r_s = -0.39$, $p < 0.05$). While baseline ACTH values were not different, the blunted ACTH response was associated with increased basal cortisol concentrations. A positive correlation was found between GHRH-evoked GH release and ACTH responses to CRH across the entire sample ($r_s = 0.45$, $p < 0.05$). This relationship persisted only in the patient group when AD patients and controls were tested separately ($r_s = 0.67$, $p < 0.05$). GH and cortisol responses were not correlated. In the patients with AD a correlation was found between GH responses to GHRH and severity of dementia as assessed by the BCRS ($r_s = -0.56$, $p < 0.05$) and the FMMS ($r_s = 0.50$, $p < 0.1$). The magnitude of GH responses to GHRH and CRH-induced ACTH/cortisol secretion did not differ between men and women and was not significantly correlated with age and BMI.

Discussion

Our data indicate that GH responses after GHRH administration are attenuated in patients with early-onset AD as compared to normal control. The impaired pituitary response could have resulted from diurnal GH hypersecretion, previously demonstrated in Alzheimer-type dementia (Christie et al., 1987). Although exaggerated GH output may reflect increased release of endogenous GHRH (Lesch et al., 1989), it is more likely to be due to a cholinergic and/or SRIH-ergic deficiency. This hypothesis is supported by the observation that cholinergic and SRIH-ergic deficits are particularly pronounced in patients with early-onset AD (Rossor et al., 1984). Since feedback inhibition of GH secretion is mediated by peripheral GH-dependent IGF-I production and by a direct effect of GH, diminished GHRH-induced GH responses may be a consequence HPS hyperactivity. Although there is no compelling evidence of diurnal GH hypersecretion in our patient sample, as reflected by elevated basal GH levels and/or increased IGF-I concentrations, the diminished GH responses to GHRH may indicate a pathologic process at the level of the pituitary and/or hypothalamus.

The discordance in ACTH and cortisol responses to CRH stimulation in AD patients indicates that the pituitary corticotroph cells respond to the negative feedback of increased baseline cortisol and that the abnormal HPA activity reflects a suprapituitary disturbance. Accordingly, the apparent normal functioning of the ACTH-secreting cells and the presumed integrity of the glucocorticoid-dependent negative feedback regulation suggests that mild hypercortisolism in patients with AD reflects an abnormality at or above the level of the hypothalamus that may lead to a hypersecretion of endogenous CRH (and/or vasopressin). Hyperactivity of CRH-containing neurons may be due to alterations in the hypothalamic, limbic and other (sub-)cortical structures where cholinergic, monoaminergic, and peptidergic interaction is known to have important modulatory influences on hormonal release (Lesch et al., 1989). These results, therefore, provide functional evidence of disturbed HPS and HPA axis regulation in AD that appears to be related to underlying brain dysfunction. Further study of neurotransmitter-mediated HPS and HPA dysregulation using selective pharmacological challenge agents combined with peripheral measures may allow a clearer understanding of pathophysiology of neuroendocrine dysfunction in AD.

Despite the limitations and low specifity of neuroendocrine challenge paradigms, GHRH-induced GH responses and ACTH/cortisol secretion following CRH may play a role in identifying early AD cases and high-risk populations, such as first-degree relatives of patients with AD. Thus, identification of peripheral endocrine abnormalities linked to neurotransmitter and/or -modulator deficiencies could help develop accessible antemortem markers and may contribute to the clarification of AD pathophysiology.

References

Balldin J, Gottfries CG, Karlsson I, et al (1983) Dexamethasone suppression test and serum prolactin in dementia disorders. Br J Psychiatry 143:277–281

Bissette G, Reynolds GP, Kilts CD, Widerlöv E, Nemeroff CB (1985) Corticotropin-releasing factor-like immunoreactivity in senile dementia of the Alzheimer type. J Am Med Assoc 254:3067–3069

Christie JE, Whalley LJ, Bennie J, Dick H, Blackburn IM, Fink G (1987) Characteristic plasma hormone changes in Alzheimer's disease. Br J Psychiatry 150:674–681

Davies P, Terry RD (1981) Reduced cortical somatostatin-like immunoreactivity in cerebral cortex from cases of Alzheimer's disease and senile dementia of the Alzheimer type. Neurobiol Aging 2:9–14

Davis BM, Levy MI, Rosenberg GS, Mathe A, Davis KL (1982) Relationship between growth hormone and cortisol and acetylcholine: a possible neuroendocrine strategy for assessing the cholinergic deficit. In: Corkin S, Davis KL, Growdon JH, Usdin E, Wurtman J (eds) Alzheimer's disease: report of progress in research. Raven, New York, pp 9–14

Davis KL, Davis BM, Greenwald BS, et al (1986) Cortisol and Alzheimer's disease: basal studies. Am J Psychiatry 143:300–305

De Souza EB, Whitehouse PJ, Kuhar MJ, Price DL, Vale WW (1986) Reciprocal changes in corticotropin-releasing factor (CRF)-like immunoreactivity and CRF receptors in cerebral cortex of Alzheimer's disease. Nature 319:593–595

Folstein MF, Folstein SE, McHugh PR (1975) "Mini-Mental State": a practical method for grading the cognitive state of patients for the clinician. J Psychiatr Res 12:189–198

Greenwald BS, Mohs RC, Davis KL (1983) Neurotransmitter deficits in Alzheimer's disease: criteria for significance. J Am Geriatr Soc 31:310–316

Lesch KP, Müller U, Rupprecht R, Kruse K, Schulte HM (1989) Endocrine responses to growth hormone-releasing hormone, thyrotropin-releasing hormone and corticotropin-releasing hormone in depression. Acta Psychiatr Scand 79:597–602

McKhann G, Drachman D, Folstein M, et al (1984) Clinical diagnosis of Alzheimer's disease: report of the NINCDS-ADRDA work group under the auspices of the Department of Health and Human Services Task Force on Alzheimer's disease. Neurology (NY) 34:485–490

Perry EK (1987) Cortical neurotransmitter chemistry in Alzheimer's disease. In: Meltzer H (ed) Psychopharmacology: the third generation of progress. Raven, New York, pp 568–575

Raskind M, Peskind E, Rivard MF, et al (1982) Dexamethasone suppression test and cortisol circadian rhythm in primary degenerative dementia. Am J Psychiatry 139:1468–1471

Reisberg B, London E, Ferris SH, Borenstein J, Scheier L, de Leon MJ (1983) The brief cognitive rating scale: language, motoric and mood, concomitants in primary degenerative dementia (PDD). Psychopharmacol Bull 19:702–708

Rossor MN, Iversen LL, Reynolds GP, Mountjoy CQ, Roth M (1984) Neurochemical characteristics of early and late onset types of Alzheimer's disease. Br Med J 288:961–969

Thienhaus OJ, Zemlan FP, Bienenfeld D, Hartford JT, Bosmann HB (1987) Growth hormone response to edrophonium in Alzheimer's disease. Am J Psychiatry 144:1049–1059

Wood PL, Etienne P, Lal S, Gauthier S, Cajal S, Nair NPV (1982) Reduced lumbar CSF somatostatin levels in Alzheimer's disease. Life Sci 31:2073–2079

Correspondence: Dr. K.-P. Lesch, Department of Psychiatry, University of Würzburg, Füchsleinstrasse 15, D-8700 Würzburg, Federal Republic of Germany.

Urinary excretion of salsolinol enantiomers and 1,2-dehydrosalsolinol in patients with degenerative dementia

P. Dostert[1], M. Strolin Benedetti[1], V. Bellotti[1], G. Dordain[2], D. Vernay[2], and D. Deffond[2]

[1] Farmitalia Carlo Erba, Research and Development-Erbamont Group, Milan, Italy
[2] Hôpital Nord, Service de Neurologie, Clermont-Ferrand, France

Summary

The daily urinary excretion of total (R)- and (S)-salsolinol and 1,2-dehydrosalsolinol (free + sulfoconjugate) was measured in patients with degenerative dementia. (R)-Salsolinol was detectable in the urine of 5 out of 12 patients and 1,2-dehydrosalsolinol in the urine of 10 patients, whereas (S)-salsolinol was not detectable. The urinary excretion of 1,2-dehydrosalsolinol appears to be similar in demented and parkinsonian patients, suggesting that the biosynthesis of salsolinol precursors may be similar in the two groups.

Introduction

Salsolinol (1-methyl-1,2,3,4-tetrahydro-6,7-isoquinolinediol) is a tetrahydroisoquinoline alkaloid which exists as R and S enantiomers (Dostert et al., 1988). There is increasing evidence that, in humans, the biosynthesis of salsolinol occurs by condensation of dopamine (DA) with pyruvic acid. Salsolinol was detected for the first time in the urine of parkinsonian patients (Sandler et al., 1973). Later on, salsolinol was also found in urine and cerebrospinal fluid of healthy volunteers (Collins et al., 1979; Sjöquist et al., 1981; Dordain et al., 1984) as well as in brain from non-alcoholic subjects (Sjöquist et al., 1982). In non-treated parkinsonian patients, the urinary excretion of total (R + S) salsolinol was shown to be significantly lower than in controls (Dordain et al., 1984). We have recently established that, under normal conditions, the R enantiomer of salsolinol predominates or is the only isomer present in the urine of healthy subjects (Dostert et al., 1987).

In biochemical studies of Alzheimer's disease and senile dementia of the Alzheimer's type, the dopaminergic system has received little attention. Findings on DA in Alzheimer's patients with histological confirmation are controversial; most studies, however, suggest that DA neurons are not affected (for review see Reinikainen et al., 1988). In this study, the urinary concentrations and daily excretion of (R)- and (S)-salsolinol and 1,2-dehydrosalsolinol, a potential bioprecursor of salsolinol, were measured in patients with degenerative dementia as an indirect evaluation of brain DA concentrations.

Material and methods

Study design

Twelve patients, 4 men and 8 women, aged 75.4 ± 8.8 years (mean $\pm$ S.D.) participated in the study (Table 1). The diagnosis of primary degenerative dementia was made by documenting that each patient had a clinical history and current mental status consistent with that diagnosis and by ruling out other possible causes of dementia. Patients had a history of progressive cognitive impairment ranging from a few months to 8 years. Each patient's current mental status and behavioral abnormalities at the time of entry into the study were assessed with the Global Deterioration Scale of Reisberg (Reisberg et al., 1982). Patients received standard food from the hospital and the 24-h urines were collected in the presence of $NaHSO_3$ and semicarbazide.

Table 1. Characteristics of the patients

Subjects	Sex	Age (years)	Weight (kg)	Reisberg's scale (grade)	Duration of the disease (years)
1	f	63	55	5	2.5
2	m	62	63	4	6
3	m	66	62	5	3
4	m	77	66	4	<1
5	f	87	55	6	8
6	f	82	–	5	2
7	f	76	37	5	3
8	m	83	59	5	–
9	f	76	40	4	1
10	f	87	58	4	<1
11	f	78	62	6	4–5
12	f	68	–	5	2

–: data not available

Determination of the urinary concentrations of (R)- and (S)-salsolinol and 1,2-dehydrosalsolinol

Determination of the urinary concentrations of total (R)- and (S)-salsolinol was performed as described by Dostert et al. (1989). 1,2-Dehydrosalsolinol urine concentrations were determined as follows: 5 ml samples of urine brought to pH 7.5 were hydrolyzed (37 °C, 16 h) with 50 μl of a solution prepared by diluting 0.1 ml arylsulphatase (Sigma S-1629) with 0.9 ml of 0.2 M Tris buffer pH 7.5. Aliquots corresponding to 1 ml urine were extracted with $CHCl_3$ first at pH 4 and then at pH 7.4. The aqueous phase brought to pH 8.5 was loaded into a phenylboronic acid cartridge. 1,2-dehydrosalsolinol was eluted with CH_3CO_2H in CH_3OH. After evaporation to dryness, the residue was dissolved in 200 μl mobile phase. Aliquots of 50 μl were injected into the HPLC system. This was essentially that used by Dostert et al. (1989) with some modifications: column: μ-Bondapak C-18, mobile phase: $CH_3OH/75$ mM KH_2PO_4 in H_2O brought to pH 3 + sodium octylsulphate (20/80, v/v), flow rate: 0.8 ml/min; limit of detection: 8.5 pmol/ml urine.

Results

The S enantiomer of salsolinol was not detectable in the urine of all the patients. The R enantiomer was clearly detectable in the urine of two patients (Table 2); in the urine of three other patients, though below the

Table 2. Urinary concentrations (U.C.: pmol/ml) and daily urinary excretion (D.U.E.: nmol/day) of (R)-salsolinol [(R)-Sal] and 1,2-dehydrosalsolinol (DSal) in 12 patients with degenerative dementia

Subjects	(R)-Sal		DSal		24-h urine excretion (ml)
	U.C	D.U.E.	U.C.	D.U.E.	
1	–	–	24.4		+
2	18.3	7.3	23.8	9.5	400
3	–	–	40.3	30.2	750
4	–	–	–	–	500
5	–	–	20.5	7.4	360
6	(10.4) *	4.7	(7.0) *	3.2	450
7	(8.5) *	9.4	–	–	1100
8	14.5	18.1	30.6	38.3	1250
9	–	–	21.8	19.6	900
10	–	–	25.1	6.3	250
11	–	–	24.3	17.3	700
12	(6.0) *	3.6	23.8	14.3	600

–: not detectable; * value determined although below the limit of detection retained;
+ data not available

limit of detection retained (14 pmol/ml), the concentrations of (R)-salsolinol were found to be quantifiable owing to the absence of interference peaks.

1,2-Dehydrosalsolinol (free + sulfoconjugate) was detectable in the urine of 10 patients out of 12; the daily excretion ranged from 3.2 to 38.3 nmoles. It is worth noting that in this study urine was not tested for the presence of the glucuroconjugate of 1,2-dehydrosalsolinol.

Discussion

Using the same methodology, neither the R nor the S enantiomer of salsolinol was detectable in the urine of 9 parkinsonian patients in the absence of L-dopa therapy (Dostert et al., 1989), whereas, in this study, (R)-salsolinol was detectable in the urine of 5 out of 12 demented patients. With respect to the degree of clinical deterioration, the two groups of demented and parkinsonian patients were different. Most of the demented patients had rather long clinical history and scored high in Reisberg's scale, whereas most of the parkinsonian patients were "de novo" parkinsonians with a rather low score in Hoehn and Yahr's scale.

In healthy subjects, the urinary salsolinol concentrations are generally found to be above the limit of detection retained (14 pmol/ml). Therefore, the urinary concentrations of (R)-salsolinol in demented patients seem to be closer to those of healthy subjects than those of parkinsonians. However, the mean daily urinary excretion of 1,2-dehydrosalsolinol (free + sulfoconjugate) was similar in demented patients (this study) and parkinsonians (Dostert et al., unpublished results), being 13.3 ± 12.3 and 16.1 ± 16.0 nmoles (mean $\pm$ S.D.), respectively. That the two groups of patients had similar mean values of daily urinary output of 1,2-dehydrosalsolinol, whereas (R)-salsolinol formation seems to be totally impaired in parkinsonians, might indicate that the reduction step 1,2-dehydrosalsolinol $\rightarrow$ salsolinol is missing or is not, or poorly, functional in parkinsonians but still working in demented patients. Determination of the urinary concentrations of 1,2-dehydrosalsolinol glucuronide in demented and parkinsonian patients should allow the comparison of the urinary concentrations of total (free + sulfo + glucuroconjugate) 1,2-dehydrosalsolinol of these patients with those of controls and, thus, to get some indirect insight into DA brain concentrations in patients with degenerative dementia with respect to controls and parkinsonian patients. However, additional studies should include healthy volunteers and recently-diagnosed parkinsonian and demented patients of the same age range before drawing any conclusion.

References

Collins MA, Nijm WP, Borge GF, Teas G, Golfarb C (1979) Dopamine-related tetrahydroisoquinolines: significant urinary excretion by alcoholics after alcohol consumption. Science 206:1184–1186

Dordain G, Dostert P, Strolin Benedetti M, Rovei V (1984) Tetrahydroisoquinoline derivatives and Parkinsonism. In: Tipton KF, Dostert P, Strolin Benedetti M (eds) Monoamine oxidase and disease. Prospects for therapy with reversible inhibitors. Academic Press, London, pp 417–426

Dostert P, Strolin Benedetti M, Dedieu M (1987) Ratio of enantiomers of salsolinol in human urine. Pharmacol Toxicol 60 [Suppl 1]:13

Dostert P, Strolin Benedetti M, Dordain G (1988) Dopamine-derived alkaloids in alcoholism and in Parkinson's and Huntington's diseases. J Neural Transm 74:61–74

Dostert P, Strolin Benedetti M, Dordain G, Vernay D (1989) Enantiometric composition of urinary salsolinol in Parkinsonian patients after Madopar. J Neural Transm (P-D Sect)1:269–278

Reinikainen KJ, Paljärvi L, Halonen T, Malminen O, Kosma V-M, Laakso M, Riekkinen PJ (1988) Dopaminergic system and monoamine oxidase-B activity in Alzheimer's disease. Neurobiol Aging 9:245–252

Reisberg B, Ferris SM, De Leon MJ, Crook T (1982) The global deterioration scale for assessment of primary degenerative dementia. Am J Psychiatry 139:1136–1139

Sandler M, Bonham Carter S, Hunter KR, Stern GM (1973) Tetrahydroisoquinoline alkaloids: in vivo metabolites of L-dopa in man. Nature 241:439–443

Sjöquist B, Borg S, Kvande H (1981) Salsolinol and methylated salsolinol in urine and cerebrospinal fluid from healthy volunteers. Subst Alcohol Actions Misuse 2:73–77

Sjöquist B, Eriksson A, Windblad B (1982) Salsolinol and catecholamines in human brain and their relation to alcoholism. Prog Clin Biol Res 90:57–67

Correspondence: Dr. P. Dostert, Farmitalia Carlo Erba, Research and Development, Via C. Imbonati 24, I-20159 Milan, Italy.

Clinics

Alzheimer's disease – one, two or several?

C. G. Gottfries, K. Blennow, B. Regland, and A. Wallin

Department of Psychiatry and Neurochemistry, Gothenburg University, Gothenburg,
Sweden

Summary

In the seventies, the two disorders, Alzheimer's disease (AD) and senile dementia of the Alzheimer type (SDAT) were sampled into one group, the dementias of the Alzheimer type. This sampling cannot be justified from a scientific point of view. The phenomenology of the two disorders differentiate. In AD a symptomatology of parietal and temporal lobe dysfunction is present, while in SDAT the clinical picture is more diffuse and indicative of not only a cortical but also a subcortical dysfunction.

Brain imaging techniques indicate white matter low attenuation and central atrophy in SDAT, while in AD the cortical atrophy is predominant.

Laboratory data show that the SDAT group includes patients with vitamin B12 deficiency and increased platelet monoamine oxidase activity. In the CSF of patients with AD there are decreased concentrations of homovanillic acid.

While AD is considered a small, relatively homogeneous group of presenile dementias, SDAT is considered a heterogeneous group including several subgroups.

When Alzheimer in 1906 first described "a characteristic disease of the cerebral cortex" he had a 51-year-old women in mind. Kraepelin (1910) in the paper "Senile and Presenile Dementias" created the concept of Alzheimer's disease, and in his paper he states: "While the anatomical findings suggest that we are dealing with a particularly serious form of senile dementia, the fact that this disease sometimes starts already around the age of 40 does not allow this supposition. In such cases we should at least assume a "senium praecox", if not perhaps a more or less age-independent unique disease process." Later on in the European textbooks Alzheimer's disease was considered a form of presenile primary dementia. However, as neuropathologists have found Alzheimer lesions, e.g. senile plaques and fibrillary tangles, also in brains from patients who have a late onset primary degenerative dementia, usually named senile dementia, these forms have been called senile dementia of the Alzheimer type (SDAT). Especially in the United States these two forms have been sampled together into one form of Alzheimer dementia (AD/SDAT).

The main reasons for sampling AD and SDAT are as follows:

— similar clinical phenomenology;
— Alzheimer lesions are present in both disorders;
— there are significant correlations between Alzheimer lesions and behavioral disturbances.

None of these reasons are valid for sampling the two disorders. Although it is evident that there are similarities between the symptomatology in the two disorders, there are also differences. These differences will be discussed below.

It is evident that senile plaques and fibrillary tangles are found in AD, as well as in SDAT, and the only differences between the two disorders are quantitative. However, it has not been documented to what extent Alzheimer lesions delimit a homogeneous group from an etiological or clinical point of view. It is well known that Alzheimer lesions are found in normal aging, Down's syndrome and other types of dementia, e.g. vascular dementia, parkinsonism with dementia and dementia pugulistica. Alzheimer lesions delimit an Alzheimer encephalopathy which may, however, be the common pathway for different degenerative disorders.

It is also true that statistically significant correlations between Alzheimer lesions and behavioural disturbances have been demonstrated (Blessed et al., 1968). However, from these statistical correlations, it is not possible to draw any conclusions about a causal relationship.

Should the presenile form of AD be separated from the senile form of SDAT?

Differences in symptomatology

The clinical diagnosis of AD and SDAT depends upon a syndrome of impairment of memory and other cognitive functions, including attention, language, praxis, visuospatial abilities, problem-solving, and abstract reasoning. Emotional disturbances, such as depression, anxiety, fear-panic and restlessness may also be present (Merriam et al., 1988). It is usually assumed that the symptomatology in AD is similar to that in SDAT, although the SDAT symptomatology is somewhat milder. In some studies (Gustafson and Nilsson, 1982; Bråne et al., 1989), it was found that there were differences in the amount of irritability, anxiety, confusion and restlessness. These kinds of symptoms dominated in the SDAT group, possibly indicating subcortical involvement in this group.

In a prospective study at our own institute, preliminary results indicate that aphasia (anomia, comprehension deficit and fluent aphasia with increased or normal verbal output, and empty stereotyped language), apraxia (inability to carry out purposive voluntary movements, when it cannot be accounted for in terms of paresis, sensory loss or comprehension deficit, the patient is for example unable to dress) and also a visual agnosia (inability to name objects by sight which are readily identified by touch or hearing) and a more severe spatial disability (impaired abstract conception of space e.g. inadequate three-dimensional figure copying, tendency to get lost even in familiar surroundings) are more frequent in AD than in SDAT. In SDAT patients with equal severity of disease, there are more non-regional symptoms, such as reduced orientation, calculating ability, logical/analytical ability (inability to interpret proverbs and to find similarities), and also a more pronounced tendency to confusion (in-alertness, in-attention, impaired concentration, disorganized thinking) perceptual disturbances (delusions/hallucinations), anxiety and increased or diminished motor activity. As is evident, this symptomatology indicates a focal, cortical disturbance in AD, while the disturbance in SDAT is more general (see Fig. 1).

	AD	SDAT
APHASIA	++	+
APRAXIA	++	+
VISUAL AGNOSIA	++	+
SPATIAL DISABILITY	++	+
REDUCED ORIENTATION	+	++
REDUCED CALCULATING ABILITY	+	++
REDUCED LOGICAL ANALYTIC ABILITY	+	++
CONFUSION	+	++
"FRONTAL LOBE" SYMPTOMS	(+)	+
"SUBCORTICAL" SYMPTOMS	(+)	+

Fig. 1. Differences in symptomatology in presenile Alzheimer's disease (AD) and senile dementia of the Alzheimer type (SDAT)

Table 1. White matter low attenuation, central and cortical atrophy for the diagnostic categories

| Diagnosis | WMLA [a] | | Atrophy | | | |
| | | | Central [a] | | Cortical [b] | |
	0	+	0	+	0	+
AD	15 (88.2%)	2 (11.8%)	16 (94.1%)	1 (5.9%)	10 (58.8%)	7 (41.2%)
SDAT	6 (20.0%)	24 (80.0%)	11 (36.7%)	19 (63.3%)	11 (36.7%)	19 (63.3%)

White matter low attenuation (WMLA): 0, not significant; +, mild-marked. Central or cortical atrophy: 0, ordinary for age; +, mild or moderate atrophy for age. AD, early onset Alzheimer's disease; SDAT, late onset Alzheimer's disease. Differences (AD vs SDAT) according to chi-square test:
[a] p<0.0001; [b] NS

Brain imaging techniques

Brain imaging techniques have improved greatly in recent years, and the localization and the degree of morphological changes in the brain, when present, can be relatively clearly described. In an investigation by Wallin et al. (1989 a), morphological changes according to CT scans are described. White matter low attenuations (WMLA) were found in a higher frequency in the SDAT group than in the AD group. In fact, the SDAT group did not differ significantly from the group of vascular dementias. The frequency of central atrophy in the SDAT group was also significantly higher than in the AD group. There were no significant differences between the AD and SDAT groups with regard to cortical atrophy. The lack of subcortical findings on CT, as well as WLMA in AD patients, suggests that AD is mainly a cortical disorder, while in SDAT there is also an involvement of subcortical regions in the degenerative process (see Table 1).

Laboratory findings

When peripheral markers are studied, some differences have also emerged in the diagnosis of AD and SDAT. Regland et al. (1988 a) studied a group of patients including AD and SDAT. In this clinical sample the concentration of vitamin B12 in serum and the activity of MAO in platelets were investigated. In the AD group, the vitamin B12 levels were normal, while in the SDAT group 23% were found to have significantly reduced vitamin B12 concentrations (below 130 pmol/l) (Fig. 2). In the sample it was found that the demented patients with low vitamin B12 levels also had significantly increased concentrations of MAO in platelets, the latter finding indicating presence of immature platelets, possibly due to mild peripheral deficiency of vitamin B12.

When studying the cerebrospinal fluid (CSF) some differences have also emerged between AD and SDAT patients. As has been shown by several groups (see Gottfries, 1983; Bråne et al., 1989), significantly lower concentrations of homovanillic acid (HVA) were found in the CSF of AD patients as compared with SDAT patients and controls. It is of interest that there was a significant increase of 3-methoxy 4-hydroxy-phenylglycol (HMPG) in SDAT patients as compared with controls. No such increase of HMPG was found in the AD patients (Table 2).

Table 2. Age, age at onset, duration of the disorder and monoamine metabolites (nmol/l) in CSF in a group of AD (onset <65 yrs), and a group of SDAT (onset ≥65 yrs) patients compared with a group of controls (means±SD). Significances of group differences were calculated using Student's t-test and Wilcoxon rank sums tests

Variables	Controls (n=26)	AD (n=13)	SDAT (n=28)
Age	64± 7	64± 8	+++ 78± 6 [c]
Age at onset	–	59± 5	+++ 73± 5
Duration	–	6± 5	5± 3
MHPG	48± 10	52±11	58± 22 [a]
HVA	276±105	153±40 [c]	+252±139
5-HIAA	137± 47	109±61	130± 45

[a] $p < 0.05$ and [c] $p < 0.005$ for differences to controls.
+ $p < 0.05$ and +++ $p < 0.005$ for differences between AD and SDAT groups (after Bråne et al., 1989)

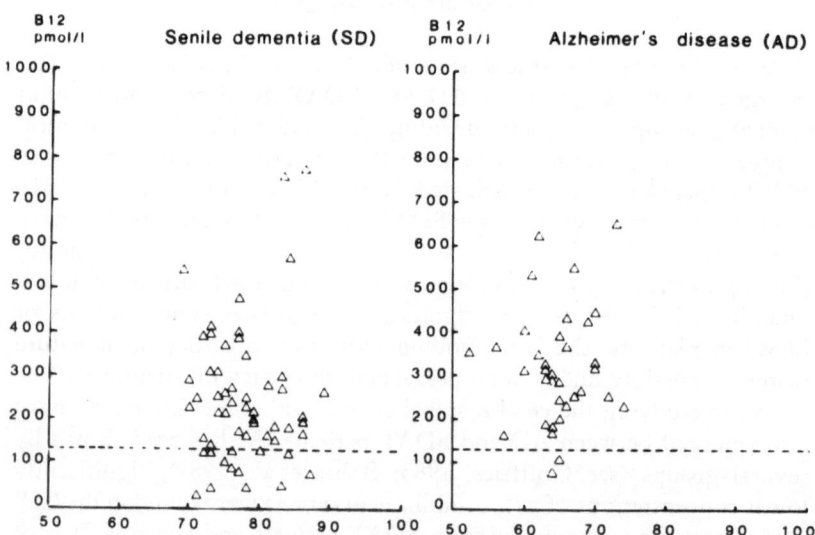

Fig. 2. Vitamin B12 concentrations (pmol/l) in serum from patients with presenile Alzheimer's disease (AD) and senile dementia of the Alzheimer type (SDAT). 130 pmol/l is considered the lower limit of normal distribution

Postmortem human brain investigations

Neurochemical differences between AD and SDAT have also been stressed in postmortem human brain investigations (Rossor et al., 1984; Gottfries, 1985). Usually the neurotransmitter disturbances seen in early onset dementia are more severe than those seen in late onset dementia. Potmortem studies of white matter changes (Brun and Gustafson, 1976; Gottfries et al., 1985; Svennerholm et al., 1988) have also shown that the recorded white matter changes are more severe in the group of demented patients with late onset of the disorder, which is in line with the CT scan findings mentioned above.

Epidemiological investigations

According to the epidemiological investigations by Sjögren et al. (1952) and Sourander and Sjögren (1970), the two disorders AD and SDAT behave like two from each other different disorders. There is a four-fold increased risk for SDAT in siblings of SDAT patients. No instances of AD were found in the 2675 family members investigated. In other studies of familial aggregation in the AD/SDAT complex, Heston (1977) found that relatives of AD patients ran a markedly increased risk of getting AD. The highest risk among relatives of getting dementia was found in siblings of autopsied Alzheimer patients with comparatively early onset of the disorder. In this group the cumulative risk was 50:50. The lowest risk was found in siblings of patients who had developed a dementia disorder after the age of 70 years.

When the symptomatology, the laboratory findings, the result from brain imaging, the postmortem findings and the epidemiological findings are taken together it is obvious that there are several arguments for distinguishing AD from SDAT.

Is AD a homogeneous group?

According to Jorm et al. (1987) and Gottfries et al. (1986), there is a strong relation between the frequencies of dementias and age. In the age group 65–70, less than 1% of the total population is demented, indicating that AD is a fairly uncommon disorder. In the age group above 80 years, the frequency of dementias is 20% or higher.

It is natural to assume that the rather malignant early onset AD is a small homogeneous group. However, in an investigation by Mayeux et al. (1985) it was shown that patients with AD (n=110) could be subdivided into four groups: patients with extrapyramidal symptoms, patients with myoclonic symptoms, patients with a benign form of dementia, and the typical form of AD. The subdividing of the dementias was not related to the age at onset. Extrapyramidal symptoms were found in 21% of the individuals investigated, and these patients had a higher prevalence of psychotic behaviour and a greater mental reduction. According to Kaye et al. (1988), a subgroup of AD with extrapyramidal features also have reduced concentrations of HVA in CSF to a greater extent than other AD patients.

In the study by Mayeux et al. (1985), myoclonus was found in 9.9%. Seven patients of the total number, 110, had a benign form of dementia, yet those patients met the criteria for dementia at the assessment. It was of interest that the benign form did not progress to the same extent as the other.

The above-mentioned findings indicate that the clinical diagnosis of AD may include subgroups of dementias.

Is SDAT a homogeneous group?

In fact, SDAT is a dementia disorder that is delimited by the age at onset and the absence of evident explanations to the mental impairment. It is an exclusion diagnosis and therefore it may very well be a heterogeneous group. Some subgroups of SDAT are suggested below.

Normal aging

Several investigations have shown that in normal aging there is an age-related decrease in the activity of neurotransmitter systems (see Gottfries, 1985). This decrease seems to have its onset in the 60s but, due to reserve capacities of the human brain, behavioural disturbances in the aging process do usually not become evident until the age of 80–85 years. It can be assumed, however, that after the age of 85 years the normal aging process, possibly in combination with negative environmental factors, may cause such cognitive impairment that a dementing disorder is diagnosed. From a clinical point of view, this type of high-age dementia cannot be separated from SDAT. It can, indeed, also be questioned whether it is possible to differentiate these disorders by histological investigations of the brain, as the amount of senile plaques and fibrillary tangles increase with age.

Secondary dementias

According to investigations by Regland et al. (1988) vitamin B12 deficiency is known to be associated with SDAT. In patients with AD there were normal concentrations of vitamin B12. It is well known that vitamin B12 deficiency goes with changes in the spinal cord, which has also been reproduced in animal experiments on rhesus monkeys (Scott et al., 1981). A brain lesion in pernicious anaemia has been described which is not different from that which occurs in the spinal cord and which consists essentially of degeneration of white matter in the cerebrum. In the study by Regland (1990) patients were also investigated for the platelet content of MAO. It was shown that in the group diagnosed as SDAT, the frequency of pathological low vitamin B12 levels was 23%, and there was also significantly higher platelet MAO activity in the group with low B12 levels. As also shown by Regland (1990), the patients with SDAT and low vitamin B12 levels had atrophic gastritis, as marked by reduced pepsinogen I concentrations in serum. When patients with low vitamin B12 levels were treated with vitamin B12, the high MAO activity in platelets was significantly reduced (Regland, 1990). As the cobalamin molecule is actively transported over membranes, the reduced concentrations of vitamin B12 in the group of late onset dementias can possibly be considered a marker of reduced membrane capacity. It can thus be assumed that other essential nutrients, such as folic acid, vitamin B6, tryptophane, tyrosine, zinc, selenium etc. are also transported insufficiently over membranes, giving rise to brain dysfunction. It is obvious that, although these patients were originally diagnosed as SDAT, the findings of reduced vitamin B12 indicate that the diagnosis of secondary dementia may be more adequate.

Immunological factors

Amyloid is often seen in diseases with dysfunction of the immune system. Due to this, immunological disturbances have been considered as etiological or pathogenetic factors in AD and SDAT. Ishii and Haga (1976) reported the presence of immunoglobulin in amyloid of senile plaques and also in the walls of small vessels. Singh and Fudenberg (1986) found antibodies binding to rat brain tissue in 57% of sera in patients with AD, compared with none in aged healthy controls. In a study on CSF in patients with AD Williams et al. (1980) also found banding in the gamma-globulin region in 5 out of 8 patients with AD. In a study at our institute (Blennow et al., 1989), a quantitative determination of albumin, IgG and IgM in serum and CSF was made. The combined group of AD and SDAT showed significantly higher IgG and IgM index values than the control group. Within the AD/SDAT group 12 out of 44 patients (26%) had an intrathecal synthesis of immunoglobulins, while there was none in the controls. None of the patients with a synthesis of immunoglobulins had signs of blood-brain barrier damage. Nor were there laboratory data indicative of virus infections. It has to be considered that a degenerative disease process through the production of tissue damage may lead, secondarily, to the development of an autoimmune process. However, the patients with an intrathecal immunoglobulin synthesis were not more severely demented than those with a normal CSF.

In an investigation by McRae Degueurce et al. (1988), antibodies in the CSF in AD and SDAT patients were shown to recognize neuronal populations and components in the medial septum and the spinal motor neurons in rodents infused with a chemical mixture that fixes small neurotransmitter molecules. In view of the results presented, it is tempting to discuss the hypothesis that morphological changes (plaque formations), as well as a decrease in presynaptic cholinergic markers in AD and SDAT, may be triggered by abnormalities in the immune system.

Confusion states

In DSM III R it is stated that a diagnosis of dementia should be made only in the absence of severe confusion. Ratings of patients with senile dementia at our institute have shown that a mild confusion is a very common symptom in patients with SDAT (Andersson, 1989). Preliminary data indicate that some of the patients at present diagnosed as SDAT, in fact, have a mild confusion in which the cognitive impairment fluctuates around the clock. The "sundown syndrome" is common in these patients. The slowly progressive course of this disorder imitates the clinical picture of SDAT. The question whether these patients have a cortical dementing disorder or a subcortical disturbance with confusion as the main symptom is being discussed. It seems obvious, however, that the group of SDAT includes a subgroup of patients with mild confusion.

Vascular non-MID patients

According to Wallin et al. (1989 b), some patients with vascular risk factors have a dementia of a non-MID type, and this is named probable vascular dementia (PVD). It may have insidious onset and slowly progressing course. Sometimes patients with PVD may have had stroke attacks, but if brain imaging is not done, these attacks may remain unidentified. Erroneous diagnoses of SDAT cannot be excluded for these cases.

Conclusions

At present, cumulative data indicate that the presenile form of Alzheimer's disease should be separated from senile dementia of the Alzheimer type.

The presenile form of Alzheimer's disease seems to be a relatively homogeneous group, although some clinical features indicate subgroups.

Senile dementia of the Alzheimer type, diagnosed according to DSM III R, is an exclusion diagnosis, and the concept is used for primary degenerative disorders with late onset (after the age of 65 years) and when no other possible explanation of the dementia can be given. The group is assumed to include several subgroups. High age (> 85 yrs) and negative environmental factors may go with cognitive impairment that is difficult to separate from SDAT. The deficiency of essential nutrients due to degenerative changes in the mucus membrane of the gut may also cause a brain dysfunction that is not possible to differentiate from SDAT. A subgroup of AD and SDAT patients also show disturbances of the immunological system, which is indicative of autoimmune disorders in AD and SDAT. Thus, a relatively large part of the SDAT group can possibly be secondary dementias.

The sampling of AD and SDAT into one large group named Alzheimer type dementia and "the disease of the century" has, indeed, created great interest in dementing illness. Politicians, as well as doctors and researchers, have displayed a keen interest. As a consequence, the status of patients and their relatives has risen substantially. However, from a scientific point of view, the sampling of AD and SDAT has not been to advantage. On the contrary, it has been a step back. AD and SDAT must be distinguished, and at least SDAT, which is an exclusion diagnosis, can be assumed to be a heterogeneous group, including several different forms of dementing illness. The question whether senile plaques and fibrillary tangles occur in the brains of patients included in this subgroup comes naturally. In a prospective study we will try to find an answer. However, the question is not particularly relevant. If Alzheimer lesions are assumed to be of a secondary nature, eventually taking place in all dementing illness, there seems to be no reason to press these clinical diagnoses into the boxes of the neuropathology. It might, in fact, be hindering. The clinical diagnosis should also be related to neurotransmitter deficits, disturbances of the calcium homeostasis or disturbed glucose metabolism.

References

Andersson M (1989) Institutionalization of the elderly and characteristics of nursing home patients. In: Andersson M (ed) Elderly patients in nursing homes and in home care. Reports from the Department of Psychiatry and Neurochemistry, St. Jörgen's Hospital, Gothenburg University, No. 18. Doctoral thesis

Blennow K, Wallin A, Davidsson P, Fredman P, Gottfries CG, Karlsson I, Svennerholm L (1989) Intrathecal synthesis of immunoglobulins in patients with Alzheimer's disease. Submitted

Blessed G, Tomlinson BE, Roth M (1968) The association between quantitative measures of dementia and of senile change in the cerebral grey matter of elderly subjects. Br J Psychiatry 114: 797–811

Bråne G, Gottfries CG, Blennow K, Karlsson I, Lekman A, Parnetti L, Svennerholm L, Wallin A (1989) Monoamine metabolites in cerebrospinal fluid and behavioural ratings in patients with early and late onset of Alzheimer dementia. Alz Dis Assoc Disord 3:148–156

Brun A, Gustafson L (1976) Distribution of cerebral degeneration in Alzheimer's disease. A clinicopathological study. Arch Psychiatr Nervenkr 223:15–33

Gottfries CG (1983) Biological changes in blood and cerebrospinal fluid. In: Reisberg B (ed) Alzheimer's disease. The standard reference. The Free Press, Collier MacMillan, London, pp 122–130

Gottfries CG (1985) Definition of normal aging, senile dementia and Alzheimer's disease. In: Gottfries CG (ed) Normal aging, Alzheimer's disease and senile dementia. Aspects on etiology, pathogenesis, diagnosis and treatment. Proceedings of two symposia held at the 14th CINP Congress, June 22–33, 1984, Florence, Italy. L'Université de Bruxelles, Brussels, pp 11–17

Gottfries CG, Karlsson I, Svennerholm L (1985) Senile dementia – a "white matter" disease? In: Gottfries CG (ed) Normal aging. Alzheimer's disease and senile dementia. Aspects on etiology, pathogenesis, diagnosis and treatment. Proceedings of two symposia held at the 14th CINP Congress, June 22–23, 1984, Florence, Italy. L'Université de Bruxelles, Brussels, pp 11–17

Gottfries CG, Bartfai T, Carlsson A, Eckernäs SÅ, Svennerholm L (1986) Multiple biochemical deficits in both gray and white matter of Alzheimer brains. Prog Neuropsychopharmacol Biol Psychiatry 10:405–413

Gustafson L, Nilsson L (1982) Differential diagnosis of presenile dementia on clinical grounds. Acta Psychiatr Scand 65:194–209

Heston LL, Mastri AR, Anderson E, White J (1981) Dementia of the Alzheimer type: clinical genetics natural history and associated conditions. Arch Gen Psychiatry 38:1085–1090

Ishii T, Haga S (1976) Immuno-electron microscopic localization of immunoglobulins in amyloid fibrils of senile plaques. Acta Neuropathol (Berl) 36: 243–249

Jorm AF, Korten AE, Henderson AS (1987) The prevalence of dementia: a quantitative integration of the literature. Acta Psychiatr Scand 76: 465–479

Kaye JA, May C, Daly E, Atack JR, Sweeney DJ, Luxenberg JS, Kay AD, Kaufman S, Milstein S, Friedland RP, Rapoport SI (1988) Cerebrospinal fluid monoamine markers are decreased in dementia of the Alzheimer type with extrapyramidal features. Neurology 38: 554–557

Kraepelin E (1910) Senile and pre-senile dementias. Johann Ambrosius Barth, Leipzig, pp 533–554, 593–632

Mayeux R, Stern Y, Spanton S (1985) Heterogeneity in dementia of the Alzheimer type. Evidence of subgroups. Neurology 35: 453–461

McRae Degueurce A, Serney B, Haglid K, Rosengren L, Karlsson JE, Karlsson I, Wallin A, Svennerholm L, Gottfries CG, Dahlström A (1988) CSF from Alzheimer's patients contain antibodies recognizing cholinergic neurons in the rodent central nervous system. Submitted

Merriam A, Aronson MK, Gaston T, Wei SL, Katz I (1988) The psychiatric symptoms of Alzheimer's disease. J Am Geriatr Soc 36: 7–12

Regland B, Gottfries CG, Oreland L, Svennerholm L (1988) Low B12 levels related to high activity of platelet MAO in patients with dementia disorders. A retrospective study. Acta Psychiatr Scand 78: 451–457

Regland B (1990) Abnormalities of vitamin B-12 and other essential nutrients in dementia of Alzheimer type, and their clinical relevance. In: Fowler Ch J, Carlson LA, Gottfries CG, Winblad B (eds) Biological markers in dementia of Alzheimer type. Proceedings of the Stiftelsen Gamla Tjänarinnor Symposium on aging and aging disorders No. 1. Smith Gordon and Company Limited, London, pp 215–219

Rossor MN, Iversen LL, Reynolds GP, Mounjoy CQ, Roth M (1984) Neurochemical characteristics of early and late onset types of Alzheimer's disease. Br Med J 288: 961–964

Scott JM, Dinn JJ, Wilson P, Weir DG (1981) Pathogenesis of subacute combined degeneration: a result of methyl group deficiency. Lancet ii: 334–337

Singh VK, Fudenberg HH (1986) Detection of brain autoantibodies in the serum of patients with Alzheimer's disease but not with Down's syndrome. Immunol Lett 12: 277–280

Sjögren T, Sjögren H, Lindgren AGH (1952) Morbus Alzheimer and morbus Pick. A genetic, clinical and patho-anatomical study. Acta Psychiatr Neurol Scand [Suppl 82]

Sourander P, Sjögren H (1970) The concept of Alzheimer's disease and its clinical implications. In: Wolstenholme FEW, O'Connor M (eds) Alzheimer's disease and related conditions. Churchill, London, pp 11–36

Svennerholm L, Gottfries CG, Karlsson I (1988) Neurochemical changes in white matter of patients with Alzheimer's disease. In: Serlupi Crescenzi G (ed) A multidisciplinary approach to myelin disease. Plenum, New York, pp 319–328

Wallin A, Blennow K, Gottfries CG, Långström G, Uhlemann C (1989 a) White matter low attenuation on computed tomography in Alzheimer's disease and vascular dementia. Acta Neurol Scand 80:518–523

Wallin A, Alafuzoff I, Carlsson A, Eckernäs S, Gottfries CG, Svennerholm L, Winblad B (1989 b) Neurotransmitter deficits in a non-multi-infarct category of vascular dementia. Acta Neurol Scand 79: 397–406

Williams A, Papadopoulos N, Chase TN (1980) Demonstration of CSF gamma-globulin banding in presenile dementia. Neurology 30: 882–884

Correspondence: Prof. C. G. Gottfries, Department of Psychiatry and Neurochemistry, Gothenburg University, St. Jörgen's Hospital, S-422 03 Hisings Backa, Sweden.

Diagnostic criteria of Alzheimer's disease

L. Gustafson[1], A. Brun[2], A. Johansson[1], and J. Risberg[3]

Departments of [1] Psychogeriatrics, [3] Psychiatry, and [2] Institute of Pathology,
Department of Neuropathology, University of Lund, Sweden

Summary

Alzheimer's disease (AD) with presenile onset is characterized by a consistent pattern of cognitive, emotional and neurological symptoms, in accordance with the accentuated degeneration of temporo-parietal and temporal limbic grey matter. The clinical picture is influenced by the age at onset and the duration of the disease and may be confounded by co-existent white matter changes. Differentiation between AD, vascular dementia and dementia with fronto-temporal degeneration of non-Alzheimer type, is possible by the use of clinical methods, e.g. special rating scales, which are described and validated.

Introduction

Since the first presentation in 1907, Alzheimer's disease (AD) has been described with increasing precision. Aloys Alzheimer, in the first case report mentioned several of the core symptoms of presenile AD, such as memory failure, spatial disorientation, language disturbances and dyspraxia. Other early studies added epileptic seizures and extrapyramidal signs to the clinical description and discussed differential diagnosis against arteriosclerotic brain disease (Perusini, 1910). The clinical significance of a common pattern of mental dysfunctions has been agreed upon in several studies (Sjögren, 1952; Brion, 1966; Lauter, 1970; Brun and Gustafson, 1976) and a three stage model has been suggested to describe the order in which these symptoms appear during the course of the disease (Sjögren, 1952). The present paper will focus upon the clinical criteria for diagnosis of presenile AD and the validity of ante mortem diagnosis based on these criteria.

Material and methods

The results are mainly based on a longitudinal prospective study of presenile and senile dementia that started twenty years ago. The patients have been studied repeatedly with psychiatric examinations, psychometric testing, regional cerebral blood flow (rCBF) measurements and EEG. Neuroradiology was performed in most cases. The psychiatric assessment was based on a rating scale with 77 items, describing various cognitive, emotional and personality changes in organic dementia (Gustafson and Hagberg, 1975). We managed to follow the patients from the very early to the terminal stage of different dementing disorders and to compare early clinical findings and the tentative diagnoses with the neuropathological findings. Neuropathological data have been obtained in about 80% (more than 250 cases) of the deceased cases. The brain pathology has been analysed with a technique that makes it possible to compare clinical features with the distribution of lesions in the brain.

For comparison three groups of patients with dementia and Alzheimer encephalopathy verified postmortem were selected:

1. Twenty-one patients with presenile AD. The mean age at onset of dementia was 56.7 (range 45–64) years, the mean age at death was 67.3 (range 56–75) years and the mean duration of the dementia was 10.6 (range 5–16) years.
2. Twenty patients with senile dementia of Alzheimer type (SDAT). The mean age at onset of dementia was 75.6 (range 65–85) years. The mean age at death was 81.5 (range 67–92) years and the duration of the disease was 6.0 (range 2–10) years.
3. Down's syndrome dementia (DSD). Nine patients with the clinical characteristics of Down's syndrome and chromosome 21 trisomy. The mean age at death was 53.9 (range 37–66) years and the mean duration of the dementia was 5 years.

The results from these groups will be compared with those from postmortem verified cases with vascular dementia and cases with frontal lobe degeneration of non-alzheimer type (FLD) belonging to the wider entity fronto-temporal dementia (FTD), which also includes cases with Pick's disease (Brun, 1987; Gustafson, 1987).

Diagnostic criteria

The clinical picture in presenile AD shows a number of symptoms and signs with more or less specific relationship to the distribution and the severity of the disease. The selection of diagnostic criteria has to consider both frequency and specificity of these symptoms and symptom clusters. The majority of presenile AD cases develop dysmnesia, dysphasia, dysgraphia, dysgnosia, dyspraxia and spatial disorientation. It seems justified to relate most of the cognitive defects to the focally accentuated degeneration of the temporo-parietal association cortex and temporal-limbic structures (Brun and Gustafson, 1976). There are important, individual variations concerning the severity and the order of appearance of the symptoms, although most of them were manifested at the second stage of the disease. Dysphasia was the first reported symptom in about 25% of our presenile AD cases and also changes of handwriting appeared early. The memory failure, appearing early in presenile AD, affects spatial as well as verbal functions and short-term as well as long-term memory. Logoclonia observed in one third of the cases, usually in the last stage was occasionally an early symptom. About 80% of the presenile AD cases developed an increase of muscle tone, that might lead to a misdiagnosis of Parkinson's disease. Tremor was observed in one third of the cases comparatively late, as also cog-wheel phenomena. The extrapyramidal signs could not be explained by neuroleptic treatment and their relationship to the degenerative brain changes is still an open question. Epileptic seizures of grand mal type and myoclonic twitchings with late onset were observed in about 50% of our cases. The age dependency of symptoms in AD was seen as a decreasing focalization of the temporo-parietal symptom pattern and a lower frequency of the neurological symptoms in the SDAT group, confirming the findings of Lauter (1970).

The cognitive and the neurological symptoms in presenile AD form a rather consistent symptom pattern, while the emotional changes are more unspecific. Alzheimer (1907) described how initially the primary defect symptoms may partly be overshadowed by secondary adaptive, insufficiency and psychotic reactions such as anxiety and paranoia. Under optimal circumstances, however, the patients' capacity for emotional and social contacts seems to be relatively preserved. This has been related to a sparing of precentral and anterior cingulate cortex, at least at an earlier stage of presenile AD (Brun and Gustafson, 1976). A minority of cases also present a frontal symptom pattern but the differential diagnosis against FTD is possible. The clinical picture in FTD is dominated by early personality changes and progressive dynamic aphasia, while spatial functions are comparatively spared (Gustafson, 1987). Certain symptoms such as fainting spells, severe dizziness and fluctuations of the clinical course which usually arouse a suspicion of an ischemic brain disorder, are not uncommon in presenile AD. It might be justified to relate these observations to the selective incomplete white matter infarctions (SIWI) described by Brun, in this volume.

There is a strong similarity both neuropathologically and clinically between presenile AD and DSD (Table 1). Memory failure, disorientation and reduction of language functions were reported in both diseases and also late onset epilepsy and myoclonia. There were often in Down's syndrome, a connection in the time between onset of epilepsy and the

Table 1. Symptoms in presenile Alzheimer's disease (AD) and in Down's syndrome dementia (DSD)

	Presenile AD (N=21)	DSD (N=9)
Age at death	67.3 + 5.2	53.9 + 9.6
Dysmnesia	+ +	+ +
Dysphasia	+ +	+ +
Dyspraxia	+ +	+ +
Spatial disorientation	+ +	+ +
Epileptic seizures	+(+)	+ +
Myoclonic twitchings	+(+)	+(+)
Syncopal attacks	+	+(+)
Increase of muscle tone	+ +	?

Frequency of symptoms: + = 20–39%; +(+) = 40–69%; + + = 70–100%; ? = reliable information not available

first signs of cognitive deterioration. The significance of the parietal lobe involvement in AD is supported by concurrent findings in a prospective study of ageing and dementia in 22 Down cases. A reduction of visuo-spatial functions was an early indication of the cognitive decline in elderly Down patients as also an rCBF decrease in the temporo-parietal cortex in contrast to the previous normal rCBF results. EEG is almost pathological at an early stage of presenile AD in contrast to the 'normal' EEG often reported in early FTD (Gustafson, 1987).

Diagnostic rating scales

A standardised procedure for diagnosis of AD was based on three diagnostic rating scales developed for recognition of AD, FTD and multi-infarct dementia respectively. The general idea was that a clinical diagnosis should be based on positive criteria for the disease and not only on exclusion of other possible causes. The three rating scales were used in combination and the scoring profile for each patient formed the basis for a tentative clinical diagnosis.

The rating scale for diagnosis of AD, was developed from data on confirmed Alzheimer cases (Gustafson and Nilsson, 1982). Twelve items (Table 2) were selected from the full psychiatric scale with 77 items used in the longitudinal study (Gustafson and Hagberg, 1975). The selection and the scoring of the items were based on item analysis according to Likert. The maximum AD score is 17 points. A score greater than 5 and

Table 2

Rating scale for diagnosis of Alzheimer's disease		Rating scale for diagnosis of Dementia with frontotemporal degeneration	
Item	Scores	Item	Scores
Slow progression	1	Slow progression	1
Early loss of insight	1	Early loss of insight	2
Early amnesia for remote events	2	Early signs of disinhibition	2
Early spatial disorientation	2	Irritability, dysphoria	1
Apraxia, aphasia, agnosia	2	Confabulation	1
Logoclonia	2	Logorrhea	1
Logorrhea	1	Progressive reduction of spontaneity of speech	1
Progressive reduction of spontaneity of speech	1	Echolalia, mutism, amimia	2
Epileptic seizures of late onset	1	Klüver-Bucy syndrome	1
Increased muscular tension	2		
Myoclonic twitchings	1		
Klüver-Bucy syndrome	1		

especially exceeding 8 points, suggests an Alzheimer diagnosis. The AD score correlates significantly (p < 0.001) with the duration of dementia in presenile AD cases.

The rating scale for diagnosis of FTD is based even more on the clinical picture of the early stage of the disease (Gustafson and Nilsson, 1982). The scoring of the 9 items is based on item analysis and a score above 5 points suggests fronto-temporal cortical degeneration. There was no correlation between FTD score and the duration of dementia in FTD cases.

The ischemic score (IS) is used with the original scoring developed by Hachinski et al. (1975). IS correlates negatively (p < 0.001) with AD score and FTD score in patients with different types of dementia (n = 121). The validity of the diagnoses based on the three rating scales have been evaluated by comparison with neuropathological diagnoses and rCBF findings.

The relationship between the diagnostic scores and the neuropathological diagnoses in 85 demented patients is shown in Fig. 1. The scoring profile for each diagnostic group is characteristic and the agreement between individual scoring profiles and neuropathological diagnosis is also satisfactory. The diagnostic decisions are, however, difficult in some individual cases such as patients with mixed vascular and Alzheimer type dementia and in advanced FTD (Brun and Gustafson, 1988).

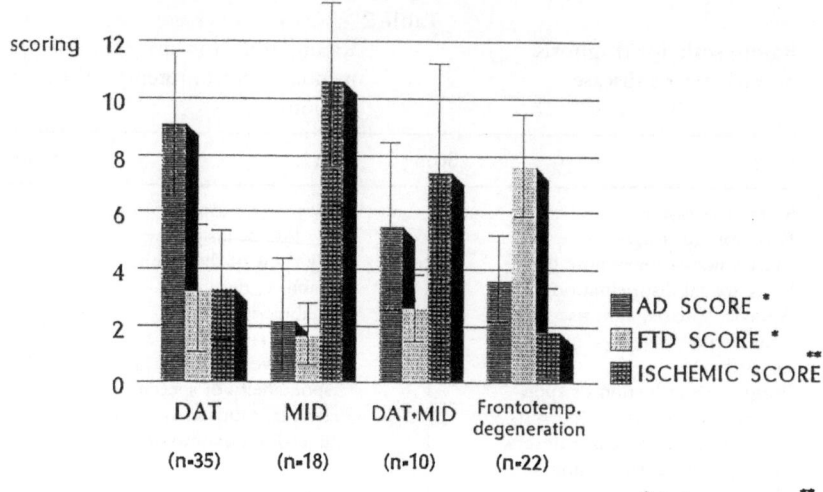

Fig. 1. Diagnostic scoring in post-mortem verified dementia (n = 85)

The diagnostic scoring was also compared with the rCBF findings in 121 patients with dementia (Risberg and Gustafson, 1988). rCBF was measured with the ^{133}Xe-inhalation technique using a 32 detector equipment. Groups with tentative clinical diagnosis of presenile AD (n = 28), SDAT (n = 27), FTD (n = 22) and vascular dementia (n = 44) showed significantly different regional patterns of rCBF pathology. Cases presumed to represent a mixture of vascular dementia and AD were excluded. 60 patients have died during the follow-up period and post-mortem examinations in 49 cases have confirmed the main diagnosis in all but one case: a case diagnosed as vascular dementia turned out to have a Creutzfeldt-Jakob disease.

Discussion

Most clinico-pathological studies of AD with presenile onset agree, regarding the existence of a consistent pattern of cognitive, emotional and neurological symptoms, several of which are found already at an early stage of the disease. The symptoms, especially the cognitive defects, are closely related to the structural and functional consequences of the underlying brain disease, as shown by autopsy and brain imaging techniques. The pathoanatomical correlates to extrapyramidal signs, epileptic seizures and logoclonia have not been clarified to the same extent. Frontal lobe symptoms, which are less dominating in presenile AD, tend to increase with the patients age at onset and the duration of the disease. The frontal lobe symptoms might partly be due to SIWI and associated with other symptoms of brain ischemia. Differential diagnosis against FTD is possible, since the symptom pattern at an early stage of presenile AD is almost contrary to that of early FTD.

The diagnostic procedure based on three rating scales described here, is one of several possible ways to improve the clinical differentiation between different types of dementia. The criteria for dignosis of AD were selected among a large number of clinical features which were evaluated repeatedly in patients with progressing dementia. The selection was based on frequency and intercorrelations between symptoms and their relationship to the distribution of brain degeneration in AD. The validity of the diagnostic procedure was strongly supported by postmortem diagnoses and rCBF findings. The positive correlation between AD score and duration of the disease indicates a limitation concerning early diagnosis based on this scale. This might be partly compensated for by using the scoring profile based on all three rating scales. The evaluation of a limited number of symptoms seems to be a simple way to improve the differentiation between the presenile dementias. A prerequisite for the present diagnostic procedure is, however, access to reliable clinical information. Early diagnosis of AD on clinical grounds remains a difficult task and clinical criteria have to be used in combination with other diagnostic tools such as psychometric testing, rCBF and other brain imaging techniques.

Acknowledgements

This study was supported by the Swedish Medical Research Council (grant No. 3950) and Greta and Johan Kock Foundation.

References

Alzheimer A (1907) Über eine eigenartige Erkrankung der Hirnrinde. Allg Z Psychiatr 64: 146–148

Brion S (1966) Démences par atrophie cérébrale primitive. Le Concours medical 15-I-88-3, Dossier EPU 31: 313–324

Brun A, Gustafson L, Englund E (1990) Morphology of white matter subcortical dementia in Alzheimer's disease. In: Maurer K, Riederer P, Beckmann H (eds) Alzheimer's disease. Epidemiology, neuropathology, neurochemistry and clinics. Springer, Wien New York, pp 79–83 (Key Topics in Brain Research)

Brun A, Gustafson L (1976) Distribution of cerebral degeneration in Alzheimer's disease. A clinico-pathological study. Arch Psychiat Nervenkr 223: 15–33

Brun A, Gustafson L (1988) Zerebrovaskuläre Erkrankungen. In: Kisker KP, et al (eds) Organische Psychosen. Psychiatrie der Gegenwart 6, 3. Aufl. Springer, Berlin Heidelberg New York Tokyo, S 253–295

Gustafson L (1987) Frontal lobe degeneration of non-Alzheimer type. II. Clinical picture and differential diagnosis. Arch Gerontol Geriatr 6: 209–223

Gustafson L, Hagberg B (1975) Dementia with onset in the presenile period. Acta Psychiatr Scand [Suppl 257]: 1–71

Gustafson L, Nilsson L (1982) Differential diagnosis of presenile dementia on clinical grounds. Acta Psychiatr Scand 65: 194–209

Hachinski VC, Iliff LD, Zilkna E, Boulay GH du, McAllister VL, Marshall J, Ross Russell RW, Symon L (1975) Cerebral blood flow in dementia. Arch Neurol 32: 632–637

Lauter H (1970) Über Spätformen der Alzheimerschen Krankheit und ihre Beziehung zur senilen Demenz. Psychiat Clin 3: 169–189

Perusini G (1910) Über klinisch und histologisch eigenartige psychische Erkrankungen des späteren Lebensalters. Histol Histopath Arb 3: 297–351

Risberg J, Gustafson L (1988) Regional cerebral blood flow in psychiatric disorders. In: Knezewic S, et al (eds) Handbook of regional cerebral blood flow. Lawrence Erlbaum, Hillsdale, New Jersey, pp 219–240

Sjögren H (1952) Clinical analysis of Morbus Alzheimer and Morbus Pick. Acta Psychiat Neurol [Suppl 82]: 69–139

Correspondence: L. Gustafson, M.D., Department of Psychogeriatrics, University Hospital, PO Box 638, S-220 06 Lund, Sweden.

Clinical diagnosis of Alzheimer's disease: DSM-III-R, ICD-10 – what else?

H. Lauter, A. Kurz, M. Haupt, B. Romero, and **R. Zimmer**

Psychiatrische Klinik der Technischen Universität München,
Federal Republic of Germany

Summary

Four sets of diagnostic criteria for Alzheimer's disease were applied in parallel to 150 patients with suspected or manifest but not severe mental deterioration and were validated against the final diagnosis derived from the subsequent course within a 12 to 24 months follow-up period. DSM-III-R criteria for Primary Degenerative Dementia of the Alzheimer Type and NINCDS-ADRDA criteria for Probable Alzheimer's Disease were superior to ICD-10 criteria for Dementia in Alzheimer Disease in terms of sensitivity, specificity, and relative predictive value of positive classification.

Several investigations have recently demonstrated that the clinical diagnosis of Alzheimer's disease can be confirmed in a high percentage of cases by brain biopsy or by postmortem neuropathological examination (Joachim et al., 1988; Morris et al., 1988; Thierney et al., 1988; Wade et al., 1988; Boller et al., 1989). Validity measures of 80 to 90 percent attained in most of these studies might however been biased by the fact that patients had been observed longitudinally and had reached fairly advanced stages of the disorder, which obviously could have facilitated the recognition of different forms of dementing processes. Therefore the question arises how often a clinical diagnosis of Alzheimer's disease can be correctly made on the basis of cross-sectional data and in patients whose mental capability may have not severely deteriorated.

During past decades the area of psychoorganic syndromes and dementing disorders has been characterized by a confused terminology and by a variety of vague and conflicting concepts (Lauter, 1988). The

world-wide interest in the nosology of mental disorders has brought
about several new classificatory systems which have already been used
in clinical practice and international research, or will be introduced in
near future. These systems provide clear inclusional and exclusional
criteria as well as definite diagnostic algorithms, which are helpful in the
decision on whether or not an individual case can be classified within a
distinct syndromal or diagnostic category of mental disorder. This state
of the art has provided us with straightforward rules which can be
applied by psychiatrists of various countries and with different orientation.

For the diagnosis of Alzheimer's disease, however, at least four
different sets of criteria have been proposed (Gustafson and Nilsson,
1982; McKhann, 1984; American Psychiatric Association, 1987; World
Health Organization, 1987).

In spite of many similarities, these systems differ in reference to the
diagnostic criteria involved. Which of these competing systems or which
combination of criteria should be preferred?

Material and methods

In order to explore these questions, we examined 150 elderly individuals in
our special dementia clinic, which is part of the outpatient service of the Department of Psychiatry at the Technische Universität in München. Patients were
referred between 1985 and 1987 from various medical institutions because of
suspected or manifest symptoms of dementia. Cases of severe mental deterioration were not included in the study. The initial assessment covered an extensive
clinical interview with the patient and with a well-informed relative, a careful
physical examination, various laboratory tests, several internationally accepted
screening instruments such as Mini Mental State Examination (Folstein et al.,
1975), Information-Memory-Concentration Test and Dementia Scale (Blessed
et al., 1968) and additional neuropsychological tests covering the areas of learning, memory, language, visuo-spatial functions, and intelligence. Certain cut-offs
on these variables were defined in advance to determine whether or not a
particular diagnostic criterion was met.

Following a polydiagnostic approach, each individual was classified by the
sets of criteria listed in Table 1.

For the validation of the different diagnostic approaches, biopsy or autopsy
results would have been preferable. But as brain biopsy was not justified for
ethical reasons and because life expectancy was quite long in most of the patients,
neuropathological confirmation was not available. Therefore we used prospective validation as the second best method.

The patients were re-examined 12−24 months after the initial assessment
(Fig. 1). Independently of the criteria which were applied on initial examination,
the diagnosis of Alzheimer's disease was made when impairments of memory

Table 1. Criteria for the diagnosis of Alzheimer's disease

1. DSM-III-R:
   Primary Degenerative Dementia of the Alzheimer Type

2. ICD-10 diagnostic criteria for research:
   Dementia in Alzheimer Disease

3. NINCDS-ADRDA work group: with dementia syndrome
   Probable Alzheimer's Disease   defined according to
                                       DSM-III-R
4. Combination of Alzheimer-,   or ICD-10
   Pick-, and Ischemic Scores

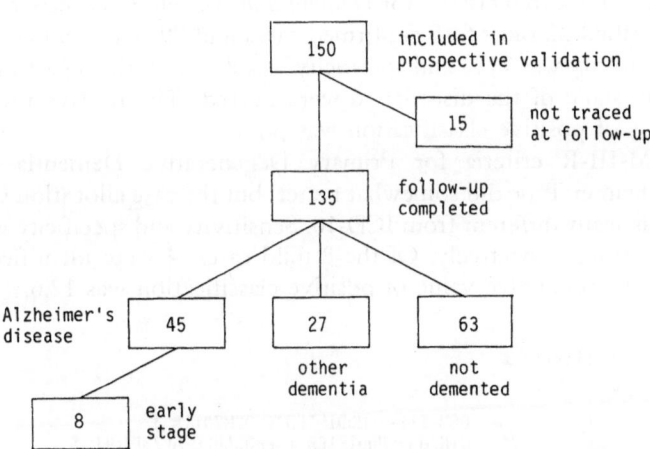

Fig. 1. Prospective validation

and intelligence had shown a continuous progression, when the interference of these symptoms with activities of daily living had increased, when no evidence of any other etiological factor had been found during the observational period, and when cognitive changes had continued to predominate over alterations of personality. If any of these requirements was not met, the diagnosis of Alzheimer's disease was rejected and the patient was categorized within other classes of mental illness.

Of the 150 individuals originally included in the investigation, 15 cases could not be traced on follow-up, thus 135 were re-examined. The diagnosis of the syndrome of dementia was rejected in 63 patients. Two patients were demented; in 45 cases the diagnosis of Alzheimer's disease was established at the end of the observational period. In 8 of these individuals the degree of cognitive impairment had been very mild at the initial assessment. In 27 cases the syndrome of dementia could be attributed to other etiological factors.

The main issue of our examination was to determine which set of criteria applied at the initial examination proved to be most predictive of the final diagnosis of Alzheimer's disease at the end of the observational period. Validity was expressed in terms of sensitivity, specificity, and in the relative predictive value of positive criteria for Alzheimer's disease. The latter measure shows how much more frequent a confirmed diagnosis of Alzheimer's disease is in the cases where the criteria were positive than in those where the criteria were negative.

Results

The essential results of the study are shown in Fig. 2.

1. ICD-10 research criteria for Dementia in Alzheimer Disease correctly identified 32 out of 45 confirmed cases and 79 out of 90 non-cases. Sensitivity was 71% and specificity was 88%. Of the 8 patients in an early stage of the disorder, 6 were missed. The relative predictive value of positive classification was poor.

2. DSM-III-R criteria for Primary Degenerative Dementia of the Alzheimer Type did somewhat better, but the case allocation was not statistically different from ICD-10. Sensitivity and specificity were 84 and 91%, respectively. Of the 8 mild cases, 4 were identified. The relative predictive value of positive classification was 12.6.

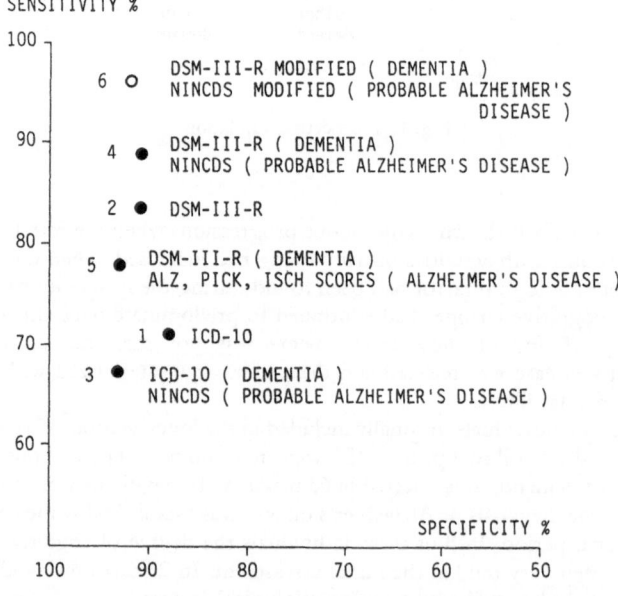

Fig. 2. Prospective validity of different diagnostic criteria for Alzheimer's disease

3. When ICD-10 criteria for Dementia were used together with NINCDS-ADRDA criteria for Probable Alzheimer's Disease, a highly specific (94%) but insensitive (67%) classification ensued.

4. When the NINCDS-ADRDA criteria for Probable Alzheimer's Disease were combined with DSM-III-R criteria for the dementia syndrome, 40 out of 45 confirmed cases and 82 out of 90 non-cases were correctly predicted, giving sensitivity and specificity values of approximately 90%. Compared to ICD-10 criteria, the increase in validity was statistically significant. 4 out of 8 early cases were still missed, and the relative predictive value of positive classification was 17.8.

5. The application of the Alzheimer-, Pick-, and Ischemic Scales to a DSM-III-R diagnosis of Dementia resulted in validity measures that were intermediate between ICD-10 and DSM-III-R criteria for Alzheimer's disease.

6. An additional though not statistically significant improvement of diagnostic validity resulted from minor modifications of the combined DSM-III-R and NINCDS-ADRDA criteria. According to these modifications an impairment of both short-term and remote memory – as required by DSM-III-R criteria – was not obligatory, as long as one of the two was present. For the diagnosis of Alzheimer's disease the additional requirement was introduced that cognitive symptoms had to predominate over personality changes. By this re-definition of criteria, sensitivity and specificity rose above 90%, and all but one patient in an early stage of the disorder were correctly identified.

Discussion

The poor performance of the ICD-10 research criteria was unexpected. It may be explained by the fact that the World Health Organization adopts a relatively restrictive concept of dementia which adheres to the European tradition and requires, in addition to memory impairment, a decline of intelligence and a deterioration of emotional control, social behavior, or motivation. This narrow concept of the dementia syndrome is combined with a wide definition of Alzheimer's disease, consisting only in the exclusion of any demonstrable cause.

The crucial influence of the restrictive concept of dementia for diagnostic validity can be shown when it is combined with the more elaborate criteria of the NINCDS-ADRDA work group for the diagnosis of Alzheimer's disease. By adopting this combination, specificity increased but sensitivity remained unsatisfactory.

In DSM-III-R, the American Psychiatric Association refers to a much broader concept of dementia. This concept may be criticized as over-inclusive: the diagnosis of dementia can be made when a decline of memory and an alteration of personality are present, while impairment of intelligence can be lacking. On the other hand, within the DSM-III-R classification, Alzheimer's disease is defined not only by the exclusion of other underlying causes but also by its typical onset and course. Our results suggest that DSM-III-R criteria for Primary Degenerative Dementia of the Alzheimer Type are superior to ICD-10 research criteria for Dementia in Alzheimer's disease in terms of sensitivity and specificity.

The NINCDS-ADRDA criteria for Probable Alzheimer's Disease are very similar to the DSM-III-R definition of Primary Degenerative Dementia of the Alzheimer Type. Thus no significant differences in validity can be expected when, in both instances, the syndrome of dementia is assessed according to DSM-III-R. As already mentioned, the NINCDS-ADRDA criteria performed much worse when the ICD-10 definition of dementia was used.

We were surprised to see that an acceptable validity of clinical diagnosis was also obtained from the application of three simple diagnostic scales. These symptom lists, however, refer to fairly advanced stages of dementing illness and therefore are less appropriate for the recognition of mild cases.

The minor modifications to the DSM-III-R criteria for the dementia syndrome and to the NINCDS-ADRDA criteria for Probable Alzheimer's Disease were aimed at the detection of early stages where impairment of remote memory may be difficult to ascertain, and at the identification of the typical symptom profile of the disorder which distinguishes it from Pick's disease and from other forms of frontotemporal degeneration. These modifications of diagnostic criteria resulted in the best validity measures that have been observed in our study.

We are aware of the fact that our attempts to compare the validity of several set of diagnostic criteria for Alzheimer's disease are only preliminary. They will have to await the final test of postmortem verification.

References

American Psychiatric Association (1987) Diagnostic and statistical manual of mental disorders, 3rd edn, revised. American Psychiatric Association, Washington

Blessed G, Tomlinson BE, Roth M (1968) The association between quantitative measures of dementia and of senile change in the cerebral grey matter of elderly subjects. Br J Psychiatry 114: 797-811

Boller F, Lopez OL, Moossy J (1989) Diagnosis of dementia: clinicopathologic correlations. Neurology 39: 76-79

Folstein MF, Folstein SE, McHugh PR (1975) "Mini Mental State". A practical method for grading the cognitive state of patients for the clinician. J Psychiatr Res 12: 189–198

Gustafson L, Nilsson L (1982) Differential diagnosis of presenile dementia on clinical grounds. Acta Psychiatr Scand 65: 194–209

Joachim CL, Morris JH, Selkoe DJ (1988) Clinically diagnosed Alzheimer's disease: autopsy results in 150 cases. Ann Neurol 24: 50–56

Lauter H (1988) Die organischen Psychosyndrome. In: Kisker KP, Lauter H, Meyer JE, Müller C, Strömgren E (Hrsg) Psychiatrie der Gegenwart, 3. Aufl, Bd 6. Organische Psychosen. Springer, Berlin Heidelberg New York Tokyo, S 3–56

McKhann G, Folstein M, Katzman R, Price D, Stadlan EM (1984) Clinical diagnosis of Alzheimer's disease: report on the NINCDS-ADRDA work group under the auspices of the Department of Health on Human Services; Task Force on Alzheimer's disease. Neurology 34: 939–944

Morris JC, McKeel DW, Fulling K, Torack RM, Berg L (1988) Validation of clinical diagnostic criteria for Alzheimer's disease. Ann Neurol 24: 17–22

Thierney MC, Fisher RH, Lewis AJ, Zorzitto ML, Snow WG, Reid DW, Nieuwstraten P (1988) The NINCDS-ADRDA work group criteria for the clinical diagnosis of probable Alzheimer's disease: a clinicopathologic study of 57 cases. Neurology 38: 359–364

Wade JPH, Mirsen TR, Hachinski VC, Fisman M, Lau C, Merskey H (1987) The clinical diagnosis of Alzheimer's disease. Arch Neurol 44: 24–29

World Health Organization (1987) ICD-10 – Organic, including symptomatic, mental disorders. Diagnostic criteria for research. WHO document number 0645d, Geneva

Correspondence: Prof. Dr. H. Lauter, Psychiatrische Klinik der Technischen Universität München, Ismaningerstrasse 22, D-8000 München 80, Federal Republic of Germany.

Clinical aspects and terminology of dementing syndromes

F. Gerstenbrand, G. Birbamer, and J. Rainer

University Clinic of Neurology, Innsbruck, Austria

Summary

As people become increasingly older dementias have been more often diagnosed during the last decade. Difficulties arise in classifying the different types of demential syndromes and in reaching a clear terminology. Modern diagnosis are reached by clinical history, neuroimaging, electrophysiological examinations, laboratory and neurophysiological tests. Following this investigations a variety of demential syndromes are differentiated. An attempt is made to classify demential syndromes into forms of organic dementia, the diffuse organic brain syndrome and the local organic brain syndrome.

Introduction

During the last few years Poeck (1982, 1985, 1988) has pointed out several times that the terminus "dementia" which is used up to now in the terminology of research and in the clinical practice should be replenished. The mainly used scales, like the score for ischemia and score for dementia is not satisfactory to classify the disintegration of brain function, mainly when it concerns the epidemiological research and concerning the different approaches towards therapy. Poeck (1988) also postulates a profile of the separate forms of dementia including the evaluation of the deficiencies as well as the preserved abilities of the performance. This goal should be an absolute one, mainly concerning the various possible therapy approaches and therefore a principal clarification of the classification groups is necessary. Primarily the irreversible diffuse damage of cortical and subcortical brain structures should be separated from the temporary diffuse lesions of the brain caused by various reasons. In both groups the clinical symptoms show a reduction of the higher and highest brain functions. These disturbances are progressive or static in patients with a manifest brain damage, in patients with a temporary brain dysfunction a partial or full remission is possible.

Therefore the task of this review should be to point out the necessity to differentiate clearly between the progredient or fixed brain damage and the transient brain dysfunction despite of the global homogeneous clinical aspects of both, as the latter one has the possibility of partial or full remission.

In the Western European countries there are 14% of over 64 years old that means a trebling of this age group in respect to the whole population within the last 90 years (Mahendra, 1987; Christie, 1987). This fact has lead to a dramatic increase in the number of patients suffering from demential syndromes with consecutive sociomedical problems. 30% of this patients with symptoms of the demential syndrome are suffering according to Mumenthaler (1986) from a so-called reversible dementia. Other authors give these figures as much as 6.5% (Smith, 1981) and 32% (Harrison, 1977).

In 1987 the definitions given in DSM-III have been revised and so became compatible with the NINCDS-ADRDA work group criteria (Wilcock, 1989). This means that we have got a tool for a more distinct clinical specification. Since the diagnostic techniques mainly the neuroimaging methods are progressing, the differential diagnoses have been widely improved (Aichner, 1988; Hershey, 1987).

Dementia is essentially an irreversible state with all the signs of impairment of the higher and highest brain functions and should be defined as the organic dementia (OD), whereby the so-called reversible dementia should be named "organic brain syndrome" (OBS). Different types of OBS according to the various clinical pictures and etiologies based on the definition by E. Bleuler (1916) and later by M. Bleuler (1967) have to be differentiated.

Classification and terminology

Derived from different etiological factors of diffuse brain dysfunction there are main groups to be defined; the organic dementia (OD) and the organic brain syndrome (OBS). The clinical syndromes related to hemispheric alterations are summarized in Table 1, subdivided into organic brain damage and transient brain dysfunctions. Based on clinical neurological manifestation the classification does not consider etiological factors.

In general three different forms of brain syndromes have to be differentiated (Gerstenbrand, 1988, 1989). The organic dementia (OD), the

Table 1. Different types of organic brain syndrome

---

1. Organic dementia (OD)

1.1. Progressive form (OD-PR)

1.2. Static form (OD-ST)

2. Organic brain syndrome (OBS)

2.1. Diffuse organic brain syndrome (DOBS)

2.2. Local organic brain syndrome (LOBS)

---

irreversible form of a demential syndrome, shows either a progressive or a static course. The two others, generally reversible brain syndromes are the diffuse organic brain syndrome (DOBS) and the local organic brain syndrome (LOBS) of which the latter one can furthermore be subdivided into the frontobasal, the frontoconvex, the temporobasal and the amnestic syndrome.

The organic dementia (OD)

The OD is defined as a mental disorder resulting from an irreversible diffuse brain damage in two different forms, the progressive and the static one.

The clinical symptoms of organic dementia consists of disturbance of highest and higher brain functions, showing a diminished psychic reactions, memory impairment, mainly affecting the short-term memory, disorientation of location and time, disturbance of judgement and apprehension, disturbances of the control of affection, emotion instinct and mood and in addition impairment of speech, calculation, reading, writing, constructional abilities, partial apraxia and agnosia.

The progressive form shows a continuous deterioration, beginning with the symptoms like dementia type DAT, in the worst form resulting in an apallic syndrome.

The organic brain syndrome (OBS)

The OBS is characterized by mental disorders which are a sequel of a reversible brain disease (inflammatory, vascular, metabolic, toxic etc.). The clinical picture of a diffuse form can be similar to the symptomatology of organic dementia, but mostly with less marked symptoms.

The diffuse organic brain syndrome (DOBS): The DOBS is characterized by reversible mental disorders. The clinical picture is similar to the one of the organic dementia, but mostly shows less marked symptoms. Nevertheless the course differs: it might present as a reversible one with a complete remission, a partially reversible one with a defective state showing symptoms of an organic dementia, or a partially reversible but with a defective state, may be later on developing into a progredient organic dementia.

The local organic brain syndrome (LOBS): The LOBS also shows mental disorders as a sequel of cerebral damage. The course of the LOBS may be reversible with a complete remission, partially reversible, but leading into a defective state with symptoms of frontal or temporal lobe lesions, partially reversible but with a later onset of a progressive course leading into a defective state. This syndrome may be differentiated into four subgroups (Table 2):

Table 2. Different forms of local organic brain syndrome

1. Local organic brain syndrome
1.1. Frontobasal syndrome (OBS-FBS)
1.2. Frontoconvex syndrome (OBS-FCS)
1.3. Temporobasal syndrome (OBS-TBS)
1.4. Amnestic syndrome (OBS-AS)

The frontobasal syndrome (FBS): The FBS has first been described by Kretschmer (1949) and occasionally has been termed the "orbital syndrome". The mental symptoms show an increase of drive, loss of emotional control, tendency for motoric overactivity, diminution of cognitive functions and overreaction on fright. Moreover, a frequent occurrence of primitive motor patterns such as oral automatism and grasping-reflexes may be observed.

In addition to these symptoms vegetative disturbances are found such as a tendency to vascular hypertonia as well as disturbances of the circadian rhythm. This syndrome can etiologically be linked with the lesion of frontobasal structures, mainly caused by traumata.

The frontoconvex syndrome (FCS): The FCS, described by Gerstenbrand (1967) and Schmieder (1968) is characterized by a lack of drive (apathy, reduction of activity), a decrease of emotional response, an inhibition of instincts and an impairment of cognitive functions. Primitive motor patterns such as grasping reflexes and less often oral automatisms are observed. This condition is due to a lesion of the convexity of the frontal lobe as well as the related structures and is mainly posttraumatic.

The temporobasal syndrome (TBS): The TBS which has been differentiated corresponds to the PSEUDOPSYCHOPATHIC SYNDROME, described by Peters (1969), consists of an increase of drive (which lessens after a short period), euphoria, distractability, reduced reaction to fright, a decrease of emotional response, hypersexuality, disturbances of social adjustment, primitive oral motoric patterns, vegetative dysregulations. In some cases temporal seizures may be observed. Lesions of the temporal lobe, mostly as a sequel of inflammatory or traumatic damage are the etiological causes of this clinical picture.

The amnestic snydrome (AS): The AS, also termed as dysmnestic syndrome by E. Bleuler (1916) mainly shows a disturbance of the memory, intellectual perception, orientation (location and time), impairment of thinking, affection with different accents. A bilateral lesion of the temporoconvex region is assumed, mostly caused by intoxications (alcohol, poisons, metals etc.) and metabolic as well as vascular disorders.

References

Aichner F, Felber S, Willeit J (1988) Magnetic resonance imaging and computed tomography in diagnosis and assessment of dementia. In: Agnoli A, et al (eds) Senile dementias, II. Internat. Symp. John Libbey Eurotext, Paris, pp 213–217

Diagnostic and Statistical Manual of Mental Disorders, 3rd edn (revised) (1987) American Psychiatric Association, Washington pp 103–123

Gerstenbrand F, Birbamer G, Rainer J (1988) Terminology in dementia. In: Agnoli A, et al (eds) Senile dementias. John Libbey Eurotext, Paris, pp 29–33

Gerstenbrand F, Birbamer G, Rainer J (1989) Clinical aspects of the dementias: a classification. In: Battistin L, Gerstenbrand F (eds) Aging brain and dementia. Wiley-Liss, New York, pp 31–37

Harrison MJG, Marsden CD (1977) Progressive intellectul deterioration. Arch Neurol 34: 199–203

Hershey LA, Modic MT, Greenough PG, Jaffe DF (1987) Magnetic resonance imaging in vascular dementia. Neurology 37: 29–36

Kokmen E (1984) Dementia – Alzheimer type. Mayo Clin Proc 59: 35–42

Kretschmer E (1949) Das Orbitalhirn- und Zwischenhirnsyndrom nach Schädelbasisfraktur. Arch Psychiat Nervengr 182: 454–579

McKhann G, Drachmann D, Folstein M, Katzman R, Price D, Stadlan EM (1984) Clinical diagnosis of Alzheimer's disease. Neurology 34: 939–944

Mumenthaler M (1986) Neurologische Gesichtspunkte in der Altersforschung unter besonderer Berücksichtigung der therapierbaren Demenzen. Triangel [Suppl 2] 25: 23–27

Peters UH (1985) Das pseudopsychopathische Affektsyndrom der Temporallappenepileptiker. Nervenarzt 72: 75–85

Poeck K (1982) Das sogenannte psychoorganische Syndrom, „hirnlokales Psychosyndrom", „endokrines Psychosyndrom". In: Poeck K (Hrsg) Klinische Neuropsychologie. Thieme, Stuttgart, S 204–209

Poeck K (1985) Neuropsychologische Aspekte hirnorganischer Psychosyndrome. In: Bente D, Coper H, Kanowski S (Hrsg) Hirnorganische Psychosyndrome im Alter, II. Springer, Berlin Heidelberg New York, S 24–29

Poeck K (1988) Das Problem der Demenz aus der Sicht der Neurologie. Thieme, Stuttgart New York (Akt Neurol 15: 1–5)

Smith JS, Kiloh LG (1981) The investigation of dementia: results in 200 consecutive admissions. Lancet i: 824–827

Schmieder F (1968) Psychiatrische Rehabilitation traumatischer Hirnschäden. Wien Med Wschr 38: 779–782

Tomlinson BE, Blessed G, Roth M (1970) Observations on the brains of demented old people. J Neurol Sci 11: 205–242

Wetle T (1985) Social epidemiology of aging in the United States. Abstracts of the First Sino-American Symposium: Perspectives of Aging, Beijing/Xian/Shanghai

Wilcock GK, Hope RA, Brooks DN, Lantos PL, Oppenheimer C, Reynolds GP, Rossor MN, Davies MB (1989) Occasional review: recommended minimum data to be collected in research studies on Alzheimer's disease. J Neurol Neurosurg Psychiatry 52: 693–700

Correspondence: Prof. Dr. F. Gerstenbrand, M.D., University Clinic of Neurology, Anichstrasse 35, A-6020 Innsbruck, Austria.

Symptoms of depression in the course of multi-infarct dementia and dementia of Alzheimer's type

W. Danielczyk[1]**, P. Fischer**[2]**, M. Simanyi**[1]**, G. Gatterer**[1]**, and A. Marterer**[2]

[1] Ludwig Boltzmann Institute of Aging Research,
Working Group: Alzheimer's-Dementia-Research, Vienna
[2] Neurological Institute, University of Vienna, Austria

Summary

Symptoms of depression are associated with both Multi-Infarct dementia (MID) and Dementia of Alzheimer's type (DAT). We investigated anxiety and depressed mood in 37 MID-patients and 55 DAT-patients of various stages of dementia. Diagnosis was carried out using the commonly accepted criteria of the DSM-III-R and the NINCDS-ADRDA criteria for "probable DAT", respectively. Severity of dementia was rated by the help of the Mini-Mental State examination.

We found comparable rates of depression in mild-to-moderate MID and DAT, respectively. Severely demented DAT-patients were far less anxious and depressed than equally demented MID-patients. Progression of dementia reduced depression in DAT but favoured depressive symptoms in MID.

Dementia of Alzheimer's type (DAT) is the most frequent type of dementia in the elderly having twice the incidence of Multi-Infarct Dementia (MID). DAT and MID account for about 90% of dementias in a geriatric population as demonstrated by neuropathological examination (Tomlinson, 1970; Jellinger, 1990). Rates of depression in DAT and MID, respectively, vary enormously between studies (Reifler, 1982; Knesevich, 1983; Reding, 1985; Cummings, 1987; Gurland, 1987; Lazarus, 1987; Siegel, 1987; Zubenko, 1988; Wragg, 1989). Estimates of depression in DAT range from 0 to 87 percent (Wragg, 1989), that in MID range from 19 to 27 percent (Reding, 1985; Cummings, 1987). Two major reasons contribute to this variability. The first is the way in which depression is defined and measured. The second reason concerns patients' selection (in- or out-patients; presenile – senile; female – male; stage of dementia; diagnostic criteria).

According to the commonly accepted ischemic score, depressive symptomatology in coexisting dementia favours the diagnosis of MID (Hachinski, 1975). But this item of the ischemic score has been excluded after neuropathological validation of the scale (Rosen, 1980; Fischer, 1989). Up to now three studies directly compared depression in MID with that in DAT. Danielczyk (1984) described a higher rate of antidepressive treatment in early DAT than in early MID. Reding (1985) found depression to be more common in DAT; Cummings (1987) found it to be more common in MID.

Depression in demented patients can either be defined as *depressive disorder* (i.e. major affective disorder of the DSM-III-R) or it can be described at the level of *depressive symptoms* (i.e. depressed mood, anxiety). Especially in the elderly, *depression-like symptomatology* interferes with the diagnosis of depression according to diagnostic criteria. For instance lack of appetite resulting in weight loss or insomnia are particulary common in old age without coexisting depressive disorder (Siegel, 1987).

Rating scales for depression, standardized and validated in younger patients, are not applicable in older individuals (Weiss, 1986): they do not address the symptoms reported to be more common among depressed elderly. The Hamilton Depression Scale (Hamilton, 1960) and the Zung self-rating depression scale (Zung, 1965) are inappropriate in old demented patients. New geriatric depression scales have been developed but have not yet been established (Brink, 1982; Yesavage, 1983; Sheikh, 1986; Sunderland, 1988).

Pathogenesis of major depressive disorder may be quite different from that of depression in old age and particularly from that in dementing illness. Because of the etiological heterogeneity of depression in dementia, it is better to use only single symptoms of depression instead of calculating sumscores of depressive symptoms, which perhaps emerge from different sources. It is these depressive symptoms which burden the care-givers and which should be treated with antidepressive drugs (Cummings, 1987; Lazarus, 1987; Reisberg, 1987).

We investigated the two most important depressive symptoms of old age, depressed mood and anxiety, in demented in-patients. We tried to answer the following questions:

1. Is there a difference regarding depressive symptoms between equally demented MID- and DAT-patients?
2. Do depressive symptoms increase or decrease during the course of MID and DAT, respectively?

Patients and methods

We investigated depressive symptoms in 37 MID and 55 DAT patients. Patients were in-patients of the neurological department of a geriatric hospital in Vienna, called the "Pflegeheim Lainz" and had been hospitalized for at least 1 month before entering the study. They took part in a prospective-longitudinal study on dementias in the elderly (Fischer, 1988).

Dementia was diagnosed according to the Diagnostic and Statistical Manual of Mental Disorders, 3rd edition-revised DSM-III-R (1987). Grading of severity of dementia was carried out by the help of the Mini-Mental State examination (MMS; Folstein, 1975). According to Folstein (1983), only patients with MMS-scores of less than 24 were called "demented". Mild dementia was defined by MMS scores ranging from 16 to 23. Moderate dementia was defined by MMS scores ranging from 6 to 15. Severely demented patients scored less than 6 on the MMS (Fischer, 1988).

Every patient diagnosed as DAT or MID, respectively, fulfilled the diagnostic criteria of the DSM-III-R (1987). DAT patients fulfilled the criteria for "Probable Dementia of Alzheimer's Type" of the NINCDS-ADRDA Work Group (McKhann, 1984). Diagnosis was based on a complete medical and neuropsychiatric examination including history, physical examination, electrocardiography, electroencephalography, computed tomography of the brain, investigation of the ocular fundi, blood count, biochemistry (electrolytes, liver and kidney function, folic acid, Vit B_{12}), thyroid function, and serology for syphilis and AIDS.

Helplessness of the patients was rated by nurses (HELP): abilities of eating, bathing, dressing, walking, incontinence, and the type of bed/chair demanded; each item scored 0 to 4; the maximum score of 24 meant maximal helplessness. This help-score is a German modification of the Physical Self-Maintenance Scale (Lawton, 1969).

The Hamilton Depression Scale (HAM) was rated by an independent rater (M.S.) in order to compare our data with that of Cummings (1987). Depressed mood and anxiety was rated using 2 items of the Sandoz Clinical Assessment-Geriatric SCAG (Shader, 1974). These two items represent depression in the elderly according to factor analysis (Gaitz, 1977) (Table 1).

Results

Characteristics of patients are listed in Table 2. MID-patients were slightly younger than DAT-patients. Dementia was more severe in DAT than in MID, which was due to only 4 severely demented MID-patients opposed to 21 equally demented DAT-patients. Nevertheless helplessness was comparable in both diagnostic groups. The Hamilton-score was higher in DAT than in MID, and this difference nearly reached significance.

Table 1. Sandoz clinical assessment geriatric (SCAG) scale; symptoms of depression

1 not present; 2 very mild; 3 mild; 4 mild to moderate; 5 moderate; 6 moderately severe; 7 severe

Item 5:	Mood depression: dejected, despondent, helpless, hopeless, preoccupation with defeat or neglect by family or friends, hypochondriacal concern, functional somatic complaints, early waking. Rate on patient's statements, attitude and behavior.
Item 6:	Anxiety: worry, apprehension, overconcern for present or future, fears, complaints of functional somatic symptoms, e.g. headache, dry mouth, etc. Rate on patient's own subjective experience and on physical signs, e.g. trembling, sighing, sweating, etc., if present.

Table 2. Patients

	N (F/M)	Age mean (SD) range	MMS mean (SD) range	Helplessness mean (SD) range	HAM mean (SD) range
MID	37 (28/2)	76.6 (7.0) 59−85	14.6 (6.5) 0−23	15.1 (6.7) 5−24	10.3 (2.5) 3−19
DAT	55 (48/7)	79.4 (8.8) 58−93	11.5 (9.1) 0−23	14.7 (8.7) 1−24	12.4 (5.8) 2−27
T-tests:		$T = 1.63$ $df = 91$ $p = 0.107$	$T = 1.81$ $df = 91$ $p = 0.073$	$T = 0.23$ $df = 89$ $p = 0.817$	$T = 1.97$ $df = 91$ $p = 0.052$

Anxiety and depressed mood were the same in MID and DAT (anxiety in MID: 3.6 ± 1.5; in DAT: 3.8 ± 1.5; $T = 0.870$; $p_T = 0.387$; depressed mood in MID: 4.2 ± 1.4; in DAT: 4.1 ± 1.7; $T = 0.293$; $p_T = 0.753$). Table 3 shows these depressive symptoms at different stages of MID and DAT, respectively. Two-way analysis of variance shows that anxiety does neither depend on severity of dementia nor on type of dementia. Depressed mood depended on severity of dementia but did not differ between MID and DAT. The interaction between severity of dementia and type of dementia was found to be highly significant regarding both symptoms, anxiety and depressed mood. Thus, during the course of the illness these two symptoms changed in a different way in MID and DAT, respectively.

The latter result was demonstrated by nonparametric rank-correlation coefficients between depressive symptoms and the MMS-score

Table 3. Depressive symptomatology at different stages of dementia of Alzheimer's type and multi-infarct dementia

	MID			DAT		
	Mild $n=21$	Moderate $n=12$	Severe $n=4$	Mild $n=22$	Moderate $n=12$	Severe $n=21$
Anxiety	3.14 (1.11)	3.75 (1.60)	5.25 (1.71)	4.18 (1.18)	4.17 (1.19)	3.30 (1.75)
Depressed mood	4.14 (1.24)	4.00 (1.65)	5.25 (1.71)	4.86 (1.55)	4.42 (1.31)	3.05 (1.61)

Two-way ANOVA

Anxiety:	Main effects:	Diagnosis	(df=1):	$F/F=1.319$	$p_F=0.254$
		Severity	(df=2):	$F/F=0.626$	$p_F=0.537$
	Two-way interaction		(df=2):	$F/F=5.872$	$p_F=0.004$
Depressed mood:	Main effects:	Diagnosis	(df=1):	$F/F=0.191$	$p_F=0.663$
		Severity	(df=2):	$F/F=4.231$	$p_F=0.018$
	Two-way interaction		(df=2):	$F/F=5.055$	$p_F=0.008$

Table 4. Spearman rank correlation coefficients between MMS and depressive symptoms

	Anxiety	Depressed mood	HAM
MID	$r=-0.340$ $p=0.020$	$r=-0.203$ $p=0.114$	$r=-0.171$ $p=0.279$
DAT	$r=+0.235$ $p=0.042$	$r=+0.495$ $p=0.000$	$r=+0.099$ $p=0.347$

(Table 4). Anxiety and depressed mood correlated negatively with MMS-scores in MID, but did so with positive sign in DAT. Correlations were found to be non-significant using the Hamilton-scale.

Discussion

We found depressive symptoms in the course of both MID and DAT. Anxiety and depressed mood were slightly higher in DAT than in MID at the stages of mild and moderate dementia. But severely demented DAT-patients were far less anxious and depressed than equally demented MID-patients. The latter group only contained 4 patients. Highly significant nonparametric correlations between cognitive decline and depressive symptoms showed that progression of dementia reduced depression in DAT but favoured depressive symptoms in MID.

Our results are in accordance with those of Reding (1985) in so far as depression in early DAT was more pronounced than in early MID. Our results differ from that of Cummings (1987), who investigated MID and DAT-patients with a degree of dementia comparable to our patients. DAT-patients in this study reached an average MMS-score of 10.7 (11.5 in our study), MID-patients reached a mean MMS-score of 15.9 (14.6 in our study). Cummings did not investigate depression in relation to severity of MID and DAT, respectively. Mean HAM-score of MID-patients was 12.7 in Cumming's study (10.3 in our MID-patients). But we found a significant higher HAM-score of DAT-patients (12.4 compared to 6.8). This could be a sample artefact: for instance relatively more DAT-patients of Cummings could have been severely demented.

Cummings accepted HAM-scores of 9 or above as indicative of some degree of depression and found 17% of DAT- but 60% of MID-patients in the depressed range. If we adopt this criterion, we find 70% of our MID-patients and 79% of our DAT-patients in the depressed range. One possibility to explain the higher rate of depression in our investigation is, that we investigated chronic in-patients. In-patients are said to show a higher degree of depression (Wragg, 1989).

Using the stricter HAM-cut-off of 13/14, which is used to exclude depression in subjects with age-associated memory impairment (Crook, 1986), we found 27% of our MID-patients and 40% of our DAT-patients in the depressed range. But, as stated before, the HAM is not designed for studying depression in old demented patients.

Taken together our results indicate that the significance of depressive symptoms in the diagnosis of MID and DAT, respectively, depends on severity of dementia. Only severely demented DAT-patients are less depressed than equally demented MID-patients. Future research will have to show to what extent depression in MID and DAT is explained either by organic factors (Zubenko, 1988) and/or by psychological (reactive) processes. Nevertheless, unmodified depressive symptoms may be one important factor in the decision to institutionalize a patient. And these depressive symptoms are frequent. We, therefore, recommend to treat depressed patients vigorously even in the presence of dementia.

Acknowledgements

This research was supported by the "Fonds zur Förderung wissenschaftlicher Forschung der Stadt Wien (grant 86/449)", by the Ludwig Boltzmann Institute of Aging Research, Working Group: Alzheimer's Dementia, and by a grant from the Austrian Academy of Sciences: Institute of Brain Research ("Aktuelle Fragen der Hirnforschung").

References

American Psychiatric Association (1987) Diagnostic and statistical manual of mental disorders, 3rd edn (revised). APA, Washington, DC

Brink TL, Yesavage JA, Lum O, Heersema PH, Adey M, Rose TL (1982) Screening tests for geriatric depression. Clin Gerontologist 1:37–43

Crook T, Bartus RT, Ferris SH, Whitehouse P, Cohen GD, Gershon S (1986) Age-associated memory impairment: proposed diagnostic criteria and measures of clinical change – Report of a National Institute of Mental Health Work Group. Dev Neuropsychol 2:261–276

Cummings JL, Miller B, Hill MA, Neshkes R (1987) Neuropsychiatric aspects of multi-infarct dementia and dementia of the Alzheimer type. Arch Neurol 44:389–393

Danielczyk W (1984) Psychiatrische Komplikationen beim Morbus Parkinson: Krankheitssymptome, Folge der Alterung oder der Therapie? Neuropsychiatr Clin 3:145–155

Fischer P, Gatterer G, Marterer A, Danielczyk W (1988) Nonspecificity of semantic impairment in dementia of Alzheimer's type. Arch Neurol 45:1341–1343

Fischer P, Jellinger K, Gatterer G, Marterer A, Danielczyk W (1989) Neuropathological validation of the Hachinski Scale. J Neural Transm (P-D Sect) 1:57

Folstein MF, Folstein SE, McHugh PR (1975) "Mini-Mental State". A practical method for grading the cognitive state of patients for the clinician. J Psychiatr Res 12:189–198

Folstein MF (1983) The mini-mental state examination. In: Crook TH, Ferris S, Bartus R (eds) Assessment in geriatric psychopharmacology. Mark Powley, Connecticut, pp 47–51

Gaitz CM, Varner RV, Overall JE (1977) Pharmacotherapy for organic brain syndrome in late life. Arch Gen Psychiatry 34:839–845

Gurland B, Toner J (1987) The epidemiology of the concurrence of depression and dementia. In: Altmann HJ (ed) Alzheimer's disease. Plenum Press, New York, pp 45–58

Hachinski VC, Iliff LD, Zilhka E, DuBoulay GH, McAllister VL, Marshall J, Russell RWR, Symon L (1975) Cerebral blood flow in dementia. Arch Neurol 32:632–637

Hamilton M (1960) A rating scale for depression. J Neurol Neurosurg Psychiatry 23:56–62

Jellinger K (1990) Histopathological validation of dementia disorders in the aged. In: Battistine L (ed) Aging brain and dementia: new trends in diagnosis and therapy. Liss, New York, pp 79–97

Knesevich JW, Martin RL, Berg L, Danzinger W (1983) Preliminary report of affective symptoms in the early stages of senile dementia of the Alzheimer type. Am J Psychiatry 140:233–235

Lawton MP, Brody EM (1969) Assessment of older people: selfmaintaining and instrumental activities of daily living. Gerontologist 9:179–186

Lazarus LW, Newton N, Cohler B, Lesser J, Schweon C (1987) Frequency and presentation of depressive symptoms in patients with primary degenerative dementia. Am J Psychiatry 144:41–45

McKhann G, Drachman D, Folstein M, Katzman R, Price D, Stadlan EM (1984) Clinical diagnosis of Alzheimer's disease: report of the NINCDS-ADRDA Work Group under the auspices of Department of Health and Human Services Task Force on Alzheimer's disease. Neurology 34:939–944

Reding M, Haycox J, Blass J (1985) Depression in patients referred to a dementia clinic. A three-year prospective study. Arch Neurol 42:894–896

Reifler BV, Larson E, Hanley R (1982) Coexistence of cognitive impairment and depression in geriatric outpatients. Am J Psychiatry 139:623–626

Reisberg B, Borenstein J, Salob SP, Ferris SH, Franssen E, Georgotas A (1987) Behavioral symptoms in Alzheimer's disease: phenomenology and treatment. J Clin Psychiatry 48:9–15

Rosen WG, Terry RD, Fuld PA, et al (1980) Pathological verification of ischemic score in differentiation of dementias. Ann Neurol 7:486–488

Shader RI, Harmatz JS, Salzman C (1974) A new scale for clinical assessment on geriatric populations: Sandoz clinical assessment geriatric (SCAG). J Am Geriatr Soc 22(3):107–113

Sheikh JI, Yesavage JA (1986) Geriatric depression scale (GDS). Recent evidence and development of a shorter version. Clin Gerontologist 5:165–173

Siegel B, Gershon S (1987) Dementia, depression, and pseudodementia. In: Altman HJ (ed) Alzheimer's disease problems, prospects, and perspectives. Plenum Press, New York, pp 29–44

Sunderland T, Alterman IS, Yount D, Hill JL, Tariot PN, Newhouse PA, Mueller EA, Mellow AM, Cohen RM (1988) A new scale for the assessment of depressed mood in demented patients. Am J Psychiatry 145:955–959

Tomlinson BE, Blessed G, Roth M (1970) Observations on the brains of demented old people. J Neurol Sci 11:205–242

Weiss IK, Nagel CL, Aronson MK (1986) Applicability of depression scales to the old person. J Am Geriatr Soc 34:215–218

Wragg RE, Jeste DV (1989) Overview of depression and psychosis in Alzheimer's disease. Am J Psychiatry 146:577–587

Yesavage JA, Brink TL (1983) Development and validation of geriatric depression screening scale: a preliminary report. J Psychiatr Res 17:37–49

Zubenko GS, Moossy J (1988) Major depression in primary dementia. Clinical and neuropathologic correlates. Arch Neurol 45:1182–1186

Zung WWK (1965) A self-rating depression scale. Arch Gen Psychiatry 12:63–70

Correspondence: Prof. Dr. W. Danielczyk, Ludwig Boltzmann Institute for Aging Research, Working Group: Alzheimer's Dementia Research, Neurological Department, Geriatric Hospital Lainz, Versorgungsheimplatz 1, A-1130 Wien, Austria.

Cognitive deterioration and dementia outcome in depression: the search for prognostic factors

F. M. Reischies, P. v. Spieß, and J.-P. Hedde

Department of Psychiatry, Freie Universität Berlin, Berlin

Summary

The time course of recovery of depressed patients' cognitive deficits and the eventual outcome in dementia was studied in patients (N = 59) over 45 years of age. In only two patients could a dementing illness be verified in the follow-up. For the other patients, after a slight recovery of cognitive performances parallel to the improvement of the depressed mood, a further decline to the level of the depressed state was observed after a period of 3 years. Age (pos.), severeness of depression (neg.) and the temporal/occipito-temporal perfusion (neg.) measured by 133 Xenon SPECT were found as probable prognostic factors, related to the decline of cognitive performance.

Introduction

The three years follow-up of cognitive performance was examined in patients admitted because of a depressive syndrome. The aim of the study was to find possible predictive factors for the time course of the cognitive performance. This is of particular interest as it is known on the one hand, that a depressive syndrome is a common initial finding of a progressive deterioration in the dementia of Alzheimer type (Lauter, 1968; Reiffler, 1986). On the other hand, a cognitive decline, which should be reversible, accompanies a depressive psychosis, especially in old age (depressive pseudodementia, Wells, 1980; Rabins et al., 1984; Reischies, 1988, 1990).

Patients

We investigated 59 patients over 45 years of age, who were suffering from a depressive syndrome. The data regarding the patients which could be reinvestigated after an average period of about three years, and who were not suffering from a relapse of the depression, are shown in Table 1.

Methods

The cognitive performance was assessed by a battery of tests, which are listed in Table 1. The clinical dementia evaluation was rated on the Global Deterioration Scale (Reisberg, 1982). The depressive psychopathology was assessed by the Hamilton Depression scale and also by self-rating VAS scales. 38 patients were investigated by 133 Xenon SPECT in the depressed state (Reischies et al., 1988, 1989). The catamnestic interview was done according to the clinical psychiatric interview of Goldberg et al. (1970).

Table 1. Population data and test results for the depressed state and the catamnesis three years later

	Depression	Follow up	Control
N	33		18
Age	58.5 (8.4)		59.8 (7.3)
Education	8.78 (10.81)		8.99 (10.02)
Sex (% female)	73		67
Depressed mood VAS (%)	60.45 (29.01)	23.48 (23.16)	14.44 (21.14)
Hamilton DS	17.38 (4.28)	5.40 (5.04)	
GDS	2.19 (0.97)	2.37 (0.83)	
Medicated (N)	14	25	
Short term memory			
averbal (LGT-3, Plan)	10.27 (4.12)**	10.42 (4.28)*	13.28 (6.20)
verbal (LGT-3, Bau)	3.36 (3.51)*	3.03 (2.73)***	5.33 (3.82)
Psychomotor speed (sec.)	56.48 (24.11)	60.57 (22.98)**	47.61 (18.33)
(Reit. Trail M. a)			
Fluency			
averbal (5-Point T.)	24.41 (9.93)***	25.16 (8.22)***	31.89 (8.80)
verbal ("FAS")	30.21 (10.26)****	31.15 (11.85)***	40.72 (15.59)
Vocabulary (max. 37)	29.78 (4.83)*	30.48 (4.69)	32.06 (3.69)
(MWT-B)	(n=24)		
Serial digit learning (max. 24)	13.00 (9.41)**	14.87 (8.78)	18.24 (4.45)
	(n=16)		

(SD); depressed state or follow up vs. Contr.

* p<0.05, ** p<0.025, *** p<0.01, **** p<0.005

Results

After three years only one patient could not be contacted, three had moved to another town, seven refused, one investigation was incomplete and 12 patients died (4 suicides; no one died demented).

Of the rest, only two patients developed a dementia (presumably of Alzheimer type). The one suffered from a psychoorganic syndrome with disturbed orientation already present in the depressed state, without improvement of cognitive functioning parallel to the improvement of the depressed mood. The other one demonstrated pronounced cognitive deficits when compared to a superior vocabulary performance in the depressed state. She showed some improvement in test-performance at the time of leaving the hospital in a remitted state, but later suffered from a further decline in cognitive performance, accompanied by a disturbance of orientation.

The result of tests on admission in the depressed state and in remitted state after three years is shown for the battery of tests for the first and final testing times on Table 1.

Even excluding patients with the dementing time course a slight deterioration in cognitive performance was found between the remitted state at the discharge from the clinic and the three years follow-up, significantly for verbal fluency (t = 3.79; p < 0.001) and verbal memory (t = 2.17; p < 0.04).

The correlations, which in an exploratory analysis consistently demonstrate a trend toward deterioration or improvement for those

Table 2. Correlations of putative predictors for the time course of cognitive performance (correlation with increase of performance)

	ST Memory		Psychom.	Vocabul.	Fluency	
	averbal	verbal	speed (−)		averbal	verbal
Age			−0.44*		−0.67**	
Hamilton DS	0.44*					
AMP depression				0.43*		
Feeling of guilt		0.40*				
RDC MDD						0.43*
Relative rCBF						
occipito-temporal le	0.52*					
temporal le			0.54*		0.52*	

* p < 0.05
** p < 0.01

patients which did not develop a dementia or a relapse of the depression, are shown on Table 2. The correlations for the relative cerebral blood flow are reported in Table 2; the absolute value of cerebral blood flow during depression was found to be negatively related to the degree of improvement of mood during the total follow-up period (CBF and improvement of the rating or self-rating of depressed mood on VAS: $r = -0.64$, $p < 0.005$ and $r = -0.55$, $p < 0.01$).

Conclusions

Only a few depressed patients developed a progressive dementia during a three years follow-up, all of which had at least some performance deficits at the first time of testing – this result confirms earlier outcome studies (Post, 1962).

There seems to be, however, a slight cognitive deterioration in three years for the depressed patients, if one considers the testing time in the remitted state before leaving the clinic. The data of Rabins et al. (1984), which indicate an almost complete recovery of cognitive functions over a 2 year follow-up, could not be confirmed.

Our result may on the one hand be related to a more pronounced age associated cognitive impairment or benign senescent cognitive decline in depression of – especially – old age, which is slightly progressive, or on the other hand be related to a final outcome in a progressive dementia, as Kral (1982) proposed. Further follow-up studies are required to answer this question.

From the possible predictive factors the age of the patient seems to be related to a less favourable time course of the cognitive performance and the intensity of the depressive syndrome, on the contrary, to a more favourable time course – perhaps indicating a more pure endogenous depression without psychoorganic admixture (Post, 1962). Interestingly the occipito-temporal and temporal hypoperfusion, which has been found to be a characteristic finding in the dementia of Alzheimer type (see Reischies et al., 1987; metabolism investigations: Friedland et al., 1983), seems to be related to a less favourable time course of the cognitive performance in patients with a depressive syndrome. This finding should also be confirmed in further follow-up studies, before speculation about the possibility to detect very early stages of Alzheimer's disease in depressed patients is allowed.

References

Friedland RP, Budinger TF, Yano Y, Huesman GH, Knittel B, Derenzo SE, Koss B, Ober BA (1983) Regional cerebral metabolic alterations in Alzheimer type dementia. J Cereb Blood Flow Metab 3 [Suppl 1]:510–511

Goldberg DP, Cooper B, Eastwood MR, Kedward HB, Shepherd M (1970) A standardized psychiatric interview for use in community surveys. Br J Prev Soc Med 4:18–23

Kral VA (1982) Depressive Pseudodemenz und senile Demenz vom Alzheimer Typ. Nervenarzt 53:284–286

Lauter H (1968) Zur Klinik und Psychopathologie der Alzheimer'schen Krankheit. Psychiatr Clin 1:85–108

Post F (1962) The significance of affective symptoms in old age. Oxford University Press, London

Rabins PV, Merchant A, Nestadt G (1984) Criteria for diagnosing reversible dementia caused by depression: validation by 2-year follow-up. Br J Psychiatry 144:488–492

Reifler BU, Larson E, Hanley R (1982) Coexistence of cognitive impairment and depression in geriatric outpatients. Am J Psychiatry 139:623–626

Reischies FM (1988) Neuropsychologische Befunde bei der Depression im Involutionsalter und Senium und ihre Beziehung zur regionalen Hirndurchblutung. In: Oepen G (Hrsg) Psychiatrie des rechten und linken Gehirns. Deutscher Ärzteverlag, Köln, S 187–197

Reischies FM, Hedde J-P, Gutzmann H (1987) The investigation of dementia syndromes by 133 Xenon dynamic single photon emission computer tomography. Neurosurg Rev 10:105–108

Reischies FM, Hedde J-P, Drochner R (1989) Cerebral blood flow in the time course of depression – clinical correlates. Psychiatry Res 29:323–326

Reischies FM, Spieß P v, Stieglitz R-D (1990) The symptom pattern variation of unipolar depression during life span. (submitted)

Wells CE (1980) The differential diagnosis of psychiatric disorders in the elderly. In: Cole JO, Barrett JE (eds) Psychopathology of the aged. Raven Press, New York, pp 19–35

Correspondence: Dr. F. M. Reischies, Department of Psychiatry, Freie Universität Berlin, Eschenallee 3, D-1000 Berlin 19.

Diagnostic significance of language evaluation in early stages of Alzheimer's disease

B. Romero[1], A. Kurz[1], M. Haupt[1], R. Zimmer[1], H. Lauter[1], F. Pulvermüller[2], and V.M. Roth[3]

[1] Psychiatrische Klinik der Technischen Universität München
[2] Abteilung für Neuropsychologie, Städtisches Krankenhaus München-Bogenhausen
[3] Philosophische Fakultät, Universität Konstanz, Federal Republic of Germany

Summary

Recent studies using formal linguistic analysis of self-generated speech material have shown that an impairment of language is present in most patients with Alzheimer's disease (AD) even at early stages. For the assessment of these disturbances in clinical routine, however, no appropriate methods are available. In this report we propose a rating scale for conversational ability in AD. This instrument was used in 32 patients with mild AD and in 14 patients with disorders which are difficult to distinguish from AD. An impairment of conversational ability was present in the majority of patients with AD but was rarely observed in non-Alzheimer cases. These results suggest that the assessment procedure may be useful for the early and differential diagnosis of AD.

Introduction

Impairment of language is a second-rate symptom of Alzheimer's disease (AD) according to current diagnostic criteria. Recent studies using formal linguistic analysis have demonstrated, however, that language abnormalities are present in most if not all patients (Bayles, 1982; Skelton-Robins and Jones, 1984; Hier et al., 1985; Hart et al., 1988). They show a typical pattern. Deficits of a pragmatic-cognitive nature become apparent early in the course of the disease, including impoverishment of communicable content, frequent irrelevancies, and intrusions (Blanken et al., 1987; Bayles and Kaszniak, 1987; Ulatowska et al., 1988). Semantic aspects of language such as word finding and word selection may also be affected at an early stage (Cummings and Benson, 1983; Bowles et al., 1987; Flicker and Ferris, 1987; Shuttleworth and Huber, 1988). On the other hand, the phonematic and syntactic structure of language remains relatively well-preserved in mild and moderate AD. For research purposes, the assessment of discourse ability has been based on a formal linguistic analysis of self-generated speech material derived from conversation, from a description of pictures or of everyday activities, or from similar tasks. These complex and time-consuming procedures are not applicable in clinical diagnostic routine. In this report we attempt to show that a brief and practical rating of conversational ability is sufficient to identify language impairment in mild and moderate AD, and that this method may be of significant diagnostic value, particularly when used in combination with a naming test.

Methods

The study refers to 46 patients who were seen in a special out-patient service because of suspected or manifest symptoms of dementia. Diagnostic criteria of the NINCDS-ADRDA workgroup for Probable Alzheimer's disease and DSM-III-R criteria for Dementia and for Major depression were applied to the data collected at the initial assessment by conventional methods such as Mini Mental State Examination (MMSE), Dementia Scale, and Global Deterioration Scale (GDS). All patients were re-examined after a follow-up period of 9 to 24 months.

In 6 patients (Group A) formal diagnostic criteria for Alzheimer's disease were not met at the first examination but were fulfilled by the end of the observational period. In 26 cases the diagnosis of Alzheimer's disease was made at initial assessment and was confirmed at follow-up. 12 of these patients (Group B) showed mild cognitive impairment (MMSE ≥ 20) and 14 (Group C) were more markedly impaired (MMSE < 20). Intermediate (GDS $= 6$) and advanced (GDS $= 7$) stages of Alzheimer's disease were excluded.

Group D, consisting of 14 patients, was diagnostically heterogeneous. In 6 cases memory impairment of unknown origin but not dementia was present at first assessment as well as on re-examination. In 3 patients the cause of dementia remained unclear at first assessment, and the diagnosis of Alzheimer's disease could not be established at folllow-up. A diagnosis of Major depression was made in 5 patients and was confirmed prospectively.

Conversational ability was rated on a special scale (see appendix) from a 10-minute tape-recorded sample by two linguistic experts who were blind to the diagnosis.

The scale was designed to cover phonematic, syntactic, semantic, and pragmatic-cognitive aspects of language behaviour. Operational criteria were used to define six degrees of severity within each of these areas. The criteria adhered to the assessment of spontaneous speech as included in the Aachener Aphasie Test (Huber et al., 1983), but modifications were introduced in order to adjust the instrument to the type of language impairment that was expected to occur in

Alzheimer's disease. Ratings were made separately on each dimension of language behaviour. An impairment of conversational ability was defined by a disturbance in at least one area. The interrater reliability of the procedure was acceptable (R = 0.5).

Naming of objects and colours was taken from the Aachener Aphasie Test; the patients' responses were classified into right and wrong.

Results

Impairment of conversational ability was rated present in 27 out of 32 patients (84%) with beginning, mild, or moderate AD (Fig. 1). These deficits were associated with impaird naming in 23 (72%) of these patients (Fig. 2). As expected, the frequency of language disturbance increased with the severity of the disease.

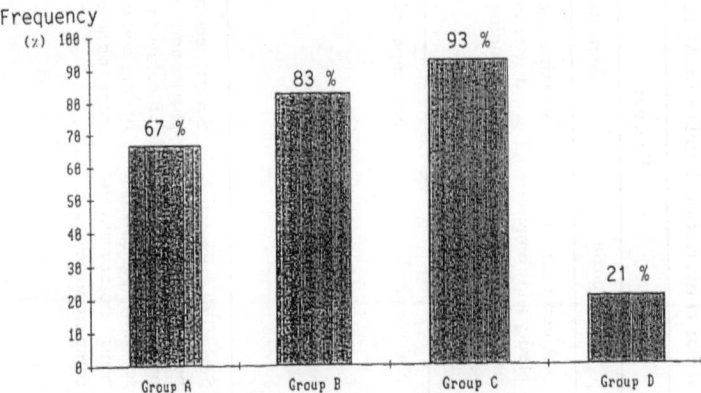

Fig. 1. Impairment of conversational ability

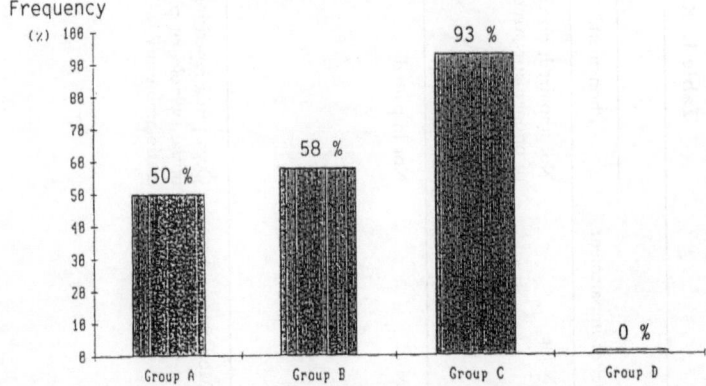

Fig. 2. Impairment of naming and conversational ability

Table 1. Rating scale for conversational ability in Alzheimer's disease

Level of impairment	Function				
	Phonematic	Syntactic	Semantic	Pragmatic-conceptual	
0 None	Not impaired; no change of phonematic structures	Not impaired; complexity and completeness of sentence composition conforms to colloquial standards	No impairment of word selection and of word finding	Language behaviour is directed towards communicating information. Patient sustains the topic of conversation and responds adequately to questions	
1 Mild	Not impaired	Not impaired	Not impaired	Mild impairment in response to questions, in planning and structuring of longer utterances. Communication of irrelevant information, while the point may be missed and relevant information may not be given	
2 Moderate	Occasional repetition of initial syllable, or phonematic ambiguity	Several or many interpositions of sentences or doubling of partial sentences (paragrammatisms) and/or many incomplete sentences	Several semantic paraphasias and/or several difficulties in word finding and/or many empty phrases	Moderate impairment in planning and structuring of longer utterances. Speech is circumlocutory and/or the thread of conversation is frequently lost and/or topics are repeated and/or unexpected switching of topic. Communication of relevant information is possible for the most part	

Table 1 (continued)

Level of impairment	Function				
	Phonematic	Syntactic	Semantic	Pragmatic-conceptual	
3 Severe	Occasionally imprecise articulation/occasional phonematic paraphasias	Many paragrammatic sentences and/or very many incomplete sentences	Marked difficulty in word finding and word selection. Utterances are chiefly vague and empty of content. Stereotyped utterances and perseveration of words of phrases may occur	Severe impairment in planning and structuring of utterances. Paradoxical switching of topic may occur. Utterances frequently lose substantial connection with the topic of conversation. Only little information can be communicated. Mild or moderate echolalia may be present	
4 Very severe	Frequent phonematic paraphasias, palilalia may occur	Fragmented sentences and paragrammatical utterances predominate	Almost entirely meaningless sequence of words and phrases. Perseverations and stereotyped utterances may appear	The substantial connection of utterances with the topic of conversation is hardly recognizable. Utterances have almost no meaningful content. A minimum of relevant information can still be communicated. Severe echolalia may occur	
5 Extremely severe	Language is no longer at the patient's command. The rare attempts of verbal expression are poorly articulated. On occasion, however, an utterance may be fluent and well-articulated				

Of the 14 patients in group D, 3 showed an impairment of conversational ability (Fig. 1) but a combination with naming deficit was not observed (Fig. 2).

If the presence of conversational impairment and naming deficit had been used to distinguish between Alzheimer and non-Alzheimer cases in the present sample, 80% of the patients would have been correctly classified.

Discussion

According to the method described, impairment of conversational ability was found in the majority of patients with early Alzheimer's disease. Even in 4 out of 6 cases which would have been missed by formal diagnostic criteria at the initial examination, conversational disturbances were demonstrable.

These results suggest that the newly developed rating procedure for the assessment of conversational ability is useful to identify even mild degress of language impairment. As the sensitivity of the scale depends on the user's competence, it might be increased by an additional rater training.

In a group of patients with disorders that are difficult to distinguish from Alzheimer's disease, impairment of conversational ability was rarely observed, and the combination with naming deficits was completely absent.

Further investigation is required to determine the value of the method for the early and differential diagnosis of Alzheimer's disease, particularly with respect to age-related memory problems, to cognitive loss in depression and to dementias which are difficult to classify on clinical grounds.

References

Bayles KA (1982) Language function in senile dementia. Brain Lang 16:265–280
Bayles KA, Kaszniak AW, Tomoeda CK (1987) Communication and cognition in normal aging and dementia. Taylor and Francis, London
Blanken G, Dittmann J, Haas JC, Wallesch CW (1987) Spontaneous speech in senile dementia and aphasia. Implications for a neurolinguistic model of language production. Cognition 27:247–274
Bowles NL, Obler LK, Albert ML (1987) Naming errors in healthy aging and dementia of the Alzheimer type. Cortex 23:519–524
Cummings JL, Benson DF (1983) Dementia: a clinical approach. Butterworths, Boston

Flicker C, Ferris SH, Crook T (1987) Implications of memory and language dysfunction in the naming deficit of senile dementia. Brain Lang 31:187–200

Hart S, Smith CM, Swash M (1988) Word fluency in patients with early dementia of Alzheimer type. Br J Clin Psychol 27:115–124

Hier D, Hagenlocker K, Shindler AG (1985) Language disintegration in dementia: effects of etiology and severity. Brain Lang 25:117–133

Huber W, Poeck K, Weniger D, Willmes K (1983) Aachener Aphasie Test (AAT). Hogrefe, Göttingen

Shuttleworth EC, Huber SJ (1988) The naming disorder of dementia of Alzheimer type. Brain Lang 34:222–234

Skelton-Robinson M, Jones S (1984) Nominal dysphasia and the severity of senile dementia. Br J Psychiatry 145:168–171

Ulatowska HK, Allard L, Donnell A, Bristow J, Haynes SM, Flower A, North AJ (1988) Discourse performance in subjects with dementia of Alzheimer type. In: Whitaker HA (ed) Neuropsychological studies of nonfocal brain damage. Dementia and trauma. Springer, Berlin Heidelberg New York Tokyo

Correspondence: Dr. B. Romero, Psychiatrische Klinik der Technischen Universität München, Möhlstrasse 26, D-8000 München 80, Federal Republic of Germany.

The Alzheimer patient in the family context: how to help the family to cope

J. Selmes van den Bril and M. Selmes Antoine

Fundación Ciencia y Medicina, Madrid, Spain

Summary

Alzheimer's disease is the commonest type of dementia. At the present time, there are no medical or surgical measures for curing it or for preventing its appearance – and there will be none in the near future.

During the years that elapse between the onset of the disease and the death of the patient (average 8.1 years), a 'continuum of care' responding to specific criteria – adaptation to the patient's condition, flexibility in relation to the course of the disease, easy and free access to health services – must be provided.

Apart from the purely medical aspects, this continuum of care represents an extremely heavy social burden. This is borne essentially by two structures:

- the society, or the nation, as a whole, which bears the economic burden;
- the family, in whose bosom, traditionally, the person with this disease is provided with care for as long as possible.

Since the 1980s a movement, operating at a level between the above two structures and composed of people who do not find in the existing institutions the help or the type of help that they are seeking, has been playing a leading role in providing a continuum of care. Based on the principle of self-help, this movement has 3 principal aims: (1) to increase the awareness of and to educate the public; (2) to provide the patients and their families with material, psychological and social help; and (3) to incite the political and health authorities to find socio-medical solutions.

Before looking at these 3 aims in more detail, it is important to know why the social aspects of the disease have been arousing such great interest in the last few years.

From problem to crisis

If the socio-economic aspects of Alzheimer's disease seem to constitute an important problem today, then by the year 2000 they will very likely have given rise to a real crisis. Several factors account for this pessimistic outlook:

- the ageing of the population in the industrialized countries is going to increase the number of patients with this disease, and the increase in the duration of life indicates that the number of severe forms of the disease will increase. Let us remember that the Spanish population over the age of 64 years will increase by 61% between 1980 and the year 2000, and that the number of persons over 80 years will be almost 1 million at the end of the century;
- the family unit is undergoing profound changes from the sociological viewpoint. On the one hand, the availability of the largest group of people providing family care, namely wives and children, is less than in former times, basically because of the increase in the number of women entering the work market. On the other hand, because of the decreasing birth rate and the ever-increasing divorce rate, the family is often reduced to the simplest of units;
- and finally, the individual, in the assurance of his 'right to health', is becoming more and more aware of the problem posed by the dementias; he is demanding a quality of care that will take into greater and greater consideration the dignity of these patients; he is organizing himself into pressure groups in order to incite the health authorities to provide new health benefits and to lighten, by the participation of the state, the very heavy economic burden associated with caring for a patient with Alzheimer's disease in the family environment.

Thus, all the facts indicate that the socio-economic burden of this disease will increase in the very near future.

Society and the economic burden of Alzheimer's disease

It is recognized that the economic burden of Alzheimer's disease on a nation is heavy, and that in the coming years it will increase in direct proportion to the demographic evolution of the population.

But estimating the exact economic cost remains very uncertain: any estimation can only be a rough approximation, which will vary according to the health systems of each country.

We are going to refer to the American studies and, in particular, to the report published in April 1987 by the Office of Technology Assessment of the United States Congress.

The total cost represented by the dementias (diagnosis, treatment, long-term care, family care, indirect costs, etc.) amounted to 38 000 million dollars in 1983, according to a study from the National Institute of Aging, and ranged from 24 000 to 48 000 million dollars in 1985, according to a study from the Battelle Memorial Institute. As Alzheimer's disease constitutes 60% of the dementias, the total cost of this disease to the American nation amounted to 23 000 million dollars in 1983, and between 15 000 and 29 000 million dollars in 1985.

Apart from these economic considerations, an increase of 50% in the number of patients suffering from Alzheimer's disease is predicted, in the United States, for the year 2000.

In addition to the studies on the estimation of total costs, several studies have focused on estimating the specific costs associated with different categories.

Although it is not possible to extrapolate from these studies, they are of interest for three reasons:

- they provide a systematization of costs;
- they reveal the methodological difficulties involved in attempting to estimate costs;
- and they permit the breaking down of total costs into the specific costs associated with the different categories.

Thus the following categories can be distinguished:

1) Costs associated with diagnosis. They are estimated on the basis of 200 000 new cases of severe dementia per year, taking into account the fact that the diagnoses may be made in either ambulatory or hospitalized patients. A reasonable estimate would put the cost at between 500 and 1000 million dollars per year.
2) The cost of medicines and of medical services, once the diagnosis has been established, would be 10 000 million dollars according to a study carried out with reference to the year 1983.
3) The cost of residential care is a matter of pure speculation. It varies according to the samples chosen: some studies mention that 3% of the population in residential care are patients with dementia, whereas in other studies the proportion is between 40 and 60%. And so it is easy to see why estimations of the cost range from 10 000 to 33 000 million dollars per year, depending on the study.
4) The cost of care provided in the home – in particular, by the family – should include the costs associated with a number of parameters that are not easily evaluated, for example: the economic loss resulting from the disappearance of income from work (either of the patient, if he was professionally active, or of the care-provider), the health costs resulting from stress in the care-provider or providers, the cost of re-organizing the living space in the home, etc.

Based on the need for an annual expenditure of 11 700 dollars (in 1983), for each person involved, to provide home care for a person with dementia, and extrapolating to the level of the American nation, a figure of 26 700 million dollars per year is reached.

A comparison of the costs relating to the different categories shows clearly that the predominant share of the total cost is derived from long-term care in either an institutional environment or the family unit. This category represents a cost that is from 2.5 to 5 times greater than the cost of diagnosis, drugs and purely medical services.

The magnitude of the problem explains the efforts instigated in many countries to combat the economic cost, both present and predicted, of this disease. At present, within the Spanish Society of Neurology, a study group is working with the available economic data on this problem.

The seriousness of the problem can be broadly outlined as follows: if the 230 000 patients with Alzheimer's disease in Spain had to be admitted to residential care, (on the basis of a daily cost of 10 000 pesetas) it would cost 2300 million pesetas per day, that is to say, it would require an annual budget of 840 500 million pesetas.

The family and the patient suffering from Alzheimer's disease

The family plays a fundamentally important role in the continuum of care required by a patient with Alzheimer's disease. Traditionally, the family looks after the patient for as long as possible; the decision to place the patient in an institution is taken on the basis of not only the course of the pathological process and the degree of worsening, but also the moral and psychological strength of those providing family care.

To examine the social role of the family is, at the same time, to answer the following questions:

– What type of care does the patient maintained in the family environment require?
– Who takes the responsibility for and charge of providing this care?
– What are the stress factors that make this a difficult and stressful task?

1 – The type of care

It is a question of 'informal' care which, most of the time, is not strictly medical. Although the type of care may vary considerably depending on the degree of deterioration, there is one constant feature – supervision 24 hours of the day, which is well expressed by the title of a guide-book available to American families: 'The 36-hour day'.

At the onset of the illness, the family has to take a whole series of decisions concerning the financial, legal, and practical problems – from driving a car to signing cheques. With time, the family will have to take charge of the patient's personal tasks: dressing, washing, eating. Then, as the condition worsens, it will have to deal with problems of incontinence, excitement, insomnia, aggressive and dangerous behaviour, etc., in a situation where communication with the patient is becoming increasingly difficult, indeed, even completely impossible.

If the patient has to be institutionalized, the family often experiences not a feeling of freedom but of guilt, arising out of their inability to maintain the sick person in his family environment.

2 – Profile of the care-providers

In general, the task of providing care falls on to the married couple. They constitute 30 to 50% of the providers of family care, followed by the children, mainly daughters or daughter-in-laws, who represent 25 to 30% of care-providers. The overall picture is one where the burden falls more on women than on men, perhaps because of the former's traditional role in the family or the fact that they are often younger than their husbands.

According to a 1982 Long-Term Care Survey, the average age of caregivers is 57 years, with 10% aged 75 or over, which may be unable to meet the physical demands of caregiving.

3 – Stress factors

Three groups of stress factors that make the task facing families more difficult can be distinguished:

– Factors directly related to the disease

The behaviour of a patient with Alzheimer's disease in the family environment includes a whole series of features that are considered by his close relatives to be stress factors. According to the surveys carried out on families, the following can be mentioned in order of decreasing frequency:

- nocturnal agitation;
- wandering;
- incontinence;
- aggressive behaviour;
- the successive accumulation of minor problems, leading to a feeling of an unbearable burden;
- and secondary effects of prescribed medication.

In this respect, a further feature may be added: the emotional shock to the care-provider when he/she is no longer recognized as spouse, daughter or daughter-in-law by the patient.

— At the level of the health professionals, factors related to unawareness of the problems

The families suffer as a result of not being able to find, at the level of the nonspecialist health-care professionals, the support and the advice that they believe they have a right to expect.

Lack of knowledge of the disease is particularly evident at the level of the general practitioners or family doctors, when they are consulted about pathological problems that are not directly related to dementia.

Where else can the families find the answers to the practical problems concerning diet, organization of living space, safety, etc., with which they are faced every day?

The complexity of the social and administrative services that should be able to provide help gives the families a feeling that they have been abandoned or left alone with their problems; part of the problem is the difficulty the families have in understanding a specialized and, indeed, confusing vocabulary.

— Factors related to the repercussions that the presence of a sick person can have on the family itself

Anxiety and fear of the future increase in the providers of family care as a result of:

- the effects of the situation on their physical and mental health;
- the absence of possibilities to participate in social activities and the feeling of being confined to the home throughout the 24 hours of the day;
- the impossibility of making changes in the way of life;
- and the financial insecurity; indeed the care-provider may have to give up a stable job in order to look after the patient.

What can be done to lighten the burden of families that very often appear as the secondary victims of the dementing illness?

Both surveys of and personal interviews with the providers of family care have given rise to a definition of their needs. These can be grouped under 3 main headings: (1) alleviate the burden of constant 24-hour surveillance; (2) provide access to information; and (3) make available services that are specifically adapted to this type of illness.

1 — Alleviate the burden of 24-hour surveillance. Care-providers have expressed the essential importance of having at their disposal, for several hours or 24 hours (including the night) per week, a paid person who can substitute for them, thus allowing them to rest, to participate in social activities, and to attend the meetings of support groups.

They also stress the need for:

- help in the home with the patient's personal tasks: dressing, washing, eating;
- short-stay facilities in hospital or residential care;
- and day centres where the patient can participate in activities compatible with his mental state

2 – The need for information. The families need information on:

- the disease and its course;
- the services and organizations providing care and
- the experience of other families facing the same problem.

3 – The existence of services specifically adapted to this type of patient, in particular, home care. The SELF-HELP movement arose out of these needs and the inability of the existing health and social structures to respond adequately to them.

The third way: the SELF-HELP movement

Originating in the United States with the creation of ADRDA (Alzheimer Disease and Related Disorders Association) in 1979, this movement has given rise to an international organization (Alzheimer's Disease International) which groups together the national organizations in 20 countries.

This movement, based on private initiative, brings together – in a varying manner, depending on the country – health professionals and people who are facing or who have faced the problem of Alzheimer's disease in the form of a family member. The aims of this movement are to increase awareness of the problem of dementia among the general public, to provide the patients and their families with help and information, and to pressurize the political and health authorities to study and to set in motion adequate solutions.

Without being a substitute for the existing structures, this movement constitutes a new dimension in the endeavour to solve the social problems of Alzheimer's disease by promoting:

- the sharing of personal experiences;
- the diffusion of informative material;
- the provision of moral, psychological and, at times, financial support for the care-providers;

– and the setting up of a collective force that can play the role of a pressure group by acting on public opinion and the political powers.

Conclusion

In view of the considerable economic burden that Alzheimer's disease is exerting, and will continue to exert, on society, the social aspects of the disease are of prime importance.

It is a pity that the research effort in the area of the social aspects of the disease is not as well developed – and, indeed, lags far behind – as that in the areas of aetiology, epidemiology, biochemistry, clinical medicine and pathology. The types of care required, the techniques, the creation of specific services, and the search for financial solutions require the participation of everybody: doctors, nurses, social workers, families, and politicians.

Finally, is not helping the family to care for the patient within his own environment one of the solutions to the agonizing economic problem posed by Alzheimer's disease?

References

Battelle Memorial Institute (1984) The economics of dementia. Contract report prepared for the Office of Technology Assessment. US Congress

Frank Father D, Smith M, Caro F (1981) Family care of the elderly. Lexington Books, Toronto

Hu TW, Huang LF, Cartwright WS (1988) Evolution of the costs of caring for the senile demented elderly. A pilot study. Gerontologist 26:158–163

Huang LF, Hu TW, Cartwright WS (1986) The economic cost of senile dementia in the United States, 1983. Contract report prepared for the National Institute of Ageing No 1-AG-3-2123

Losing of million minds (1987) VS Congress, Office of Technology Assessment, OTA-BA-323 (Washington, DC: US Government Printing Office)

Male NL, Rabins PV (1981) The 36-hour day. John Hopkins University Press, Baltimore London

Stone R, Caffarata G, Sanglel J (1986) Care givers of the frail elderly: a national profile. US Department of Health and Human Services, National Center for Health Services Research

US Department of Health and Human Services (1984) "Alzheimer's disease". Report of the Secretary's Task Force on Alzheimer's Disease. DHHS 84-1323, Washington DC, Sept 1984

Correspondence: Dr. J. Selmes van den Bril, Fundación Ciencia y Medicina, Alberto Alcocer 33,6 c, 28036 Madrid, Spain.

Towards a clinically specific profile of severe senile primary degenerative dementia of the Alzheimer type (PDDAT)

H. Stuckstedte, C.-M. Abrahams, and **G. Ulmar**

Psychiatrisches Landeskrankenhaus, Wiesloch, Federal Republic of Germany

Summary

DSM-III-R-criteria for diagnosis, type, and degree of dementia are applied to 65 psychogeriatric in-patients.

In order to establish a clinically valid profile for PDDAT the degree of severity as well as the distinction from other types of dementia are taken into consideration. This two-dimensional approach to PDDAT seems to account better for the complexity of its clinical phenomenology. Patients with severe PDDAT are compared with a moderate PDDAT-group as well as a severe MID-group.

Abbreviations

AGP	The AGP System (Guy et al., 1985)
DS	Dementia Scale (Blessed et al., 1968)
DSI	Depression Status Inventory (Zung, 1972)
DSM-III-R	Diagnostic and Statistical Manual of Mental Disorders, Third Edition, Revised (American Psychiatric Association, 1987)
IMCT	Information-Memory-Concentration Test (Blessed et al., 1968)
IS	Ischemic Score (Hachinski et al., 1975)
MID	Multi-infarct Dementia
MMS	Mini-Mental State (Folstein et al., 1975)
PDDAT	Primary Degenerative Dementia of the Alzheimer Type (DSM-III-R)
***	significance: $p < 0.001$
**	significance: $p < 0.01$
*	significance: $p < 0.05$
T	tendency: $p < 0.10$

Introduction

It is of great importance in clinical research to distinguish between PDDAT and other types of dementia.

Prior to DSM-III-R, even important contributions to the clinical discrimination of different etiological groups of dementia do not consider the degree of severity to be an integral part of the diagnosis (Hachinski et al., 1975; Gustafson and Nilsson, 1982; Brun and Gustafson, 1988).

On the other hand, most studies dealing with different degrees of severity do not take into account different types of dementia (Reisberg, 1986; Erkinjuntti et al., 1988).

In order to ascertain clinical differences and find common aspects, our study takes into consideration the type of dementia as well as the degree of severity.

A major part of our patients is severely demented. We compare groups of elderly patients with severe PDDAT, moderate PDDAT, and severe MID, respectively.

Patients and methods

65 demented psychogeriatric in-patients (49 women, 16 men; age range 65–95 yrs., M = 80.5). Symptoms existed for about 2 yrs.

DSM-III-R-criteria for diagnosis, type, and degree of dementia. Severe PDDAT: 35 patients, moderate PDDAT: 11 patients, severe MID: 19 patients.

Methods applied

Dementia Scale DS (Blessed et al., 1968), Mini-Mental State MMS (Folstein et al., 1975), Information-Memory-Concentration Test IMCT (Blessed et al., 1968), Ischemic Score IS (Hachinski et al., 1975); evaluation of selected partial scores and single items; single items of the AGP System (Guy et al., 1985). Comparison of frequencies: G-Test (Woolf), two-tailed, continuity-adjusted according to Yates.

Results

Severe vs. moderate PDDAT

As anticipated, no differences are found with regard to the course of the disease.

"Changes in performance of everyday activities" (1) as well as "Changes in habits" (2) occur more often in the severe PDDAT-group (94% vs. 0% ***; 89% vs. 18% ***).

The severe PDDAT-group is more severely impaired in general cognition (3, 4) (80% vs. 27% **; 94% vs. 36% ***). The same holds true for "Abstract thinking" (5) (94% vs. 64% *), "Judgment" (6) (97% vs. 64% *); "Orientation to time" (7) (66% vs. 18% *), "Orientation to place" (8) (74% vs. 27% **) as well as "Language-related tasks" (9) (57% vs. 9% *). The two groups do not differ with regard to "Impaired concentration". Patients with severe PDDAT are more severely impaired in "Immediate recall" (10) (83% vs. 45% *) and tend to be more severely impaired in "Short-term memory" (11) (91% vs. 64% T) and "Personal long-term memory" (12) (43% vs. 9% T). In both groups "Non-personal long-term memory" is impaired to an equivalent degree. In the moderately as well as in the severely demented group personal data can be remembered better than non-personal data. The degree of "Indecisiveness" is equivalent in both groups. This also holds true for "Paranoic symptoms".

In general, an increase in the severity of symptoms can be observed in advanced stages of dementia. Rather unexpectedly, several disorders of affect occur more often in the moderate PDDAT-group: these patients are more irritable (13) (45% vs. 9% *) and depressed (14) (73% vs. 29% *) and complain more often about somatic disturbances (15) (45% vs. 9% *).

The "Major depressive syndrome" of DSM-III-R can be found significantly more often in the moderately demented as compared to the severely demented PDDAT-group (16) (64% vs. 20% *).

With regard to "Emotional incontinence", "Anxiety", "Euphoria", and eight DS-items relating to changes in personality, no differences are found between the two groups. The same holds true for five IS-items concerning somatic and neurological findings.

Severe PDDAT vs. severe MID

Four items of the IS relate to the course of the disease. "Abrupt onset" (17) and "Stepwise deterioration" (18) are found more often in MID-patients (26% vs. 0% **; 74% vs. 26% **), whereas "Fluctuating course" and "Nocturnal confusion" occur equally in both groups.

"Changes in performance of everyday activities" (1) occur more often in the PDDAT-group (94% vs. 68% *). "Habits" (Eating, dressing, sphincter control) (19) tend to be more severely impaired in the MID-group (58% vs. 29% T).

No differences can be found with regard to general cognition, "Abstract thinking", "Judgment", "Orientation to time", and "Concentration". "Orientation to place" (8) tends to be more severely disturbed in

PDDAT-patients (74% vs. 47% T), whereas MID-patients tend to be more severely impaired in "Language-related tasks" (9) (84% vs. 57% T).

In both groups "Immediate recall" and "Long-term memory" are impaired to an equivalent degree; "Short-term memory" (20) tends to be more severely impaired in PDDAT-patients (100% vs. 84% T).

Both groups are comparable with regard to disorders of affect and most of the personality-related DS-items.

"Increased rigidity" (21) can be found more often in the MID-group (84% vs. 49% *). Surprisingly, "Emotional incontinence" and "Relative preservation of personality" do not occur more frequently in patients with MID. As could have been expected, significantly more patients with "History of strokes" (22), "Focal neurological symptoms" (23), and "Focal neurological signs" (24) belong to the MID-group (68% vs. 3% ***; 53% vs. 11% **; 74% vs. 0% ***). There is only a tendency of "History of hypertension" (25) to occur more often in MID-patients (37% vs. 11% T), whereas "Evidence of associated atherosclerosis" can be found in both groups to an equivalent degree.

Conclusions

Our results indicate, that in order to establish a clinically valid profile for PDDAT the degree of severity as well as the distinction from other types of dementia should be taken into consideration. This two-dimensional approach to PDDAT seems to account better for the complexity of its clinical phenomenology. Diagnostic precision could be enhanced by the additional use of NINCDS-ADRDA-criteria for Alzheimer's disease (McKhann et al., 1984; Lauter et al., 1989).

With regard to the differences found and in order to widen the scope of our results it seems worth-while to include a greater number of patients with moderate and mild dementia of different types.

It should be examined, whether pathological changes in different areas depend on a difference in type of dementia, degree of severity, or both. Based on the outline of DSM-III-R the operational criteria for degrees of severity should be elaborated in more detail.

The diagnostic criteria of DSM-III-R do not permit the simultaneous diagnosis of PDDAT and MID. If patients with both diagnoses could be identified the remaining groups would be more homogeneous etiologically. Thus more distinct clinical profiles of different degrees of severity in PDDAT and MID could be established.

Notes

(1) DS partial score (max. partial score: 8; marked/severe: 4.5–8.0 vs. slight/moderate: 1.0–4.0)
(2) DS partial score (max. partial score: 9; moderate/severe: 2–9 vs. none/slight: 0–1)
(3) MMS score (max. score: 30, normal: 24–30; marked/severe: 0–15 vs. slight/moderate: 16–23)
(4) IMCT score (max. score: 37, normal: 28–37; marked/severe: 0–18.5 vs. slight/moderate: 19–27)
(5) single item (max. score: 2; severe: 0 vs. moderate: 1)
(6) AGP item 42 (markedly/severely impaired vs. slightly impaired)
(7) IMCT partial score (max. partial score: 7; severe: 0–1.5 vs. slight/marked: 2.0–6.5)
(8) MMS partial score (max. partial score: 5; severe: 0–1 vs. none/marked: 2–5)
(9) MMS partial score (max. partial score: 8; marked/severe: 0–5 vs. none/moderate: 6–8)
(10) IMCT partial score (max. partial score: 5, first trial only; slight/severe: 0–4 vs. none: 5)
(11) MMS partial score (max. partial score: 3; severe: 0 vs. moderate/marked: 1–2)
(12) IMCT partial score (max. partial score: 7; moderate/severe: 0–5.5 vs. none/slight: 6–7)
(13) DSI item 17 (max. score: 4; moderate/severe: 3–4 vs. none/slight: 1–2)
(14) IS item 6
(15) IS item 7
(16) Major Depressive Syndrome according to DSM-III-R-criteria
(17) IS item 1
(18) IS item 2
(19) DS partial score (max. partial score: 9; marked/severe: 5–9 vs. none/moderate: 0–4)
(20) IMCT partial score (max. partial score: 5; severe: 0 vs. marked: 1–2)
(21) DS item 12
(22) IS item 10
(23) IS item 12
(24) IS item 13
(25) IS item 9

References

American Psychiatric Association (1987) Diagnostic and statistical manual of mental disorders, 3rd edn (revised). American Psychiatric Association, Washington, DC

Blessed G, Tomlinson BE, Roth M (1968) The association between quantitative measures of dementia and of senile change in the cerebral grey matter of elderly subjects. Br J Psychiatry 114:797–811

Brun A, Gustafson L (1988) Zerebrovaskuläre Erkrankungen. In: Kisker KP, Lauter H, Meyer J-E, Müller C, Strömgren E (Hrsg) Psychiatrie der Gegenwart 6. Organische Psychosen. Springer, Berlin Heidelberg New York Tokyo, pp 253–295

Erkinjuntti T, Hokkanen L, Sulkava R, Palo J (1988) The Blessed dementia scale as a screening test for dementia. Int J Geriatr Psychiatry 3:267–273

Folstein MF, Folstein SE, McHugh PR (1975) "Mini-Mental State". A practical method for grading the cognitive state of patients for the clinician. J Psychiatr Res 12:189–198

Gustafson L, Nilsson L (1982) Differential diagnosis of presenile dementia on clinical grounds. Acta Psychiatr Scand 65:194–209

Guy W, Ban TA (1985) The AGP system. Manual for the documentation of psychopathology in gerontopsychiatry. Springer, Berlin Heidelberg New York Tokyo

Hachinski VC, Iliff LD, Zilhka E, Du Boulay GH, McAllister VL, Marshall J, Ross Russell RW, Symon L (1975) Cerebral blood flow in dementia. Arch Neurol 32:632–637

Lauter H, Kurz A, Haupt M, Romero B, Zimmer R (1989) Clinical diagnosis of dementia and Alzheimer's disease: a comparative validation of current diagnostic criteria. J Neural Transm (P-D Sect) 1:24

McKhann G, Drachman D, Folstein M, Katzman R, Price D, Stadlan EM (1984) Clinical diagnosis of Alzheimer's disease. Neurology 34:939–944

Reisberg B (1986) Dementia: a systematic approach to identifying reversible causes. Geriatrics 41:30–46

Zung WWK (1972) The depression status inventory: an adjunct to the self-rating depression scale. J Clin Psychol 28:539–543

Correspondence: Dr. H. Stuckstedte, Psychiatrisches Landeskrankenhaus, P.O. Box 1420, D-6908 Wiesloch, Federal Republic of Germany.

Sequential clinical approach to differential diagnosis of dementia

T. Wetterling[1], D. Rumpf-Höling[1], P. Vieregge[2], K.-J. Borgis[3], P. Delius[1], K.-H. Reger[1], R. Kanitz[1], and H.-J. Freyberger[1]

[1] Klinik für Psychiatrie, [2] Klinik für Neurologie and [3] Institut für Radiologie, Medizinische Universität zu Lübeck, Lübeck, Federal Republic of Germany

Summary

In this study an attempt is made to prove whether a sufficient differentiation can be obtained by the sequential application of some short scores [Blessed Dementia Scale, Mini Mental State Exam, Ischemic Score, and the Alzheimer Inventory (Cummings and Benson, 1986)], CT scan, EEG and laboratory tests. The data of 76 patients admitted with the presumptive diagnosis of dementia are presented. 20.5% of the demented patients could be diagnosed to suffer from a possibly treatable cause.

Introduction

A precise differentiation of the underlying cause of dementia is urgently necessary to separate patients with probable Alzheimer's disease from cases with other causes of dementia, especially those suffering from treatable disorders.

In this study an attempt is made to prove whether a sufficient differentiation can be obtained by the sequential application of some short scores, CT scan, EEG and laboratory tests.

Patients and methods

In this study 76 patients, 57 female and 19 male (mean age: 74.2 ± 7.8 years, range: 45–89 years), admitted consecutively for evaluation of their presumptive diagnosis of dementia are included. Education level: 8 years 77.7%; 10 years 15.8%; ≥13 years 6.5%.

Applied instruments:

1. modified Dementia-Scale (DS) (Blessed et al., 1968) and Mini Mental State Exam (MMSE) (Folstein et al., 1975);
2. Hamilton Depression Scale (HDS) (Hamilton, 1960);
3. laboratory tests consisting of: thyroid hormones (T3/T4/TSH), vitamine B12, folate, treponemal agglutination test and in some cases CSF protein analysis
4. CT scan;
5. EEG;
6. Ischemic Scale (IS) (Hachinski et al., 1975), and
7. Alzheimer Inventory (AI) (Cummings et al., 1986).

Results

Discrimination of demented from nondemented patients

According to the MMS and DS 13 cases (17.1%) are judged as severely (MMS <10 and/or DS <20), 9 (11.8%) as moderately (MMS >10 and <18, DS ≥20 and <35), and 22 (28.9%) as mildly demented (MMS >18 and <25, DS ≥35 and <50), while 32 patients (42.1%) are not demented. The correlation between DS and MMS is good ($r = 0.94$; $p < 0.000$).

Differentiation of the demented patients

To differentiate the most common types of dementia, Alzheimer's disease (AD) and multiinfarct-dementia (MID), some clinical scores have been developed. The Alzheimer Inventory (AI) pretends to differentiate AD from other causes of dementia, while the more established Ischemic Score (IS) is considered to discriminate MID from degenerative demen-

Table 1

Ischemic score	Alzheimer inventory	
	<15	>15
≤4	20 5/11/4/0	5 2/2/0/1
5–6	6 3/0/1/1	2 2/0/0/0
≥7	11 5/1/4/1	0

tia. If applied together with IS the AI can only classify 18 (44%) cases sufficiently. Seven cases with a IS <7 and AI >15 can be classified as probable AD-patients, while 11 with IS ≥7 and AI <15 can be judged as MID-patients. No confusing overlap (cases IS >7 and AI >15) can be found. But the agreement with the corresponding CT-findings (given as cases with WML/normal CT or atrophy/infarcts/intracerebral mass in the second line) is only poor.

Separation of possibly treatable causes of dementia

Out of the 22 patients with mild dementia 8 (36.4%) and 2 moderately (22.2%), but none of the severely demented cases have a HDS score >18. These patients are assumed to suffer from major depression too. Five (50%) of these cases improved by an antidepressant treatment.

Extensive physical examinations and laboratory tests reveal in 2 patients a possibly curable cause of dementia [renal failure, chronic cardiac failure (ECG findings: Lown IV b)]. The CSF analysis (not done in all cases) reveals in 2 cases a blood-CSF barrier dysfunction. Two further patients can be separated by CT bearing a brain metastasis or a frontal meningeoma, respectively.

CT- and EEG-findings

Nine demented patients (20.5%) show multiple infarcts, 17 (38.6%) diffuse white matter lesions, located predominately periventricularly (WML) 9 (20.5%) atrophy in CT, while 5 (11.6%) have normal CT-findings. The EEG shows in 50% of the demented cases an alpha rhythm, in 20% a general slowing and in 30% focal changes. The agreement of CT- and EEG-findings is good (data not shown).

Discussion

The various approaches to discriminate different types of dementia by i.e. IS, AI, CT or EEG yielded, as our data (Table 1) show, in rather confusing results. Therefore a sequential procedure is considered to allow a more sufficient differentiation. The diagnostic value of the applied instruments is still a matter of discussion in literature (i.e. Schröder et al., 1989).

Assuming infarcts seen in CT are the strongest indication for a vascular cause of dementia, the following sequential approach to differential diagnosis of dementia is suggested (Fig. 1).

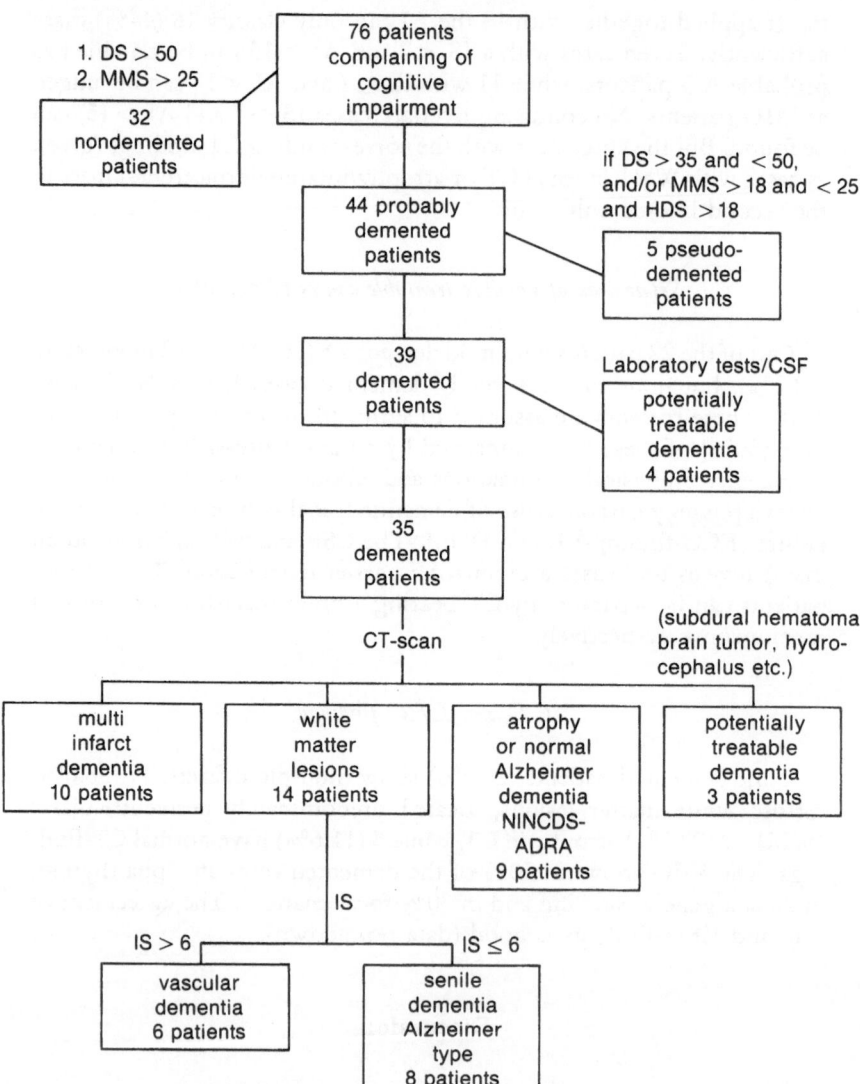

Fig. 1. Decision-tree for differential diagnosis of dementia

In sum, 20.4% of the demented patients can be diagnosed to suffer from a possibly treatable disease. 60% of these cases are considered to suffer from depressive pseudodementia due to their improvement by antidepressant therapy. 10 cases (22.7%) are diagnosed as MID by the CT-finding of multiple infarcts. The classification of 14 patients having WML in CT is difficult. Although 6 of them have an IS score >6, suggesting they suffer from MID, most of the WML cases have an Alzheimer-like course (insidious onset and progressive deterioration). Gottfries (1989) reported too, that many senile AD cases show WML in CT. But according to the NINCDS-ADRDA criteria (McKhann et al., 1984) neither these 8 cases nor 2 further cases (due to their CSF findings) can be classified as probable AD. Thus, only 9 cases (20.5% of all demented cases) fulfill all NINCDS-ADRDA criteria for AD. Our results show that the NINCDS-ADRDA criteria are very specific.

The sequential application of some short scores, CT, EEG and laboratory tests allow a sufficient clinical differential diagnosis of demented people. The rating can be done within 1 hour. Therefore even outpatients, suspected to suffer from dementia, might be diagnosed by this procedure.

References

Blessed G, Tomlinson BE, Roth M (1968) The association between quantitative measures of dementia and of senile change in the cerebral grey matter of elderly subjects. Br J Psychiatry 114:797–811

Cummings JL, Benson DF (1986) Dementia of the Alzheimer type. An inventory of diagnostic clinical features. J Am Geriatr Soc 34:12–19

Folstein M, Folstein S, Mc Hugh PR (1975) Mini-Mental state: a practical for grading the cognitive state of patients for the clinican. J Psychiatr Res 12:189–192

Gottfries CG (1989) Alzheimer's disease – one, two, or several? J Neural Transm (P-D Sect) 1:22

Hachinski VC, Illif LD, Zilhka E, du Boulay GH, Mc Allister VL, Marshall J, Russell RWR, Symon L (1975) Cerebral blood flow in dementia. Arch Neurol 32:632–637

Hamilton M (1960) A rating scale for depression. J Neurol Neurosurg Psychiatry 23:56–62

McKhann G, Drachman D, Folstein M, Katzman R, Price D, Stadlan EM (1984) Clinical diagnosis of Alzheimer's disease. Neurology 34:939–944

Schröder J, Haan J, Dickmann E (1989) Computerized tomography (CT) in multi-infarct (MID) and dementia of Alzheimer's type (DAT). J Neural Transm (P-D Sect) 1:127

Correspondence: Dr. T. Wetterling, Klinik für Psychiatrie, Medizinische Universität zu Lübeck, Ratzeburger Allee 160, D-2400 Lübeck, Federal Republic of Germany.

Do old patients with Down's syndrome develop premature brain atrophy?

D. Ehrmann and **P. Stoeter**

Stiftung Liebenau and Elisabethen-Kankenhaus, Ravensburg, Federal Republic of Germany

Summary

Evaluation of the clinical course and of CT examinations of 20 patients with Down's syndrome over 40 years of age showed a secondary clinical deterioration in cerebral performance in 5 patients, with significant cerebral atrophy in CT in 4 of them. In the others, the only significant difference to a control group of 40 patients of the same age were age-dependent atrophic changes of the temporal lobes. These findings which are similar to the early stage of Alzheimer's disease support the idea of a common genetic defect of both conditions.

A common genetic defect is discussed in trisomy 21 and Alzheimer's disease because of similar degenerative changes on post-mortem brain examinations and a reduced activity of the neurotransmitter system. But not every patient with Down's syndrome (DS) becomes demented and CT studies were not able to show significant or increasing cerebral atrophy up to the age of 40 (Ieshima et al., 1984). In the present study, the degree of cerebral atrophy in DS of older age is correlated to the clinical course.

Patients and CT examinations

Clinical evaluation and CT examinations (Somatom DR) are carried out in 2 groups of 10 DS of 40–49 and 50–63 years of age (7 female, 13 male) and 2 age-matched control groups of 20 patients without neurological deficit (15 female, 25 male). The DS have been hospitalised at 2 centres for mental retardation (Stiftung Liebenau and Anstalt Stetten) for mostly over 30 years. Changes of daily living skills, social behavior, and personality, neurological signs, seizures, and deterioration to dementia could be jugded without difficulty.

Clinical findings

The clinical evaluation of 20 patients with DS showed the following course:

subgroup 1: no changes: 6; slight deterioration: 9.
subgroup 2: severe dementia: 3; psychosis/epilepsy: 2.

CT-findings

Significant and age-dependent differences between patients with DS and the control groups are found only in the temporal lobes (enlargement of temporal horn and Sylvian fissure, Figs. 1, 2) whereas other significant changes (size of cisterna magna and cerebral peduncles) are similar in both or found in one age group only (size of the 3rd ventricle and number of the fissures of the superior cerebellar vermis, Fig. 3). The other ventricular parameters (Fig. 4) and the frequency of calcifications of the basal ganglia show a non-significant trend towards atrophic changes.

The subgroup of the 5 severely deteriorated patients with DS however, has a more pronounced brain atrophy "on inspection" in 4 cases with significant enlargement of the CSF spaces (Fig. 5) as compared to their non- or mildly affected collegues (Fig. 6, Table 1).

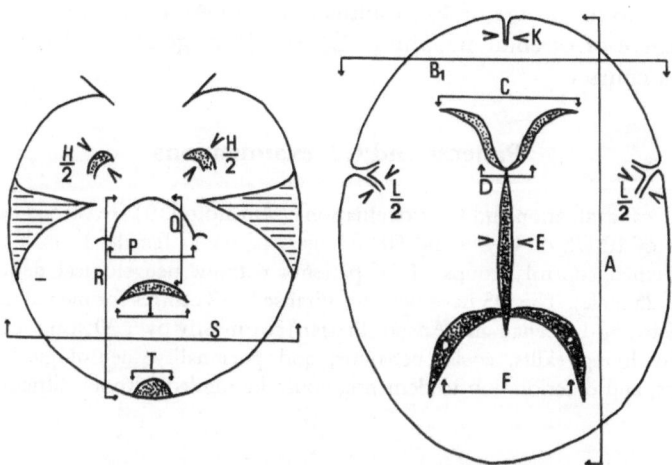

Fig. 1. CT diagram. Measurement of distances as indicated below

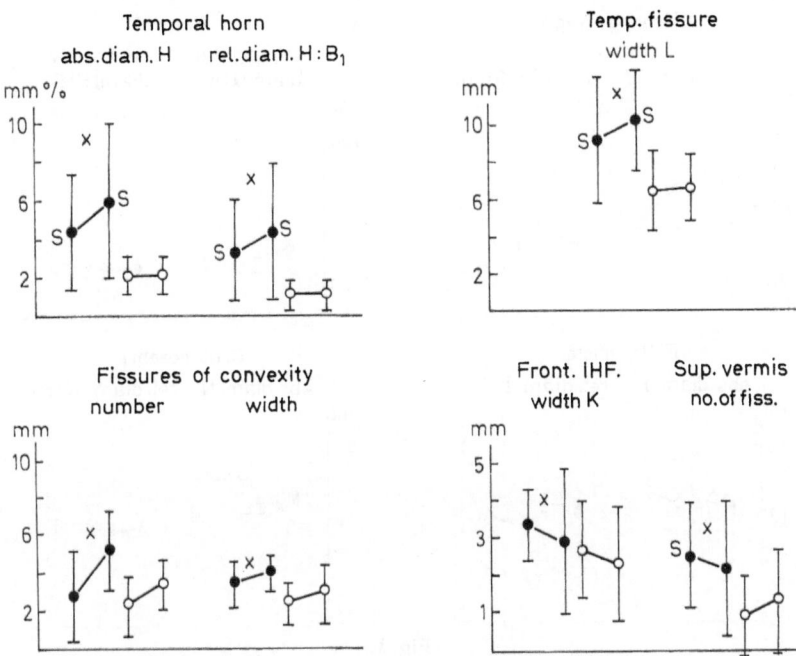

Fig. 2. Size of temporal horn and fissure, number/size of fissures of convexity, frontal interhemispheric fissure, and fissures of sup. cerebellar vermis. DS (all patients) (o); DS (severely deteriorated subgroup 2) (×); control group (●). Age group 40−49 years: left symbol. Age group 50−63 years: right symbol. Significant difference to control group: S. *Significant differences:* diameter of temporal horn and width of temporal fissure. Both parameters are age-dependent and significantly more pronounced in the severely deteriorated subgroup and therefore probably represent degenerative changes. The widening of the sulci of the cerebellar vermis is significantly different in the younger age group only

Table 1. Cerebral atrophy and calcifications

	M. Down			Control group	
	40−49 y n=10	50−63 y n=10	subgroup 2 n=5	40−49 y n=20	50−63 y n=20
Cerebral atrophy on inspection	2	2	4*	0	0
CSF-cranial ratio in %, acc. to (2)	7.2±2.6	8.5±3.2	11.1*	5.8±1.8	6.8±1.3
Calcification of basal ganglia	6	3	2	3	2

* Significant difference (α=99%) of severely deteriorated subgroup 2 to other 15 DS (subgroup 1) acc. to χ²-test resp. Student's t-test

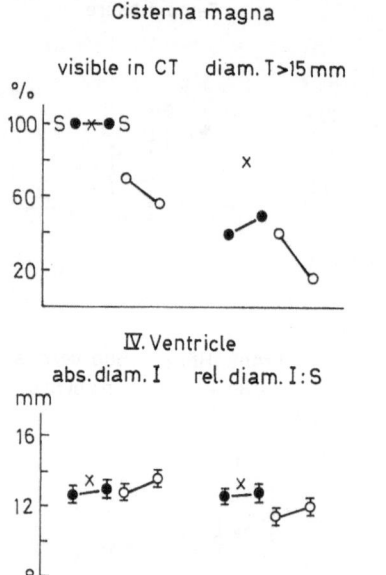

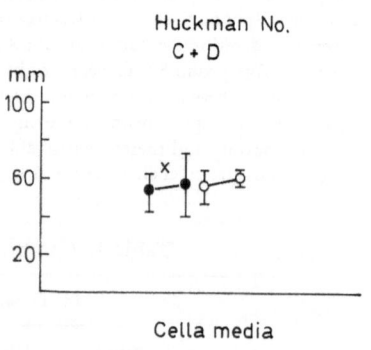

Fig. 3

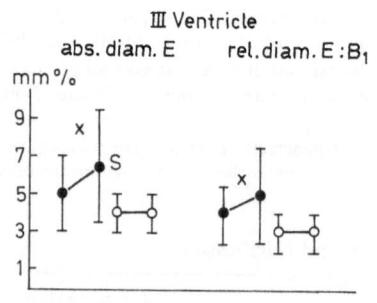

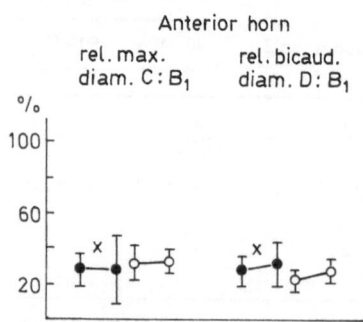

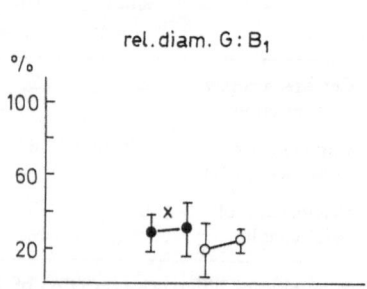

Fig. 4

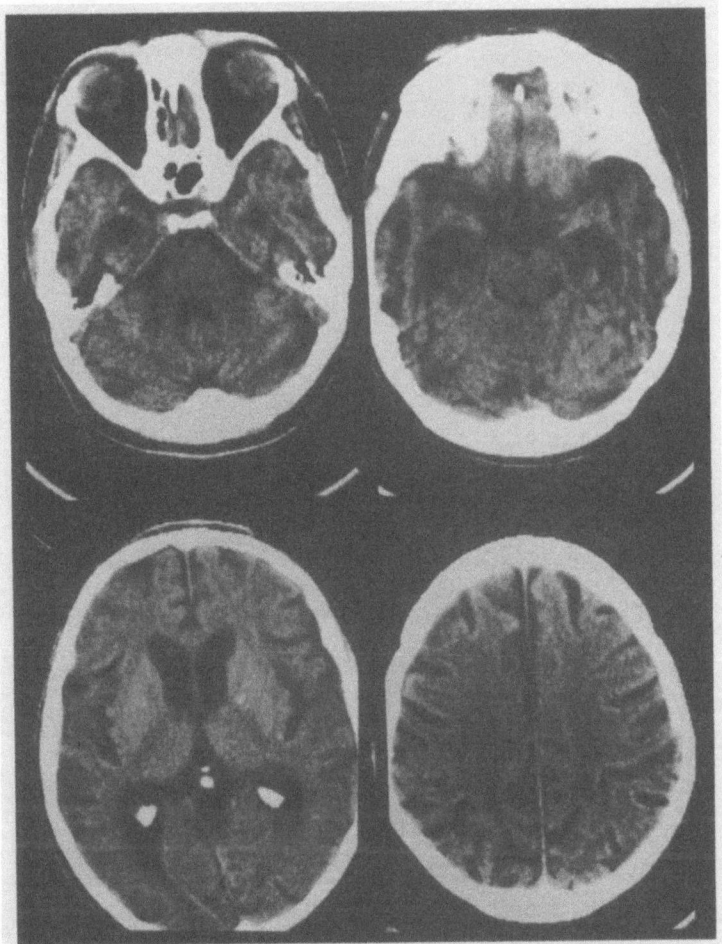

Fig. 5. M. Down, M. 49 years old. Severely demented. CT: Severe general cortical atrophy. Moderate general ventricular enlargement, but important dilatation of the temporal horns. Slight calcification of left basal ganglia

◄───

Fig. 3. Size of cisterna magna, pons, 4th ventricle, and cerebral peduncles. Symbols as in Fig. 2. *Significant differences:* Size of cisterna magna and absolute diameter of the cerebral peduncle. These parameters do not change with age and their deviation probably is due to developmental reasons

Fig. 4. Size of 3rd ventricle and lateral ventricles. Symbols as in Fig. 2. *Significant differences:* enlargement of 3rd ventrile in the DS group of older age. Dilatation of the anterior horn and cella media which is significantly larger in the deteriorated subgroup of DS. These changes probably are due to degeneration

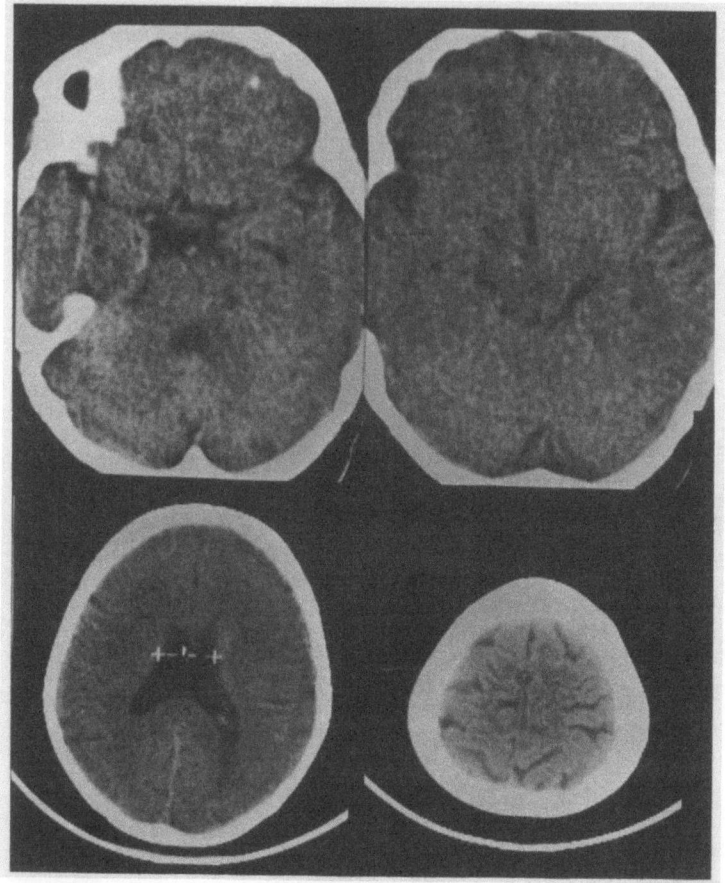

Fig. 6. M. Down, F. 57 years old. No clinical deterioration. CT: Slight widening of
Sylvian fissures

Discussion and conclusion

Our results show that only a subgroup (20%) of patients with DS
suffer from secondary clinical cerebral deterioration similar to presenile
dementia after the age of 40. Only these patients present significant
general cerebral atrophy in CT. The rest of the Down patients only show
significant and age-dependent atrophic changes of the temporal lobes.
Similar histological findings of a high quantity of plaques and tangles as
in Alzheimer's disease (Wisniewski et al., 1985) support the idea of a
common genetic defect of both diseases, and recently the – in DS –
triplicated chromosome 21 has been shown to carry the genes responsi-

ble for the increased formation of amyloid in Alzheimer's dementia (St. George-Hyslop et al., 1987).

Acknowledgment

We thank Dr. Letzkus, Anstalt Stetten, for the data of her 4 patients.

References

Ieshima A, Kisa T, Yoshino K, Takashima S, Takeshita K (1984) A morphometric CT study of Down's syndrome showing small posterior fossa and calcification of basal ganglia. Neuroradiology 26:493–498

Nagata K, Basugi N, Fukushima T, Tango T, Suzuki I, Kaminuma T, Kurashina S (1987) A quantitative study of physiological cerebral atrophy with aging. A statistical analysis of the normal range. Neuroradiology 29:327–332

St George-Hyslop PH, Tanzi RE, Polinsky RJ, et al (1987) The genetic defect causing familial Alzheimer's disease maps on chromosome 21. Science 235:885–890

Wisniewski KE, Wisniewski HM, Wen GY (1985) Occurrence of neuorpathological changes and dementia of Alzheimer's disease in Down's syndrome. Ann Neurol 17:278–282

Correspondence: D. Ehrmann, Stiftung Liebenau, D-7996 Meckenbeuren-Liebenau, Federal Republic of Germany.

Results of EEG brain mapping and neuroimaging methods in Senile Dementia of Alzheimer's Typ (SDAT) and Vascular Dementia (VD)

H. Lechner, K. Niederkorn, Ch. Logar, R. Schmidt, F. Fazekas, S. Horner, and H. Valetitsch

Universitäts-Nervenklinik, Graz, Austria

Summary

In order to determine the impact of neuroimaging methods and EEG brain mapping on the differential diagnosis of dementia 11 patients with SDAT and 11 patients with VD were studied. All SDAT patients scored 2 or less on the Rosen-Hackinski scale, the vascular dementia patients 4 or more. The Mini Mental and Mattis scale scores were similar. MRI showed cerebral infarcts in 82% of the VD group versus 0% in the SDAT group. Basal ganglia hyperintensities in early confluent and confluent white matter lesions occurred more frequently in the VD group SPECT revealed cortical minor perfusions mainly in patients with VD, diffuse frontal and diffuse parietal temporal cortical minor perfusions were detected exclusively in patients with SDAT. Duplex scanning of the carotid arteries showed no significant difference of the status of the extracranial carotid arteries between the 2 dementia groups.

EEG brain mapping clearly demonstrate the slowing down of the basis rhythm in the SDAT patients, the main advantage in comparison with routine EEG lies in the statistical possibilities of this method.

It could be shown that the use of neuroimaging methods and EEG brain mapping adds significantly to the capability of clinical scores to differentiate between degenerative and vascular etiology of dementia, improving the possibilities of treatment and prevention.

Introduction

Alzheimer's disease and Vascular Dementia are two pathologically well-defined diseases, that often cannot be separated reliably enough by the clinician (Meyer et al., 1988, 1989). Neuropsychological testing and clinical rating scales allow a differentiation of both diseases to some extent. The value of conventional EEG in the differential diagnosis of dementias has been described in the 1970s, especially by Obrist (1978) and van der Drift and Kok (1972). Considerable progress has been made by the introduction of EEG brain mapping, which allows computerized data analysis of the amplitudes and topographic distribution of EEG frequencies. These distributions may be imaged and averaged values of different patient groups can be compared statistically (Duffy et al., 1979). Difficulties with this method arise from the absence of a really well defined and investigated normal control group and from fluctuations, which may occur in normal individuals. To overcome these problems the use of additional neuroimaging methods like MRI, SPECT and ultrasound investigations is necessary.

The objective of the present study is to investigate the significance of EEG brain mapping for the differential diagnosis of Alzheimer's disease versus vascular dementia in comparison to the results on MRI, SPECT and ultrasound methods in these patient groups.

Patients and methods

We studied 22 demented patients (mean age 67 years, 10 male, 12 female). The diagnosis of dementia and the differential diagnosis between SDAT and vascular dementia was based on the SDM III, the NINCDS-ADRDA criteria and on the Rosen-Hachinski scale. The severity of dementia was assessed by the Mini Mental State Examination and the Mattis dementia rating scale. 11 patients were identified as SDAT cases (mean age 66 years, 4 males, 7 females), 11 patients were diagnosed as having vascular dementia (mean age 68 years, 6 males, 5 females).

All SDAT patients scored 2 or less on the Rosen-Hachinski scale (mean 0.91), the vascular dementia patients 4 or more (mean 9.6, $p = <0.001$, T-Test). The SDAT patients had an average score of 17.2 and 78.6 on the Mini Mental and Mattis scale respectively versus 22.9 and 104.5 for the vascular dementia patients (n.s.).

For the EEG brain mapping we used the "Brain Atlas III" (Bio-logic Systems Corp., Mundelein, IL, U.S.A.). Data from 16 scalp electrodes following the 10:20-system and 2 mandibular reference electrodes were worked up. EEG data segments of 2 seconds duration were subjected to spectral analysis by means of fast Fourier transformation (FFT). With only 16 measured values the computer calculated the data necessary for the imaging by means of linear interpolation. Statistical analysis (t-test) was used to demonstrate the areas where single patients or clinical subgroups, whose data have been averaged differ from a normal control group or groups with other clinical diagnoses.

MRI was performed using a superconducting magnet at a field strength of 1.5 T (Gyroscan 15, Philips, Eindhoven, The Netherlands) and the spin echo technique. A repetition time of 1800–2500 msec and two echo times of 30 and

60 msec were used to generate mixed-intensity and T2-weighted images in the axial plane. Slices were 5–6 mm thick.

SPECT was carried out using the Tomomatic 564 (Medimatic, Copenhagen, Denmark) 20–30 min after injecting 15–25 mCi Tc99-HMPAO.

Extracranial Doppler and Duplex sonography of the carotid and vertebral arteries (Vingmend CFM 700, Vingmed, Oslo, Norway) was performed in all patients.

Results

The comparison of the SDAT and vascular dementia groups showed a higher incidence of hypertension, diabetes mellitus, coronary heart disease and also of hypercholesteremia in the VD patients, although the significance level was not reached (see Table 1). A neurological deficit was more frequent (8/11 versus 2/11) in patients with vascular dementia than in those with SDAT. This difference did not reach significance. The status of the extracranial arteries as diagnosed by Duplex scanning was similar in both groups (Table 2).

MRI revealed cerebral infarctions in 9 of 11 (82%) of the VD cases versus of 0/10 patients with SDAT (p = 0.012, Fisher's Exact Test). The complete MRI results are shown in Table 3.

When comparing the SPECT results of SDAT and vascular dementia patients, focal cortical minor perfusions were significantly more present

Table 1. Risk factors for stroke in SDAT and VD (in %)

	SDAT (n = 11)	VD (n = 11)	
Hypertension	27	90	N.S.
Diabetes mellitus	27	55	N.S.
Coronary heart disease	20	82	N.S.
Hypercholesterolemia (> 250 mg %)	12	30	N.S.
Smoking	9	0	N.S.*

* Fisher's exact test

Table 2. Extracranial vessel wall damage in SDAT and VD patients (in %)

	SDAT (n = 11)	VD (n = 11)
Normal	0	0
ASKL < 50% sten.	91	82
ASKL ≥ 50% sten.	9	18

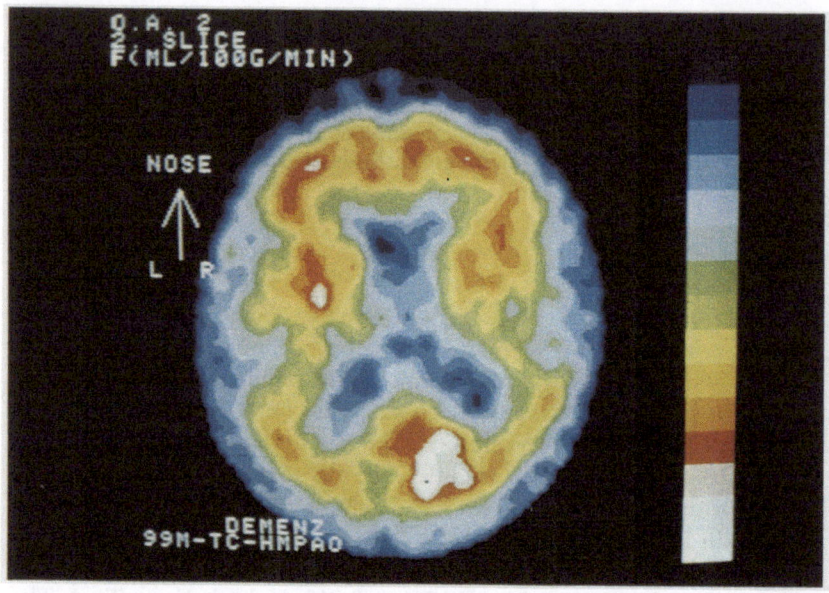

Fig. 1. Tc99-HMPAO SPECT of a 70 year old female patient with SDAT. Parietotemporal cortical minor perfusions are visible

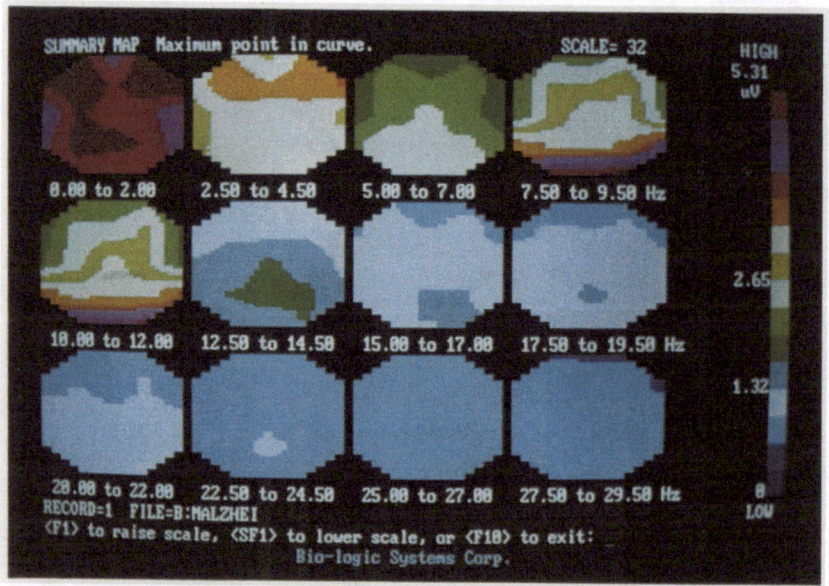

Fig. 2. The averaged EEG brain maps of 11 patients with SDAT

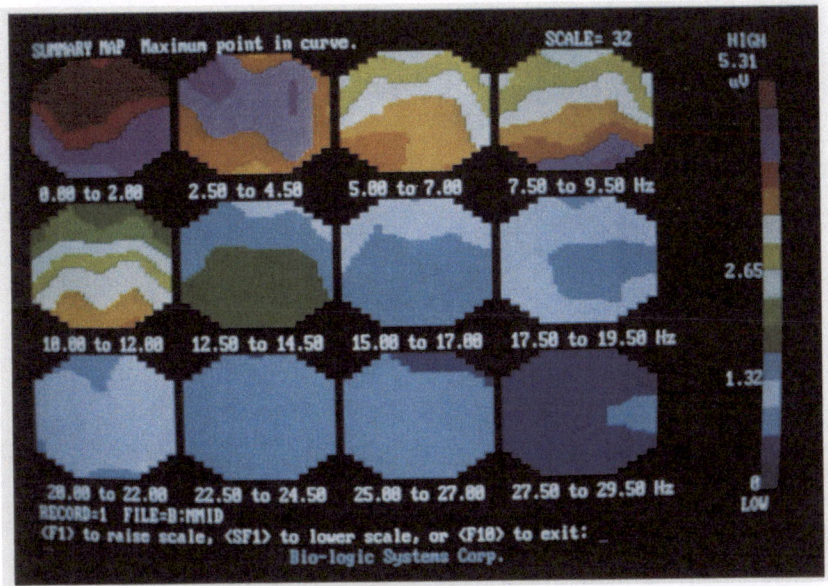

Fig. 3. The averaged EEG brain maps of 11 patients with vascular dementia

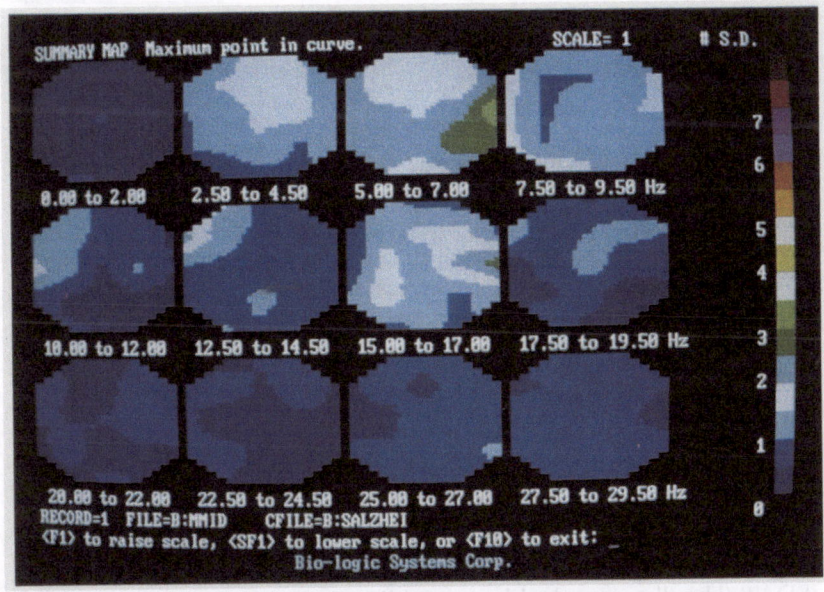

Fig. 4. The comparison of SDAT and VD groups reveales significant differences in the 5–7 Hz range

Table 3. MRI findings in SDAT and VD (in %)

	SDAT (n = 10)	VD (n = 11)	P* =
Cerebral infarcts	0	82	0.012
Basal ganglia hyperintensities	20	82	N.S.
Basal ganglia lacunes	0	27	N.S.
White matter lesions-punctate	20	18	N.S.
WML-early confluent + confluent	30	64	N.S.
Cortical hyperintensity (Insule, Parahippocampus)	60	20	N.S.
Cortical atrophy	100	89	N.S.
Ventricular atrophy	80	89	N.S.

* Fisher's exact test

Table 4. Comparison of spect results in patients with SDAT and VD (cortical minorperfusion, in %)

	SDAT (n = 10)	VD (n = 9)	P* =
None	0	22	N.S.
Focal lesion	0	56	0.047
Patchy	0	22	N.S.
Diffuse frontal	70	0	0.030
Diffuse parietotemporal	80	0	0.020

* Fisher's exact test

in patients with vascular dementia. Diffuse frontal and diffuse parietotemporal cortical minorperfusion was significantly more frequent in patients with SDAT (Table 4). A typical example is shown in Fig. 1.

On EEG brain mapping, focal changes were the dominant finding in the vascular dementia group, whereas Alzheimer patients showed reduced amplitudes of the basis rhythm. Statistical analysis of a comparison of the averaged results of both groups showed significant differences predominantly in the 5 to 7 Hz range (see Figs. 2–4).

Discussion

The present study describes the results of EEG brain mapping and various neuroimaging methods in patients with vascular dementia and SDAT. The diagnoses had been established on the basis of a battery of psychometric tests and clinical scores. When comparing risk factors stroke (hypertension, diabetes mellitus, coronary heart disease, hyper-cholesteremia, smoking) no significant difference between groups was observed. There were also no significant differences in the status of the extracranial carotid artery vessel wall. MRI showed cerebral infarcts in 82% of the vascular dementia group versus 0% of the SDAT group. Basal ganglia hyperintensities and early confluent and confluent white matter lesions occurred more frequently in the VD group, but there were no significant differences.

SPECT was very helpful and revealed significantly more focal corti-cal minor perfusions in patients with vascular dementia, whereas diffuse frontal and diffuse parietal temporal cortical minor perfusions occurred exclusively in patients with SDAT. The use of EEG mapping showed a slowing down of the basis rhythm as the most striking characteristic for separation of SDAT and VD group, a finding that has also been ob-served by classical EEG. The advantage of EEG brain mapping lies predominantly in the possibility to. statistically compare and evaluate defined patients groups. Fluctuation problems may be solved by increas-ing the registered period of artefact free EEG. MRI and SPECT, cover-ing morphologic and metabolic aspects of the dementia process add significant information for and allow improved differential diagnosis (Lechner et al., 1990).

References

Van der Drift JH, Kok NK (1972) The EEG in cerebrovascular disorders in relation to pathology. In: Remond A (ed) Handbook of EEG and clinical neurophysiology, vol 14, part A. Elsevier, Amsterdam, pp 12–64

Duffy HF, Buchfield JL, Lombroso CT (1979) Brain electrical activity mapping (BEAM): a method for extending the clinical utility of EEG and evoked potentials data. Ann Neurol 5:309–321

Lechner H, Niederkorn K, Logar C, Schmidt R, Fazekas F (1990) EEG brain mapping in patients with SDAT and vascular dementia – results and correla-tions with MRI, SPECT and TCD. In: Battistin L (ed) Aging brain and dementia: new trends in diagnosis and therapy. Alan R Liss, New York, pp 337–348

Meyer JS, Lechner H, Marshall J, Toole JF (eds) (1988) Vascular and multi-in-farct dementia. Futura Publishing Co, Mount Kisco, NY

Meyer JS, Lechner H, Reivich M, Toole JF (eds) (1989) Cerebral vascular diseases, 7. Excerpta Medica, Amsterdam New York Oxford

Obrist WD (1978) Electroencephalography in aging and dementia. In: Katzman R, Terry RD, Blick KL (eds) Alzheimer's disease: senile dementia and related disorders. Raven Press, New York (Aging, vol 7)

Correspondence: Dr. H. Lechner, Universitäts-Nervenklinik, Auenbruggerplatz 22, A-8036 Graz, Austria.

Regulation of EEG delta activity by the cholinergic nucleus basalis

**P. J. Riekkinen[1], P. J. Riekkinen Jr.[1], H. Soininen[1],
K. J. Reinikainen[1], V. Laulumaa[1], V. S. J. Partanen[2], T. Halonen[1],
and L. Paljärvi[3]**

Departments of [1] Neurology, [2] Neurophysiology, and [3] Pathology, University of Kuopio,
Kuopio, Finland

Summary

To study the relationship between cholinergic pathology and EEG deterioriation, spectral EEG and CSF neurotransmitter markers were examined in 23
probable AD patients. CSF acetylcholinesterase was correlated with delta power
and alpha/delta ratio. In an autopsy study, cortical ChAT activity and the nucleus
basalis cell number were lower and delta power higher in 4 demented patients
than in 6 age-matched controls. In the group of demented patients, the lowest
ChAT activities were associated with the highest delta power.

Introduction

Alzheimer's disease (AD) is characerized by alterations of the normal
electrical activity of the brain: a decrease has been observed in both the
alpha power and in the dominant EEG frequency (Soininen et al., 1982).
Also an increase in a slow frequencies (Coben et al., 1983) and a decrease
in beta power (Duffy, 1984) has been shown to occur in AD patients.
Further, several studies have demonstrated that the EEG deterioriation
is correlated with the degree of cognitive decline (Kazsiak et al., 1979),
and the duration of the illness (Coben et al., 1985) in AD patients.
However, a recent follow-up study did show that only in 50% of early
AD cases EEG worsening could be detected during a period of 1 year

(Soininen et al., 1989). Interestingly, in the subpopulation of AD patients with deterioriating EEG, cortical cognitive functions declined more during a 3-year follow-up than in the AD patient group with stable EEG (Helkala et al., unpublished).

Though many neurochemical abnormalities occur in the brains of patients with AD, the most consistent and severe deficiency is the reduction in cortical cholinergic markers (Reinikainen et al., 1988; Bird et al., 1983), and degeneration in the nucleus basalis of Meynert (NBM) (Whitehouse et al., 1981), a cholinergic nucleus located in the basel forebrain sending projections to all the areas of the neocortex.

The cholinergic system has been shown to play an important role in the regulation of neocortical electrical activity. Administration of muscarinic antagonist, scopolamine or atropine, abolish normal EEG activity and produce large amplitude slow waves (Vanderwolf, 1975; Wikler, 1952). The role of NBM in the activation of cortical EEG has been shown to be crucial; lesioning the basal forebrain cholinergic cells increases slow delta waves (Buzsaki, 1988; Stewart, 1984). However, the brainstem cholinergic systems located in the pedunculopontine and laterodorsal tegmental nuclei, Hallanger (1987), may also contribute to the regulation of electrocorticogram, but no studies exist examining the effects of restricted destruction of these structures on EEG formation.

Because cholinergic cells in the NBM function in desynchronization of the electrocorticogram, abnormalities in these cells may underlie, at least to a certain degree, the EEG deterioriation associated with AD. However, to date no studies have examined the possible relationship between the cholinergic pathology and changes in quantitative EEG parameters in patients with AD. Thus, in the present study the relationship between the EEG alterations and the decline of markers of the cholinergic system were studied. The CSF study was done in order to obtain a deeper understanding of the possible correlation between the cholinergic marker (acetylcholinesterase, AChE) and EEG slowing. The number of other subcortical structures innervating the neocortex which could regulate neocortical EEG are few; the dorsal raphe (Moore et al., 1978), the locus coruleus (Lindvall et al., 1974), and the ventral tegmental area. Because in AD these systems also degenerate (Palmer et al., 1987; Rossor and Iversen, 1986; Reinikainen et al., 1988), and thus might contribute to the EEG slowing, CSF markers of these systems were studied and correlated with the EEG parameters measured. In addition, four demented patients (2 cases with stable EEG and 2 with rapidly deterioriating EEG) and 6 controls were included in a preliminary autopsy study that was conducted to investigate the possible relationship between the increase in delta power and loss of presynaptic cholinergic markers in the cortex as well as loss of NBM cells.

Patients and methods

CSF study

The CSF study group consisted of 23 patients (16 women and 7 men) with probable AD according to the criteria of the NINCDS-ADRDA Work Group. The mean age of the patients was 69 years, S.D. ± 7, the mean age at the onset of the disease was 66 years, S.D. ± 6, and the mean duration of the disease 2 years, S.D. 1. The patients were examined using a clinical neurological investigation, neuropsychological evaluation, routine blood and CSF chemical studies, EEG and brain CT scan. The severity of the dementia was estimated using the Clinical Dementia Rating scale of Hughes et al. (1982). According to the CDR scale, 13 patients had mild, 8 moderate, and one severe dementia. Nineteen patients were without CNS active medication; one patient was receiving benzodiazepine, one patient an antidepressant, and three patients were receiving neuroleptic drugs in small doses.

At 8–10.00 a.m. a standard sample of lumbar CSF was obtained from the patients as they were lying in a recumbent position. The samples were frozen immediately and stored at $-80°C$ until assayed. Cholinesterase (ChE) activity was determined using a modified method of Ellman et al. (1961). A specific AChE inhibitor (BW284C51) was used to inhibit true AChE, and the residual activity was subtracted from the ChE activity to obtain the AChE activity. The levels of the monoamine metabolites HVA, 5-HIAA, and MHPG were measured with HPLC using electrochemical detection.

Pathology and cell counts

At autopsy, the brain was halved sagittally. The right half was immersed in 10% buffered formalin for several weeks. Diagnostic samples were taken from twelve standard cortical and subcortical areas and processed into hematoxylin-eosin, Congo red and Bielscowsky sections as described previously (Mölsä et al., 1987).

The sample for cell counts was taken at the level of the anterior commisure and included the cholinergic Ch4a area (Rinne et al., 1987). Three 20 µm coronal sections were taken at 60 µm intervals and stained with cresyl violet. The cell group of interest was marked with India ink, and – using an ocular grid with an area of 0.408 mm² (25 × objective) – a systematic sampling was done comprising one sixth of the total area. Nucleolated neurons having a minor perikaryonal diameter of 15 µm or more were counted in each field. In this report we use the cell counts of the fields with maximum cell densities as measures of nucleus basalis degeneration. The results of four demented patients were compared with six samples from patients (aged 60–83 years) having no history or pathological findings of neurological disease.

The left half of the brain was stored at $-70°C$ until dissection and biochemical assay. At dissection, the brains were sectioned coronally into serial sections 5–10 mm thick, starting at the frontal pole. Gray matter from the frontal cortex

(Brodmann area 9) and from the temporal cortex (Brodmann area 21) were taken for the biochemical analysis.

Noradrenaline and dopamine were analysed in brain areas by high pressure liquid chromatography (HPLC) as described by Mefford et al. (1981) with the exception that a coulometric detector with an oxidation voltage of 0.40 V was used instead of an amperometric detector. Amine metabolites were measured by HPLC method with an amperometric detector. ChAT was analysed by the method of Fonnum (1975) and AChE by the method of Ellman (1961).

EEG values

A conventional EEG recording was performed with a 10- or 16-channel Siemens Minograph apparatus using the International 10−20 system in attaching the Ag-AgCl-electrodes. The EEG from derivation T6-O2 was amplified and filtered by a DISA 15C01 amplifier with the bandpass set at 0.5−50 Hz, and digitized at 125 samples/s with a 12-bit multichannel analyzer (Tracor-Northern 1710). The EEG was visually inspected for artefacts. Four artefact-free epochs of 8.192 s were recorded. A Fast Fourier Transform (FFT) was computed on a series of half-overlapping sections for each of the four epochs. The first section comprised the first half, the second section the second and third quarters, and the third section the second half of the epoch. To reduce leakage and time truncation errors, the time-domain waveform of each section was multiplied by a cosine (Hannig) window before the FFT operation. FFTs from a total of 12 sections were then averaged to obtain the frequency spectrum of the whole EEG sample of 32.772 s. The relative powers (percentage of total power) in the delta (1.46−3.91 Hz), theta (4.15−7.32 Hz), alpha (7.57−13.92 Hz) and beta (14.16−20.02 Hz) bands were calculated. The total range of analysis was 1.46−20.02 Hz. Several power ratios: alpha/theta, alpha/delta, alpha/(delta + theta) and (alpha + beta)/(theta + delta) were calculated, as well as the peak and mean frequency values. The subjects were instructed to relax and shut their eyes. Verbal communication was used, if necessary, to maintain alertness of the subjects.

Results

Spearman's correlation coefficients were used to examine correlations between the spectral EEG variables and the CSF values.

Correlations between the CSF ChE, AChE, HVA, 5-HIAA and MH-PG values and spectral EEG variables are shown in Table 1. The CSF AChE values correlated significantly with the delta power ($r = -0.57$; $p = 0.002$) and alpha/delta ratio ($r = -0.44$; $p = 0.017$), but not with other spectral variables. Also, the total ChE correlated significantly with the same EEG parameters (delta; $r = -0.59$, $p = 0.002$ and alpha/delta; $r = 0.46$, $p = 0.013$). In addition, low ChE activity correlated with the mean frequency in the combined alpha-theta range ($r = 0.043$; $p = 0.019$).

Table 1. Correlations between spectral EEG variables and CSF total cholinesterase (ChE), acetylcholinesterase (AChE), homovanillic acid (HVA) 5-hydroxy-indoleacetic acid (5-HIAA), and 3-methoxy-4-hydroxyphenylglycol (MHPG) in patients with Alzheimer disease

	ChE n = 23	AChE n = 23	HVA n = 23	5-HIAA n = 23	MHPG n = 23
Beta %	0.31	0.21	0.32	−0.06	−0.10
Alpha %	0.16	0.15	0.00	−0.02	0.08
Theta %	−0.22	−0.13	0.03	0.14	0.08
Delta %	−0.59 [c]	−0.57 [c]	−0.19	−0.06	0.06
Alpha/theta ratio	0.22	0.17	0.03	0.00	−0.08
Alpha/delta ratio	0.46 [c]	0.44 [c]	0.05	0.00	−0.04
Alpha/theta + delta	0.28	0.25	0.07	0.07	0.05
Alpha + beta/theta + delta	0.30	0.26	0.16	0.10	0.09
Frequency peak Hz	0.15	0.11	0.07	−0.25	−0.24
Frequency mean Hz	0.43 [c]	0.34	0.20	−0.20	−0.30

Correlations are calculated with Spearman's correlation test, [a] $p < 0.05$; [b] $p < 0.02$; [c] $p < 0.01$

Table 2. Mean of the frontal and temporal noradrenaline (NA), dopamine (DA), homovanillic acid (HVA), 5-hydroxyindoleactic acid (5-HIAA), cholineacetyltransferase (ChAT), nucleus basalis (NBM) cell density and delta power

	Controls	Case 1	Case 2	Case 3	Case 4
Frontal and temporal NA (pmol/g)	− −	2.9 7.7	3.4 7.8	29.8 39.4	36.9 43.1
Frontal and temporal DA (pmol/g)	− −	10.1 3.8	17.8 8.0	− 63.7	129.0 36.9
Frontal and temporal HVA (nmol/l)	127.0 189.0	252.0 23.0	173.0 246.0	− 171.0	129.0 248.0
Frontal and temporal 5-HIAA (mg/wet weight)	134.0 164.0	196.0 75.0	− −	74.0 98.0	52.0 142.0
Frontal and temporal ChAT (nmol/mg prot/min)	94.5 −	5.9 11.6	22.4 24.4	81.6 71.4	70.9 88.8
NBM cell density (mm⁻²)	195.0	70.0	64.0	108.0	168.0
Delta power (μV²)	13.1	55.9	65.2	12.9	30.0

In contrast, no systematic correlation could be found between the slowing of EEG and the levels of CSF monoamine metabolites.

The results of the biochemical analysis, nucleus basalis cell counts, pathological diagnosis and absolute delta powers are shown in Table 2. The mean ChAT levels were lower in the frontal and parietal cortex in the demented patients than in the controls. The mean levels of 5-HIAA were lowered in the frontal cortex as well as in the temporal cortex in the demented patients. Mean HVA values were not changed either in the frontal or temporal cortex. Interestingly, the lowest levels of ChAT as well as noradrenaline were observed in the two demented patients with largest delta power.

Discussion

In the present study of a group of AD patients, the lowered level of CSF AChE activity correlated with the relative delta power and a decrease in the alpha/delta ratio of the spectral EEG. No correlation could be detected between the CSF markers of the aminergic systems and EEG variables. Further, in a preliminary autopsy study the patients with the lowest level of cortical ChAT activity as well as NBM cell density had the highest amount of delta power, thus highlighting the importance of the cholinergic deficit in EEG slowing.

The CSF AChE activity as a marker of the degeneration of the basal forebrain cholinergic neurons in AD patient has many limitations. First, the origins of lumbar CSF AChE and the relative amounts of lumbar CSF AChE produced by different brain regions are not exactly known. Second, AChE is not only found in cholinergic cells but may be produced by non-cholinergic neurons located in the locus coruleus and dorsal raphe (Greenfield, 1984), two subcortical areas that also degenerate in AD (Palmer et al., 1987; Rossor and Iversen, 1986). Third, the spinal cord may also contribute to lumbar CSF AChE activity (Appleyard et al., 1987). Taking the above consideration into account, it is important to note that it has consistently been shown that in neuropathologically verified AD cases the AChE activity is reduced (Reinikainen et al., 1988; Arendt et al., 1984). Further, the two cases in the present study that had the lowest density of NBM cells also had the lowest levels of CSF AChE activity. Interestingly, following the lesioning of the basal forebrain cholinergic neurons in rats the Cisterna Magna CSF AChE levels were reduced, thus giving new empirical support for the validity of the CSF AChE as a marker indicating the loss of cholinergic cells in the basal forebrain (Riekkinen, 1990a).

Despite the above limitations of the CSF AChE activity as a marker for the cholinergic neurons, both the correlation observed between the CSF AChE and EEG delta power as well as the preliminary autopsy evidence relating the loss of both the ChAT activity in neocortex and neurons in nucleus basalis support the assumption that the cholinergic deficit is responsible, at least partly, for the slowing of the EEG in AD patients. This result is interesting in view of the recent experiments demonstrating an increase in delta power following lesioning of the NBM (Buzsaki et al., 1988). Indeed, the increase of delta power following the nucleus basalis lesion is correlated with reduced ChAT activity (Riekkinen, 1990 b). Anti-muscarinic drugs also increase slow delta power and the change induced by cholinolytics is smaller in the NBM lesioned rats than in control animals, thus further underscoring the importance of the cholinergic component of the EEG slowing induced by lesioning of the NBM. Further, in aged rats the frontal delta power is correlated with the reduction of AChE activity underscoring the importance of cholinergic pathology in age-related EEG deterioriation (Sirviö, 1989).

Though the other aminergic systems studied (noradrenaline, serotonin, dopamine) may play an important role in the regulation of neocortical EEG, it is important to note that no correlation could be found between the CSF markers of these systems and EEG slowing. The lack of correlation could be explained by the less consistent and marked degeneration of other aminergic systems compared with the cholinergic (Palmer et al., 1987; Davies, 1983). Based only on the CSF study it is premature to say that only the cholinergic system and not any of other systems studied would contribute to the EEG changes in AD patients. Indeed, in the autopsy study, the two patients with rapidly deterioriating EEG had not only low levels of ChAT but also low levels of noradrenaline in the cortical samples measured.

In the present study, the observed relationship between the reduction in CSF AChE activity as well as the low amount of cortical ChAT activity and the NBM cell bodies with increased EEG delta power support the view that the cholinergic pathology contributes significantly to the EEG slowing. However, this relationship as well as the role of the other aminergic systems should be further studied in a more extensive autopsy study.

References

Appleyard ME, Smith AD, Berman P, Wilcock GK, Esiri MM, Neary D, Bowen DM (1987) Cholinesterase activities in cerebrospinal fluid of patients with senile dementia of Alzheimer type. Brain 110:1309–1322

Arendt T, Bigl V, Walther F, Sonntag M (1984) Decreased ratio of CSF acetyl-cholinesterase to butylcholinesterase activity in Alzheimer's disease. Lancet ii:173

Bird TD, Stranahan S, Sumi SM, Raskind M (1983) Alzheimer's disease: choline acetyltransferase activity in brain tissue from clinical and pathological sub-groups. Ann Neurol 14:284–293

Buzsaki G, Bickford RG, Ponomareff G, Thal LJ, Mandel R, Gage FH (1988) Nucleus basalis and thalamic control of neocortical activity in the freely moving rat. J Neurosci 8:4007–4026

Coben LA, Danziger WL, Berg L (1983) Frequency analysis of the resting awake EEG in mild senile dementia of Alzheimer type. Electroencephalogr Clin Neurophysiol 55:372–380

Coben LA, Danziger WL, Storand M (1985) A longitudinal EEG study of mild senile dementia of Alzheimer type: changes at 1 year and 2.5 years. Electroencephalogr Clin Neurophysiol 61:101–112

Davies P (1983) An update on the neurochemistry of Alzheimer's disease. In: Mayeux R, Rosen WG (eds) The dementias. Raven Press, New York, pp 75–86

Duffy FH, Albert MS, McAmilty G (1984) Brain electrical activity in patients with presenile and senile dementia of the Alzheimer type. Ann Neurol 16:439–448

Ellman GL, Courtney DK, Anders V, Featherstone RM (1961) A new and rapid calorimetric determination of acetylcholinesterase activity. Biochem Pharmacol 7:88–95

Fonnum F (1975) A rapid radiochemical method of the determination of choline acetyltransferase. J Neurochem 24:407–409

Greenfield SA (1984) Acetylcholinesterase may have novel functions in the brain. Trends Neurosci 7:364–368

Hallanger AE, Levey AI, Lee HJ, Rye DB, Wainer BH (1987) The origins of cholinergic and other subcortical afferents to the thalamus in the rat. J Comp Neurol 262:105–124

Hughes CP, Berg L, Danzinger (1982) A new clinical scale for staging dementia. Br J Psychiatry 140:566–572

Kaziak AW, Garron DC, Fox JH, Bergen D, Huckman M (1979) Cerebral atrophy, EEG slowing, age, education and cognitive functioning in suspected dementia. Neurology 29:1273–1279

Lindvall O, Björklund A, Nobin A, Stenevi U (1974) The adrenergic innervation of the rat thalamus as revealed by the glyoxylic acid fluorescence method. J Comp Neurol 154:314–348

Mefford IN (1981) Application of high performance liquid chromatography with electrochemical detection to neurochemical analysis: measurements of chatecholamines, serotonin and metabolites in rat brain. J Neurosci Methods 3:207–224

Moore RY, Halaris AE, Jones BE (1978) Serotonin neurons of the midbrain raphe: ascending projections. J Comp Neurol 180:417–438

Mölsä P, Säkö E, Paljärvi L, Rinne JO, Rinne UK (1987) Alzheimer's disease: neuropathological correlates of cognitive and motor disorders. Acta Neurol Scand 75:376–384

Palmer AM, Wilcock GK, Esiri MM, Francis PT, Bowen DM (1987) Monoaminergic innervation of the frontal and temporal lobes in Alzheimer's disease. Brain Res 401:231–238

Reinikainen K, Paljärvi L, Halonen T, Malminen O, Kosma VM, Laakso M, Riekkinen PJ (1988) Dopaminergic system and monoamine oxidase-B activity in Alzheimer's disease. Neurobiol Aging 9:245–252

Reinikainen KJ, Riekkinen PJ, Paljärvi L, Soininen H, Helkala EL, Jolkkonen J, Laakso M (1988) Cholinergic deficit in Alzheimer's disease: a study based on CSF and autopsy data. Neurochem Res 13:135–146

Riekkinen P Jr, Miettinen R, Rummukainen J, Pitkänen A, Paljärvi L, Riekkinen P (1990a) The effects of lesioning the basal forebrain neurones on CSF AChE activity. Neurosci Res 6:37–43

Riekkinen P Jr, Sirviö J, Riekkinen P (1990b) Relationship between cortical ChAT content and EEG delta power. Neurosci Res (in press)

Rinne JO, Paljärvi L, Rinne UK (1987) Neuronal size and density in the nucleus basalis of Meynert in Alzheimer's disease. J Neurol Sci 79:67–76

Rossor MN, Iversen LL (1986) Non-cholinergic neurotransmitter abnormalities in Alzheimer's disease. Br Med Bull 42:70–74

Sirviö J (1989) The cholinergic system in aging and dementia. Academic Diss, University Printing Office, University of Kuopio

Soininen H, Partanen VJ, Helkala EL, Riekkinen PJ (1982) EEG findings in senile dementia and normal aging. Acta Neurol Scand 65:59–70

Soininen H, Partanen J, Laulumaa V, Helkala EL, Laakso M, Riekkinen PJ (1989) Longitudinal EEG spectral analysis in early stage of Alzheimer's disease. Electroencephalogr Clin Neurophysiol 72:290–297

Stewart DJ, MacFabe DF, Vanderwolf CH (1984) Cholinergic activation of the electrocorticogram: role of substantia innominata and effects of atropine and quinnuclidinyl benzylate. Brain Res 322:219–232

Vanderwolf CD (1975) Neocortical and hippocampal activation in relation to behavior. Effects of atropine, eserine, phenothiazines and amphetamine. J Comp Physiol Psychol 88:300–323

Whitehouse PJ, Price DL, Clark AW, Coyle JT, DeLong MR (1981) Alzheimer's disease: evidence for selective loss of cholinergic neurons in the nucleus basalis. Ann Neurol 10:122–126

Wikler A (1952) Pharmacologic dissociation on behavior and EEG sleep patterns in dogs: morphine, N-allylnormorphine and atropine. Proc Soc Exp Biol Med 79:261–265

Correspondence: Prof. P. Riekkinen, Department of Neurology, University of Kuopio, P.O.B. 6, SF-70211 Kuopio, Finland.

EEG- and cognitive changes in Alzheimer's disease – a correlative follow-up study

M. Streifler[1], M. Simanyi[2], P. Fischer[3], and W. Danielczyk[2,4]

[1] Sackler Faculty of Medicine, Tel Aviv University, Israel
[2] Neurology Department, Geriatric Hospital-Lainz, Vienna
[3] Neurological Institute, University of Vienna
[4] Ludwig Boltzmann Institute for Aging Research,
Workgroup Alzheimer-Dementia Research, Vienna, Austria

Summary

Electroencephalograms (EEG) obtained in a two years follow up period from 39 patients with Alzheimer's disease – presenile (AD) and senile (SDAT) – were compared with psychomental ratings (MMS and OMDS) and with EEG records from two age-matched normal control groups, comprising 33 subjects. Normal EEGs were obtained 7 times more often in controls than in Alzheimer patients and alpha was mildly to severely disturbed in close to 68 percent of the latter.

The EEG-differences between controls and patients were significant for global rating, alpha activity and diffuse slowing. Beta activity was significantly reduced in SDAT. Alpha and beta activity were significantly more impaired in SDAT than in AD, but their changes with progression of disease did not differ significantly. Close to 10% of initial records were and remained normal.

The study of the sequential EEGs showed a good correlation between global rating, slowing of basic rhythm and irregularity of alpha activity and degrees of dementia.

This study, in consistency with most of reports in the literature shows, that sequential recording of EEG is a valuable tool in the confirmation of Alzheimer's disease in doubtful cases and in the differentiation of this disease from normal ageing.

Introduction

Electroencephalography (EEG) holds an important position in the diagnosis of dementing disorders in the aged. As for senile dementia of the Alzheimer type (SDAT) the diagnostic value of EEG is claimed to surpass that of the CT-scan (86% versus 84%), while their combined use can increase the extent of correct classification up to 90 percent (Soininen et al., 1982 a). Moreover, normal records are rarely found in this condition (around 10 percent), even in the early stages of its clinical manifestation.

In addition to this diagnostic efficiency EEG qualifies itself as a painless, non-invasive and low-priced technique. Easily performable, it is free of the effects of repetition ("practice effect") and motivation, encountered in psychomental evaluations.

The study of the possible contribution of electroencephalography in the field of the dementias, started by Berger (1932) close to sixty years ago, has led to a respectable number of publications. To those based on conventional EEG techniques, more elaborated ones, like frequency and power-spectral analysis and finally also topographical mapping of brain potentials have been added in recent years.

In all these studies the hallmark of EEG pathology in Alzheimer's disease has been shown to be a progressive slowing of brain potentials, with an early bilateral reduction of occipitoparietal rhythmic activity. This alpha (8–12 c/s) rhythm is progressively replaced by diffuse theta (4–7 c/s), theta-delta and finally by delta (0.5–3 c/s) activity alone. Other changes affect beta rhythm (14–30 c/s), responsivity to intermittent light stimulation and to overbreathing, as well as the appearance of paroxysmal activity, of sharp waves and rarely also of triphasic waveforms.

The main problem in the diagnosis of dementia in ageing people is the differentiation of Alzheimer's disease from depression with cognitive retardation, from a variety of organic or functional cerebral disorders, especially multi-infarct dementia (MID) and, in early stages also from normal ageing.

Senescence in itself implies a continuance of slowly appearing EEG alterations, the significance of which is still being debated. Regarding the anomalies found in about one third of records from aged people, good concensus exists on a) a slight decrease of alpha presence and frequency b) a slight increase in theta and delta and c) an increase in beta activity. Some authors claim additional changes like the appearance of paroxysms and of slow temporal activity (especially left), while others point to the reduced response to hyperventilation and to intermittent photic stimulation (Guggenheim and Karbowski, 1979; Soininen et al., 1982 b; Busse, 1983).

While qualitatively similar to those of the healthy elderly the basic EEG changes in demented aged persons are much more pronounced and their progress is remarkably accelerated. This stresses the need for follow-up in suspected and established cases of dementia. Such studies, while undoubtedly of diagnostic importance, if properly designed, should help to further the understanding of the disruptions of the cerebral functions, which lead to dementia and of the significance of concomitantly occurring changes in the electrical brain signals.

Assuming a satisfactory knowledge of how to measure severity of EEG abnormality and degree of dementia, the question evolves, whether, how and to what extent are impairments in memory, cognition and in the behavioural adaptabilities of Alzheimer's disease mirrored and/or measurable by EEG-techniques.

Serial EEG studies in Alzheimer's disease, which seem to have started with that of Gordon (1968), initially aimed only at the changes of the electrical potentials per se, have recently also taken up the problems of their correlation with psychometric data and even with quantitative histopathological findings.

This report deals with a two-years follow-up of EEGs of two groups of Alzheimer patients, namely that of (presenile) early onset (AD) versus that of senile (late) onset (SDAT). We studied the intercorrelation of their EEG-parameters also with respect to the degree of dementia.

Methods

The following findings are derived from ongoing studies on dementia in the aged, conducted by the Alzheimer-Dementia Research Group of the Ludwig Boltzmann Institute of Geriatric Research. All Alzheimer cases were in-patients of the neurological department of the Lainz-Vienna Geriatric Hospital (Vienna, Austria). As shown in Table 2, 33 volunteers served as age-matched healthy controls: 15 subjects 69 years and below (range 55–69 years, average 63.6 ± 3.4) and 18 in the age range of 74–97 years, average 81.1 ± 6.7. These subjects had one to two EEG's each.

In all the 39 Alzheimer patients the disease had presumptively started two to thirteen years before the first EEG and in all, except for one patient, two more EEGs were recorded within 24 months of follow-up. Eleven of these subjects belong to the early onset group (AD) and 28 to the senile type (SDAT).

The diagnosis of Alzheimer's disease was based on clinical criteria, on those of the DSM-III-R and the Hachinski Ischemic Score (Hachinski, 1975); on neuroimaging methods (CT in all patients) and on neuro-psychological testing: Mini-Mental Scale of Folstein (MMS) (Folstein et al., 1975), the Organic Mental Disorder Scale (OMDS) (Danielczyk, 1986) – Table 1 – and the Hamilton depression scale (Hamilton, 1967). The degrees of dementia in most patients listed from moderate to marked.

Medication, wherever necessary consisted in minor tranquilizers, cardiac fortificants, vitamin-mineral supplements and occasional light hypnotics.

The EEGs were recorded in the waking state, visually analyzed by two independent observers and globally classified – Table 1 – as: 0 = normal;

Table 1. Grading of dementia: Organic Mental Disorder Scale (OMDS) and of EEG-parameters

0 = No cognitive and intellectual impairment
1 = Mild memory and concentration disorders, irresoluteness, mild organic personality syndrome
2 = Obvious impairment of memory and concentration, reduced capability of criticism
3 = Additionally: temporary mental disorientation, confusion and aggressivity, severe impairment in capability of criticism, no insight into illness
4 = Persistent disorientation, disturbed communication, incontinence, psychological testing impracticable

Grading of EEG-parameters
0 = Normal
1 = Mildly disturbed
2 = Moderately disturbed
3 = Severely disturbed

Table 2. Demographic data

	AD*	SDAT*	Controls <70 yrs	Controls >70 yrs
	n = 11	n = 28	n = 15	n = 18
Gender f/m	6/5	27/1	13/2	17/1
Age in years at the time of last EEG				
Average (SD)	65.1 (5.6)	84.4 (5.5)	63.6 (3.4)	81.1 (6.7)
Range	60–76	76–97	55–69	74–97
Age at onset				
Average (SD)	59.5 (3.6)	77.7 (7.1)	–	–
Range	53–64	66–91		
Duration of disease at the time of last EEG				
Average (SD)	6.0 (3.3)	6.2 (3.4)	–	–
Range	2–13	1–17		
Last OMDS				
Average (SD)	3.5 (0.68)	3.3 (0.82)	–	–
Range	2–4	1–4	0	0

* *AD* Alzheimer's disease (onset before age 65)
 SDAT Senile Dementia of Alzheimer's Type (onset after age 65)

1 = mildly irregular; 2 = moderately irregular and 3 = severely irregular. Features especially looked for were basic rhythm, global rating, alpha activity, beta activity, paroxysmal activity, focal changes, myoclonic discharges, sharp waves, triphasic waves and reaction to hyperventilation and intermittent light stimulation. Only the first five features lent themselves to a meaningful inter-comparison and to a correlation with levels of dementia.

For statistical analysis the Student-t-test and the methods of Kruskal-Wallis, Mann-Whitney and Spearman-Brown were used.

Results

During the years 1985–1988 seventy Alzheimer patients had been included in the study but only 39, fulfilling all criteria are presently discussed. In 13 patients the clinical diagnosis of Alzheimer's disease was confirmed by post-mortem studies.

As shown in Table 2 the average age at disease onset was 59.5 ± 3.6 years (range: 53–64 years) in the AD group and 77.7 ± 7.1 years (range: 66–91 years) for the SDAT patients. The average duration of disease was about six years in both groups at the time of inclusion in the study.

Since, as seen in Table 5 the performance of the patients on the Mini-Mental-Scale (MMS) correlated well with their rating in the Organic Mental Disorder Scale (OMDS) – which presents a fairly comprehensive portrayal of the patients psychomental and social abilities, only the statistical correlations of the latter with EEG parameters are shown in the Tables.

The OMDS ratings, 3.26 ± 0.87 for AD and 3.14 ± 0.94 for SDAT at the onset declined to 3.5 ± 0.68 and 3.3 ± 0.82 respectively during the two-years period. The difference between these two deteriorations is not significant. Of the 33 EEGs obtained from the controls – Table 3 – none was rated as globally severely disturbed, in contrast to more than half of the patients records. The comparison of the two control groups (A versus B) shows a significant slowing of "basic rhythm" in the older group while the other parameters are not significantly different.

When comparing controls with patients (A versus C and B versus D), differences relating to global evaluation and to alpha reactivity are highly significant, attesting to the marked deterioration in the patients. For basic rhythm, the highly significant difference, shown between the older cohorts is not found between the younger ones. As for beta activity we find a significant reduction in the SDAT-group. Differences regarding paroxysmal activity and presence of sharp waves (not shown in Table 3) are of no significance in any of the comparisons.

Comparison of the two patient groups (C versus D) reveals a significant worsening of alpha activity in the presenile group, while beta activity is reduced in the senile contingent. The deterioration of the basic rhythm only borders on significance.

In Table 4 a the EEG development in the two patient groups is shown by comparing the last records of each individual. The mean time interval between two compared records is equal in the two groups ($p = 0.590$)

Table 3. Comparison of age-matched controls and demented patients on EEG-parameters

	Global				Alpha-reaction				Paroxy		Beta		Basic-rhythm	OMDS 0–4
	0	1	2	3	0	1	2	3	no	yes	no	yes	mean (SD)	mean (SD)
A) Controls <70 yrs n=24	15	4	5	0	8	16	0	0	24	0	13	11	11.0 (1.0)	0.0
B) Controls >70 yrs n=22	17	4	1	0	13	9	0	0	22	0	11	11	10.3 (1.3)	0.0
C) AD n=31	5	5	4	17	1	13	12	5	29	2	16	14[a]	9.0 (3.8)	3.26 (0.87)
D) SDAT n=70	6	13	11	40	11	32	19	8	65	5	55	15	7.8 (1.6)	3.14 (0.94)
Mann-Whitney U-test between														
A and C:	z=4.340 p=0.000				z=4.597 p=0.000				n.s.*		n.s.		n.s.	—
B and D:	z=6.240 p=0.000				z=4.388 p=0.000				n.s.		z=2.582 p=0.009		z=5.689 p=0.000	—
C and D:	n.s.				n.s.				n.s.		z=2.536 p=0.011		n.s.	n.s.
A and B:	n.s.				n.s.				n.s.		n.s.		z=2.079 p=0.037	—

[a] beta presence questionable in one AD-patient
* not significant at level 0.05

Table 4a. Comparison of two sequential EEG-records (early versus late onset Alzheimer's disease)

	Global					Alpha-reaction			Paroxy			Intervall months	OMDS
	-2	-1	0	+1	+2	-1	0	+1	-1	0	+1	mean (SD)	mean (SD)
AD n=14*	0	1	9	2	2	1	10	3	3	10	1	12.4 (6.3)	3.546 (0.688)
SDAT n=32*	0	1	22	9	0	3	20	9	2	18	12	13.1 (5.7)	3.214 (0.876)
U-test:	z=0.101 p=0.919					z=0.083 p=0.934			z=2.342 p=0.019			z=0.539 p=0.590	z=1.280 p=0.260

* number of EEG-pairs compared

Range of interval between EEGs: 6–24 months. 0, no change; +, improvement; –, worsening

Table 4b. Changes in routine EEG in Alzheimer's disease (AD and SDAT) during 24 months of follow up (N = 53 comparisons)

	Global					Alpha-reaction			Paroxysmal activity		
	+2	+1	0	−1	−2	+1	0	−1	+1	0	−1
N	0	3	35	13	2	3	39	10*	0	21	32
%	0	5	66	25	4	6	75	19	0	40	60

* only 52 comparisons possible
Range of interval between EEGs: 6−24 months mean time interval: 12.7 ± 5.8 months.
0, no change; +, improvement; −, worsening

Table 5. Rank correlation coefficients between EEG and degree of dementia and age; presenile and senile patients; 101 records (significance in brackets)

	Global	Alpha-reaction	Paroxy	Beta	Basic-rhythm
OMDS [a]	0.4331 (0.001)	0.4512 (0.000)	0.2172 (0.057)	0.0067 (0.481)	−0.2767 (0.020)
MMSE*	−0.4526 (0.000)	−0.5604 (0.000)	−0.2312 (0.050)	0.0028 (0.492)	0.2882 (0.020)
Age	0.165 (0.024)	0.0750 (0.197)	0.0080 (0.462)	0.2004 (0.008)	−0.3876 (0.000)

[a] Organic Mental Disease Scale
* Mini Mental State Examination

and also the difference between the degree of dementia (OMDS) of the two groups is not significant (p = 0.260) at the time of the last record. The differences of changes in global rating (p = 0.919) and in alpha activity (p = 0.934) in the two patient cohorts (AD/SDAT) also are non-significant. This shows that the rate of deterioration was globally equal in both Alzheimer varieties.

The statistically almost significant reduction of paroxysmal activity in SDAT, compared to AD (p = 0.019) is an unexpected finding. It should be mentioned that Soininen et al. (1982a) had reported paroxysmal bifrontal episodic runs of delta waves, of paroxysmal connotation, 10 times more often in demented than in healthy old people.

Table 4b shows that within one half to two years, average 12.7 ± 5.8 months, the majority of EEGs have remained practically unchanged with respect to global rating (66%) and alpha reactivity (75%). A further, mainly moderate deterioration, was found in less than one third of Alzheimer (DAT) patients, while a moderate improvement occurred in six percent.

Based on 101 records of presenile and senile DAT patients (Table 5) the calculation of rank correlation coefficients shows a significant correlation between grades of dementia, as established by OMDS and MMS and global EEG rating and alpha activity. The correlation with slowing of basic rhythm borders on significance. The psychometric tests do not correlate with absence or presence of beta and of paroxysmal activity.

Age on the other hand correlates significantly with beta presence and with slowing of the basic rhythm. Occipito-parietal rhythmic activity i.e. alpha, however does not correlate with age. As already shown before (Table 3), alpha is significantly reduced in both Alzheimer cohorts, compared to age matched controls.

These findings permit the conclusion that good alpha activity pleads against the diagnosis of Alzheimer's disease.

About 11% of the patient records, see Table 3, were classified as "globally" normal, compared to average 70% normal global ratings for both aged control groups. This incidence of "normalcy" in the present study is consistent with that reported in other studies.

The corresponding figure for "severely disturbed" global ratings, zero in the controls, amounts to 56% in the patients.

Discussion

In the Alzheimer patients 90 percent of the records were abnormal, displaying a severe disturbance in more than half, while in the aged control person about 70% were normal and mildly to moderately disturbed records were encountered in about 30 percent. None of the control records was found to be severely disturbed. In the patients close to 10 percent of the records were classified as normal (Table 3).

The differences between patient and control records with regard to the EEG parameters global assessment and alpha reactivity are highly significant and are close to significance also for paroxysmal activity and basic rhythm (Table 3). Beta presence is statistically reduced in the late onset patient group (SDAT) when compared both with aged normal controls and the presenile patient cohort (AD) (Table 3).

These findings are in line with those reported in the majority of pertinent studies (Coben et al., 1983; Kaszniak et al., 1979; Letemendia and Pampiglione, 1959; Deisenhammer and Jellinger, 1972; Pillhatsch, 1985) and attest to the diagnostic value of routine EEG in the diagnosis of Alzheimer's disease.

Comparing EEG with degree of dementia, Table 5, we find a highly significant correlation between the psycho-mental evaluations and the EEG-parameters of global rating and alpha reactivity and borderline significance also for basic rhythm (see also Johanneson et al., 1979; Rae-Grant et al., 1987; Soininen and Partanen, 1988). The differences between the initial versus final psycho-mental test scores, however, showed no significance, which seems to portray a rather parallel and equal rate of mental decline in our two patient cohorts during the observation period.

Intercorrelating EEG course and mental developments, the findings of the present follow-up study can be compared with those of other longitudinal studies, published during the past 20 years (Gordon, 1968; Johanneson et al., 1977, 1979; Coben et al., 1983, 1985; Soininen and Partanen, 1988; Giannitrapani and Collins, 1968). Some of these were based on EEG spectral analysis. Findings, common to both modes of analysis, visual and computerized, are slowing of dominant frequency, especially of the occipito-parietal alpha activity, as well as the episodic increase in theta and delta. These changes, which seem to attest to a progressive loss of the capacity for rhythmic potential generation, can indeed be expected to appear and progress alongside with the decrease in intellectual functioning.

Also beta activity, which increases with ageing, has been shown to decrease with the progression of dementia (see also Coben et al., 1983; Berg et al., 1984; Visser et al., 1985).

While well in line with these findings, the present study, as outlined previously shows that close to two thirds of our Alzheimer patients, in spite of disease progression, maintained their EEG pattern and grading during the two years of follow-up. Only 29 percent deteriorated. In about 5 percent a slight improvement was scored.

This lack of conspicuous progression of EEG deterioration during follow-up studies has already be pointed out by Letemendia and Pampiglione (1959) "in a number of cases" and by Gordon (1968). Johanneson et al. (1977) reported in three out of seven patients no EEG deterioration in a four-years follow-up. Rae-Grant et al. (1987), stressed this "even greater tendency for the (EEG) findings to remain the same" in a one to four year study period. "Indeed," – 12 percent –, "of the EEGs from patients with dementia showed improvement from initial to final study and in three patients this was apparent over a four-year period".

Coben et al. (1985) had, however observed significant EEG changes already during one year of follow-up, while they encountered several demented patients with normal EEG and no worsening during one year.

Soininen and Partanen (1988) recently reported a significant deterioration within one year in 12 out of 24 DAT patients. "There were (however) 12 patients (50%!) with a quite normal EEG at baseline and with minor or no changes between baseline and one-year recordings. Four of the patients with unchanged EEG had moderate dementia and 8 had mild dementia."

As for the study of Rae-Grant (1987) and the present one it should be kept in mind that a considerable proportion of patients, at the time of their inclusion in the study, had been sick for several years and had already presented moderate to marked initial grades of mental decline and EEG perturbation.

Although there is no one-to-one correlation and no parallelism in course between slowing of EEG activity and decline in intellectual performance, this study shows that EEG, and particularly the test-retest method of sequential recording constitutes a valuable tool for the early recognition and confirmation of doubtful Alzheimer's disease. This simple procedure is also particularly helpful in the differentiation of this disorder from senility and from pseudo-demential depression.

Acknowledgement

This study was supported by Bürgermeister-Fonds der Stadt Wien, Grant Nr. 449.

References

Berg L, Danziger WL, Storandt M, Coben LA, Gado M, Hughes CP, Knesevich JW, Botwinick J (1984) Predictive features in mild senile dementia of the Alzheimer type. Neurology 34:563–569

Berger H (1932) Über das Elektroenzephalogramm des Menschen. Fünfte Mitteilung. Arch Psychiat Nervenkr 98:231–254

Busse EW (1983) Electroencephalography. In: Reisberg B (ed) Alzheimer's disease. Free Press, New York, pp 231–236

Coben LA, Danziger WL, Berg L (1983) Frequency analysis of the resting awake EEG in mild senile dementia of the Alzheimer type. EEG Clin Neurophysiol 55:372–380

Coben LA, Danziger WL, Storandt M (1985) A longitudinal EEG study of mild senile dementia of Alzheimer type. Changes at one year and 2.5 years. EEG Clin Neurophysiol 61:101–112

Danielczyk W (1986) Akinetische Krisen, akinetische Endzustände und Sterbealter bei hospitalisierten Parkinson Patienten. In: Fischer PA (Hrsg) Spätsyndrome der Parkinson-Krankheit. Edit Roche, Basel, S 89–98

Deisenhammer E, Jellinger K (1972) Korrelation elektroenzephalographischer Befunde bei präsenilen und senilen Demenzen. In: Kanowski S (Hrsg) Gerontopsychiatrie 2. Janssen Symposien, Berlin, S 62–83

458   M. Streifler et al.

Folstein MF, Folstein SE, McHugh PR (1975) Mini-Mental State. A practical
    method for grading the cognitive state of patients for the clinician. J Psychi-
    atr Res 12:189–198
Giannitrapani D, Collins J (1968) EEG differentiation between Alzheimer's and
    Non-Alzheimer's dementias. J Psychiatr Res 26–41
Gordon EB, Sim M (1967) The EEG in presenile dementia. J Neurol Neurosurg
    Psychiatry 30:285–291
Gordon EB (1968) Serial EEG studies in presenile dementia. Br J Psychiatry
    114:779–780
Guggenheim P, Karbowski K (1979) EEG-Befunde bei 40–60jährigen gesun-
    den Probanden. Z Gerontol 12:365–375
Hachinski VC (1975) Cerebral blood flow in dementia. Arch Neurol 32:632–637
Hamilton M (1967) Development of a rating scale for primary depressive illness.
    Br J Soc Psychol 6:278–296
Johanneson G, Brun A, Gustafson J, Ingvar DH (1977) EEG in presenile
    dementia related to cerebral blood flow and autopsy findings. Acta Neurol
    Scand 56:89–103
Johanneson G, Hagberg B, Gustafson L, Ingvar DH (1979) EEG and cognitive
    impairment in presenile dementia. Acta Neurol Scand 59:225–240
Kaszniak AW, Garron DC, Fox JH, Bergen D, Huckman M (1979) Cerebral
    atrophy, EEG slowing, age, education and cognitive functioning in suspect-
    ed dementia. Neurology 29:1273–1279
Letemendia F, Pampiglione G (1959) Clinical and electroencephalographic obser-
    vations in Alzheimer's disease. J Neurol Neurosurg Psychiatry 21:167–172
Pentilla M, Partanen JV, Soininen H, Riekkinen PJ (1985) Quantitative analysis
    of occipital EEG in different stages of Alzheimer's disease. J Neurol Neuro-
    surg Psychiatry 60:1–6
Pillhatsch K (1985) Korrelation des organischen Psychosyndroms im Alter mit
    apparativen Befunden – Aussagekraft von EEG und CCT bei psychia-
    trischen Erkrankungen im Alter. Z Gerontol 18:330–336
Rae-Grant A, Blum W, Lau C, Hachinski VC, Fisman M, Merskey H (1987) The
    EEG in Alzheimer-type dementia. A sequential study correlating the EEG
    with psychometric and quantitative pathologic data. Arch Neurol 44:50–54
Soininen H, Partanen JV (1988) Quantitative EEG in the diagnosis and follow-
    up of Alzheimer's disease. In: Giannitrapani D, Murri L (eds) The EEG of
    mental activities. Karger, Basel, pp 42–49
Soininen H, Partanen JV, Puranen M, Riekkinen PJ (1982a) EEG and computed
    tomography in the investigation of patients with senile dementia. J Neurol
    Neurosurg Psychiatry 45:711–714
Soininen H, Partanen JV, Helkala EL, Riekkinen PJ (1982b) EEG findings in
    senile dementia and normal ageing. Acta Neurol Scand 65:59–70
Visser SL, Van Tilburg W, Hooijer C, Jonker C, DeRijke W (1985) Visual
    evoked potentials (VEPS) in senile dementia (Alzheimer type) and in non-or-
    ganic behavioural disorders in the elderly: comparison with EEG parame-
    ters. Electroencephalogr Clin Neurophysiol 60:115–121

Correspondence: Prof. M. B. Streifler, M.D., M.Sc., 53, David Hamelekh Blvd.,
    64237 Tel Aviv, Israel.

Decreased hippocampal metabolic rate in patients with SDAT assessed by positron emission tomography during olfactory memory task

M. S. Buchsbaum, C. Cotman, P. Kesslak, G. Lynch, H. Chui, J. Wu, N. Sicotte, and E. Hazlett

Department of Psychiatry and Human Behaviour, University of California, Irvine, CA, U.S.A.

Summary

Positron emission tomography with F-18 deoxyglucose was used to assess cortical metabolic rate during an olfactory memory task in six patients with senile dementia of the Alzheimer's type and six healthy, age-matched controls. Decreases in metabolic rate were observed in the anterior portion of the medial temporal cortex, especially on the left in the patients. This region is known to receive a large olfactory input, and to have been implicated in the encoding of human memory. Our results are consistent with earlier reports of temporal lobe decreases in metabolic rate, extending them by examining areas salient to the behavioral loss.

Introduction

Patients with Alzheimer's disease have been reported to have a loss of cells in the entorhinal and subicular areas, located along the medial-ventral surface of the temporal lobe (Hyman et al., 1984). The entorhinal-hippocampal-subicular complex has been implicated as a key structure in the encoding of human memory. Entorhinal cortex is also known to receive a large olfactory input via the lateral olfactory tract which originates in the olfactory bulbs. Olfactory areas also have been reported to show more marked neurofibrillary tangles and neuritic plaques than visual cortex or other sensory areas (Pearson et al., 1985). Consequently, one would predict that Alzheimer's disease patients would be deficient in the performance of an olfactory memory task. While such a deficit might be demonstrable as a fall in psychophysical performance scores, deficits in attention, motivation, cooperation, and olfactory sensitivity could also contribute. These factors could also appear in other dementias, blurring the distinction between Alzheimer's and non-Alzheimer's patients. Direct assessment of entorhinal metabolic rate with positron emission tomography (PET) during performance of an olfactory memory test would provide a neuroanatomically based functional evaluation of these brain regions and might provide a test to facilitate the early detection and diagnosis of Alzheimer's disease.

PET studies have revealed decreases in glucose metabolic rate ratios of frontal to temporal regions (Friedland et al., 1983, 1985) and general decreases in metabolic rate (Friedland et al., 1983). More marked parietal and temporal but not frontal decreases are shown in the data of Duara et al. (1986) and McGeer et al. (1986) and diagram of Tamminga et al. (1987). Recently Miller et al. (1987) obtained PET scans during a verbal memory task. While widespread cortical reduction in metabolic rate was observed in patients during baseline, only the temporal lobe showed a differential task effect. Normal individuals activated their left temporal lobe with the task, whereas patients activated the right side. Several limitations of these studies as a test of hippocampal memory deficit in Alzheimer's disease should be noted. First, the Friedland et al. (1983), de Leon et al. (1983), McGeer et al. (1986), and Duara et al. (1986) studies did not use a memory task. The regions of interest in all of these studies were either lateral temporal or whole temporal lobe, rather than limited to the hippocampus. Definition of hippocampal metabolic rate is also diminished because the scans were obtained parallel to the canthomeatal line while the hippocampus itself is tilted with respect to this plane. Nevertheless, the studies were not inconsistent with hippocampal deficits. In the current study, we report decreased metabolic rates in patients carrying out an olfactory memory task during FDG uptake with the use of small hippocampal regions of measurement and slices tilted 15 degrees to parallel the hippocampal plane.

Methods

Patients

Six patients with senile dementia of the Alzheimer's type, mean age 67.8, SD = 11.1, two men and four women, served as subjects. They were recruited from the Alzheimer's Disease Diagnostic and Treatment Center of the University of Southern California at Rancho Los Amigos Medical Center. All six met NINCDS criteria for probable Alzheimer's disease (McKhann et al., 1984) and

DSM-III criteria (American Psychiatric Association, 1980) for primary degenerative dementia. The clinical evaluation included complete histories, physical and neurological examinations, and the following laboratory studies: complete blood count, electrolytes, liver, kidney and thyroid functions, calcium, phosphorus, B12, folic acid, and syphilis serology. All patients had Hachinski Ischemia (Hachinski et al., 1975) Scores <4. No focal abnormalities were noted in computerized tomographic scans of the head. Electroencephalography showed mild slowing in four patients. The mean duration of the dementia symptoms was 5.8 years and mean age of onset was 65 years. Mean Mini-Mental State Exam score (Folstein et al., 1975) was 15.2 (range 11–21 out of 30). Mean score on the Everyday Activities and Habits sections of the Blessed-Tomlinson-Roth Dementia Scale (Blessed et al., 1968) was 5.2 ± 2.1 (17 indicating greatest disability). Patients were not taking psychoactive medications.

Five of the patients were enrolled in a double-blind, placebo controlled cross-over study of intraventricular bethanechol chloride (Harbaugh et al., 1984) after completion of the PET scan. In these patients, cerebral cortical biopsies were obtained from the right frontal lobe at the time of intraventricular catheter placement. In all five subjects, plaques and tangles were found in Bielschowsky silver-stained formalin-fixed sections, thereby confirming the diagnoses of Alzheimer's disease.

Six normal subjects, three men and three women, mean age 67.8, SD = 5.2, served as controls. They received a medical and psychiatric history, physical exam and screening blood chemistry; subjects with significant medical or psychiatric illness, head injury, epilepsy, family history of psychiatric illness or Alzheimer's disease were excluded.

Task

An olfactory match to sample test was administered for 30 minutes following FDG injection. The patient performed the test in a dimly lit room sitting in an office chair. The examiner sat in front of him across a small table. The patient was instructed not to speak during uptake. At the time of FDG injection a series of 30 match-to-sample trials were started, each trial lasting approximately one minute. Each trial was as follows: the patient was first asked to smell a target odor for 2–3 seconds. Following a 10-second delay, the patient was presented with three other odors, one at time. The patient was then asked which one of the three "best matched" the target and was instructed to raise either one, two or three fingers on his right hand to signal his answer. A small penlight was used to illuminate his right hand for viewing. The presentation of the matched samples was randomized, and each stimulus was presented approximately an equal number of times as either a target or potential match.

Positron emission tomography

Regional brain activity changes were imaged as glucose metabolic rate using sterile, pyrogen-free ^{18}F-2-deoxyglucose, prepared at the Crocker Nuclear Laboratory, University of California, Davis. Before PET scanning, an individually molded, thermosetting plastic head holder was made for each subject to minimize head movement. The same head holder was used for MRI scans.

For the PET procedure, subjects were seated in a darkened isolation room and IV lines inserted as described elsewhere (Buchsbaum et al., 1987). At 2–3 minutes before the ^{18}F-2-deoxyglucose injection, room lights were extinguished and the olfactory task was begun; the stimuli continued for 30–35 minutes after isotope injection. Subjects were not spoken to during uptake, and all remained quiet and cooperative. Subjects received 4 to 5.2 millicuries of FDG. After 30–35 minutes of FDG uptake, the subject was transferred to the adjacent scanning room. Three planes (CTI NeuroECAT) at 10 mm increments and at 15 degrees of extension relative to the canthomeatal line (CM) were done between 45 and 100 minutes after FDG injection. Scans were performed with both shadow and septa shields in, a configuration with measured in plane resolution of 7.6 mm and 10.9 mm resolution in the Z-dimension. Typical total plane counts were in the range of 800,000–1,200,000 counts.

Scans were transformed to glucose metabolic rate according to the model of Sokoloff et al. (1977), using kinetic constants and the lumped constant from Phelps et al. (1979). Following a method common in cerebral blood flow analysis and in reports on Alzheimer's disease by de Leon et al. (1983), Friedland et al. (1983) and McGeer et al. (1986), we expressed data as a fraction of whole slice metabolic rate to correct for individual differences in brain metabolism unrelated to regional activation.

Magnetic resonance procedure

Fourteen horizontal planes were obtained at 10 mm increments at 15 degrees extension to the canthomeatal line (Fig. 1). The 15-degree angle was chosen because it would provide a plane parallel to the entorhinal cortex (Salamon and Huang, 1980). Due to logistical issues, MRI were not obtained on all patients and normals. However, the MRI on the first patient was used as a template for locating regions of interest on all 14 subjects as described below.

A diagrammatic representation of the template MRI is shown in Fig. 1. In the lower section, the pons and cerebellum can be visualized in the center of the brain, anterior to the suprasagittal sinus and dorsal of the optic tract and optic chiasm. The parahippocampal gyrus/entorhinal cortex appears lateral to the cerebellum and medial from the collateral sulcus. The hippocampus was visualized on the cut 10 mm dorsal. At this upper level, the ventral portion of the frontal cortex is clearly seen. The dorsal portion of the cerebellum is also apparent. Lateral ventricles can be seen in the posterior half of the scan. The hippocampi are located immediately lateral to the cerebellum and hypothalamus. The two PET slices best matching these anatomical images were chosen from among the 3–4 obtained on each subject. One subject did not have a suitable upper slice, and was excluded from upper slice and combined slice ANOVA.

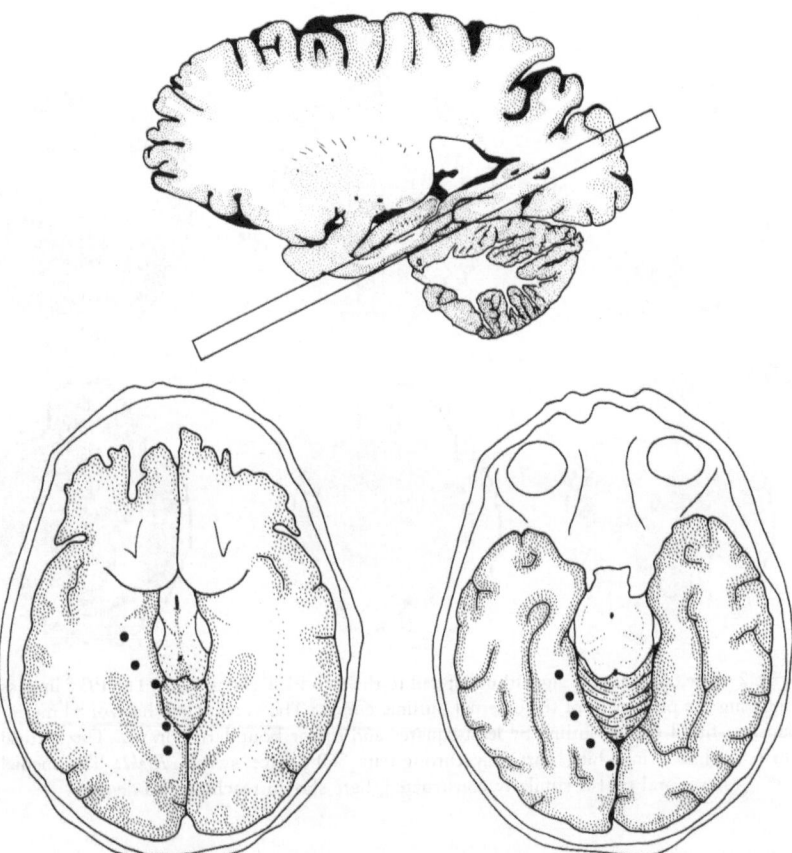

Fig. 1. *Above:* Sagittal view of human brain adapted from Roberts and Hanaway (1970) showing the approximate plane of the PET scans (tilted at 15 degrees). The two planes (*below*) were drawn from MRI with details adapted from the Salamon atlas (Salamon and Huang, 1980) to develop a typical example of structures seen in the two planes at 30 and 20 mm above the CM line

Region of interest location

The box method used small 3 × 3 pixel square regions of interest. Five boxes were placed along the medial temporal cortex in areas corresponding to the parahippocampal gyrus in an anterior to posterior direction. The proportional locations on the PET scan were located using a computer program (Fig. 1).

The peel method (Fig. 2) used a computer algorithm described elsewhere (Buchsbaum et al., 1984). In a second step, an area swept at a distance of 18 mm from the surface with a thickness of 6 mm was obtained, to assess medial temporal cortical measurement. The cortical strip from each hemisphere was

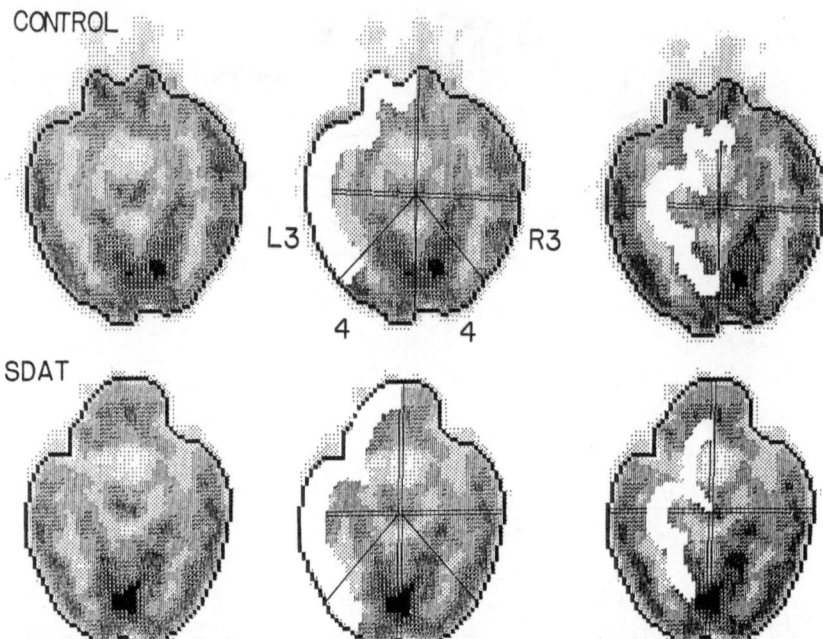

Fig. 2. The cortical peel algorithm applied to the two PET slices. *Left:* The PET images showing the placement of the external outline. *Center:* The vertical and horizontal meridians are fitted to the outline by least-squares and sector boundaries drawn. The cortical strip is shown in white, part way through its 360-degree sweep. *Right:* The medial temporal peel is similarly constructed, here shown reaching 180 degrees

divided into four sectors from front to back at the 45-, 90-, and 135-degree positions. Eight values (four anteroposterior sectors for each hemisphere) were obtained from each slice. With this method, the second and third sectors lie within the temporal lobe, both on the outer and inner peel segments.

Statistical methods

The main hypothesis predicted diminished relative glucose metabolism in the most anterior regions of the medial temporal cortex in patients with Alzheimer's disease while performing an olfactory memory task. This was examined with a t-test on the hippocampus obtained from the peel (sector 2) and box (box 1) methods selected on this a priori basis. We also used repeated measures ANOVA to examine the difference in metabolic rate between anterior, more olfactory regions, and posterior, more visual regions. These analyses used BMDP 2V (Dixon, 1981) with independent groups (normals, patients) and repeated measures dimensions of hemisphere (right, left), anteroposterior position, slice level, and for the peel analysis, medial and lateral cortex. The a priori assumption is that a group by region interaction would be significant since decreases in the hippocampus but sparing in visual cortex was hypothesized.

T-tests on other regions are presented on an exploratory basis. Two other statistical evaluations were performed to match the right minus left score variance differences between normals and patients reported by Friedland et al. (1985) and Duara et al. (1986), and a right greater than left abnormal asymmetry during a memory task reported by Miller et al. (1987); both of these replication findings are tested with t-tests, one-tailed.

Results

Stereotaxic box method

Decreased relative metabolic rate was confirmed in the anterior portion of the left anterior medial temporal cortex (Tables 1 and 2) at both levels and on the right side for the upper level. We considered the most anterior box in the upper and lower slices as representing the best estimate of the olfactory memory area of the hippocampus. Combining across levels and hemispheres for the anterior-most box yielded significantly higher values for normals (1.01, SD = 0.15) than for patients (0.81, SD = 0.10) by t-test (t = 2.44, d.f. = 9, p = 0.04). Using 1.00 as a dividing criterion identifies four of five patients and five of six normals correctly. An ANOVA with diagnostic group (Alzheimer's, controls), and repeated measures for anteroposterior position (five boxed from anterior hippocampus to the margin of the visual cortex), hemisphere (right, left) and level (upper slice, lower slice) confirmed the critical group by anteroposterior position interaction for our major hypothesis (Table 1); post-hoc t-tests confirmed the difference in the most anterior region, most closely corresponding to the olfactory region, and showed no difference in the posterior-most box, on the border of the occipital cortex.

Normals had higher metabolic rates on the left than on the right in boxes 1, 2, 4, and 5 on the upper slice and 1, 2, 3, and 4 on the lower slice. This asymmetry was statistically confirmed for box 2 in the upper slice (t = 2.77, p < 0.01) and box 1 in the lower slice (t = 2.31, p < 0.05). Patients with Alzheimer's disease had higher values on the right than the left for boxes 1, 2, 4, and 5 on the upper slice and boxes 3 and 5 on the lower slice; only box 1 on the upper slice differed significantly between normals and patients in asymmetry [unpaired t contrasting difference scores, t = 2.13, p = 0.03, 1-tailed replication of Miller et al. (1987)].

A focal asymmetry in metabolism not favoring either hemisphere was tested for by comparing the variances of left minus right metabolic

Table 1. Relative hippocampal metabolic rate; 30 mm and 20 mm slices combined

	Controls n = 6	SDAT n = 5
1. Hippocampus	1.01 ± 0.16	0.82 ± 0.11 *
2. Ant. parahippocampus	0.96 ± 0.14	0.95 ± 0.11
3. Medial parahippocampus	1.02 ± 0.23	1.16 ± 0.16
4. Post. parahippocampus	1.17 ± 0.20	1.46 ± 0.24
5. Visual cortex	1.36 ± 0.24	1.49 ± 0.25

* Different from controls by 2-tailed t-test, df = 9, 2, p = 0.04.
ANOVA is group (control, sdat) by anteroposterior box position (1, 2, 3, 4, 5), F = 4.24, Huyhn-Feldt adhusted df = 1.87, 16.82, p = 0.0345

Table 2. Relative hippocampal metabolic rate

	Controls n = 6			SDAT n = 5		
+ 30 mm	L	R	Comb.	L	R	Comb.
1	1.06 ± 0.18	0.96 ± 0.07	1.01	0.80 ± 0.07 **	0.85 ± 0.08 *	0.82 *
2	1.06 ± 0.18	0.90 ± 0.13	0.98	0.86 ± 0.22	0.93 ± 0.16	0.90
3	0.88 ± 0.25	0.93 ± 0.22	0.90	0.97 ± 0.21	0.90 ± 0.16	0.94
4	1.20 ± 0.27	1.06 ± 0.31	1.13	1.35 ± 0.30	1.44 ± 0.26	1.40
5	1.38 ± 0.39	1.28 ± 0.28	1.33	1.53 ± 0.44	1.62 ± 0.22	1.58
+ 20 mm		n = 6			n = 6	
1	1.14 ± 0.22	0.89 ± 0.31	1.02	0.80 ± 0.21 **	0.80 ± 0.32	0.80
2	0.94 ± 0.27	0.92 ± 0.21	0.93	0.92 ± 0.26	0.92 ± 0.26	0.90
3	1.14 ± 0.25	1.11 ± 0.29	1.13	1.19 ± 0.43	1.28 ± 0.44	1.23
4	1.22 ± 0.20	1.19 ± 0.26	1.20	1.48 ± 0.31	1.42 ± 0.29	1.45
5	1.35 ± 0.18	1.42 ± 0.27	1.39	1.34 ± 0.36	1.45 ± 0.25	1.39

Areas 1–5 are as listed in Table 1.
t-test, * $p < 0.05$, ** $p = 0.01$.
Anteroposterior position main effect, F = 25.0, df = 1.84, 16.8, $p < 0.0001$. Position by group effect, F = 4.24, p = 0.03; see collapsed means, Table 1. Slice level by anteroposterior position, F = 7.09, df = 3.2, 28.8, p = 0.0009. Slice by position by group, F = 2.38, p = 0.08

difference or quotient scores between patients and controls. Using the same statistical method, we found 8 of 10 left minus right difference score variances to be higher in the patients, with box 2 on the upper slice and box 3 on the lower slice reaching statistical significance (F > 5.05, $p < 0.05$). However, in one of the boxes showing patient/normal differences in metabolic rate (box 1, left hemisphere, upper slice), this variance was nonsignificantly smaller in the patients.

Cortical peel method

Consistent with the box results, decreased relative metabolic rate was observed in the anterior portion of the left medial temporal cortex peel (upper level) of patients with Alzheimer's disease (0.78, SD = 0.09) in contrast to normal controls (0.86, SD = 0.05). This was confirmed by t-test (see Table 3). Lateral temporal cortex also showed a significant decrease, confirmed both by t-test and by ANOVA (Table 3). A similar left-sided reduction was seen at the lower level (Table 3) in the medial region but not in the lateral cortex. Examination of micromole metabolic rate data yielded similar findings; the left hemisphere medial peel most anterior segment was significantly reduced (t = 1.95, p = 0.046, 1-tailed) and the anterior lateral peel sectors on the left at the upper level (t = 2.72, p = 0.012, 1-tailed) and lower level (t = 2.09, p = 0.034, 1-tailed) also reached significance.

The ANOVA revealed a significant anteroposterior position by group interaction for the lateral temporal cortex peel but not for the medial peel, unlike the box analysis (Table 3). A five-way ANOVA with diagnostic group, anteroposterior position, lobe position (medial, lateral), hemisphere (right, left) and level (upper and lower slice) did not confirm a group by lobe position by anteroposterior position interaction or any other diagnostic group effect or interaction term with diagnostic group (main effects of anteroposterior position, lobe position, and level as well as interactions with hemisphere not including group were all significant). This ANOVA was also carried out on micromole metabolic rate data with a similar lack of significance, indicating that whole temporal lobe metabolic rate was not significantly decreased in our patients.

In the anterior lateral temporal lobe, normals had higher relative metabolic rates on the left and patients on the right as found with box values. This difference was confirmed by unpaired t-test on right minus left difference scores at the upper slice level (t = 2.85, p = 0.02). The asymmetry in normals was small, but in patients the right was much higher than the left (right minus left scores in patients, paired t = 4.4, p = 0.01, see Table 3). Right-left asymmetries were not found significant for medial cortex peel or at the lower level. Analysis of glucose metabolic rate in micromoles glucose/100 g/min revealed similar findings, although of mildly diminished statistical power, again with only the anterior lateral cortical peel sector showing a significant normal patient difference in asymmetry [right minus left scores, t = 1.91, p = 0.045, 1-tailed replication of Miller et al. (1987)].

Table 3. Upper and lower slice inner and outer peel

| | Controls n=6 | | | | SDAT n=5 | | | |
| | Inner | | Outer | | Inner | | Outer | |
	L	R	L	R	L	R	L	R
Upper Slice								
2	0.86±0.05	0.86±0.08	0.98±0.03	0.97±0.10	0.78±0.09**	0.81±0.26*	0.78±0.10	0.93±0.09
3	1.07±0.11	1.04±0.08	1.13±0.05	1.15±0.07	0.98±0.22	1.01±0.21*	0.86±0.23	1.03±0.18
4	1.39±0.23	1.35±0.24	1.16±0.08	1.13±0.07	1.46±0.12	1.45±0.08	1.22±0.21	1.21±0.19
Lower Slice								
2	0.96±0.20	0.91±0.23	0.94±0.16	1.00±0.09	0.82±0.06	0.90±0.16	0.82±0.08	0.94±0.13
3	1.14±0.22	1.11±0.25	1.06±0.10	1.08±0.07	1.17±0.08	1.23±0.13	0.93±0.18	1.02±0.17
4	1.35±0.16	1.40±0.22	1.17±0.12	1.15±0.13	1.48±0.18	1.55±0.13	1.13±0.11	1.18±0.09

* $p=0.09$, 2-tailed t-test; ** $p<0.05$, 2-tailed t-test.

Analysis of variance for outer peel: Group by quadrant, $F=4.16$, $df=3, 27$, $p=0.0152$, other interactions ns. Inner peel ANOVA ns.

For outer peel (upper and lower slices), ANOVA group by quadrant: $F=3.25$, Huyhn-Feldt adjusted df 3, 27, $p=0.0371$. Other interactions ns. The approximate anatomical areas are: Outer peel, upper slice, 2=anterior temporal pole, 3=middle temporal, 4=occipital cortex; lower slice, 2=temporal pole, 3=middle temporal cortex, 4=occipital cortex.

Inner peel, upper slice, 2=temporal cortex/hippocampus/amygdala, 3=parahippocampal gyrus, 4=occipital cortex (inner layer); lower slice, 2=temporal cortex/hippocampus, 3=parahippocampal gyrus, 4=occipital cortex

Discussion

Our finding of a relative reduction of metabolic rate in the temporal lobe of patients with senile dementia of the Alzheimer's type is consistent with earlier reports (de Leon et al., 1983; Friedland et al., 1983; Friedland et al., 1985; Duara et al., 1986; Miller et al., 1987) and extends them in demonstrating the reduction in an area of the temporal lobe specifically related to the memory deficit and to cholinergic function. Friedland et al. (1983) used a region of interest similar to our third sector in antero-posterior cortical location but narrower (about 10 mm), higher above the CM line (temporoparietal margin), and expressed as a frontal lobe/temporal lobe ratio rather than as a ratio to whole slice. They report a valaue of 1.34 in normals and 1.04 in patients (left side), expectedly higher than our values but showing a similar normal/patient difference to our 1.13 in normals and 0.86 in patients (Table 1). They also investigated hemispheric asymmetry (Friedland et al., 1985), finding an absence of significant right-left differences, but noting a greater tendency for either right or left to be decreased leading to an increased variance in left minus right difference scores. We confirmed this finding as did Duara et al. (1986) with analysis of variances of $R-L/R+L$ scores and McGeer et al. (1986) with analyses of absolute values of left/right scores.

Greater parietal and temporal than occipital decreases were observed by Duara et al. (1986) and McGeer et al. (1986). Our results similarly showed decreases in temporal but not occipital cortex. Of the studies reporting metabolic rates for the four lobes of the brain only, de Leon et al. (1983) found all areas showing approximately equal decreases in patients with dementia. The pattern of temporal and parietal but not occipital metabolic decreases is consistent with the distribution of neu-rofibrillary tangles and neuritic plaques reported by Pearson et al. (1985). They observed changes in association areas of temporal and parietal cortex with motor, somatic sensory and primary visual areas relatively unaffected. They also noted "the invariable and severe involvement of the olfactory areas of the brain . . . in striking contrast to the minimal changes in the somatic sensory and primary visual areas raises the possibility that the olfactory pathway may be initially involved". In order to obtain the maximal resolution for olfactory areas in planes containing the hippocampus and other temporal lobe structures with our single-ring tomograph, we did not obtain slices surveying the entire cortex. Our findings were not uniform throughout the temporal lobe, but decreases of metabolic rate in patients were restricted to the anterior portions, both medial and lateral and were absent in visual cortex, consistent with the pathological changes reported. This is demonstrated by the finding of actually higher relative metabolic rate values in posterior hippocampus and visual regions in patients (Tables 1 and 2) and by the significant group by region interaction in the ANOVA on box data. While the ANOVA did not show a significant group by anteroposterior position by hemisphere interaction, we examined right-left differences by t-test on the strength of the lateral asymmetry results of Miller et al. (1987) during a verbal memory task. Our results are consistent with theirs, finding normal individuals with left greater than right temporal lobe metabolic rates during a memory task but patients with right greater than left. In the study of Miller et al. (1987), the slices were cut parallel to the CM line at the basal ganglia and centrum semiovale level; the temporal lobe region of interest extended from lateral cortex to the lateral ventricles medially. Our results found the effect limited to anterior temporal cortex, both medial and lateral. It should be noted that the results of Friedland et al. (1985) of increased variance in right minus left scores are not necessarily contradictory, and suggest a range of severity within the patient group beyond the normal variation in the controls.

Statistical analysis of relative and absolute metabolic rate data produced very similar results, with relative metabolic rates tending to yield stronger regional interactions in ANOVA and larger asymmetries. Both measures find that it is the decrease in left temporal lobe rather than an increase in the right which characterizes the patient group. While our study did not have a baseline non-memory control task, Miller et al. (1987) found no significant group differences in asymmetry in subjects resting with eyes open. Further, temporal lobe metabolic rate showed a similar percent decrease in patients compared to controls (−27%) to those found in parietal (−27%) and frontal regions (−22%). Thus, the activation condition reveals a relatively specific functional regional deficit related to the behavioral deficit in memory observed clinically. No PET study to date has included contrast groups of demented patients of other etiologies than Alzheimer's, but these results suggest that memory tasks may be important in developing a diagnostic test with specificity.

References

American Psychiatric Association (1980) DSM-III: diagnostic and statistical manual, 3rd edn. Washington DC, APA

Blessed G, Tomlinson BD, et al (1968) The association between quantitative measures of dementia and of senile change in the cerebral grey matter of elderly subjects. Br J Psychiatry 114:797–811

Buchsbaum MS, De Lisi LE, et al (1984) Anteroposterior gradients in cerebral glucose use in schizophrenia and affective disorders. Arch Gen Psychiatry 41:1159–1166

Buchsbaum MS, Wu J, et al (1987) Positron emission tomography assessment of effects of benzodiazepines on regional glucose metabolic rate in patients with anxiety disorders. Life Sci 40:2393–2400

De Leon MJ, Ferris SH, et al (1983) Positron emission tomographic studies of aging and Alzheimer's disease. Am J Neuroradiol 4:568–571

Dixon WJ (1981) BMDP statistical software. University of California Press, Berkeley, pp 388–412

Duara R, Grady C, et al (1986) Positron emission tomography in Alzheimer's disease. Neurology 36:879–887

Folstein M, Folstein S, et al (1975) The mini-mental state examination. J Psychiatr Res 12:189–198

Friedland RP, Budinger TF, et al (1983) Regional cerebral metabolic alterations in dementia of the Alzheimer type: positron emission tomography with [18F]fluorodeoxyglucose. J Comput Assist Tomogr 7:590–598

Friedland RP, Budinger TF, et al (1985) Alzheimer's disease: anterior-posterior and lateral hemispheric alterations in cortical glucose utilization. Neurosci Lett 53:235–240

Hachinski VC, Iliff LD, et al (1975) Cerebral blood flow in dementia. Arch Neurol 32:632–637

Harbaugh RE, Roberts DW, et al (1984) Preliminary report: intracranial cholinergic drug infusion in patients with Alzheimer's disease. Neurosurgery 15:514–518

Hyman BI, van Hoesen GW, et al (1984) Alzheimer's disease: cell-specific pathology isolates the hippocampal formation. Science 225:1168–1170

McGeer PL, Kamo H, et al (1986) Positron emission tomography in patients with clinically diagnosed Alzheimer's disease. Can Med Assoc J 134:597–607

McKhann G, Drachman D, et al (1984) Clinical diagnosis of Alzheimer's disease. Neurology 34:939–944

Miller JD, de Leon MJ, et al (1987) Abnormal temporal lobe response in Alzheimer's disease during cognitive processing as measured by ^{11}C-2-Deoxy-D-Glucose and PET. J Cereb Blood Flow Metab 7:248–251

Pearson RCA, Esiri MM, et al (1985) Anatomical correlates of the distribution of the pathological changes in the neocortex in Alzheimer's disease. Proc Natl Acad Sci USA 82:4531–4534

Phelps ME, Huang SC, et al (1979) Tomographic measurement of local cerebral glucose metabolic rate in humans with (F-18)2-fluoro-2-deoxyglucose: validation of method. Ann Neurol 6:371–388

Roberts M, Hanaway J (1970) Atlas of the human brain in section. Lea & Febiger, Philadelphia

Salamon G, Huang YP (1980) Computed tomography of the brain. Springer, Berlin Heidelberg New York, p 132

Sokoloff L, Reivich M, et al (1977) The [14C]Deoxyglucose method for the measurement of local cerebral glucose utilization: theory, procedure, and normal values in the conscious and anesthetized albino rat. J Neurochem 28:897–916

Tamminga CA, Foster NL, et al (1987) Alzheimer's disease: low cerebral somato-statin levels correlate with impaired cognitive function and cortical metabolism. Neurology 37:161–165

Correspondence: Dr. M. S. Buchsbaum, Department of Psychiatry and Human Behaviour, University of California, Irvine, CA 92717, U.S.A.

PET criteria for diagnosis of Alzheimer's disease and other dementias

W.-D. Heiss, B. Szelies, R. Adams, J. Kessler, G. Pawlik, and K. Herholz

Max-Planck-Institut für Neurologische Forschung und Universitätsklinik für Neurologie, Cologne, Federal Republic of Germany

Summary

At present, PET is the only technology affording the quantitative, three-dimensional imaging of various aspects of brain function. Since function and metabolism are coupled, and since glucose is the dominant substrate of the brain's energy metabolism, studies of glucose metabolism by PET of 2(^{18}F)-fluoro-2-deoxy-D-glucose (FDG) are widely applied for investigating the participation of various brain systems in simple or complex stimulations and tasks. In focal or diffuse disorders of the brain, functional impairment of affected or inactivated brain regions is a reproducible finding.

While glucose metabolism is decreased slightly with age in a regionally different degree, in most types of dementia severe changes of glucose metabolism are observed. Degenerative dementia of the Alzheimer type is characterized by a metabolic disturbance most prominent in the parieto-occipito-temporal association cortex and later in the frontal lobe, while primary cortical areas, basal ganglia, thalamus, brainstem and cerebellum are not affected. By this typical pattern Alzheimer disease can be differentiated from other dementia syndromes, as e. g., Pick's disease (with the metabolic depression most prominent in the frontal and temporal lobe), multi infarct dementia (with multiple focal metabolic defects), Huntington's chorea (with metabolic disturbance in the neostriatum) and other diseases leading to cognitive impairment with more or less typical metabolic patterns. A ratio calculated from CMRGl of affected (temporo-parieto-occipital and frontal association cortex) and non-affected brain regions (primary cortical areas, brainstem, cerebellum) was able to separate clearly AD patients from age-matched controls and permitted discrimination of patients suffering from cognitive impairment of other origin in 82%. The discrimination power can be further improved by specific activation studies. In demented patients PET can also be used to assess treatment effects on disturbed metabolism. Such studies demonstrated an equalization of metabolic hetero-geneities in patients responding to muscarinergic cholinagonists and diffuse increase of metabolism during treatment with piracetam and phosphatidylserine. The therapeutic relevance of such metabolic effects, however, must be proved in controlled clinical trials.

Introduction

Since disturbances of cerebral function are followed by changes in metabolism and blood flow, and pathological impairments of blood supply and energy metabolism themselves lead to functional deficits, in many diseases of the CNS these parameters are measurably altered with-out it being possible to draw conclusions with respect to etiology. Dementias, which are clinically manifested primarily as non-localizable disturbances of cerebral function, can hardly be diagnosed by conven-tional supplementary neurological investigations, which detect mainly localized morphologic lesions. Although regional structural cerebral injuries can be demonstrated in many forms of secondary dementia, the degree of dementia often depends on functional disorders of cerebral regions not primarily affected by the disease. The primary (degenerative) diseases leading to dementia are accompanied by atrophic changes in the brain visible at CT only in the late stages. Progressive cell loss and reduced cell and synaptic activity lead to a reduction of metabolism and blood flow which can be visualized with the aid of functional imaging techniques. Since glucose is the most important substrate of cerebral energy metabolism, studies of glucose metabolism are currently the best method of detecting and quantifying functional disturbances of the brain. The glucose metabolic rate can be determined regionally and three-dimensionally in the brain by means of positron emission tomog-raphy.

Glucose metabolism in healthy subjects

The different rates of glucose metabolism in various regions of the brain depending on their functional activity have been determined in a number of studies (review by Heiss et al., 1984). The overall metabolic level depends to a great degree on internal (anxiety, vigilance) and external (illumination, environmental noise) conditions (Mazziotta et al., 1982) so that the resting conditions for the investigations must be defined. In our studies, which were carried out on subjects with their eyes closed in a darkened room with low levels of noise from equipment and manipulation, the mean glucose turnover rate of 42 normal subjects (age 43 ± 19.1 years, 14 women, 28 men) was 34.6 ± 3.83 µmol/100 g cerebral tissue/min. Highly significant regional differences (Fig. 1 a) with values between 40 and 50 µmol/100 g/min were detected in the striatum, upper limbic system, insula, frontal cortex and primary visual cortex, between 35 and 40 µmol/100 g/min in the other gray structures of the hemispheres, between 30 and 35 µmol/100 g/min in the cerebellum and hippocampal structures and below 20 µmol/100 g/min in the medullary layer. Our studies confirm additionally a certain age dependency: the global rate of cerebral glucose metabolism showed a decline with advancing age which, although statistically significant ($P < 0.05$) represented less than 2% per decade (Fig. 1 b). A detailed analysis revealed that the individual brain regions were affected symmetrically but to rather differing degrees ($P < 0.0001$). After the subjects were divided into three age groups each comprising 14 normal subjects the greatest age dependent changes were detected in the frontal cortex, the insula and the upper part of the limbic system, as well as in the parieto-temporal region and to a lesser extent also in the perirolandic region and medullary layer. No appreciable age-dependency could be demonstrated for the metabolism of the subcortical gray structures.

Metabolic disturbances in dementia syndromes

Dementias are a very heterogeneous group of diseases, common to all of which are deterioration of intellectual function and memory and a degeneration of the personality which are clearly differentiable from the usual age-related changes – slight forgetfulness, change in the intelligence structures – and greatly exceed them in scope. Clinical symptoms include impairment of learning ability and memory and reduction of attention, orientation, critical faculty and judgement. These are frequently associated with disturbances of visual-spatial orientation, speech functions and apraxie.

Primary degenerative dementias

Primary degenerative dementias of the Alzheimer type (AD) which are accompanied by a loss of cortical neurons for unknown reasons (Terry et al., 1981), disturbance of various transmitter systems (Rossor et al., 1982) but also by selective reduction of specific projection systems [especially the cholinergic system (Coyle et al., 1983)] and also by typical pathological changes (plaques and fibrils), account for more than 50% of all dementia disorders. Patients with AD show a reduction of cerebral

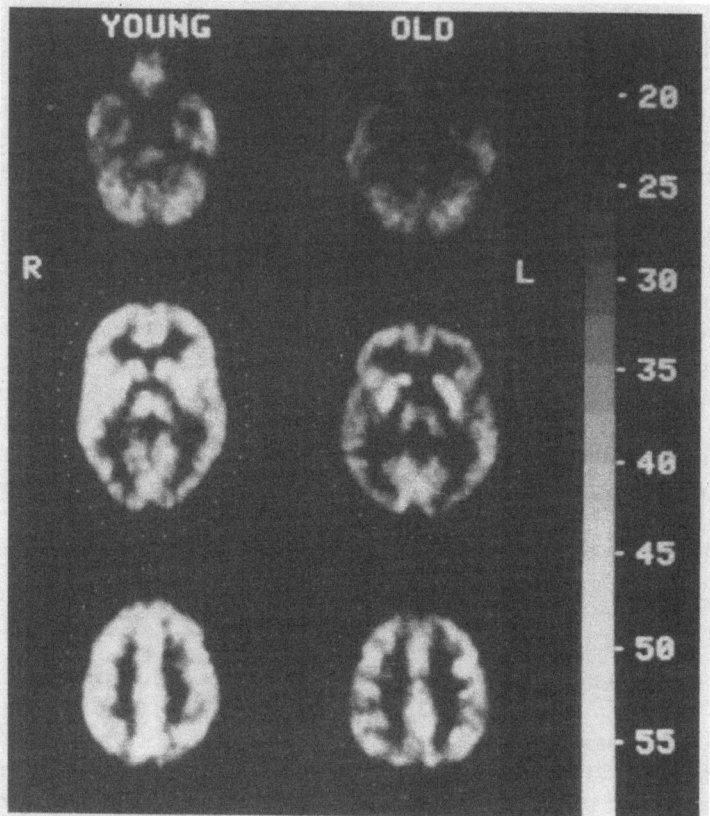

Fig. 1a. PET scans of glucose metabolism (μmol/100 g/min) (according to gray scale) in cerebral sections at the level of cerebellum, basal ganglia, thalamus and semioval centre in young (23 years) and old (67 years) healthy subjects. The individual brain structures can be differentiated according to different metabolic rates, metabolism decreases slightly in all regions in older patients

glucose metabolism [similar to oxygen utilization and blood flow (Frackowiak et al., 1981)] proportional to the severity of the dementia. The reduced metabolism is detectable (Fig. 2) before the occurrence of atrophic changes in CT and shows significant regional differences: the bilateral local reductions are especially pronounced in the parieto-temporal and frontal cortex (Fig. 2) and do not affect the primary visual and sensorimotor cortex or the subcortical structures and the cerebellum (Kuhl et al., 1983; Deleon et al., 1983; Friedland et al., 1983; Duara et al., 1986).

In studies which compared results of PET and neuropsychological tests, it was possible to demonstrate a relationship between leading

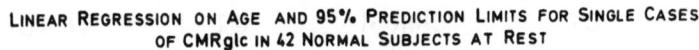

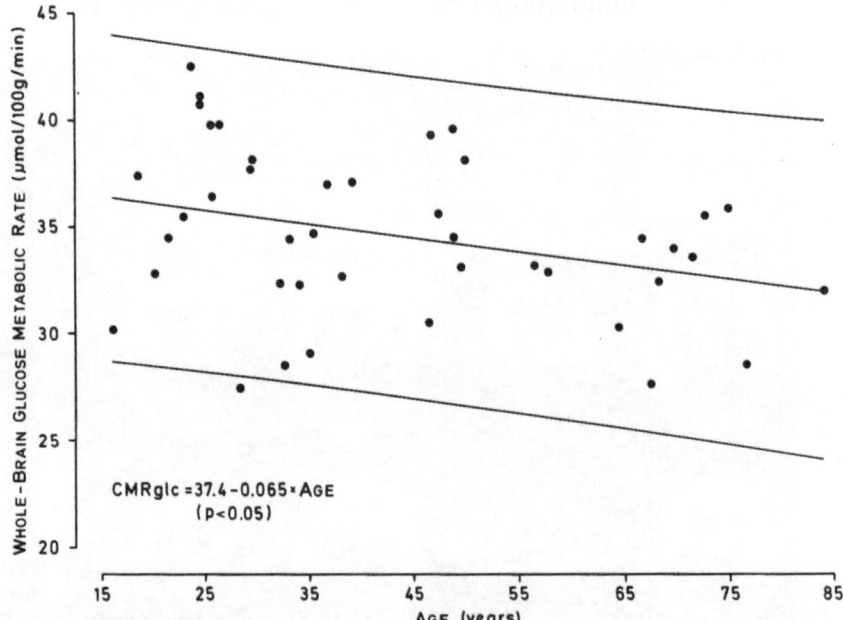

Fig. 1b. Decrease of mean global glucose metabolic rate in 42 healthy subjects with increasing age. The regression line shows a significant relationship despite the large range of variation

symptoms and the localization of especially reduced glucose metabolism: when an aphasic disorder was predominant, glucose metabolic disturbance was more pronounced in the left than in the right parietal lobe, when apraxic symptoms predominated this parameter was disturbed more on the right than the left, while on predominance of amnestic deficits no asymmetry was present (Foster et al., 1983).

In the second, but much rarer form of primary degenerative dementia, Pick's disease, the first and most marked metabolic changes – in analogy to the primary localization of pathological changes – are seen in the frontal and temporal lobe (Szelies and Karenberg, 1986). This distinctly different pattern of damage allows Pick's disease to be differentiated from AD; in moderate cases it is often not possible to make this distinction on the basis of clinical finding alone. The typical reduction of metabolism in the (frequently asymmetrically affected) atrophic frontal lobes and lower temporal lobes and in the much less changed parietal lobes as well as the basal ganglia and the thalamus is correlated with the degree of gliosis and cell depletion (Kamo et al., 1987).

CT   CMRGI   CT   CMRGI
         μmol/100g/min   μmol/100g/min

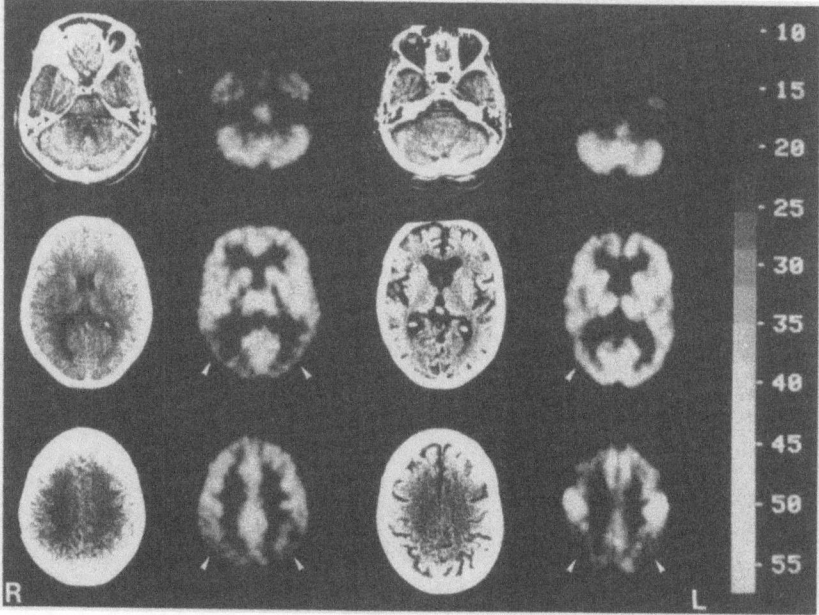

Fig. 2. CT and PET scans of glucose metabolism in patients with mild (left) and severe (right) Alzheimer's dementia. In mild AD metabolic rates in the normal CT are distinctly reduced parieto-occipito-temporal, in severe dementia there is diffuse atrophy and pronounced reduction of cortical metabolism with recessing of primary somatosensory and visual areas and of basal ganglia/thalamus and cerebellum

The pattern of metabolic disturbance is also characteristic in Huntington's chorea which in addition to the extrapyramidal-hyperkinetic syndrome is always accompanied by dementia disorders. The glucose turnover rate in the neostriatum is already significantly reduced in the early stages of this disease (Fig. 3) and as the severity and duration of the disease increases, metabolism is seen to be reduced in the nucleus caudatus and putamen, and later (according to the degree of severity of dementia) also the cerebral cortex (Kuhl et al., 1984). Since the metabolic disturbances precede the clinical manifestations of the disease, PET studies may possibly help to identify persons at risks in chorea families, and these studies as well as genetic investigations can be used to establish a prognosis for the subsequent appearance of the disease (Hayden et al., 1987; Mazziotta et al., 1987).

CT   CMRGl
                    μmol/100g/min

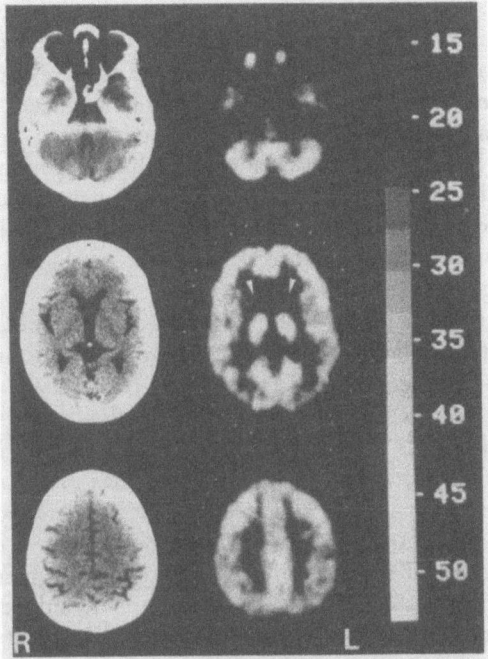

Fig. 3. CT and PET scans of glucose metabolism of patients with Huntington's chorea. In still non-pathological CT, metabolism is reduced in the neostriatum (caudate nucleus and putamen)

In Parkinson's disease, a degeneration of the dopaminergic nigrostriatal system, glucose metabolism is usually not altered, in contrast to the reduction of the dopaminergic endings in the basal ganglia demonstrated by means of PET of ^{18}F-dopa (Nahmias et al., 1985). Only on development of a dementia, a frequent concomitant of Parkinson's disease, are the metabolic changes typical of AD also in evidence (Kuhl et al., 1985).

Vascular dementias

Focal cerebral lesions caused by blood flow disturbances can induce dementia syndromes through two mechanisms in particular: multiple lesions in mostly neurologically silent, frequently subcortical regions impair cerebral function in the form of dementia (multi-infarct dementia) when they exceed a total volume that cannot be precisely defined (80 – 150 cm³). In rare cases, relatively small infarcts of critical localization can cause dementia syndrome in addition to the focally dependent neurological symptoms. Chronic inadequate blood flow in the cerebral tissue which leads to a persisting hypofunction and thus to a disturbance of intellectual function which cannot be precisely localized, is likely to be present only in exceptional cases or to be of temporary duration following transient blood flow disturbances. Such deficient perfusion syndromes probably are only very rarely a cause of dementia since in the usual forms no corresponding disproportions between blood flow and oxygen consumption or glucose metabolism could be demonstrated (Frackowiak et al., 1981; Gibbs et al., 1986).

Multi-infarct dementias (MID) together with the AD-MID mixed forms account for about 30% of all dementia syndromes. A clinical differentiation on the basis of rating scales (Hachinski et al., 1975) is often difficult, and diagnostic classification is often easier on the basis of morphological lesions demonstrated by CT or MRI. In MID patients, PET can clearly differentiate mostly multilocular metabolic reductions from the pattern typical of AD (Kuhl et al., 1983). Detection of ischemic lesions in the medullary layer in MID and Binswanger's disease can be performed with great sensitivity by means of T_2-weighted MRI (Heiss et al., 1986; Alavi et al., 1987), and the regions of reduced metabolism then correspond to the superjacent deafferentated cortical areas (Fig. 4).

Small isolated infarctions in the cerebral regions which are particularly important for the integrity of the personality lead to disturbances of behaviour, affect, mood and intellectual performance. This is especially true of infarcts in the area supplied by the anterior cerebral artery, but also for small localized infarcts in strategically important regions, for example unilaterally in the anterior centre of the thalamus or bilaterally in the median thalamus: they also lead to permanent cognitive and amnestic losses.

Dementias of other etiology

Various other causes – inflammatory diseases, as herpes encephalitis, HIV encephalopathy, Creutzfeldt-Jakob disease; posttraumatic encephalopathy; toxic affections of the central nervous system; communicating hydrocephalus – can lead to severe dementias which then must be differentiated from the more frequent etiologies. Their representation in PET is usually not following a typical pattern – as AD or Pick's disease – or metabolic changes are not restricted to small areas – as in postinfarc-

CT   CMRGl
                   μmol/100g/min

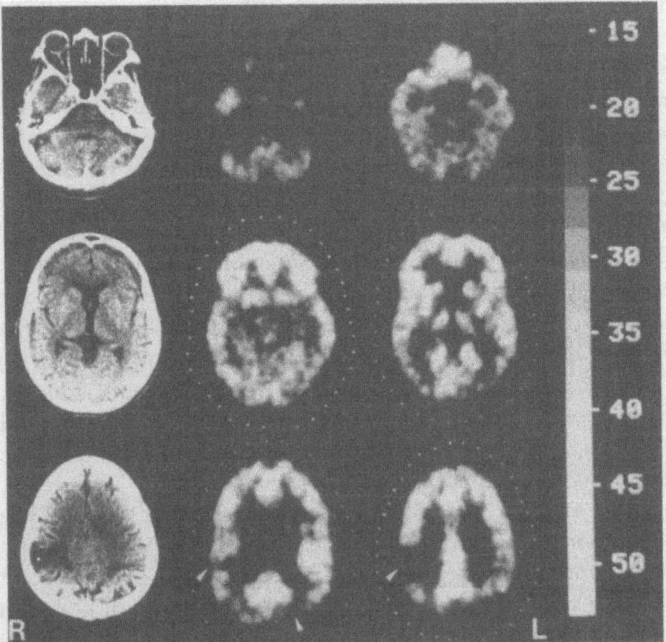

Fig. 4. CT and PET scans of glucose metabolism in patients with multiple infarcts: several morphological lesions (CT) cause regional metabolic disturbances in the infarcted area or in the superjacent deafferentated cortex

tion states. Therefore, a differentiation from the more frequently occurring form of dementia – e. g., AD and MID – can usually be achieved.

It is often difficult to differentiate the affective disorders with impairment of drive and psychomotoricity seen in depressions from similar symptoms in the early stages of dementia. The metabolic patterns observed in depressed patients are not comparable to the characteristic changes seen in dementias, particularly in AD. When the overall metabolic level is in relation to the mood (Baxter et al., 1987) there are sometimes regional differences of varying distribution; a pattern typical of depression or correlations between the metabolic values of certain regions with the severity of specific symptoms or function deficits have not so far been described.

Differentiation of AD from other dementias

In order to test the value of FDG-PET in the differential diagnosis of dementias, rCMRGl measurements of 19 patients with probable Alzheimer's disease according to the NINCDS-ARDA criteria (McKhann et al., 1984) were compared to those from 19 age-matched healthy subjects and 22 patients with cognitive impairment due to other diseases. In comparison to the 19 healthy normals the AD patients (age 49 to 71 years, cognitive deterioration for $1-5$ years, average mini mental status 14.5 ± 7.3, average global deterioration scale 4.9 ± 0.9) were characterized by significantly lower global cerebral metabolic rates for glucose (30.3 ± 3.2 µmol/100 g/min) and considerable metabolic asymmetries with the most conspicious decrease of regional CMRGl (in terms of Z-transforms) found in the supramarginal and angular gyrus, the adjacent parts of the superior temporal gyrus and the medial frontal gyrus. While abnormally low metabolism ($Z < -2$) in supramarginal/angular gyrus was observed in all AD patients and is a highly sensitive indicator of the disease, abnormal metabolism in temporal and frontal association areas supports the diagnosis, but is not mandatory.

However, despite the decrease of global CMRGl in AD, some regions maintained a strikingly normal metabolism even in most severely affected patients. Those regions were the cerebellum, brainstem, primary sensorimotor cortex and the occipital cortex including the cortex around the calcarine fissure. Notably, relative cerebellar CMRGl was normal (within 2 SD of normal range) in all 19 AD patients, and even absolute cerebellar rCMRGl was normal in 17 of the 19 AD patients.

Considering the contrast between typically affected and non-affected regions, a ratio R of the rCMRGl in those regions was calculated. Its average value in normals was 1.09 (SD 0.084, range 1.01 to 1.15) vs. 0.77 (SD 0.11, range 0.60 to 1.00) in AD, thus a complete separation of the two groups was achieved.

The diagnostic criteria of AD deducted from the metabolic pattern on FDG-PET – reduced temporo-parietal metabolism, normal cerebellar metabolism, reduced CMRGl ratio of temporo-parietal and frontal association areas to primary sensory areas, brainstem and cerebellum – were fulfilled only by 4 of the 22 patients with cognitive impairment of other origin than AD. One of those suffered from dementia in the course of Parkinson's disease, one had multiple infarcts and in two the cognitive impairment was a sequel to diffuse hypoxic or hypoglycemic brain damage. The overall specificity of this diagnostic procedure in this sample was 82%. FDG-PET reaches a diagnostic sensitivity and specificity unsurpassed by other imaging modalities (see also Szelies et al., this symposium).

Since AD affects mainly and primarily parieto-temporo-occipital areas involved in processing of visual information, a test procedure was developed to further improve the diagnostic sensitivity of PET measurements. When a continuous visual recognition task (Kessler et al., 1989) is performed during metabolic studies a significant activation is observed in normal controls averaging 24% for global CMRGl with the most prominent changes (38%) in occipital regions. In AD the global increase during the performance of an adapted visual recognition task was only 8%, and the changes mainly occurred in areas primarily not affected by the disease. This finding points to the impaired functional reserve capacity of the brain in degenerative dementia of the Alzheimer type. The pattern typical for AD – low CMRGl in parieto-temporo-occipital regions, high CMRGl in primary cortical areas, brainstem and cerebellum – becomes more obvious enhancing the diagnostic contrast between AD patients and normal age matched controls or patients with non-AD dementia.

Evaluation of drug effects

The effect of therapeutic interventions in dementia syndromes is difficult to assess because the degenerative diseases often progress slowly or intermittently or are interrupted by phases without appreciable deterioration. Because of the differences in the disease dynamics, it is also very difficult to make comparisons between individuals. Because metabolic disturbances show a characteristic distribution particularly in degenerative dementias of the Alzheimer type and furthermore are correlated with the severity and duration of the disease, and because the functional activity is reflected in metabolic values, measurements of this kind may be valuable in assessing the effects of drugs. Metabolic investigations could then also be useful in providing objective evidence of therapeutic results within a relatively short time, when clinical improvements or a slowing of the progression of the deficits would not yet be apparent.

In the last few years some principles have been elaborated for therapeutic strategies aimed at improving certain clinical deficits in senile or presenile degenerative dementia of the Alzheimer type (AD). These concepts include measures for substitution of cholinergic deficiency that is presumed to be specific (Davies and Maloney, 1976; Coyle et al., 1983). This can be done by inducing a presynaptic increase in the synthesis and release of acetylcholine, by inhibiting the breakdown of acetylcholine at the synapse and by postsynaptic stimulation of the acetylcholine receptors. All these therapeutic approaches centering on the

cholinergic activity have reportedly brought about improvements in the
memory disturbances typical of AD, with inhibition of cholinesterase
with physostigmine (Davis et al., 1978) and the administration of
tetrahyodroaminoacridine in larger controlled clinical studies yielding
successful results in individual cases (Summers et al., 1986). Muscarinergic choline agonists have also been able to improve the symptoms in
cases that are not so advanced (Szelies et al., 1986). In contrast, precursors of acetylcholine administered by improving its bioavailability were
not effective if they were given as the sole form of therapy (review by
Kurz et al., 1986; Hollander et al., 1986) although specific memory disturbances were improved by a combination with nootropic substances
which stimulate cerebral metabolism (Ferris et al., 1982; Smith et al.,
1984).

The use of PET to objectivize the effects of drugs is still rare and has
so far been limited to small groups of patients: glucose metabolism was
monitored for six to twelve weeks in eight patients with AD of differing
severity undergoing therapy with the muscarinic choline agonist (RS 86
Sandoz, 2.5–3 mg/d, Szelies et al., 1986). Over this period the global
metabolic rate decreased under therapy, but there was a compensation of
the heterogeneous metabolic pattern typical of AD with a particular
reduction of the slightly elevated values (sensorimotor and visual cortex)
measured before starting treatment and there was only a slight influence
on the typically lowered parieto-occipital to temporal values. This effect
was especially pronounced in patients who became clinically stabilized
on this therapy and showed improved performance in several functions;
this group which profited from the therapy originally showed regional
glucose metabolic rates diverging relatively little from the norm and
were also those whose AD was less severe. This study therefore shows
the importance of initiating therapy at an early stage before severe cell
destruction takes place and suggests that a metabolic decoupling takes
place between different regions of the brain as the functional substrate
of the specific symptoms. Another study (Heiss et al., 1988) examined
whether piracetam, which improves memory performance when administered in combination with precursors of acetylcholine (Ferris et al.,
1982; Smith et al., 1984) has metabolic effects in AD. Of 16 patients with
dementia syndrome (DSM-III, American Psychiatric Association, 1980)
nine fulfilled the criteria for AD (McKhann et al., 1984) and the remaining seven were graded as MID or unclassifiable and used as a control
group. Between the PET investigations all patients received 6 g piracetam b.i.d. (Nootrop ®, UCB) for 14 days as a short infusion. The groups
differed significantly from each other and from a control group of
similar age with respect to regional rates of glucose metabolism, the
reductions being particularly pronounced in the parieto-temporo-occip-

ital regions for the AD groups. Under piracetam treatment the glucose metabolism values in the AD group increased in the frontal, central, parieto-occipital, visual, auditory and cingulate cortex, basal ganglia and thalamus whereas no significant changes were detected in the non-AD group. The differences in the effects of treatment between AD and non-AD groups were statistically significant (ANOVA $P < 0.02$ for interactions between regions, treatment and group); on the basis of the ANOVA, the increases in the individual regions were checked by paired t-test. The results were supported by improvements in five AD patients during the short therapy phase with respect to their clinical deficits and their performance in tests. Similar results were obtained in 8 AD patients under treatment with phosphatidylserine (FIDIA, 500 mg/d for 3 weeks) which is suggested to have an effect on membrane structure and cell function. In these patients rCMRGl increased during the treatment with the most significant effect in occipital areas ($+17\%$, $P < 0.05$). For all drugs shown to be effective on one aspect of AD – the regional disturbance of metabolism – controlled clinical studies will be needed in order to justify their clinical use.

References

Alavi A, Fazekas F, Chawluk J, Zimmerman R (1987) Magnetic resonance imaging of the brain in normal aging and dementia. In: Meyer JS, Lechner H, Reivich M, Ott EO (eds) Cerebral vascular disease 6. Excerpta Medica, Amsterdam New York Oxford, pp 191–195

American Psychiatric Association (1980) Diagnostic and statistical manual of mental disorders, 3rd edn (DSM-III). Washington, DC, pp 124–126

Baxter LR, Phelps ME, Mazziotta JC, et al (1987) Local cerebral glucose metabolic rates in obsessive-compulsive disorder – a comparison with rates in unipolar depression and in normal controls. Arch Gen Psychiatry 44:211–218

Coyle JT, Price DL, Delong MR (1983) Alzheimer's disease: a disorder of cortical cholinergic innervation. Science 219:1184–1190

Davies P, Maloney AJF (1976) Selective loss of control cholinergic neurons in Alzheimer's disease. Lancet ii:1403

Davis KL, Mohs RC, Tinklenberg JR, et al (1978) Physostigmine: improvement of long-term memory processes in normal humans. Science 201:272–274

DeLeon MJ, Ferris SH, George AE, et al (1983) Computed tomography and positron emission transaxial tomography evaluations of normal aging and Alzheimer's disease. J Cereb Blood Flow Metab 3:391–394

Duara R, Grady C, Haxby J, et al (1986) Positron emission tomography in Alzheimer's disease. Neurology 36:879–887

Ferris SH, Reisberg B, Crook T, et al (1982) Pharmacologic treatment of senile dementia: choline, L-dopa, piracetam, and choline plus piracetam. In: Corkin S, et al (eds) Alzheimer's disease: a report of progress. Raven Press, New York, pp 475–481

Foster NL, Chase TN, Fedio P, et al (1983) Alzheimer's disease: focal cortical changes shown by positron emission tomography. Neurology (Cleveland) 33:961–965

Frackowiak RSJ, Pozzilli C, Legg NJ, et al (1981) Regional cerebral oxygen supply and utilization in dementia. A clinical and physiological study with oxygen-15 and positron tomography. Brain 104:753–778

Friedland RP, Budinger TF, Ganz E, et al (1983) Regional cerebral metabolic alterations in dementia of the Alzheimer type: positron emission tomography with (18F)fluorodeoxyglucose. J Comput Assist Tomogr 7:590–598

Gibbs JM, Frackowiak RSJ, Legg NJ (1986) Regional cerebral blood flow and oxygen metabolism in dementia due to vascular disease. Gerontology 32 [Suppl 1]:84–88

Hachinski VC, Iliff LD, Zilkha E, et al (1975) Cerebral blood flow in dementia. Arch Neurol 32:632–637

Hayden MR, Hewitt J, Stoessl AJ, et al (1987) The combined use of positron emission tomography and DNA polymorphisms for preclinical detection of Huntington's disease. Neurology 37:1441–1447

Heiss WD, Hebold I, Klinkhammer P, et al (1988) Effect of piracetam on cerebral glucose metabolism in Alzheimer's disease as measured by PET. J Cereb Blood Flow Metab 8:613–617

Heiss WD, Herholz K, Böcher-Schwarz HG, et al (1986) PET, CT, and MR imaging in cerebrovascular disease. J Comput Assist Tomogr 10:903–911

Heiss WD, Pawlik G, Herholz K, et al (1984) Regional kinetic constants and CMRGlu in normal human volunteers determined by dynamic positron emission tomography of (18F)-2-fluoro-2-deoxy-D-glucose. J Cereb Blood Flow Metab 4:212–223

Hollander E, Mohs RC, Davis KL (1986) Cholinergic approaches to the treatment of Alzheimer's disease. Br Med Bull 42:97–100

Kamo H, McGeer PL, Harrop R, et al (1987) Positron emission tomography and histopathology in Pick's disease. Neurology 37:439–445

Kessler J, Adams R, Herholz K, et al (1989) Impaired metabolic activation (FDG-PET) in patients with Alzheimer's disease under stimulation by continuous recognition. Aging of the brain and dementia: ten years later. Symposium, Florenz, May 31–June 3, 1989

Kuhl DE, Metter EJ, Benson DF, et al (1985) Similarities of cerebral glucose metabolism in Alzheimer's and Parkinsonian dementia. J Cereb Blood Flow Metab 5 [Suppl 1]:S169–S170

Kuhl DE, Metter EJ, Riege WH, Markham CH (1984) Patterns of cerebral glucose utilization in Parkinson's disease and Huntington's disease. Ann Neurol 15 [Suppl]:S119–S125

Kuhl DE, Metter EJ, Riege WH, et al (1983) Local cerebral glucose utilization in elderly patients with depression, multiple infarct dementia, and Alzheimer's disease. J Cereb Blood Flow Metab 3 [Suppl 1]:S494–S495

Kurz A, Rüster P, Romero B, Zimmer R (1986) Cholinerge Behandlungsstrategien bei der Alzheimer'schen Krankheit. Nervenarzt 57:558–569

Mazziotta JC, Phelps ME, Carson RE, Kuhl DE (1982) Tomographic mapping of human cerebral metabolism: sensory deprivation. Ann Neurol 12:435–444

Mazziotta JC, Phelps ME, Pahl JJ, et al (1987) Reduced cerebral glucose metabolism in asymptomatic subjects at risk for Huntington's disease. N Engl J Med 316:357–362

McKhann G, Folstein M, Katzman R, et al (1984) Clinical diagnosis of Alzheimer's disease. Neurology 34:939–944

Nahmias C, Garnett ES, Firnau G, Lang A (1985) Striatal dopamine distribution in Parkinsonian patients during life. J Neurol Sci 69:223–230

Rossor MN, Emson PC, Mountjoy CQ, et al (1982) Neurotransmitters of the cerebral cortex in senile dementia of Alzheimer type. Exp Brain Res [Suppl 5]: 153–157

Smith RC, Vroulis G, Johnson R, Morgan R (1984) Pharmacologic treatment of Alzheimer's-type dementia: new approaches. Psychopharmacol Bull 20:542–545

Summers WK, Majovski LV, Marsh GM, et al (1986) Oral tetrahydroaminoacridine in long-term treatment of senile dementia, Alzheimer-type. N Engl J Med 315:1241–1245

Szelies B, Herholz K, Pawlik G, et al (1986) Zerebraler Glukosestoffwechsel bei präseniler Demenz vom Alzheimer-Typ – Verlaufskontrolle unter Therapie mit muskarinergem Cholinagonisten. Fortschr Neurol Psychiatr 54:364–373

Szelies B, Karenberg A (1986) Störungen des Glukosestoffwechsels bei Pick'scher Erkrankung. Fortschr Neurol Psychiatr 54:393–397

Szelies B, Wullen T, Adams R, et al (1989) Comparison between cerebral glucose metabolism and late evoked potentials in patients with Alzheimer's disease. International Symposium on Alzheimer's disease, Würzburg, June 21–24, 1989

Terry RD, Peck A, De Teresa R, et al (1981) Some morphometric aspects of the brain in senile dementia of the Alzheimer type. Ann Neurol 10:184–192

Correspondence: Prof. Dr. W.-D. Heiss, Universitätsklinik für Neurologie, Joseph-Stelzmann-Strasse 9, D-5000 Köln 41, Federal Republic of Germany.

Oxygen metabolism in the degenerative dementias

P. J. Tyrrell [1,2,3], **M. N. Rossor** [2,3], and **R. S. J. Frackowiak** [1,2]

[1] MRC Cyclotron Unit, Hammersmith Hospital
[2] National Hospital for Nervous Disease, Queen Square
[3] St Mary's Hospital, London, United Kingdom

Summary

Positron Emission Tomography (PET) is a quantitative technique which can be used to measure regional values of cerebral metabolism in the living human at rest. While structural scans (CT or MRI) may be normal in the degenerative dementias, or show generalised atrophy, functional imaging techniques allow differentiation of subtypes of dementias according to patterns of dysmetabolism. This paper describes some of the variety of alterations in patterns of oxygen metabolism that may be observed in degenerative cognitive disorders, and their correlations with clinical subtypes.

Introduction

The degenerative dementias are a group of disorders characterised by gradually progressive neuronal loss, often in association with other histopathological changes, such as the senile plaques and neurofibrillary tangles of Alzheimer's disease, or the inclusion bodies of Pick's disease. Clinically, they are typically characterised by the inexorable loss of intellectual and social skills. The most widely studied of the degenerative dementias is Alzheimer's disease, which is associated with posterior biparietal and bitemporal hypometabolism on PET scan (Frackowiak et al., 1981). However, it is becoming increasingly apparent that patients with degenerative cognitive disorders may differ considerably in the clinical features with which they present, and such clinical variation may be reflected in the anatomical site and pathological nature of the disorder.

1. Oxygen studies in dementia of the Alzheimer type

Dementia of the Alzheimer type (DAT) is associated with a character-istic pattern of posterior biparietal and bitemporal hypometabolism, associated in more severely demented cases with bifrontal hypome-tabolism (Frackowiak et al., 1981). Metabolism in the anterior temporal, parietal, and occipital cortices, and cerebellum, may be normal. As in the other degenerative dementias, this reduction in oxygen metabolism ($CMRO_2$) is matched by an equivalent and appropriate drop in cerebral blood flow (CBF), with normal oxygen extraction (OER). Correlations are observed between patterns of hypometabolism and neuropsycholog-ical features, for example, patients with dysphasia exhibit a predominant reduction in metabolism in the left hemisphere, while patients with more severe visuospatial abnormalities may have more prominent right hemi-sphere abnormalities (Foster et al., 1983). In addition, the severity of dementia correlates well with the degree of reduction in metabolism.

Although the pattern of posterior biparietal hypometabolism is typi-cal of DAT, it is not absolutely specific: other dementias, for example Creutzfeld-Jakob disease, may have a similar metabolic picture (Fried-land et al., 1984).

2. Oxygen metabolism in the non-Alzheimer dementias

Whilst DAT is the commonest of the degenerative dementias, other degenerative dementias may be associated with other patterns of hy-pometabolism. Frontal lobe hypometabolism is seen in Pick's disease (Kamo et al., 1987) where the severity of hypometabolism has been correlated with the presence of gliosis and neuronal loss seen at post-mortem. In addition, cerebral blood flow studies using Xenon have shown bifrontal hypometabolism in a group of patients with a "frontal-lobe dementia", characterised clinically by early behavioral change in addition to intellectual decline, and pathologically by spongiform degen-eration without Pick bodies (Risberg, 1987; Gustafsson, 1987; Brun, 1987). The same pattern has been observed using HMPAO and SPECT (single photon emission tomography) scanning, in a similar group of patients (Neary et al., 1988).

3. Oxygen metabolism in focal progressive cognitive disorders

Focal progressive dysphasia without dementia

Focal progressive dysphasia without dementia was first described by Mesulam in 1982, who reported six patients with a slowly progressive language disorder, which continued to deteriorate in the presence of relatively intact non-verbal intellectual skills. The length of history varied, but one patient was followed for 11 years after initial presentation. CT head scans performed in a number of the cases revealed left hemisphere or bilateral fronto-temporal atrophy. The pathology underlying this condition is unclear. However, one case with a six month history of progressive dysphasia preceding the development of a more generalised dementia was subsequently found to have Alzheimer's disease at postmortem (Pogacar and Williams, 1984), one case was found to have Pick's disease (Wechsler et al., 1982), two cases had a localised left fronto-temporal spongiform degeneration (Kirshner et al., 1987), and two Jakob-Creutzfeld disease (Yamanouchi, 1986; Mandell et al., 1989). Two patients reported by Chawluk et al. (1986) underwent PET scans, using [18]FDG, which showed left frontotemporal hypometabolism, in the absence of posterior parietal hypometabolism. We have studied six such patients, using the ^{15}O-steady state technique, to obtain regional values of $CMRO_2$. The six cases varied in length of history from two to six years, and varied in severity. The least affected patient had an isolated naming deficit, with a normal verbal and performance IQ, but poor performance on a stringent naming test (McKenna and Warrington, 1983). In contrast, the most severely affected patient, who also had the longest history, had complete obliteration of all language skills with maintenance of at least some non-verbal skills, for example continuing to drive a car and to be able to navigate in his locality without difficulty. The other patients studied had varying degrees of difficulty with verbal comprehension and expression, reflected in impaired scores on the verbal sub-tests of the Wechsler Adult Intelligence Scale (WAIS). One patient differed from the others, in that she had primarily a deficit of speech expression, associated with an orofacial dyspraxia, in the presence of intact verbal comprehension. All patients had a slowly progressive history, suggesting a degenerative rather than a vascular aetiology. MRI and or CT scans performed on these patients were either normal, or showed left hemisphere atrophy. [15]O steady state scans revealed various patterns of hypometabolism, affecting various areas of left, and in two cases, right hemispheres, correlating with the severity of the clinical deficit. Using a stereotactic localisation technique (Friston et al., 1989), based on the atlas of Talairach et al. (1967), regional values of $CMRO_2$ were obtained for different anatomical areas, including the three frontal gyri and three temporal gyri on each side, together with values for parietal and occipital cortices. The patient with the isolated naming deficit had an area of hypometabolism limited to the left superior temporal gyrus, while the most severely affected patient had severe and widespread reductions throughout the entire left hemisphere, and an additional area of right temporal lobe hypometabolism. The patient with the orofacial dyspraxia and loss of speech expression had a deficit of left fronto-temporal hypometabolism, affecting primarily the left inferior frontal gyrus. The other cases, with varying severities of language comprehension and expression, also had left fronto-temporal lobe deficits, but these were maximal in the left superior temporal gyrus (Tyrrell et al., 1990). The heterogeneity of the oxygen scans reflect the heterogeneity of the clinical features of the disorder, and may reflect heterogeneity of the underlying pathology.

Progressive occipital atrophy

Progressive cortical blindness as a presenting feature of a dementing illness has been reported in one patient who subsequently developed pathologically confirmed Alzheimer's disease (Faden and Townsend, 1976), and as an initial presentation in a number of patients who eventually developed generalised dementia (Benson et al., 1988). One patient with a 12 year history of progressive cortical blindness, with a Balint type syndrome and recent development of more widespread cognitive abnormalities, has been studied with the ^{15}O-steady state technique. This revealed widespread bilateral occipital hypometabolism, with forward projection into the posterior temporal and parietal lobes. This pattern is reminiscent of the pattern of hypometabolism commonly seen in dementia of the Alzheimer type (Frackowiak et al., 1981).

Discussion

These studies reflect the variety of clinical presentations of cortical degenerative disease, and the different patterns of dysmetabolism that may be observed. The extent of the hypometabolism seen in all these cases is greater than the area of atrophy visible on the CT or MRI scans, and indeed in some cases the structural scans were normal. Functional imaging is therefore of considerably more value than structural imaging in delineating the site and extent of the abnormal tissue in cortical degenerative disease, and allows more precise correlations between the anatomical site of a lesion, and its clinical and neuropsychological consequences. Although it is not possible to diagnose the underlying pathological disorder from the PET scan, some clues about possible aetiology may be obtained. For example, the fact that the patient with progressive occipital atrophy who is now developing widespread neuropsychological abnormalities, has bilateral posterior temporal and parietal hypometabolism suggests that he may be developing an Alzheimer type dementia. Functional metabolic scanning, coupled with clinical, neuropsychological, and eventually pathological evaluation of patients with different presentations of cortical degenerations, is likely to be of considerable value in the classification of the dementias.

Acknowledgements

We are grateful to Prof. E. K. Warrington for permission to report neuropsychological data on some of the patients reported in this paper, and to the scientific and technical staff of the MRC Cyclotron Unit, without whom these studies would not have been possible.

References

Benson DF, Davis RJ, Snyder BD (1988) Posterior cortical atrophy. Arch Neurol 33:789–793

Brun A (1987) Frontal lobe degeneration of non-Alzheimer type. I. Neuropathology. Arch Gerontol Geriatr 6:193–208

Chawluk JB, Mesulam M-M, Hurtig H, Kushner M, Weintraub S, Saykin A, et al (1986) Slowly progressive aphasia without generalised dementia: studies with positron emission tomography. Ann Neurol 19:68–74

Faden AI, Townsend JJ (1976) Myoclonus in Alzheimer's disease: a confusing sign. Arch Neurol 33:278–280

Foster NL, Chase TN, Fedio P, Patronas NJ, Brooks RA, Di Chiro G (1983) Alzheimer's disease: focal cortical changes shown by positron emission tomography. Neurology 33:961–965

Frackowiak RSJ, Pozzilli C, Legg NJ, Du Boulay GH, Marshall J, Lenzi GL, Jones T (1981) Regional cerebral oxygen supply and utilisation in dementia. A clinical and physiological study with oxygen-15 and positron tomography. Brain 104:753–778

Friedland RP, Prusiner SB, Jagust WJ, Budinger TF, Davis RL (1984) Bitemporal hypometabolism in Creutzfeld-Jakob disease measured by positron emission tomography with 18-F-2-fluorodeoxyglucose. J Comput Assist Tomogr 8:978–981

Friston K, Passingham RE, Nutt J, Heather JD, Sawle GV, Frackowiak RSJ (1989) Localisation in PET images: direct fitting of the intercommissural (AC-PC) line. J Cereb Blood Flow Metab 9:690–695

Gustafson L (1987) Frontal lobe degeneration of non-Alzheimer type. II. Clinical picture and differential diagnosis. Arch Gerontol Geriatr 6:209–223

Kamo H, McGeer PL, Harrop R, McGeer EG, Calne DB, Martin WRW, Pate BD (1987) Positron emission tomography and histopathology in Pick's disease. Neurology 37:439–445

Kirshner HS, Tanridag O, Thurman L, Whetsell WO (1987) Progressive aphasia without dementia: two cases with focal spongiform degeneration. Ann Neurol 22:527–532

McKenna P, Warrington EK (1983) Graded naming test. NFER-Nelson Pub. Co Ltd, Windsor, Berks

Mesulam M-M (1982) Slowly progressive aphasia without generalised dementia. Ann Neurol 11:592–598

Neary D, Snowden JS, Northen B, Goulding P (1988) Dementia of frontal lobe type. J Neurol Neurosurg Psychiatry 51:353–361

Pogacar S, Williams RS (1984) Alzheimer's disease presenting as slowly progressive aphasia. RI Med J 67:181–185

Risberg J (1987) Frontal lobe degeneration of non-Alzheimer type. III. Regional cerebral blood flow. Arch Gerontol Geriatr 6:225–233

Talairach J, Szikla G, Tournoux P, Prossalentis A, Bordas-Ferrer M, Covello L., et al (1967) Atlas d'Anatomie Stereotaxique du Telencephale. Masson et Cie, Paris

Tyrrell PJ, Warrington EK, Frackowiak RSJ, Rossor MN (1990) Heterogeneity in progressive aphasia due to focal cortical atrophy: a clinical and PET study. Brain (in press)

Wechsler AF, Verity A, Rosenschein S, Fried I, Scheibel AB (1982) Pick's disease: a clinical, computed tomographic and histologic study with Golgi impregnation observations. Arch Neurol 39:287–290

Yamanouchi H, Budka H, Vass K (1986) Unilateral Jakob-Creutzfeld disease. Neurology 36:1517–1520

Correspondence: Dr. P. J. Tyrrell, MRC Cyclotron Unit, Hammersmith Hospital, London W12 OHS, United Kingdom.

Positron emission tomography for differential diagnosis of dementia: a case of familial dementia

R. Adams[1], J. Kessler[1], K. Herholz[1], and A. Mackert[2]

[1] Max-Planck-Institut für Neurologische Forschung, Köln, Federal Republic of Germany
[2] Psychiatrische Klinik und Poliklinik der FU Berlin, Berlin

Summary

In a 52-year-old patient, member of a family with an accumulated number of dementias, the diagnosis of Alzheimer's disease was confirmed by measurement of cerebral glucose metabolism. Three of her first-degree relatives without clinical abnormalities were also examined and showed a normal pattern of metabolism.

Introduction

It has been demonstrated that positron emission tomography (PET) with 18F-2-fluoro-deoxyglucose (FDG) shows characteristic metabolic alterations of cerebral glucose metabolism (CMRGlu) in patients with Alzheimer's disease (AD). The typical findings include hypometabolism of the temporoparietal association areas, and in most cases, especially in more advanced stages, also of frontal association areas but preservation of normal metabolism in cerebellum, brainstem, and in most cases also of primary visual and primary sensorimotor cortex. As an example of the diagnostic use of the method, we demonstrate the findings in a patient with familial dementia.

Patients and methods

A 52-year-old female (E. K.) was admitted to a psychiatric clinic because of paranoid delusion and progressive dementia. One year ago she gradually began to loose memory, about six months before she had difficulty recalling words when speaking and was impaired in performing her household duties. Besides

Fig. 1. Neuropsychological test battery. Results of neuropsychological examinations in the 4 tested subjects

	Subjects			
	E. K.	A. D.	A. B.	S. R.
Mini-mental-state-test	8	28	30	29
Gollin's incomplete pictures (5 step version)				
1st presentation (mean of 1st identification)	naming not	2.9	2.1	2.6
2nd presentation (mean of 1st identification)	possible	1.7	1.2	1.3
Saving-score	—	26%	27.2%	33.3%
Buschke's selective reminding paradigm (percentage of recalled items in 5 trials)	26.6% DR* = 0%	58% DR = 50%	74% DR = 70%	76% DR = 80%
Corsi's tapping task	2	5	7	6
Token-test (subtest 4) (percentage of errors)	80%	10%	0%	0%
Reaction time-measuring	not possible	X = 0.6 ms S = 0.185 ms	X = 0.402 ms S = 0.041 ms	X = 0.5 ms S = 0.067 ms
Laterality-questionaire (modified after Oldfield)	right	right	right	right
Tapping task (32 sec)	le ri 119 127	le ri 209 208	le ri 174 201	le ri 168 201
Pursuit rotor	not possible	61 errors	58 errors	52 errors
Apraxia (range: 0–3)	3	0	0	0
Aphasia	transcortical sensoric	—	—	—
State-*trait* anxiety inventory from Spielberger	not possible	RS = 25 PR = 9	RS = 31 PR = 39	RS = 30 PR = 34

* Delayed recall after 30 min

of nicotine, there were no risk factors for vascular disease present. Neither was there a history of alcohol or drug abuse. – Other medical history: appendectomie 1955, extrauterine pregnancy 1969, nephrolithiasis.

Family history: The patient's mother had also suffered from a severe confusional state before her death at age 38 (suicide). The father was unknown. A 48-year-old brother was also demented, but unavailable for examination, alcohol abuse was suspected. A 58-year-old sister (A. D.) and two daughters (S. R., A. B.) did not show clinical abnormalities.

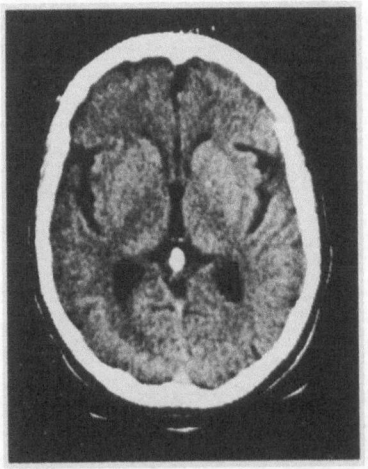 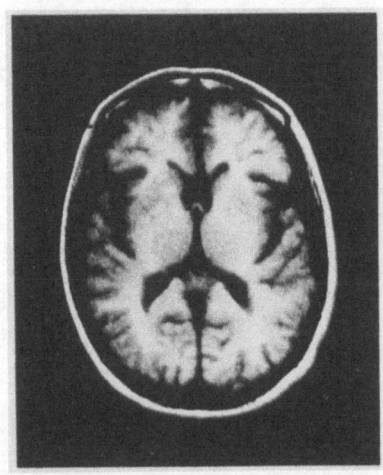

Fig. 2. CT and MR imaging of patient E. K. with mild global atrophy

General physical examination was unremarkable. – Findings from neurological examination were normal besides a dysdiadochocinesia and an inborn convergent strabismus of the right eye.

The mental status examination indicated that the patient was disorientated to place and time. Her speech and motor activity were slowed, but periods of agitation and apprehensiveness sometimes intervened. Memory was impaired severly. The affect was dull and mentation was slowed. Further there was a weakness of comprehension and impulse debilation. Delusion of persecution exsisted and suspicion of cenesthetic hallucination arised. The patient herself did not feel ill. The GDS score of Reisberg et al. (1982) was 6.

Neuropsychological examinations see Fig. 1.

The laboratory examinations, including serum electrolytes, glucose, BUN, SGOT, SGPT, ceruloplasmin, thyroid function tests, serum vitamin B 12 level and routine urinanalysis were all normal. Serological tests for syphilis and HIV were negative. Cerebrospinal fluid, ECG and chest radiographs were also normal. CT and magnetic resonance imaging (see Fig. 2) showed mild global brain atrophy. In the EEG background activity was only slightly present and slowed. Focal abnormalities did not exist.

FDG-PET findings: The 18-F-2-fluoro-2-deoxyglucose (FDG) technique, as described by Reivich et al. (1979) and Phelps (1981) was used. 30 to 50 min after an i.v.-injection of approximately 190 mbq FDG a total of 14 partially overlapping transaxial images of regional brain glucose metabolism were acquired in parallel to the canthomeatal line using a four-ring PET-Scanner (Scanditronix PC 384). The in-plane resolution was 7.8 mm at a slice thickness of 11 mm. Regions of interest (ROI) were outlined according to a standard scheme using a semiautomatic computer-assisted procedure (Herholz et al., 1985). During the measurement subjects lay quietly in supine position under dim lighting with closed eyes and unplugged ears.

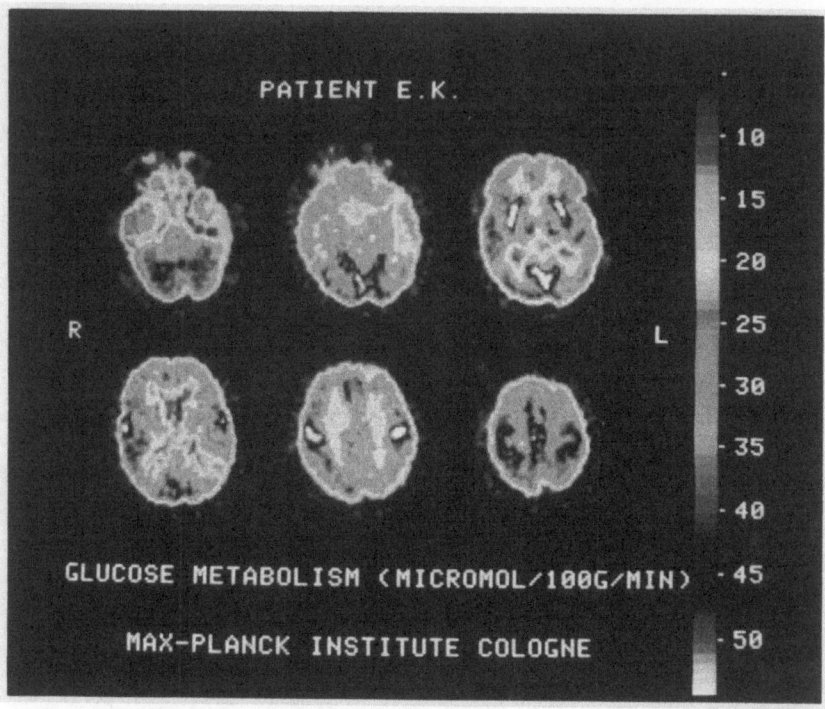

Fig. 3. Regional glucose metabolism of the patient E. K.

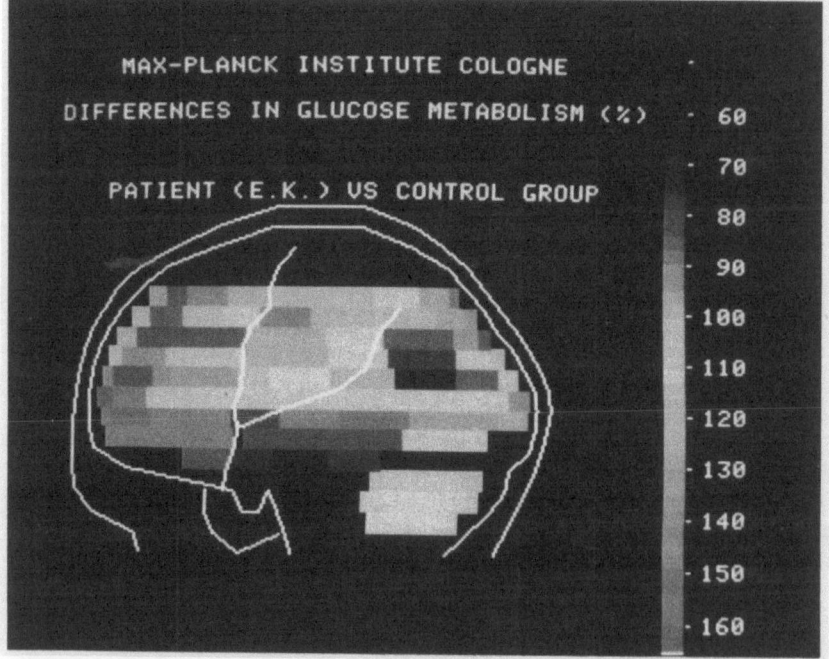

Fig. 4. A lateral view of the differences in the regional relative glucose metabolism between patient E. K. and a control group (n = 20)

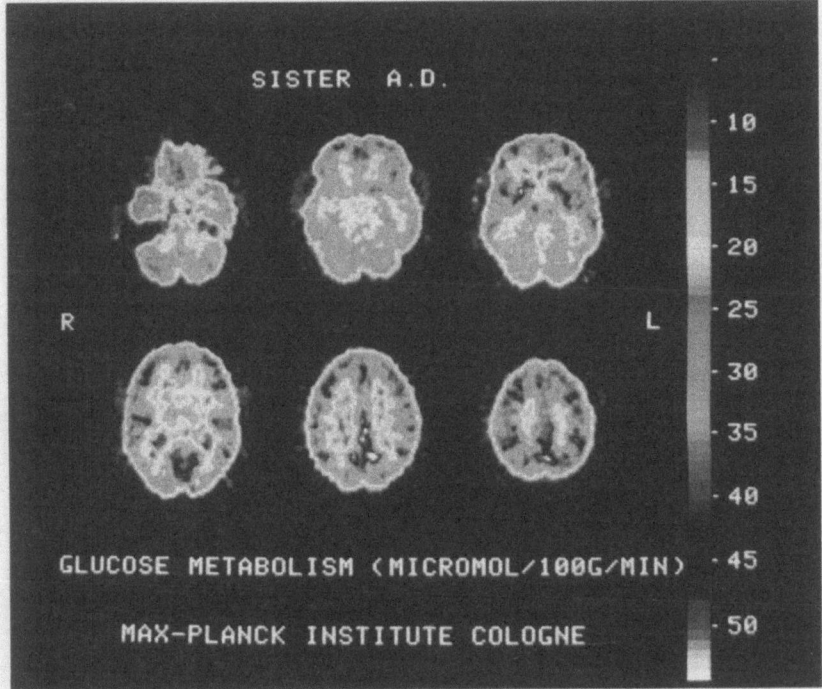

Fig. 5. Regional glucose metabolism of patient's sister A. D.

Resting FDG-PET revealed severe hypometabolism of the left temporoparietal association cortex (22 µmol/100 g/min, normal range 29.6 to 45.3) and of the frontal association cortex (left 26, right 27 µmol/100 g/min). The right temporoparietal cortex was less severely affected. CMRGlu was normal in cerebellum, visual cortex, basal ganglia, and primary sensorimotor cortex (see Figs. 3, 4). In contrast the FDG-PET examinations of the clinically unaffected sister (see Fig. 5) and the two daughters were completely normal.

Comments

- Our patient showed a typical pattern of glucose metabolism as it is known to exist in Alzheimer's disease with a hypometabolism in parieto-temporal region and in the frontal areas. Global metabolism was also decreased.
- Thus PET is able to confirm the clinical diagnosis of Alzheimer's disease not only by exclusion like other investigations. It renders

other possible causes of familial dementia unlikely, in particular metabolic disorders.

– Perhaps genetic predisposition caused the early onset of the disease. PET findings in inherited and sporadic AD seem to be similar, as reported earlier (Cutler et al., 1985; Polinsky et al., 1987; Hoffmann et al., 1989). Thus there are no differences to be expected in the patterns of familial and sporadic AD.

References

Cutler NR, Haxby JV, Duara R, Grady CL, Moore AM, Parisi JE, White J, Heston L, Margolin RM, Rapoport SI (1985) Brain metabolism as measured with positron emission tomography: serial assessment in a patient with familial Alzheimer's disease. Neurology 35:1556–1561

Herholz K, Pawlik G, Wienhard K, Heiss WD (1985) Computer assisted mapping in quantitative analysis of cerebral positron emission tomograms. J Comput Assist Tomogr 9:154–161

Hoffmann JM, Guze BH, Baxter L, Hawk TC, Fujikawa DJ, Dorsey D, Maltese A, Small G, Mazziotta JC, Kuhl DE (1989) Familial and sporadic Alzheimer's disease: an FDG-PET study. J Cereb Blood Flow Metab 9 [Suppl 1]:S547

McKhann G, Drachman D, Folstein M, Katzman R, Price D, Stadlan EM (1984) Clinical diagnosis of Alzheimer's disease: report of the NINCDS-ADRA Work Group under the auspices of the Department of Health and Human Services Task Force on Alzheimer's disease. Neurology 34:939–944

Phelps ME (1981) Positron computed tomography studies of cerebral glucose metabolism in man: theory and application in nuclear medicine. Semin Nucl Med 11:32–49

Polinsky RJ, Noble H, DiChiro G, Nee LE, Feldman RG, Brown RT (1987) Dominantly inherited Alzheimer's disease: cerebral glucose metabolism. J Neurol Neurosurg Psychiatry 50:725–757

Reisberg B, Ferris SH, deLeon MJ, Crook T (1982) The global deterioration scale for assessment of primary degenerative dementia. Am J Psychiatry 139:1136–1139

Reivich M, Kuhl D, Wolf A, Greenberg J, Phelps ME, Ido T, Casella V, Fowler J, Hoffman E, Alavi A, Som P, Sokoloff L (1979) The [18F] fluorodeoxyglucose method for the measurement of local cerebral glucose utilization in man. Circ Res 44:127–137

Correspondence: Dr. R. Adams, Max-Planck-Institut für Neurologische Forschung, Ostmerheimer Strasse 200, D-5000 Köln 91, Federal Republic of Germany.

Comparison between cerebral glucose metabolism and late evoked potentials in patients with Alzheimer's disease

B. Szelies, T. Wullen, R. Adams, M. Grond, H. Karbe, K. Herholz, and W.-D. Heiss

Max-Planck-Institut für neurologische Forschung und Klinik für Neurologie der Universität, Cologne, Federal Republic of Germany

Summary

In an ongoing 2-year-prospective study 21 patients with probable Alzheimer's disease (Global Deterioration Scale 3–6) using NINCDS-ADRDA criteria were examined by PET of glucose metabolism and P300 brain mapping to compare the power of both functional imaging methods to distinguish between patients and normals.

Data of P300 latency and topography were compared with a ratio of glucose metabolism in frontal and temporoparietal association areas over typically not affected regions. PET proved to be a method with high sensitivity (85%) and specificity (90%) while P300 mapping was less sensitive with 58% at a specificity of 90%.

Introduction

Many imaging procedures have been proposed as an adjunct in the diagnosis of Alzheimer's disease. Among them, positron emission tomography (PET) with 18-F-2-fluoro-2-deoxyglucose (FDG) emerged as the most promising method (Cutler et al., 1985; DeLeon et al., 1983; Duara et al., 1985; Foster et al., 1983; Friedland et al., 1983). Yet, PET is still very expensive and therefore restricted to few centers. P300 is reported to be a sensitive functional parameter in the diagnosis of dementia (Maurer and Dierks, 1987; Pfefferbaum et al., 1984; Gordon et al., 1986; Goodin et al., 1983; Polich et al., 1986). Computerized brain mapping is much less expensive and therefore more widely applicable.

As part of an ongoing 2-year prospective study we therefore compare the diagnostic power of the two methods and present here preliminary results obtained in 21 patients with DAT with FDG-PET and P300 brain mapping.

Methods

FDG-PET

The 18-F-2-fluoro-2-deoxyglucose (FDG) technique (PET), as described by Reivich et al. (1979) and Phelps (1981) was used in this study. A four-ring PET-scanner (Scanditronix PC 384) was used. The in-plane resolution was 7.8 mm at a slice thickness of 11 mm. Regions of interests (ROI) were outlined according to a standard scheme using a semiautomatic computer-assisted procedure (Herholz et al., 1985).

As reported in many studies before DAT patients show a reduction of glucose metabolism in the superior temporal and adjacent parietal cortex, as well as in the middle frontal areas. This pattern of hypometabolism matches the location reported for pathological and neurochemical alterations in convexity cortex. Considering the contrast between most affected (temporoparietal cortex, middle frontal cortex) and least affected regions (cerebellum, primary visual cortex, primary sensorimotor cortex) a ratio of the regional cerebral metabolic rate of glucose (rCMRGlu) in these regions was calculated. As we saw in a pilot study, this ratio is able to distinguish patients with probable Alzheimer's disease from normals and other organic brain syndromes. Nearly all DAT patients showed a ratio lower than 1.0.

P300 brain mapping

P300 was elicited with a two-tone "oddball" paradigm according to Maurer et al. (1988) using a Brain atlas system III (Biologic Corp.) (analysis time 1024 ms, pre-post point −55, stimulus rate 0.5/s, P300-ratio 5, stimulation of both ears with a 1000/2000 Hz/90 dB tone burst, raise/fall 10 ms, plateau 50 ms, number of rare tones: 30. Gain 20000, high filter 30 Hz, low filter 1.0 Hz). Surface scalp electrodes were applied according to the international 10−20 system with linked mastoid reference. Two identical runs were done, P300 latency was determined by two independent investigators. For each participant a mean P300 latency was calculated and examined individually to determine whether it was within 95% prediction limits. We also calculated a "latency ratio" between the deviation of the individual latency from the age corrected normal mean value and the difference between the age corrected normal mean value and the corresponding upper 95% prediction limit. Ratios greater than 1 indicated a latency beyond the 95% prediction limit. Group means of P300 latency were compared via Wilcoxon test. In order to assess the fronto-parietal spatial distribution a mean amplitude of the 8 pre- and 8 postcentral electrodes and the difference between them were calculated. The individual fronto-parietal differences as well as the group means were analyzed statistically.

Subjects

21 patients with probable Alzheimer's disease according to the NINCDS-ADRDA workgroup (McKhann et al., 1984) and DSM-III-R criteria were tested (14 females, 7 males, mean age 65.9 ± 7.3 years). Duration of illness was 6 months at minimum. The patients were mildly to moderately demented and scored between 3 to 6 on the GDS of Reisberg et al. (1982). (Mean GDS 4.1 ± 1.2, mean MMSE 15.0 ± 6.5).

All subjects underwent detailed neurological and psychiatric examination including extensive laboratory studies, ultrasound examination of the neck vessels, chest radiographs, ECG, MRI or x-ray CT of the brain and extensive neuropsychological testing (e.g. Buschke's selective reminding paradigm, Corsi's tapping task, Gollin's incomplete pictures, Token test, Pursuit rotor, Reaction time measurement).

The Brain mapping control group (n = 39, 18 female, 21 male, mean age 44.2 ± 16.3 years) consisted of healthy volunteers and patients of our hospital without diseases of the central nervous system. None had subjective memory complaints or abnormal EEG. The latencies of the normal participants were regressed on their ages, and the resulting linear equation, $y = 0.9 \times age + 302$, demonstrated that P300 latency increased by 0.9 msec/year. From this regression equation age dependent normal values and 95% prediction limits were derived. There was no statistically significant relationship between amplitude and age. The subjects in the PET-control group were found healthy upon medical examination and without abnormal neurological or psychiatric history. Subjective memory complaints could not be verified psychometrically. The groups were comparable in age, education and profession.

Results

A P300 potential could be identified in 19 out of 21 DAT patients, the maximal difference between the two runs was 24 msec in both groups. The mean latency was significantly greater in the demented (385 msec, SD 35.5) than in the normal subjects (342 msec, SD 31.5) (p < 0.01) (see Fig. 1 and Fig. 2) but only two demented patients had latencies that exceeded the 95% prediction limits. The mean precentral and postcentral amplitudes as well as the differences between them were different between the DAT and control group showing a lower postcentral amplitude (4.9 ± 4.3 vs 9.1 ± 3.8) (p < 0.01), a higher precentral amplitude (1.3 ± 4.3 vs 0.5 ± 4.6) (p: n.s.), and a lower difference between post and precental amplitudes (3.7 ± 5.8 vs 8.6 ± 4.2) (p < 0.01) in the demented patients. 6 (32%) of the 19 DAT patients, but no normal participant, had

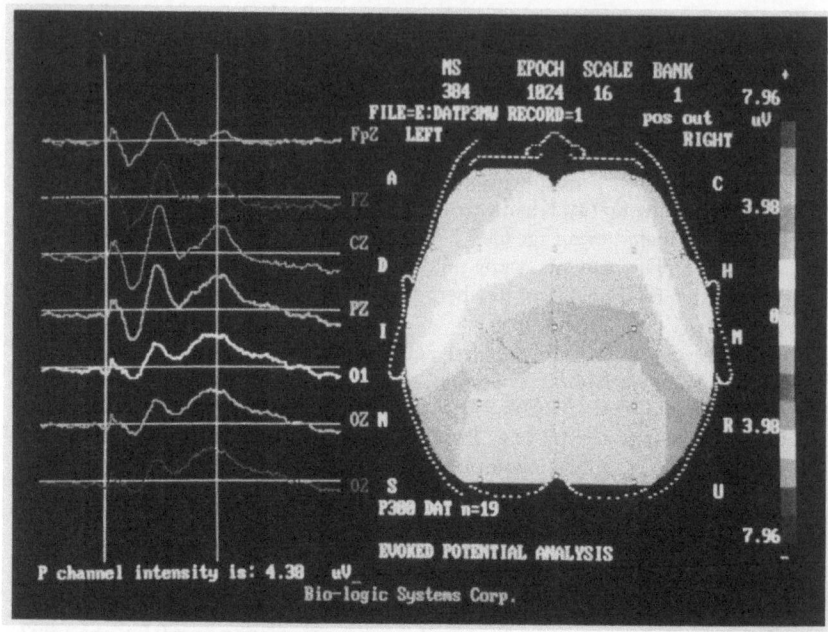

Fig. 1. Mean P300 latency of 19 patients with DAT

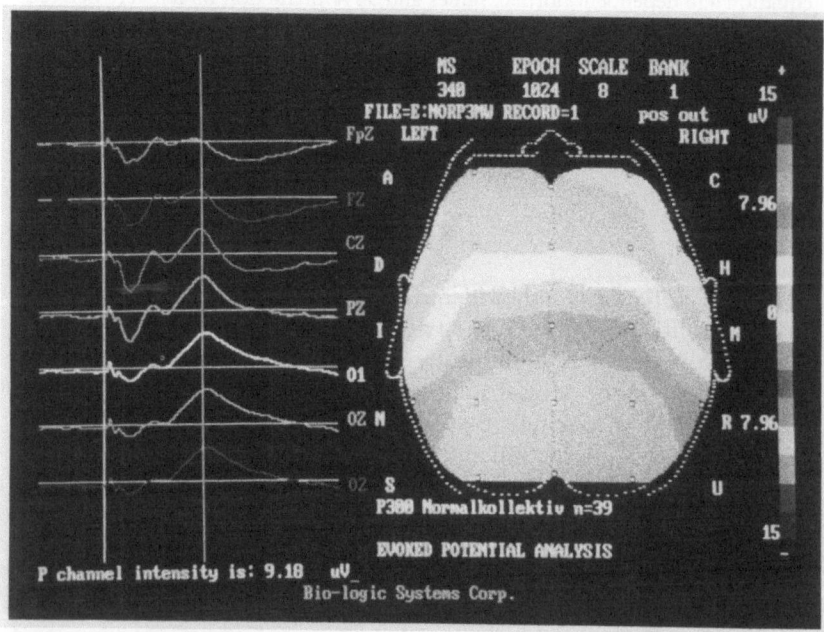

Fig. 2. Mean P300 latency of the control group (n = 39)

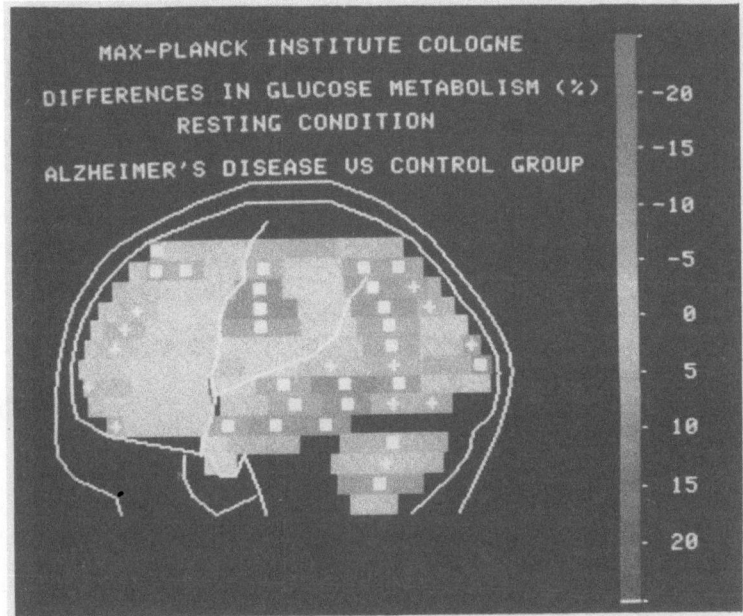

Fig. 3. A lateral view of the differences in the regional relative glucose metabolism between DAT group (n = 21) and a control group (n = 20)

a negative difference, 10 (53%) of the DAT patients had a difference lower than 2.8 but one (2.6%) normal fell in this range. 8 (42%) of the 19 DAT patients but also one normal subject (2.6%) had a latency ratio greater than 0.75. A combination of the two criteria did not add useful diagnostic information.

With PET, most DAT patients showed the typical metabolic pattern as described above (see Fig. 3). Global metabolic rate was only slightly decreased in comparison with the control group. The metabolic ratio separated DAT patients from control subjects with a sensitivity of 85% and a specificity of 90% (see Fig. 4).

Discussion

Both functional imaging methods, PET of glucose metabolism and P300 brain mapping showed significant differences between groups of clinically diagnosed mild to moderate forms of DAT and normal controls. Furthermore PET proved to be a method with high sensitivity (85%) and specificity (90%). P300 brain mapping did not reach a com-

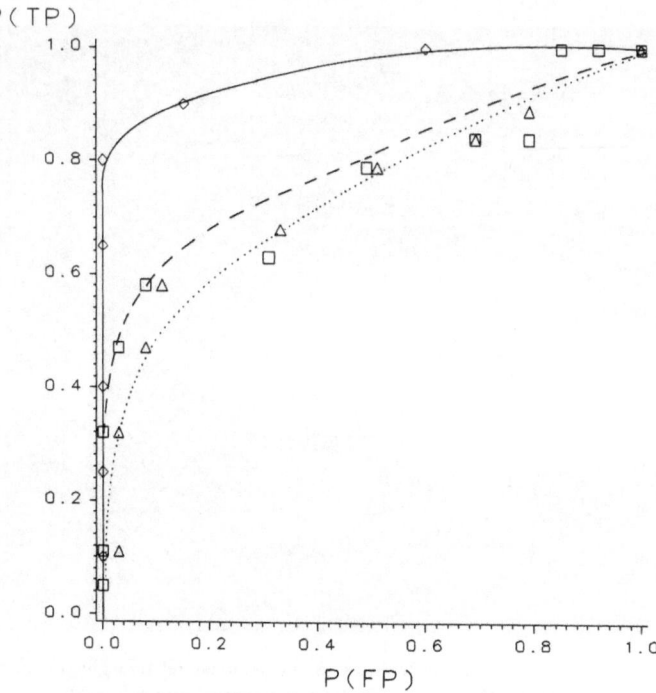

Fig. 4. ROC curve plotting probability of false-positive response P(FP) against probability of true-positive response P(TP). ◇—◇: Metabolic ratio of affected and unaffected regions. △···△: Latency ratio of P300. □---□: Pre-postcentral amplitude difference

parable power to distinguish between DAT patients and normals. Sensitivity of spatial distribution was only 58% at a specificity of 90%, or, if one would attempt to achieve a sensitivity of 85%, specificity drops to 21%. Sensitivity of latency was only 47% at a specificity of 90%.

References

Cutler NR, Haxby JV, Duara R, Cheryl L, Grady CL, Moore AM, Parisi JE, White J, Heston L, Margolin RM, Rapoport SI (1985) Brain metabolism as measured with positron emission tomography: serial assessment in a patient with familial Alzheimer's disease. Neurology 35:1556–1561

DeLeon MJ, Ferris SH, George AE, Reisberg B, Christman DR, Kricheff II, Wolf AP (1983) Computed tomography and positron emission transaxial tomography evaluations of normal aging and Alzheimer's disease. JCBFM 3:391–394

Duara R, Grady C, Haxby J, Sundaram M, Cutler NR, Heston L, Moore A, Schlageter N, Larson S, Rapoport SI (1986) Positron emission tomography in Alzheimer's disease. Neurology 36:879–887

Foster NL, Chase TN, Fedio P, Patronas NJ, Brooks RA, DiChiro G (1983) Alzheimer's disease: focal cortical changes shown by positron emission tomography. Neurology 33:961–965

Friedland RP, Budinger TF, Ganz E, Yano Y, Mathis CA, Koss B, Ober BA, Huesman RH, Derenzo SE (1983) Regional cerebral metabolic alterations in dementia of the Alzheimer type: positron emission tomography with [18F]fluorodeoxyglucose. J Comput Assist Tomogr 7:590–598

Goodin DS, Starr A, Chippendale T, Squires KC (1983) Sequential changes in the P3 component of the auditory evoked potential in confusional states and dementing illnesses. Neurology 33:1215–1218

Gordon E, Kraiuhin C, Harris A, Meares R, Howson A (1986) The differential diagnosis of dementia using P300 latency. Biol Psychiatry 21:1123–1132

Herholz K, Pawlik G, Wienhard K, Heiss W-D (1985) Computer assisted mapping in quantitative analysis of cerebral positron emission tomograms. J Comput Assist Tomogr 9:154–161

Maurer K, Lowitzsch K, Stöhr M (1988) Evozierte Potentiale AEP-VEP-SEP, Atlas mit Einführungen, 1. Aufl. Enke, Stuttgart, S 55–56

Maurer K, Dierks Th (1987) Brain mapping–topographische Darstellung des EEG und der evozierten Potentiale in Psychiatrie und Neurologie. Z EEG-EMG 8:4–12

McKhann G, Drachman D, Folstein M, Katzman R, Price D, Stadlan EM (1984) Clinical diagnosis of Alzheimer's disease: report of the NINCDS-ADRDA Work Group under the auspices of the Department of Health and Human Services Task Force on Alzheimer's disease. Neurology 34:939–944

Pfefferbaum A, Wenegrat BG, Ford JM, Roth WT, Kopell BS (1984) Clinical application of the P3 component of event-related potentials. II. Dementia, depression and schizophrenia. Electroencephalogr Clin Neurophysiol 59:104–124

Phelps ME (1981) Positron computed tomography studies of cerebral glucose metabolism in man. Theory and application in nuclear medicine. Semin Nucl Med 11:32–49

Polich J, Ehlers CL, Otis S, Mandell AJ, Bloom FE (1986) P300 latency reflects the degree of cognitive decline in dementing illness. Electroencephalogr Clin Neurophysiol 63:138–144

Reisberg B, Ferris SH, DeLeon MJ, Crook T (1982) The global deterioration scale for assessment of primary degenerative dementia. Am J Psychiatry 139:1136–1139

Reivich M, Kuhl D, Wolf A, Greenberg J, Phelps ME, Ido T, Casella V, Fowler J, Hoffmann E, Alavi A, Som P, Sokoloff L (1979) The [18F]fluorodeoxyglucose method for the measurement of local cerebral glucose utilization in man. Circ Res 44:127–137

Correspondence: Dr. B. Szelies, MPI für Neurologische Forschung, Klinik für Neurologie der Universität Köln, Joseph-Stelzmann-Strasse 9, D-5000 Köln, Federal Republic of Germany.

High resolution regional cerebral blood flow measurements in Alzheimer's disease and other dementia disorders

J. Risberg[1], L. Gustafson[2], and A. Brun[3]

[1] Departments of Psychiatry, [2] Psychogeriatrics, and [3] Institute of Pathology, Department of Neuropathology, University Hospital, Lund, Sweden

Summary

High-resolution regional cerebral blood flow (rCBF) measurements using inhalation of ^{133}xenon and a newly developed 254 detector system were performed in patients with organic dementia. Neuropathological findings were compared to rCBF in 13 deceased patients. A close relationship was found between the localization and severety of atrophy and/or infarcts situated in the cortical mantle and the rCBF-findings. Pathological changes in subcortical regions and white matter showed a less clear relationship to rCBF. The rCBF technique was found to be a useful diagnostic tool.

Introduction

One of the first functional brain imaging methods was the intra-carotid ^{133}xenon injection technique for measurement of the regional cerebral blood flow (rCBF; Lassen and Ingvar, 1961). Although the close coupling between cortical flow and function was not empirically proven until several years later (Raichle et al., 1976), were rCBF measurements used early for describing functional brain disturbances in organic dementia (Ingvar and Gustafson, 1970). The potential of rCBF data for differential diagnosis of organic dementia was evident already from the results of these studies. Soon were neuropathologically verified diagnoses available and found to compare favourably with the rCBF findings (Brun et al., 1975). The invasive nature of the intra-arterial injection technique prohibited, however, a routine clinical use of the method.

J. Risberg et al.

In the middle of the 1970-ies was the intra-arterial rCBF technique replaced by non-invasive methodology based on inhalation or i. v. injection of [133]Xe (Obrist et al., 1975; Risberg et al., 1975). Large series of patients could now be studied as well as normal control groups. We continued our prospective longitudinal study of rCBF in dementia and were soon able to demonstrate the existence of different and rather specific rCBF changes in patients with different diagnoses based on clinical criteria (Alzheimer's disease, frontal lobe dementia, multi-infarct dementia; Gustafson and Risberg, 1979). In a later study we compared autopsy verified diagnoses with diagnoses based on evaluation of the regional flow pattern and found an agreement in 90% of the patients (Risberg, 1985).

Until now the standard rCBF recording systems have contained 32 scintillation detectors covering the hemispheres with a spatial resolution of 3–5 cm. This poor resolution has limited the possibility to study flow changes localized within small areas of the normal brain (e. g. rCBF changes during mental activation) as well as in focal brain disorders. In the present paper a new high-resolution rCBF system (spatial resolution about 1 cm) is described and its potential for research and routine clinical use in organic dementia is illustrated by some preliminary findings.

Material and methods

The high-resolution rCBF system (Cortexplorer 256 HR; Scan. Detectronic Inc., Hadsund, Denmark) contains 254 scintillation detectors [10 mm × 10 mm NaI (Tl) crystals] mounted in a pneumatically controlled helmet type of holder. The system adjusts for a wide variation of head sizes and shapes providing an optimal coverage of all parts of the hemispheric cortical mantle. The measurement procedure starts with one minute of inhalation of a mixture of ^{133}Xe and air (90 MBq/l) followed by 10 minutes of breathing of ordinary air. The detectors

---

→

Fig. 1. High-resolution regional cerebral blood flow results in a normal subject obtained by 133-Xe inhalation and the Cortexplorer 256-HR system (Scan. Detectronic Inc., Hadsund, Denmark). The hemispheric average flow (ISI units of cortical blood flow) is shown in the lower part of the figure. The vertex projection shows all parts of the cortical mantle and indicates regional flow values as per cent of the average flow by means of a colour code (red shades-values above and green-values below the average). Note the normal "hyperfrontal" flow pattern

Fig. 2. RCBF results from a patient with DAT. The neuropathological investigation (upper left) showed the typical distribution of degeneration with the corresponding flow changes shown as high-resolution lateral (upper right) and vertex (lower right) maps. As a comparison are simulated low resolution results (32 detectors) shown in the lower left part of the figure. The colour coding is the same as in Fig. 1

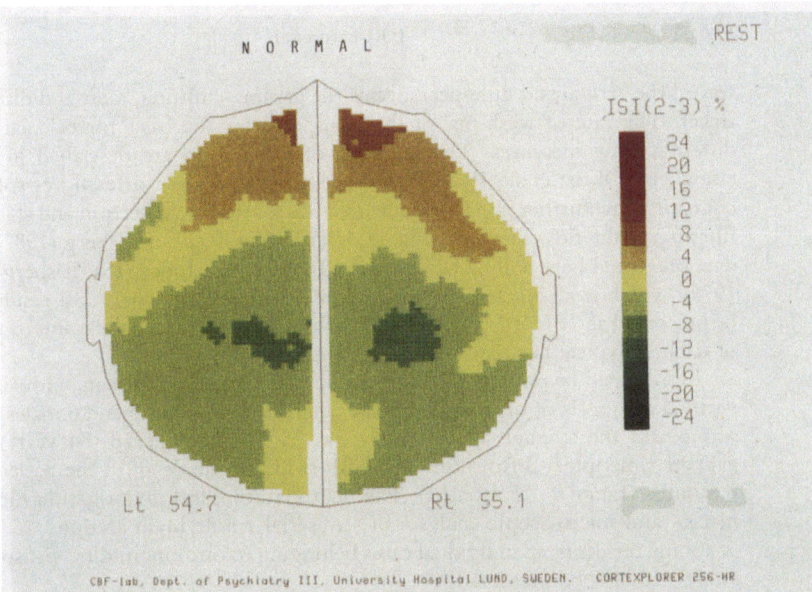

Fig. 1

NORMAL REST

ISI(2-3) %

24
20
16
12
8
4
0
-4
-8
-12
-16
-20
-24

Lt 54.7 Rt 55.1

CBF-lab, Dept. of Psychiatry III, University Hospital LUND, SWEDEN. CORTEXPLORER 256-HR

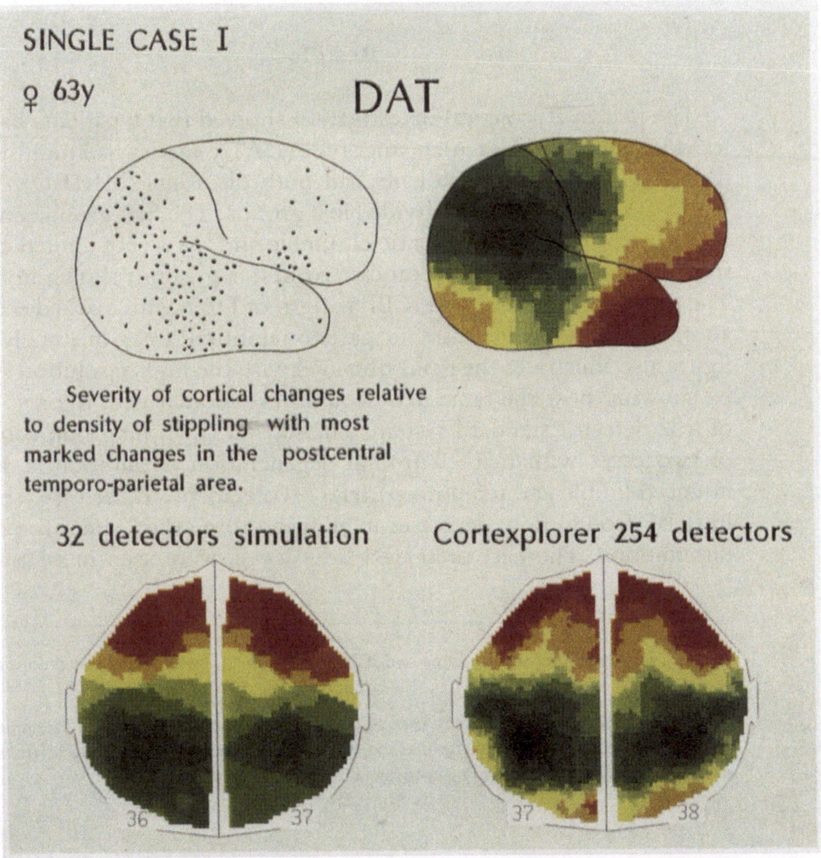

SINGLE CASE I

♀ 63y DAT

Severity of cortical changes relative to density of stippling with most marked changes in the postcentral temporo-parietal area.

32 detectors simulation Cortexplorer 254 detectors

36 37 37 38

Fig. 2

record the arrival and disappearance of the gamma-emitting, inert and diffusible tracer. The rate of wash-out of the isotope forms the basis for calculation of different flow measures. The methods of calculation are described in detail elsewhere (Obrist et al., 1975; Risberg et al., 1975). The Cortexplorer software contains some further improvements regarding artefact correction and statistical filtering of the flow maps, which are described in detail in Risberg (1987). The data presented here will be based on the Initial Slope Index (ISI; Risberg et al., 1975), which is a parameter of mainly grey matter flow. The rCBF results will be presented as "BEAM-type" of vertex flow maps with triangular interpolation of data. A typical normal flow map is shown in Fig. 1.

Results will be presented from a group of 13 deceased patients with autopsy verified diagnosis of organic dementia. There were 7 females and 6 males in the group and the average age at death was 75 years (range 58–84 years). The average time interval from the last rCBF study to the death of the patient was 8 months (range 0–15 months). The neuropathological investigation included macro- and micro-scopic analyses of semiserial whole brain sections.

Some results from individual cases belonging to our longitudinal prospective study will also be presented.

Results

The neuropathological investigation showed that 6 patients had suffered from dementia of Alzheimer type (DAT), 4 cases had multi-infarct dementia (MID) and 3 patients had both disorders (MIXED). The 6 DAT patients could be subdivided in 3 groups. The first group consisted of 3 patients with temporoparietal, limbic and, to a very limited extent, frontal degeneration. A typical case from this group is shown in Fig. 2. The patients had the typical rCBF pattern of DAT with marked accentuation of the flow decreases in parieto-temporal areas bilaterally. The figure also illustrates the gains obtained with the high-resolution system by showing how the same data would have looked with the resolution of a 32 detector standard system. The second DAT subgroup consisted of two cases with marked frontal degeneration in addition to a pronounced limbic and temporo-parietal involvement. These cases showed in rCBF frontal flow decreases in addition to marked temporo-parietal diminutions. The best preserved flow levels were seen in central and

$\longrightarrow$

Fig. 3. Neuropathological findings and rCBF results in a case of MID. The colour coding is the same as in Fig. 1

Fig. 4. High-resolution rCBF in a case of progressing tentative frontal lobe degeneration of non-Alzheimer type. Note the marked frontal flow decreases especially at the second investigation when the clinical symptoms of frontal lobe dysfunction were very prominent. Symbols as in Fig. 1

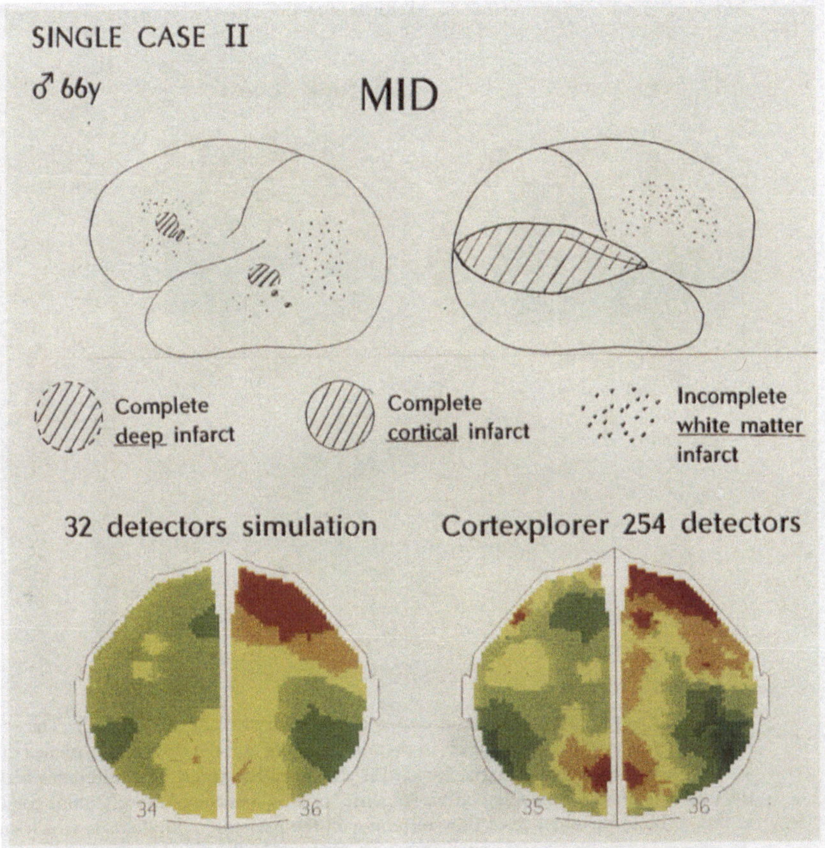

SINGLE CASE II

♂ 66y

MID

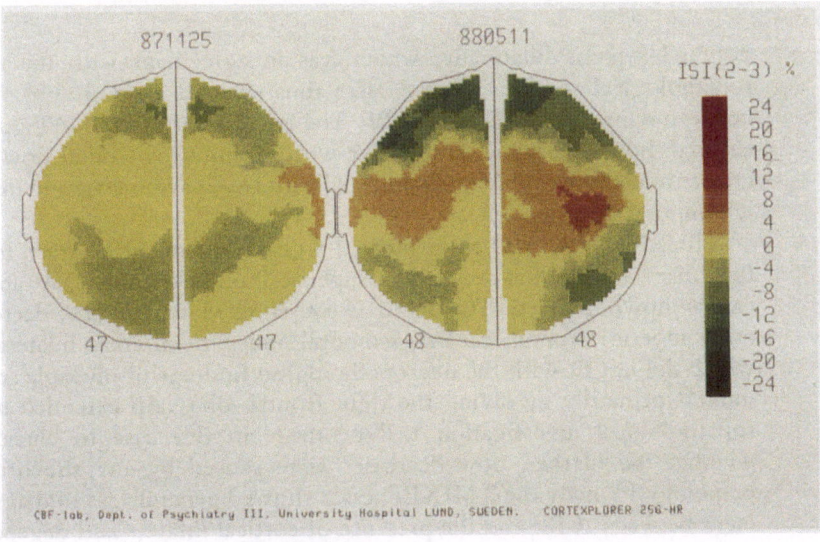

Complete **deep** infarct

Complete **cortical** infarct

Incomplete **white matter** infarct

32 detectors simulation   Cortexplorer 254 detectors

34   36   35   36

Fig. 3

871125   880511   ISI(2-3) %

24
20
16
12
8
4
0
-4
-8
-12
-16
-20
-24

47   47   48   48

Fig. 4

CBF-lab, Dept. of Psychiatry III, University Hospital LUND, SWEDEN.   CORTEXPLORER 256-HR

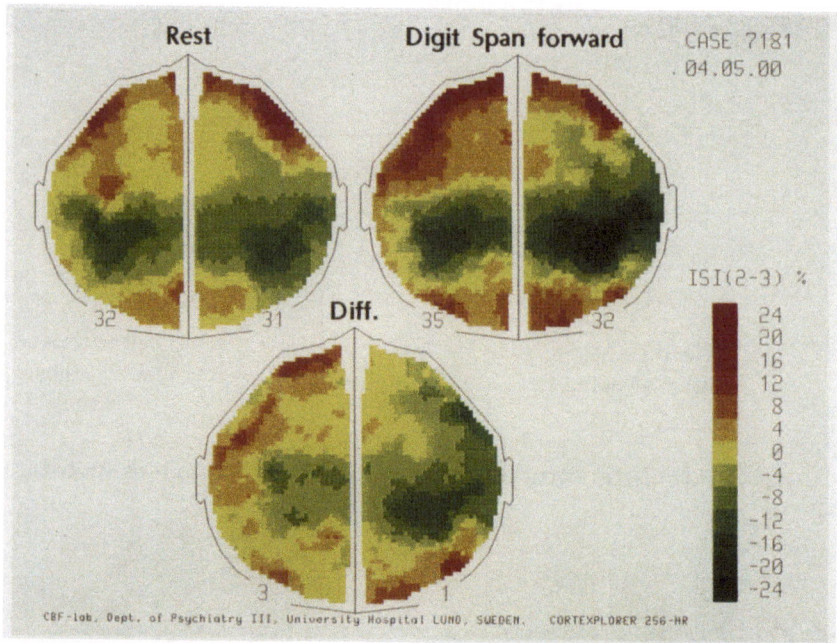

Fig. 5. High-resolution rCBF in a case with the clinical diagnosis of DAT. The resting flow pattern is shown in the upper left part of the figure and the flow map obtained during activation by repetition of digits forward is shown to the right. The difference between the two maps (the isolated activation response) is shown at the bottom. Colour coding as in Fig. 1. Note the absence of flow responses in the parietotemporal areas, which are the ones most affected in DAT

occipital regions bilaterally, which was in agreement with the neuropathological results. These 5 cases thus showed a good agreement between neuropathology and rCBF. The last patient in the DAT group showed, however, some discrepancy with pronounced frontal flow decreases in combination with an unusual pattern of asymmetric and mainly temporal atrophy.

Three of the 4 MID cases showed a good agreement between focal flow decreases and the presence of superficial cortical infarcts. A typical case is shown in Fig. 3. The last MID case showed marked flow decreases in superior frontal, central, temporal and parietal areas bilaterally, which did not fit with the neuropathological findings of multiple small infarcts primarily involving the right frontal lobe. (An extended neuropathological investigation will be made in this case to elucidate whether the marked flow decreases were caused by any subcortical pathology.) Finally the 3 MIXED cases showed generally a good agreement between rCBF and the presence of cortical infarcts and degeneration. The relationship between subcortical grey and white matter changes and rCBF was less clear.

Figure 4 shows typical rCBF findings in a case of tentative fronto-temporal degeneration of non-Alzheimer type. The patient showed a rapidly progressing dementia dominated by frontal symptoms parallel to progressive frontal flow diminutions.

Finally, Fig. 5 illustrates the enhancing effect of mental activation on the rCBF pathology in a case of tentative DAT. The digit-span test given during the second flow measurement caused flow increases only in areas known to be affected to a lesser extent by the disease.

Discussion

The present results lend further support to the clinical usefulness of rCBF measurements for the diagnosis of dementia. The existence of fairly specific topographic DAT pathology involving mainly posterior, parieto-temporal cortical areas has now been established also with several other brain imaging methods (PET, SPECT, BEAM). The ^{133}Xe inhalation technique, as described here, provides a moderately expensive (100−200 US$) and rather easily applicable technique for the demonstration of this regional functional pathology. The technique provides absolute flow values (in contrast to SPECT with ^{99m}Tc-HM-PAO) and fairly unlimited repeatability (due to rapid elimination of the tracer and low radiation dose). This makes the method suitable also for the study of flow changes induced by mental activation or drug treatment. The study of such changes of rCBF might further improve the possibility to early and correctly diagnose Alzheimer's disease and to follow the effects of treatment.

Acknowledgements

Supported by the Swedish Medical Research Council (project nr 4969) and the King Gustaf V and Queen Victoria Foundation.

References

Brun A, Gustafson L, Ingvar DH (1975) Neuropathological findings related to neuropsychiatric symptoms and regional cerebral blood flow in presenile dementia. In: Proceedings of the 7th International Congress of Neuropathology, Budapest, Hungary, Sept 1−7, 1974. Excerpta Medica, Amsterdam

Gustafson L, Risberg J (1979) Regional cerebral blood flow measurements by the 133-xenon inhalation technique in differential diagnosis of dementia. Acta Neurol Scand 60 [Suppl 72]:546–547

Ingvar DH, Gustafson L (1970) Regional cerebral blood flow in organic dementia with early onset. Acta Neurol Scand 46 [Suppl 43]:42–73

Lassen NA, Ingvar DH (1961) The blood flow of the cerebral cortex determined by radioactive krypton-85. Experientia 17:42–50

Obrist WD, Thompson HK, Wang HS, Wilkinson WE (1975) Regional cerebral blood flow estimated by 133-xenon inhalation. Stroke 6:245–256

Raichle ME, Grubb RL, Gado MH, Eichling JO, Ter-Pogossian MM (1976) Correlation between regional cerebral blood flow and oxidative metabolism. Arch Neurol 33:523–526

Risberg J (1985) Cerebral blood flow in dementias. Dan Med Bull 32 [Suppl 1]:48–50

Risberg J (1987) Development of high-resolution two-dimensional measurement of regional cerebral blood flow. In: Wade J, Knezevic S, Maximilian VA, Mubrin Z, Prohovnik I (eds) Impact of functional imaging in neurology and psychiatry. John Libbey, London, pp 35–43

Risberg J, Ali Z, Wilson EM, Wills EL, Halsey JH (1975) Regional cerebral blood flow by 133-xenon inhalation. Stroke 6:142–148

Correspondence: J. Risberg, Ph. D., Department of Psychiatry, University Hospital, S-221 85 Lund, Sweden.

Single photon emission computed tomography (SPECT) in Pick's disease: two case reports

F. Reisecker[1], E. Laich[2], F. Leblhuber[3], J. Trenkler[4], and E. Deisenhammer[2]

[1] Department of Neurology, Hospital Barmherzige Brüder, Graz, and [2] Department of Clinical Neurophysiology, [3] Department of Neurologic-Psychiatric Gerontology, and [4] Institute of Radiology, Wagner Jauregg Hospital, Linz, Austria

Summary

The diagnostic value of SPECT in Pick's disease is discussed on the basis of two cases. Personality changes predominated in both of them. Atrophy of the frontal and temporal lobes was seen on CT only in one of the two female patients, who was in a stage of advanced dementia. SPECT showed reduced frontal accumulation in both cases in analogy to PET studies. The results of the present study demonstrate that a tentative diagnosis of Pick's disease can either be excluded or substantiated with the help of SPECT.

Pick's disease has a special position among the dementias because of its rare incidence, particular epidemiology and circumscribed (lobar) atrophy. Its histological characteristics are gliosis and neuronal loss in the atrophic area and intracellular Pick bodies in various localizations (Seitlberger et al., 1983). Clinical observations of the few neuropathologically verified cases show that personality changes, emotional disorders, speech disturbances and hyperorality – similar to Kluver-Bucy syndrome – predominate in the early stages of the disease, while intellectual defects prevail only later in the illness (Constantinidis et al., 1974; Cummings and Duchen, 1981; Gustafson and Nilson, 1982; Knopman et al., 1989; Munoz-Garcia and Ludwin, 1984; Seitlberger et al., 1983). Therefore, it is often difficult to distinguish Pick's disease in its inital stages from affective psychoses and also from Alzheimer's disease in cases with marked intellectual deficits or memory impairment (Cummings and Duchen, 1981; Knopman et al., 1989).

X-ray transmission computed tomography (CT) shows frontal and fronto-temporal atrophies (Corsellis, 1977), but also global atrophies with accentuation outside the frontal lobe have been identified (Groen and Hekster, 1982; Cummings and Duchen, 1981; Knopman et al., 1989; Munoz-Garcia and Ludwin, 1984).

Szelies and Karenberg (1986) studied glucose metabolism in a demented female patient with atrophy predominantly located in the frontal lobe using positron emission tomography (PET) and found global hypometabolism which was clearly accentuated in the frontal lobes (left more than right). Similar metabolic conditions were found by Kamo et al. (1987) in a comparable patient, quantitative metabolic analysis showing high-grade correlation between metabolic decrements and frontally accentuated gliosis as well as neuronal loss. The authors conclude from their results that a premortem diagnosis of Pick's disease might be possible with the help of PET.

Single photon emission computed tomography (SPECT) with tracers that pass the blood-brain barrier (IMP, HMPAO) performed in patients with Alzheimer-type dementia showed perfusion patterns similar to those obtained with PET measurements (Deisenhammer et al., 1989). Another SPECT study (Leblhuber et al., 1989) demonstrated abnormalities similar to those detected by PET also in other forms of primary degenerative dementias.

We performed SPECT studies in two female patients suspected of suffering from Pick's disease.

Case reports

Case 1

A 67-year-old woman had been experiencing progressive personality changes for about a decade. Initially her personality profile became coarsened, she lost interest, showed signs of disinhibition, flat drive and increased suggestibility; polyphagia, polydipsia and oral mechanisms followed. Later on intellectual deficits, memory impairment, unmotivated and uncontrolled movements, states of confusion, urinary incontinence occurred.

When the patient was seen for the first time, she was alert, completely disoriented to time and place, and partly oriented to person. She showed high-grade amnestic impairment, perseveration, oral mechanisms with thumb sucking, disinhibition and undirected motor activity. The neurological examination revealed a positive snapping reflex on the left side and the presence of a snout reflex. The patient reached 8 points on the Hachinski Ischemia Scale and 12 points on the Mini-Mental Scale. Psychometric testing (Rohrschach, Raven, HAWIE, Benton) revealed deficits in abstract logical thinking, in availability of

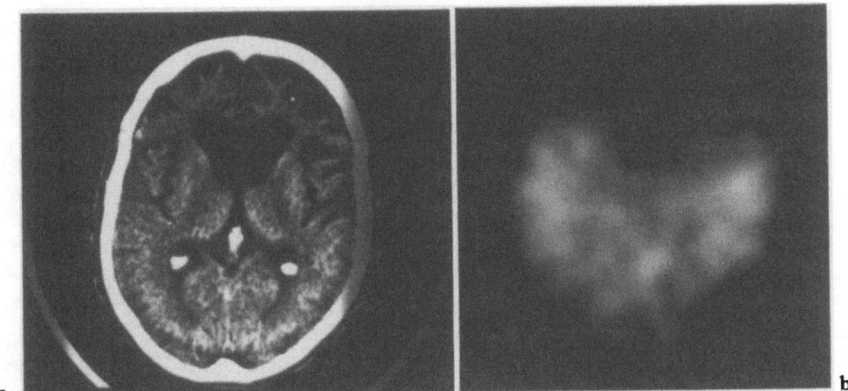

a   b

Fig. 1. CT (a) and SPECT (b) studies of case 1 (axial slices parallel to, and approximately 5 cm above cantho-meatal plane). There is a marked atrophy in the frontal and temporal lobe and a severe decrease of tracer uptake in SPECT

general knowledge and understanding of social environment, impaired short-term recall of geometric figures, medium-term memory, attention, concentration, calculation, spatial orientation and planning.

Laboratory examination showed creatinine and cholesterol to be elevated, while all other values including vitamin B12 and folic acid were normal. EEG was normal, CT showed marked, predominantly cortical atrophy in frontal and temporal lobes (Fig. 1a). SPECT studies with HMPAO revealed reduced accumulation in the frontal lobe extending to the insula and exceeding by far the magnitude of the atrophy (Fig. 1b).

Case 2

A 63-year-old woman was hospitalized because of a manic disorder with paranoid symptoms. The patient had been getting more and more forgetful for some years. She used to go shopping without any reason up to twenty times a day, imagined to receive letters from strangers, felt robbed, showed abnormal drinking behaviour and consumed up to 3 liters of coffee per day. Upon admission she was well oriented to time, place and person and was sociable. The clinical picture was that of a manic disorder with foolish euphoric behaviour, increased drive, psychomotor unrest, desultory thinking, logorrhea, lack of criticism and distance, impaired memory. Neurological findings were normal apart from the presence of a snout reflex. The patient reached 25 points on the Mini-Mental Scale and 2 points on the Hachinski Ischemia Scale. Psychometric testing (see case 1) revealed severe impairment of short-term memory, medium-term memory and recall, concentration, attention and analytical synthetic thinking. Laboratory parameters including vitamin B12 and folic acid were normal, with the exception of elevated cholesterol. The EEG showed minor general

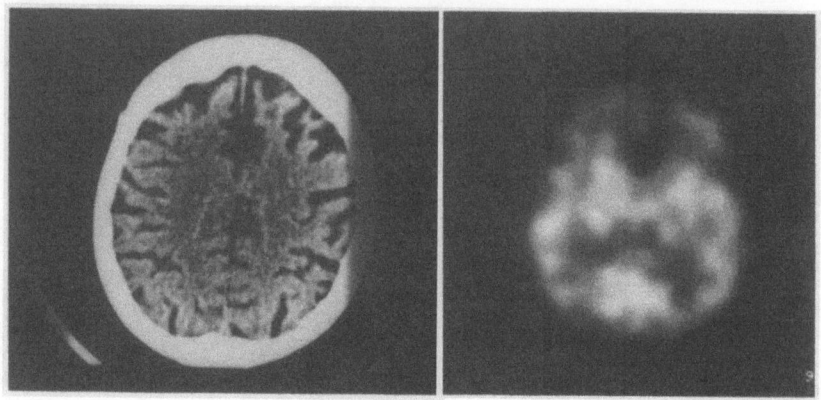

Fig. 2. CT (**a**) and SPECT (**b**) studies of case 2 (axial slices parallel to, and approximately 7.5 cm above cantho-meatal plane). There is only a mild atrophy in CT, but a marked frontal decrease of tracer uptake in the SPECT

alterations, slight enlargements of the cisterns around the midbrain and of the sylvian and frontal fissures were seen on CT (Fig. 2a). SPECT with HMPAO showed markedly reduced accumulation in the frontal lobes (Fig. 2b).

Discussion

PET studies performed in a patient with Pick's disease who subsequently died showed a high-grade correlation between the extent of hypometabolism and typical histological abnormalities, i.e. gliosis and neuronal loss (Kamo et al., 1987). The frontal accentuation of these abnormalities results in a typical hypofrontal PET image. The coupling of regional blood flow to regional metabolism (Lou et al., 1987) permits – in analogy to PET – the interpretation of blood flow patterns tomographically demonstrated by SPECT imaging as the expression of regional metabolism. Advanced dementia, hyperorality, and frontal atrophy suggested the diagnosis of Pick's disease in case one. The hypofrontal pattern of the SPECT images seems to substantiate this diagnosis. The second patient on the other hand presented with manic psychosis without proof of dementia and without lobar atrophy so that a differentiation from affective psychosis was impossible. In this case only the SPECT pattern lead to the suspicion of Pick's disease. Although histological verification is lacking we believe on the basis of our experience with typical PET-type SPECT patterns that it is possible to substantiate a tentative diagnosis of Pick's disease during the patient's lifetime –

similar to Alzheimer's disease (Deisenhammer et al., 1989) and Huntington's chorea (Leblhuber et al., 1989) – or to suspect Pick's disease in cases of affective psychoses.

References

Constantinidis J, Richard J, Tissot T (1974) Pick's disease. Histological and clinical correlations. Eur Neurol 11:208–217

Corsellis J (1977) Aging and dementias. In: Blackwood W, Corsellis J (eds) Greenfield's neuropathology. Edward Arnold, London, pp 796–848

Cummings JL, Duchen LW (1981) Kluver-Bucy syndrome in Pick disease: clinical and pathologic correlations. Neurology 31:1415–1422

Deisenhammer E, Reisecker F, Leblhuber F, Höll K, Markut H, Trenkler J (1989) Beitrag der Single-Photon-Emissions-Computer-Tomographie (SPECT) zur Differentialdiagnose der Demenz. Dtsch Med Wochenschr 114:1639–1644

Groen JJ, Hekster REM (1982) Computed tomography in Pick's disease: findings in a family affected in three consecutive generations. J Comput Assist Tomogr 6(5):907–911

Gustafson L, Nilson L (1982) Differential diagnosis of presenile dementia on clinical grants. Acta Psychiatr Scand 65:194–209

Kamo H, McGeer PL, Harrop R, McGeer EG, Calne DB, Martin WRW, Pate D (1987) Positron emission tomography and histopathology in Pick's disease. Neurology 37:439–445

Knopman DS, Christensen KJ, Schut LJ, Harbaugh RE, Reeder T, Ngo T, Frey W (1989) The spectrum of imaging and neuropsychological findings in Pick's disease. Neurology 39:362–368

Leblhuber F, Brucker B, Reisecker F, Trenkler J, Deisenhammer E (1989) Single photon emission computed tomography in probable Huntington's chorea. J Neural Transm (P-D Sect) 1:93–94

Lou HC, Edvinson L, McKenzie ET (1987) The concept of coupling blood flow to brain function: revision required? Ann Neurol 22:289–297

Munoz-Garcia D, Ludwin SK (1984) Classic and generalized variants of Pick's disease: a clinicopathological, ultrastructural, and immunocytochemical comparative study. Ann Neurol 16:467–480

Seitlberger F, Gross H, Pilz P (1983) Pick's disease: a neuropathologic study. In: Hirano A, Miyoshi K (eds) Neuropsychiatric disorders in the elderly. Igaku-Shoin, Tokio, pp 87–117

Szelies B, Karenberg A (1986) Störungen des Glukosestoffwechsels bei Pickscher Erkrankung. Fortschr Neurol Psych 54:393–397

Correspondence: Prim. Univ.-Doz. Dr. F. Reisecker, Neurologisch-Psychiatrische Abteilung, Krankenhaus der Barmherzigen Brüder, Bergstrasse 27, A-8020 Graz-Eggenberg, Austria.

In vivo studies of hippocampal atrophy in Alzheimer's disease

M. J. de Leon, A. E. George, L. A. Stylopoulous, D. C. Miller, and G. Smith

Departments of Psychiatry, Radiology and Pathology, New York University School of Medicine, New York, NY, U.S.A.

Summary

Using 5 mm negative angulation CT scans, observations on the temporal lobes and hippocampal areas were made on 113 Alzheimer (AD) patients and 72 controls. The results indicate that hippocampal changes including dilatation of the choroidal and hippocampal fissures, appear to be an early feature of AD. Over a three year interval these changes are predictive of AD type clinical deterioration in non-demented elderly subjects. The CT hippocampal changes are correlated with postmortem hippocampal changes. Hippocampal evaluations in individuals < 80 years of age are a useful addition to the diagnostic workup of AD patients.

Introduction

Severe hippocampal degeneration is often documented in AD and is occasionally observed in the absence of neocortical deficits; these findings have led to the characterization of AD as "a hippocampal dementia" (Ball et al., 1985). It has been observed recently that the first site of degeneration in Down's syndrome occurs in the hippocampus (Mann and Esiri, 1988). The structural hippocampal abnormalities documented in AD include neuronal loss, granulovacuolar degeneration, neurofibrillary tangles and neuritic plaques (Ball, 1978; Kemper, 1984; Terry et al., 1981). Numerous biochemical deficits have been reported including depletions of hippocampal acetylcholinesterase, choline acetyltransferase, norepinephrine, somatostatin and glutamate (Davies and Maloney, 1976; Powers et al., 1988; Hyman et al., 1987; for a review see Selkoe and Kosik, 1983).

Neuropathological findings in AD show a consistent topography with respect to the hippocampal formation and its efferent projections. Pyramidal cells located within layer II and superficial layer III of entorhinal cortex are the primary neuronal population involved. These are the neurons of origin of the perforant pathway. The subiculum, CA1 of the hippocampus and Layers II, III and IV of temporal association cortex are also involved. This pattern of degeneration results in the "isolation" of the hippocampal formation from its input to neocortical association areas (Hyman et al., 1984).

Structural hippocampal pathology qualitatively similar to that documented in AD also occurs in normal aging. From the seventh to tenth decade, hippocampal neuronal loss, neurofibrillary tangles, granulovacuolar degeneration and neuritic plaques have been reported (Ball, 1978; Kemper, 1984; Terry et al., 1981; Hyman and Van Hoesen, 1988). Characterization of the hippocampal lesions and their temporal relation to the development of intellectual changes and neocortical pathology may be essential for an understanding of the early stages of AD progression and the normal aging process.

Recent CT and MRI studies have also yielded valuable data on temporal lobe and hippocampal changes in AD. LeMay et al. (1986) found that subjective ratings of temporal lobe atrophy produced a high classification accuracy in identifying AD patients vs. normal controls. Our findings, using a special "negative angle" imaging protocol (de Leon et al., 1988), indicate for the first time that atrophic changes of the hippocampus and parahippocampal areas can be visualized *in vivo* with CT. Moreover, our data indicated that atrophic hippocampal changes are related to altered endocrine functioning (de Leon et al., 1988) and with good pathologic verification and anatomical specificity to the symptoms of dementia (George et al., 1990). In an important but preliminary study, Seab et al. (1988) found MRI hippocampal volumes reduced in AD relative to control. We now extend our CT results to a large cohort of 175 patients and controls. A report of this work is also found in de Leon et al. (1989).

Method

Using a GE 9800 scanner, CT evaluations were conducted using a gantry tilt that produced a scan angulation of 20° negative to the cantho-meatal plane. The slice thickness was 5 mm. All study participants received extensive research diagnostic evaluations that included medical, neurologic, psychiatric, neuropsychological, and laboratory examinations. All AD patients met NINCDS-ADRDA diagnostic criteria. Controls were volunteers, generally spouses and relatives of affected patients. The 1 to 7 scale of the GDS (Reisberg et al., 1982) was used to stage the severity of cognitive change. The CT scans of 113 AD patients (aged

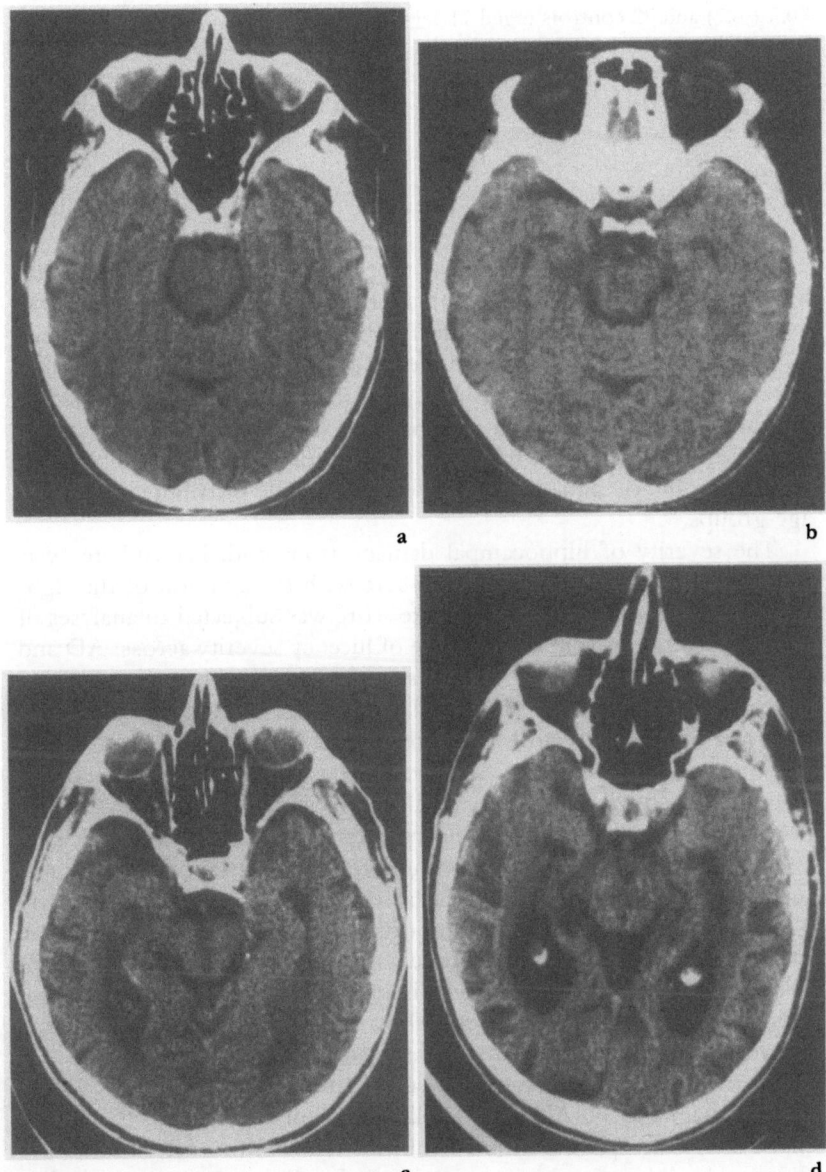

Fig. 1. Four levels of increasing atrophic change of the hippocampus (a) no atrophy (b) questionable and insufficient evidence of atrophy (c) mild but definite atrophy and (d) moderate to severe atrophy

70.3 ± 8.7) and 72 controls (aged 71.1 ± 8.0) were evaluated subjectively on the basis of hippocampal lucencies. Findings were rated on a four-point scale for each hemisphere: 0 = no lesion, 1 = questionable lesion, 2 = mild-moderate lesion and 3 = severe lesion (see Figs. 1a–1d). A 3-year longitudinal study was done on 48 non-demented individuals.

Results

Using a cut-off hippocampal rating of ≥ 2 for either hemisphere, hippocampal atrophic changes or alternatively hippocampal lucencies were significantly more prevalent in AD than in controls (77% versus 22%, p < .01). The hippocampal lucencies occurred more frequently in the advanced stages of AD (see Fig. 2). Consistent with the neuropathological literature, the normal controls with hippocampal lucencies were significantly older than those without lucencies (75.2 versus 68.9 years, see Fig. 3). In AD patients, lucencies were equally distributed across all age groups.

The severity of hippocampal damage from both hemispheres was estimated by creating a composite score with the addition of the right and left rating scores. The composite score was subjected to analyses of variance to establish the significance of lucency severity across: AD and normal samples, levels of disease severity, and in both groups, as a function of advancing age. The results indicated that hippocampal lucencies were significantly more severe relative to normal controls in each of the questionable, mild and moderate to severe AD patient groups (Fig. 4). Moreover, the severity of the hippocampal changes were significantly associated with increasing cognitive impairment. A significant age by diagnosis interaction was found for the severity of hippocampal lucencies. Specifically, from the fifth to seventh decade lucencies in AD patients were more severe than in normals. Lesions were equally severe in both groups in the ninth decade. AD patients showed a consistent relationship between cognitive impairment and the severity of hippocampal damage but did not show any evidence of an age effect. Normal controls demonstrated a clear cut age dependence.

In the longitudinal study, subjects deteriorated and received the clinical diagnosis of AD. At baseline 91% of these subjects showed hippocampal atrophy. Of the 37 subjects that did not deteriorate 81% did not show baseline hippocampal atrophy. Postmortem neuropathologic correlation with antemortem CT and postmortem MRI has also been conducted on six cases. The neuropathology protocol included coronal hippocampal samples taken at the level of the lateral geniculate nucleus. Following standard processing with paraffin, the formalin-fixed sections were stained with Luxol fast blue and with hematoxylin and

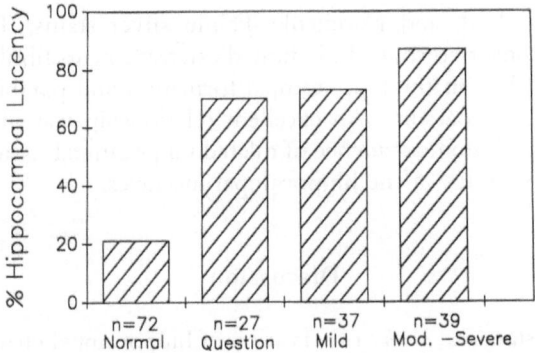

Fig. 2. The prevalence CT hippocampal lucency and cognitive functioning (n = 175)

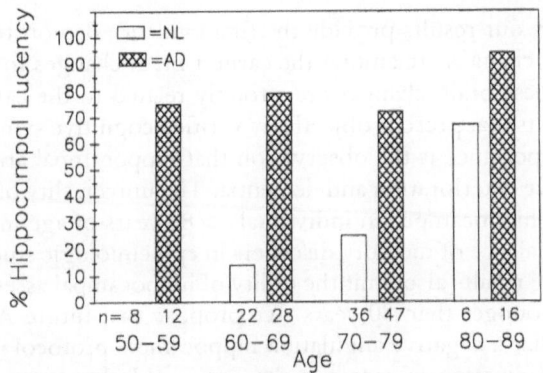

Fig. 3. The prevalence CT hippocampal lucency and age (n = 175)

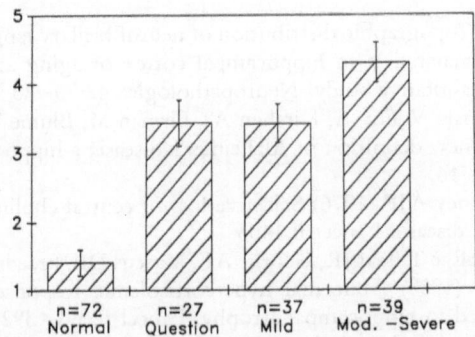

Fig. 4. The severity of CT hippocampal lucency and cognitive function (n = 175)

eosin, Congo Red, and Naomenko-Feigin silver stains. The results, compared against two controls, indicated extensive neurofibrillary tangles and neuronal loss in the hippocampal formation and parahippocampal gyrus of the AD patients. An excellent relationship was observed between neuropathologic evidence of dilated hippocampal fissures and the antemortem severity of the hippocampal lucencies.

Discussion

In conclusion, the *in vivo* observation of hippocampal changes in AD is of clinical value. Our CT data are in agreement with the neuropathological literature and our results directly show a correlation between the CT and the postmortem appearance of the choroidal hippocampal fissure complex.

Moreover, our results provide the first evidence demonstrating that hippocampal changes are among the earliest brain changes to be detected *in vivo*. These brain changes are strongly related to the early subjective complaints that precede objectively verified cognitive symptoms. Of particular importance is the observation that hippocampal atrophy predicts cognitive deterioration and dementia. The universality of the aging effect on the hippocampus in individuals > 80 years of age may explain the high prevalence of memory disorders in epidemiologic studies of the elderly. These results also limit the utility of hippocampal assessments to individuals younger than 80 years. We propose that future AD studies routinely utilize a negative angulation hippocampal protocol in the clinical brain examination of memory changes in elderly groups.

References

Ball MJ (1978) Topographic distribution of neurofibrillary tangles and granulo-vacuolar degeneration in hippocampal cortex of aging and demented patients. A quantitative study. Neuropathologica 42:73–80

Ball MJ, Hachinski V, Fox A, Kirshen AJ, Fisman M, Blume W, Kral VA, Fox H (1985) A new definition of Alzheimer's disease: a hippocampal dementia. Lancet i:14–16

Davies P, Maloney AJF (1976) Selective loss of central cholinergic neurons in Alzheimer's disease. Lancet ii:1403

de Leon MJ, McRae T, Tsai JR, George AE, Marcus DL, Freedman M, Wolf AP, McEwen B (1988) Abnormal hypercortisolemic response in Alzheimer's disease linked to hippocampal atrophy. Lancet ii:391–392

de Leon MJ, George AE, Stylopoulos LA, Smith G, Miller DC (1989) Early marker for Alzheimer's disease: the atrophic hippocampus. Lancet ii:672–673

George AE, de Leon MJ, Stylopoulos LA, Miller J, Kluger A, Smith G, Miller DC (1990) Temporal lobe CT diagnostic features of Alzheimer's disease: dilatation of the choroidal/hippocampal fissure complex. Am J Neuroradiol 11:101–107

Hyman BT, Van Hoesen GW (1988) Sites of earliest Alzheimer-type pathological changes in entorhinal cortex and hippocampus visualized by ALZ-50 immunocytochemistry. Soc Neurosci Abstr 14 (2):1084 (Abstract)

Hyman BT, Van Hoesen GW, Damasio AR (1987) Alzheimer's disease: glutamate depletion in the hippocampal perforant pathway zone. Ann Neurol 22:37–40

Hyman BT, Van Hoesen GW, Damasio AR, Barnes CL (1984) Alzheimer's disease: cell-specific pathology isolates the hippocampal formation. Science 225:1168–1170

Kemper T (1984) Neuroanatomical and neuropathological changes in normal aging and in dementia. In: Albert ML (ed) Clinical neurology of aging. Oxford University Press, New York, pp 9–52

LeMay M, Stafford JL, Sandor T, Albert M, Haykal B, Zamani A (1986) Statistical assessment of perceptual CT scan ratings in patients with Alzheimer type dementia. J Comput Assist Tomogr 10:802–809

Mann DMA, Esiri MM (1988) The site of the earliest lesions of Alzheimer's disease. N Engl J Med 318(12):789–790

Powers RE, Struble RG, Casanova MF, O'Connor DT, Kitt CA, Price DL (1988) Innervation of human hippocampus by noradrenergic systems: normal anatomy and structural abnormalities in aging and Alzheimer's disease. Neuroscience 25:401–417

Reisberg B, Ferris SH, de Leon MJ, Crook T (1982) The global deterioration scale for assessment of primary degenerative dementia. Am J Psychiatry 139:1136–1139

Seab JP, Jagust WS, Wong SFS, Roos MS, Reed BR, Budinger TF (1988) Quantitative NMR measurements of hippocampal atrophy in Alzheimer's disease. Magn Reson Med 8:200–228

Selkoe D, Kosik K (1983) Neurochemical changes with aging. In: Albert M (ed) The clinical neurology of aging. Oxford University Press, New York, pp 53–75

Terry RD, Peck A, De Teresa R, Schechter K, Horoupian DS (1981) Some morphometric aspects of the brain in senile dementia of the Alzheimer type. Ann Neurol 10:184–192

Correspondence: Dr. M. J. de Leon, Department of Psychiatry, New York University Medical Center, 550 First Avenue, New York, NY 10016, U.S.A.

Outline for the evaluation of nootropic drugs

S. Kanowski, G. Ladurner, K. Maurer, W. D. Oswald, and **U. Stein**

Hirnliga Heidelberg, Federal Republic of Germany

General recommendations

In the course of the discussion concerning proof of the clinical relevance of nootropic effects, which was started in 1975 by Kanowski, the following items have been generally adopted as a basis for agreement on the performance of clinical studies for evaluating the efficacy of nootropics:

a) Selection of the trial parameters and thus of the target criteria is to take place as directed by theory. Among other factors, as a prerequirement for this, only those areas of function can be taken into account as target criteria for which a change correlated with modifications in cerebral performance and/or syndromal modifications have been demonstrated in these fields.

b) Confirmatory analysis of data presupposes an a priori definition of the target criteria and the methods of statistical inference.

c) Evaluating the effect of the test substance is to be carried out on different planes which are independent of each other (e.g. physician, patient, nursing staff or objective test procedures, subjective assessment by third parties and everyday behaviour).

d) Complex target criteria (such as e.g. objective test procedures) are to be examined as regards the importance of therapy-induced changes for the everyday behaviour of the patient and his subjective state, that is: they are to be validated.

e) The measurement methods to be applied must meet the generally valid theoretical test requirements as to objectivity, reliability, validity and dimensionality, and be standardized or be in accordance with norms in the context of the age group to which a sample set of patients belongs.

f) Where findings are to be evaluated parametrically, the effects ought to involve approximately half of a standard deviation in order to be clinically relevant.

g) Non-parametric evaluation of the findings (e.g. in the form of the number of patients plus relevant effects) is less subject to influence through extreme values and, at the same time, makes easier recognition of the effects possible for the sample (of patients) examined.

h) Taking into account that in multicentre studies the drug effect may be conceiled or accentuated by differences in therapeutic settings, long duration of the study, social structures, meteorological influences, dietary habits etc. etc. a comparable specific effect being documented in all or at least a majority of the study centres involved would doubtlessly be accepted as an indicator for therapeutical efficacy of the drug.

General study design

To demonstrate the clinical efficacy of a nootropic, a randomized, double-blind study versus placebo (or active placebo) is a conditio sine qua non.

Comparison with a second substance known to be active is problematic in that very large patient numbers are required to represent a difference at an approximately equal potency. However it may be defined, a medicational equality is generally not accepted as a proof of efficacy.

Single vs multicentre testing

Single-centre tials present the fewest problems and should therefore be the preferred clinical testing strategy, provided that a sufficient number of patients is available. Multicenter trials may be performed by different ways.

a) Multiple, single-centre testing. This model presents the advantage of internal replication (confirmation by replication) as well as approximative assessment of the influence of varying conditions in therapeutical environment. The possibility of a sequential assessment is a further advantage which, nevertheless, must be established in the test plan prior to starting the study (Bauer, 1987). As the test protocol does not have to be exactly the same for each centre, room is still available for procedures and study methods specific to each centre.

b) Multicentre testing with one team of examiners. Quasi single-centre (single examiner model; comp. Oswald, 1987; Oswald and Oswald, 1988): minimization of the systematic error variance produced by different examiners which, in the case of evaluation scales especially, results in a marked reduction in reliability, consequently having a pronouncedly restrictive effect on both the sensitivity and the validity of test parameters.

In this model particularly, however, major logistic problems are yet to be solved.

c) Multicentre testing with multiple teams of examiners. Individual evaluation of each single centre is not possible due to restrictions in the contribuable number of patients.

This means that the impact of the variance in environment conditions depending on the centres in question cannot be estimated. To compensate for the variability of different examiners, a uniform training of the examining team and their monitoring during the entire study period is deemed obligatory.

Neurophysiological tests

The neurophysiologic and functional dynamic test methods cannot so far be used to proof the efficacy of nootropic drugs.

Nonetheless they provide valuable tools to analyse and describe various effects of these drugs. The following target variables can be determined:

EEG   a) Bioavailability
b) Determination of dosage window!
c) Assessment of vigilance
d) Speed components of cognitive information processing (e.g. P300)
e) Substance-specific action profiles

PET   a) Metabolic changes
b) Neutrotransmitterreceptor dynamics
c) Cerebral blood flow (CBF)
d) Oxygen extraction

SPECT a) Cerebral blood flow
b) Neurotransmitterreceptor dynamics

CBF   a) Cerebral blood flow

In the above procedures, all parameters measured can, in addition, be brought into relation with cognitive activation and/or deprivation of performance.

Patient selection criteria

Patient groups must exhibit the target symptoms to be improved by therapy with nootropics. In this case, patients suffering from a dementia in accordance with the criteria of the DSM IIIR or, respectively, the ICD 10 must be involved: this means that patients with a pseudodementia (depression), consciousness disturbances and primary intellectual retardation are to be excluded. The diagnostic procedure may be directed by the following decision-tree: For assuming the diagnosis of SDAT the NINCDS-ADRA criteria should be applied.

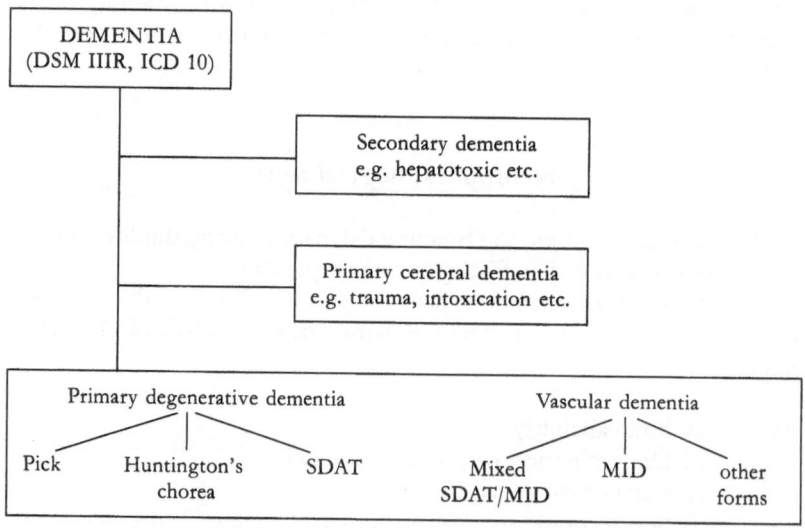

(Kanowski and Hedde, 1986)

According to the apparatus and logistics available, substrate-specific methods (such as CT, NMR and Doppler/Duplex) and functionally dynamic methods (EEG, PET, SPECT, CBF and Doppler) may also be applied over and beyond the differentiation given above.

Inclusion criteria

The diagnosis of dementia is to be undertaken clinically, whereby the extent of a cognitive disturbance and/or dementia must be defined via psychological tests. Furthermore, only patients with a primary degenerative dementia or a vascular dementia are to be included, in which case the other forms of dementia (secondary) are to be excluded either clinically or using laboratory and test apparatus methods.

As a rule, only patients with mild to medium symptoms ought to be included in clinical trials with nootropic substances, as no evidence has been presented up till now that these substances produce relevant effects in patients with more severe disturbances. Nonetheless, when defining inclusion criteria, the fact should be considered that, in the light of fundamental statistical considerations, inclusion parameters should not be applied simultaneously or evaluating therapeutic drug effects.

Patients with mild to medium forms of dementia (corresponding to the chronic organic brain syndrome) can be diagnosed, among other factors, in the context of their cognitive performance, this taking place via:

SCAG score from 40 through 90,
      with the mean value ≥ 3.5 in the items 2, 3, 6 and 8, total point value ≥ 24 in the items $1-8$, whereby none of these items have a point value of 7
SKT $9 \leq$ point value ≤ 18
BCRS $2 \leq$ point value ≤ 4, max. 5 on each scale
NAB $17 \leq$ test value ≤ 27
NAA $25 \leq$ test value ≤ 45.

Exclusion criteria

− Other severe mental conditions not in connection with dementia (major affective disorders, schizophrenia)
− Physical (non-mental) conditions which are capable of exerting, due to their degree of severity, a considerable influence on a patient's general condition, or which may result in a considerable change of the overall situation due to exacerbations or improvement.
  Steady state conditions can therefore be accepted when the above considerations are taken into account.
− The ability to communicate with the patient must be maintained as far as possible. An absence of his or her capacity to understand instructions, diseases affecting the sensory system considerably influencing communication perception or fine motor functions thus constitute exclusion criteria. In the same way, patients with aphasia, apraxia or other disturbances of the higher cerebral performance level which present major hindrance to communication, are not conducive to examination.
  Where cases of high-degree dementia are encountered, difficulties of the same kind may also arise, so that special target symptoms and

testing instrumentation have to be defined for such patients with
communication disturbances and/or high-degree dementias – if these
are to be tested in particular.

In the case of patients not capable of understanding, for example, the
Nuremberg Ageing Inventory (Nürnberger-Alters-Inventar) NAI
(Oswald and Fleischmann, 1986) and processing it without a set time
limit, the reliability and validity as well as the completeness of data
gathering cannot be guaranteed.

– Conditions not subject to control such as a patient's environment/
surroundings and life habits.

Comedication

a) Total exclusion:

   Other nootropics
   Stimulants

b) Relative exclusion:

   Hypnotics
   Sedatives
   Antidepressants
   Neuroleptics
   Tranquillizers

Insofar as those forms of medication belonging to the relative exclusion group cannot be dispensed with, the requirement must be made that
these be in a steady state throughout the entire duration of the study and
be uninterruptedly documented. This allows a stratified evaluation of
the study comparing the group of patients without any comedication to
the patients in whom comedication could not be avoided and its influence on the therapeutic results.

Observation planes

The variables aimed at in a test on efficacy should at all times include
the following areas, whereby only the first three come under consideration for an assessment of their *efficacy* at present. For the purpose of
complementation, the *effect* can be described and analyzed on the plane
of neurophysiology/functional dynamics.

– The psychopathological plane.
– The psychometric plane.

- The behavioural plane.
- The plane of neurophysiology/functional dynamics.

If significant changes are demonstrated versus placebo in one variable respectively from two of the above planes (1–3), which are to be evaluated independently (or in all target planes after corresponding alpha-adjustment), a proof of efficacy may then be assumed insofar as the study has been carried out under adequate adherence to the criteria listed under 1 a–h.

In the area of cognitive performance capacity, only those test procedures may as a result be applied which meet the requirements for internal and internal validity. This means that they must be correlated according to age, differentiated between normal and pathological changes in cerebral performance and, over and beyond this, be capable of reproducing those partial performances important for carrying out everyday activities (in the context of clinical relevance). These requirements are only fulfilled by test methods capable of being assigned to the areas listed for functions related to rapidity (of performance) and/or the secondary memory field.

Table 1. Shows, by way of example, the individual processes actually applied on the separate planes (for explanation of abbreviations see appendix, pages 539–540)

1 Psychopathologic plane		CGI AGP BCRS	SCAG MMS
2 Psychometric plane		NAS NAF BF-S ZVT-G ZS-G Multiple choice reaction	LT-G LATL BT SKT
3 Behavioural plane	a)	Activities of daily living NOSIE BGP	NAB GBB
	b)	Instrumental activities of daily living NAA	NAB
	c)	Behavioural assessment NAR	
	d)	Specific questioning AAT NAI	DT
4 Neurophysiologic/ functional dynamic plane		EEG PET Doppler	SPECT CBF

Clinical relevance

Does the clinically demonstrated therapeutical efficacy of a given nootropic produce an increase in a patient's ability to meet the demands of everyday existence? How can clinical relevance be operationalized to be taken as a basis in the clinical assessment of nootropics?

Different solutions to this problem are conceivable:

a) Agreement of therapeutical results on different planes of observations independent of each other could be considered as an indicator of clinical relevance (see Observation planes).

b) Percentage threshold value for the number of improved clinical symptoms coupled with a second percentage threshold value for the improvement of each individual symptom.

c) Number of improved patients as complementary measure.

d) Effect of active medicational treatment in long-term studies on the delay in progression of the disease concerned, using the placebo group as reference.

e) Improvement by one or two stages when using instruments such as Alzheimer staging according to Reisberg et al. (1982, 1983).

f) Improvement in competent behaviour (Quality of life), procedures already in common use as ADL and IADL activities as well as of life satisfaction (Israel et al., 1984) could thus become a starting point for initial concepts in the development of objective and subjective measurement procedures.

In this context, the question may also be asked as to whether the training of cognitive and everyday competence in effective combination with nootropic medication produces better results than each method by itself. Initial results by Yesavage et al.(1985) point in this direction and ought to stimulate further research in this direction.

Ethical and legal aspects

Under prevailing legal conditions, the following recommendations may be made:

1. The ability to understand explanations and give consent must be examined carefully in the case of each patient.
2. In spite of "dementia" being given as a diagnosis, patients capable of giving their consent may be included in therapeutical studies; here, as a rule, we are probably dealing with mild to medium cases of dementia.

3. If the ability to understand explanations and give consent is not present, and there is a legal guardianship for the corresponding duties, the person to whom such a guardianship case has been entrusted is able to give his/her assent to inclusion of the patient in a therapeutical study.
4. Patients without any ability to understand explanations and give consent and patients not subject to a legal guardianship as described cannot, unter present regulations, be included in any therapy studies (phase III and IV tests). We must here go by the assumption that legal counsellors are not, at present, willing to establish special "research care provisions" exclusively for this purpose.

Appendix

(Explanation of abbreviations of Table 1)

SCAG Sandoz Clinical Assessment Geriatric Scale
Venn et al. (1986)

SKT Syndrom-Kurztest
Erzigkeit (1986)

BCRS Brief cognitive rating scale
Reisberg et al. (1983)

CGI Clinical Global Impressions; National Institute of Mental Health; Vgl. CIPS (1986)

AGP Dokumentationssystem der Arbeitsgemeinschaft für Gerontopsychiatrie
Ciompi et al. (1986)

MMS Mini-Mental-State
Folstein et al. (1975)

NAS Nürnberger-Alters-Selbstbeurteilungs-Skala
Oswald and Fleischmann (1986)

NAF Nürnberger-Alters-Fragebogen
Oswald and Fleischmann (1986)

Bf-S Befindlichkeits-Skala
von Zerssen (1986)

ZVT-G Zahlen-Verbindungs-Test, a revised form of the Trail Making-Tests; Reitan (1958), vgl. auch Oswald and Roth (1987)

ZS-G Zahlen-Symbol-Test, a revised form of ZS,
Wechsler (1964)

LT-G Labyrinth-Test, a revised form of common Maze-Tests, e.g. von Chapuis (1959)

LATL Latentes Lernen
similar to Bromley (1958), Randt et al. (1984)

BT Bilder-Test, a revised form similar to Ferris et al. (1980), Wechsler (1945)

NOSIE Nurses' Observation Scale for Inpatient Evaluation
Honigfeld et al. (1986)
NAB Nürnberger-Alters-Beobachtungs-Skala from the NAI
Oswald and Fleischmann (1986)
BGP Beurteilungsskala für geriatrische Patienten,
Mol (1972), Kam et al. (1986), similar to BOP and BAP
GBB Geriatrischer Beurteilungsbogen
Schneider and Fisch (1981)
NAA Nürnberger-Alters-Alltags-Skala NAA from the NAI
Oswald and Fleischmann (1986)
NAR Nürnberger-Alters-Rating from the NAI
Oswald and Fleischmann (1986)
AAT Aachener Aphasie Test
Huber et al. (1982)
DT Demenztest
Markowitsch et al. (1988)
NAI Nürnberger-Alters-Inventar
Oswald and Fleischmann (1986)

References

Allard M, Signoret JL, Stalleicken D (1988) Alzheimer Demenz. Springer, Berlin Heidelberg New York Tokyo

Bauer P (1987) Sequentielle Tests von Hypothesen in aufeinanderfolgenden klinischen Studien. Biometrisches Seminar, Locarno, September 1987

Benton AL (1974) Der Benton-Test, 4. Aufl. Huber, Bern

Bromley DB (1958) Some effects of age on short-term learning and remembering. J Gerontol 13: 198–406

Chapuis F (1959) Der Labyrinth-Test. Huber, Bern

Ciompi L, Kanowski S, et al (1986) Dokumentationssystem der Arbeitsgemeinschaft für Gerontopsychiatrie. In: Collegium Internationale Psychiatriae Scalarum (ed) Internationale Skalen für Psychiatrie. Beltz Test GmbH, Weinheim

Committee for "Geriatric disease and asthenias" at BGA (1986) Impaired brain functions in old age. AMI-Heft 1/1986

Coper H, Kanowski S (1976) Geriatrika: Theoretische Grundlagen, Erwartung, Prüfung, Kritik. Hippokrates 47: 303–319

Coper H, Kanowski S (1983) Nootropika, Grundlagen und Therapie. In: Langer G, Heimann H (Hrsg) Psychopharmaka. Grundlagen und Therapie. Springer, Wien New York, S 409–430

Croog SH, Levine S, Brown B, et al (1986) Effects of antihypertensive therapy on the quality of life. N Engl J Med 14: 1657–1664

Engel RR (1987) Alzheimersche Demenz 1987. MMW 129 (26): 512–515

Erzigkeit H (1986) Syndrom-Kurz-Test, 2. überarb. Aufl. Vless, Ebersberg

Ferris SH, Crook T, Clark E, McCarthy M, Rae D (1980) Facial recognition memory deficits in normal aging and senile dementia. J Gerontol 35: 707–714

Fleischmann UM (1989) Gedächtnis und Alter. Multivariate Analysen zum Gedächtnis alter Menschen. Huber, Bern

Fleischmann UM, Oswald WD (1986) Nürnberger-Alters-Fragebogen NAF. In: Fleischmann UM, Oswald WD (Hrsg) Nürnberger-Alters-Inventar NAI. Universität Erlangen-Nürnberg, Nürnberg und Testzentrale des BDP, Stuttgart

Folstein MF, Folstein SE, McHugh PR (1975) Mini-Mental-State. A practical method for grading the cognitive state of patients for the clinician. J Psychiatr Res 12:189–198

Heimann N (1978) Medikamentöse Behandlung von zerebralen Zirkulations- und Nutritionsstörungen. Monatskurse für ärztliche Fortbildung 28:279–285

Helmchen H, Kanowski S, Koch H-G (1989) Forschung mit dementen Kranken: Forschungsbedarf und Einwilligungsproblematik. Ethik in der Medizin 1:83–98

Herrmann WM, Schuster J, Stille G (1981) Gedanken zur Phase-III-Prüfung zum Nachweis der therapeutischen Wirksamkeit von Gerontopharmaka. In: Oswald WD, Fleischmann UM (Hrsg) Experimentelle Gerontophysiologie. Beltz, Weinheim, S 23–35

Herrmann WM, Kern U (1987) Nootropika: Wirkungen und Wirksamkeit. Eine Überlegung am Beispiel einer Phase III-Prüfung mit Piracetam. Nervenarzt 58:358–364

Honigfeld G, Gillis RD, Klett CJ (1986) Nurses' observation scale for inpatient evaluation. In: Collegium Internationale Psychiatriae Scalarum (ed) Internationale Skalen für Psychiatrie. Beltz Test GmbH, Weinheim

Huber W, Poeck K, Weniger D, Willems K (1982) Der Aachener Aphasietest. Hogrefe, Göttingen

Israel L, Kozarevic D, Sartorius N (1984) Source book of geriatric assessment, vol 1 and 2. Karger, Basel München Paris

Johannson B, Berg S (1987) Memory span, ageing and terminal decline. The Third Congress of the International Psychogeriatric Association, Chicago 1987 (Book of Abstracts)

Kam P, van der Mol F, Wimmers MFGH (1986) Beurteilungsskala für geriatrische Patienten. In: Collogium Internationale Psychiatriae Scalarum (ed) Internationale Skalen für Psychiatrie. Beltz Test GmbH, Weinheim

Kanowski S (1975) Methodenkritische Überlegungen zur Prüfung von Geriatrika. Z Gerontol 5:316–322

Kanowski S (1986) Möglichkeiten und Grenzen der Therapie mit Nootropika. Hospitalis 56:400–409

Kanowski S, Hedde JP (1986) Arzneimittel für die Indikation hirnorganisch bedingter Leistungsstörungen (Nootropika). In: Dölle W, Müller-Oerlinghausen B, Schwabe U (Hrsg) Grundlagen der Arzneimitteltherapie. Bibliographisches Institut, Mannheim, S 154–171

Kanowski S, Fischhof P, Hiersemenzel R, Röhmel J, Kern U (1988) Wirksamkeitsnachweis von Nootropika am Beispiel von Nimodipin – Ein Beitrag zur Entwicklung geeigneter klinischer Prüfmodelle. Zeitschrift für Gerontopsychologie und -psychiatrie 1:35–44

Larrabee GJ, Crook T (1987) New techniques for assessing treatment effects in

Alzheimer's disease and age associated memory impairment. The Third Congress of the International Psychogeriatric Association, Chicago, 1987 (Book of Abstracts)

Lauter H, Möller HJ, Zimmer R (1986) Untersuchungs- und Behandlungsverfahren in der Gerontopsychiatrie. Springer, Berlin Heidelberg New York Tokyo

Lehmann E (1984) Practicable and valid approach to evaluate the efficacy of nootropic drugs by means of rating scales. Pharmacopsychiatrie 17: 71–75

Lehrl S (1986) Steigerung der geistigen Leistungsfähigkeit. Therapiewoche 36: 2585–2594

Lehrl S, Fischer B (1986) Steigerung der geistigen Leistungsfähigkeit im Alter. Nervenheilkunde 5: 173–181

Lindenberger U, Smith J, Baltes PB (1989) Das Altern der Intelligenz: Möglichkeiten und Grenzen. MMW 131: 93–96

Markowitsch HJ, Kessler J, Denzler P (1988) Demenztest. Hogrefe, Göttingen

Mol F (1972) Über die Anwendbarkeit und den diagnostischen Wert einer geriatrischen Meßmethodik. Gerontopsychiatrie 2: 185–210

National Institute of Mental Health (1986) Clinical global impressions. In: Collegium Internationale Psychiatriae Scalarum (ed) Internationale Skalen für Psychiatrie. Beltz Test GmbH, Weinheim

Oswald WD (1982) Alltagsaktivitäten und die Speed/Power Komponenten von Testleistungen. Z Gerontol 15: 11–14

Oswald WD (1986) Der Zahlen-Verbindungs-Test im höheren Lebensalter. In: Daumenlang K, Sauer J (Hrsg) Aspekte psychologischer Forschung. Hogrefe, Göttingen, S 377–388

Oswald WD (1987) Multizentrische klinische Prüfungen: Methodische Überlegungen. In: Coper H, Heimann H, Kanowski S, Künkel H (Hrsg) Hirnorganische Psychosyndrome im Alter III. Springer, Berlin Heidelberg New York Tokyo, S 101–108

Oswald WD (1988) Möglichkeiten und Grenzen der Psychometrie in der Psychogeriatrischen Forschung. Zeitschrift für Gerontopsychologie und -psychiatrie 1: 181–191

Oswald WD, Fleischmann UM (1985) Psychometrics in gerontological research: models, methods and results. In: Bergener M, Ermini M, Stähelin HB (eds) Thresholds in aging. Academic Press, London, pp 241–253

Oswald WD, Roth E (1987) Der Zahlen-Verbindungs-Test, 2. Aufl. Hogrefe, Göttingen

Oswald WD, Matejcek M, et al (1982) Über die Relevanz psychometrisch operationalisierter Therapie-Effekte bei der Behandlung altersbedingter Insuffizienzerscheinungen des Gehirns am Beispiel des Nürnberger-Alters-Inventars NAI. Arzneimittelforschung/Drug Research 32: 584–590

Oswald WD, Fleischmann UM (1986) Das Nürnberger-Alters-Inventar NAI. Hogrefe, Göttingen bzw. Testzentrale Stuttgart

Oswald WD, Oswald B (1988) Zur Replikation von Behandlungseffekten bei Patienten mit hirnorganischen Psychosyndromen im Multizenter-Modell als Indikator für klinische Wirksamkeit. Eine Placebo-kontrollierte Doppelblind-

Studie mit Pyritinol. Zeitschrift für Gerontopsychologie und -psychiatrie 3: 223–242

Oswald WD, Fleischmann UM (1989) Psychometric assessment in psychogeriatrics. In: Bergener M, Finkel I (eds) Clinical and scientific psychogeriatrics, vol. 2. The interface of psychiatry and neurology. Springer, New York, pp 29–44

Poon LM (1986) Handbook of the clinical memory assessment of older adults. American Psychological Association, Washington, DC

Randt CT, Brown ER, Osborne DP (1984) Randt memory test. New York University, Department of Neurology, New York

Reisberg B, Ferris SH, De Leon MJ, et al (1982) The global deterioration scale for assessment of primary degenerative dementia. Am J Psychiatry 139: 1136–1139

Reisberg B, London E, Ferris SH, Borenstein J, Scheier L, De Leon MJ (1983) The brief cognitive rating scale: language, motoric and mood concomitants in primary degenerative dementia (PDD). Psychopharmacol Bull 19: 702–708

Reitan RM (1958) Validity of the trail making test as an indicator of organic brain damage. Percept Mot Skills 8: 271–276

Satzger W, Engel RR (1987) Früherkennung dementieller Erkrankungen. MMW 129 (42): 746–749

Schneider HD, Fisch HP (1981) Der geriatrische Beurteilungsbogen (GBB): Ein Instrument zur Einschätzung des Alltagsverhaltens geriatrischer Patienten durch das Pflegepersonal. In: Oswald WD, Fleischmann UM (Hrsg) Experimentelle Gerontopsychologie. Beltz, Weinheim, S 116–130

Thomae H (1984) Lebenszufriedenheit. In: Oswald WD, Herrmann WM, Kanowski S, Lehr UM, Thomae H (Hrsg) Gerontologie. Kohlhammer, Stuttgart, S 271–275

Venn RD, Hamot HB, Shader RI (1986) SANDOZ Clinical Assessment – Geriatric. In: Collegium Internationale Psychiatriae Scalarum (ed) Internationale Skalen für Psychiatrie. Beltz Test GmbH, Weinheim

Wechsler D (1945) A standardized memory scale of clinical use. J Psychol 19: 87–95

Wechsler D (1964) Die Messung der Intelligenz Erwachsener. Huber, Bern

Yesavage JA, Westphal J, Rush L (1981) Senile dementia, combined pharmacologic and psychologic treatment. J Am Geriatr 29: 164–171

Von Zerssen D (1986) Befindlichkeits-Skala, Selbstbeurteilungs-Skala S. In: Collegium Internationale Psychiatriae Scalarum (Hrsg) Internationale Skalen für Psychiatrie. Beltz Test GmbH, Weinheim

Zimmer R, Kurz A, Lauter H (1987) Zur klinischen Relevanz nootroper Effekte. In: Coper H, Heimann H, Kanowski S, Künkel H (Hrsg) Hirnorganische Psychosyndrome im Alter III. Springer, Berlin Heidelberg New York Tokyo, S 54–61

Correspondence: Prof. Dr. S. Kanowski, Freie Universität Berlin, Reichsstrasse 15, D-1000 Berlin 19

Drug treatment of dementia

H. Coper

Department of Neuropsychopharmacology, Free University Berlin

Summary

In a brief review it is shown that historically drug treatment of Alzheimer's disease has addressed at first unspecific symptoms. The subsequent therapy with vasodilators reflected the understanding of the nature of the illness at the time. Another intellectual construction was derived from the amine deficiency hypothesis of depression (use of stimulants). The many subsequent efforts in the development of therapeutic progress based on morphological and biochemical deficits have demonstrated that all the simple cause/effect models are inappropriate. New concepts are matter for discussion.

"In line with the nature of the disease, the therapy of senile dementia usually has only a very narrow scope. The things that can be done are: physical care, supervising the often fragile and weak patients, regulating their way of life, especially regulation of diet and digestive problems, treatment of anxiety by administering small doses of opium, treatment of insomnia with bath therapy, prudent use of compresses, occasional administration of paraldehyde or Veronal. For delirious states of excitement one often needs the application of padded beds or prolonged baths as well as special diets with or without the addition of essential sedatives. On the other hand, for the calm form of dementia there is usually no need for institutionalization; the patient can be taken care of by his family or by a welfare worker".

This paragraph ends the chapter "Senile and pre-senile dementia" (Das senile und präsenile Irresein) in the textbook for medical students and physicians: *Psychiatry* written in 1910 by the famous German psychiatrist, Dr. Emil Kraepelin.

In 1986, in the summary of the book 'Treatment development strategies for Alzheimer's disease' Hollister deduces: "One can safely say 80 years after Alzheimer first described the disease that bears his eponym, that we still have very little understanding of it. To paraphrase Winston Churchill's famous statement concerning the Russians: AD is a riddle wrapped in an enigma wrapped in a mystery. This is not to say that we have not learned much about the disease, but rather that our attempts to understand its etiology and to treat it effectively still remain rather primitive" (Hollister, 1986).

But what have we learned and why, nevertheless, has the therapy remained unsatisfactory up to now?

Probably, there are many reasons for the decade-long standstill in the development of effective drugs against Alzheimer's disease. But one cause in particular seems to be worth mentioning, especially since this phenomenon has been rather the exception in the neurobiological research of the last 40 years. It is the fact that in no treatment strategies to improve impaired brain functions were outdated principles adhered to so permanently and with such remarkable perseverance as in those of organic brain syndrome and dementia. This has been the case even though the original premise could no longer be maintained.

The most impressive example of this is the use of drugs for improvement of the cerebral blood flow. From the knowledge of the fifties it was plausible that due to the development of arteriosclerosis in old age the blood supply to the brain becomes deficient and the resulting relative hypoxia affects the brain metabolism. As soon as less energy is at the brain's disposal an impairment of cognitive faculties should occur. Drugs that can improve the cerebral blood flow ought to bring about a favourable effect, especially since cerebral blood flow is in fact diminished in patients with dementia (Obrist, 1979).

However, in 1959 the premise for this was called into question (Lassen, 1959) and for nearly 20 years it has been clear that only 25% of dementia cases are of vascular origin (Tomlinson et al., 1970). In the meantime Rogers et al. (1986) could substantiate in a prospective study of seven years' duration that decreased cerebral blood flow precedes multi-infarct dementia, but follows senile dementia of Alzheimer type (see also Tachibana et al., 1984; Hoyer, 1986). Thus, drugs that should boost cerebral blood flow can be profitable only for a small group of patients. Independent of this limitation, when given orally and in the dosage recommended, none of the respective substances on the market has a therapeutic efficacy resulting from an increased cerebral blood flow (Hollister, 1986; Yesavage et al., 1979).

Later on, a number of these drugs were asserted to act by increasing brain metabolism, in particular the glucose utilisation. It would be of no benefit to present the many arguments against this interpretation (Coper et al., 1987; Duara et al., 1984; Szelies et al., 1986). But it is indisputable that the effect neither clarifies pathogenetic relationships nor substantiates the efficacy of the drugs, giving rise instead to false associations.

A look into the so-called "Red List" of 1988 – a selection of all drugs available in West Germany – shows that drugs which are intended to improve the cerebral blood flow or which use the glucose utilization as an argumentation corroborate the thesis mentioned above, that in this field there is little readiness to draw conclusions from inopportune information. Evidently, negative consequences like loss of credibility in the Pharmaceutical Industry and also in medical practice seem to be secondary. But even the sciences seem to be little worried about this untenable situation.

The use of psychostimulants was also primarily not based on the therapeutic efficacy of the drugs but rather on intellectual considerations. Referring to high fatigue in old people, their reduced drive and other mental infirmities, Jacobson (1958) attempted to utilize the cerebral adrenergic properties of stimulants as mild antidepressive supplementary medication in elderly depressive patients with decreased mental performances. This idea has been examined in manifold ways. But the asserted success could never be scientifically confirmed. All critical examinations have shown that stimulants only initially produce some positive reactions – if any at all. Subsequently, however, after six weeks, for example, there was rather an increase in depressive states, anxiety, restlessness and so on (for review see Loew and Singer, 1983). Nevertheless, in several countries Ritalin is still used for the indication discussed.

As opposed to the idle perseverance in unsatisfactory therapeutic experiences in Germany, the neurobiological research of dementia was turbulently set in motion in the early seventies in Great Britain and the USA. It was established that the amount of the enzyme cholineacetyltransferase (CAT), which is responsible for the synthesis of acetylcholine, was much lower in the brains of Alzheimer patients than in the brains of old people who have a reduced content of it as it is (Bowen et al., 1976; Bowen, 1981). The density of muscarinic receptors, however, was not found to be reduced to such an extent (Mash et al., 1985). Further findings on the structural, functional and molecular levels (Davies, 1985; Plotkin and Jarvik, 1985) have established the choline deficiency hypothesis and today nobody doubts that disturbances of cholinergic transmission have a share in causing the impairments in cognitive capacities (Bartus et al., 1985). In analogy to Parkinson's disease, it was presumed that if the signal-emitting apparatus is defective with the recognition site still intact in principle, it could be expected that direct or indirect stimulation of muscarinic receptors in the CNS could outmaneuver the impaired neurotransmission. The results with acetylcholine precursors (deanol, choline and lecithine), inhibitors of acetylcholinesterase such as physostigmine, or cholinomimetics such as areco-line, are disillusioning, but after all, not surprising (Bowen and Davison, 1981; Bartus et al., 1984; Mohs et al., 1985; Hollander et al., 1986).

1. The dopaminergic neurotransmission takes a more or less continuous course, the system is more circumscribed and in general more mod-ulating. Cholinergic neurons as an "output system", as it is termed by Davies (1985), have quite another pattern of activity. It can be characterized as more superordinating. It is impulse-dependent and therefore of a discontinuous structure. Thus, an attempt to repair the partially disturbed cholinergic transmission by stimulation of the receptor site with a simple overflow must fail and may also be the cause of the high incidence of adverse side effects and the narrow therapeutic window of some receptor agonists or inhibitors of acetylcholinesterase. Possibly in the future better cholinomimetics can be synthesized. At present it is discussed whether tetrahydroaminoacridine can be con-sidered as an advance in this field (Summers et al., 1986).
2. The second and main reason for failure of any decisive break-through arising from the cholinergic deficiency hypothesis, however, is due to oversimplification. This judgement is based on the fact that changes also in other noncholinergic transmitter or modulator systems occur with age and especially with dementia (Narang et al., 1986; Zornet-zer, 1986). However, if these alterations are again viewed in isolation they will also be unable to help to establish any significant relation-ship to particular symptoms of dementia or to be a basic idea for therapeutic strategies (Kopelman and Lishman, 1986).

With its limited possibilities for reacting by promotion or inhibition of impulse transportation for communication and selection of information, the CNS forms innumerable reaction patterns, the sum of which delin-eate a regulated balance. Therefore, the loss of several information carriers and of their mutual connections and decreased homeostatic stability have more consequences for disturbed cognitive performances than can be compensated for by substituting only one transmitter or modulator. From this it is safe to assume that, similar to neuroleptics and antidepressants, for the efficacy of drugs used in treating impaired brain functions not only one specific reaction is decisive. Instead, they must have a certain – in part even antagonistic – profile of action which is able to normalize a deviation of homeostasis.

Along these lines, studies on muscarinic receptor density (Pilch and Müller, 1988) and clinical results point to the superior effects of combi-nations e.g. of arecoline + tacrine or of piracetam with choline or with pentoxyphylline as compared to a mono-drug therapy (Bartus et al., 1981; Ferris et al., 1982; Parnetti et al., 1985; Bartus et al., 1986; Flood and Cherkin, 1988).

Furthermore, it is an accepted fact that "pure" anticholinergics such as scopolamine deteriorate cognitive performances, especially in old age (Drachmann and Leavitt, 1974; Ghoneim and Mewald, 1975; Drachmann, 1977; Davies, 1985; Broks et al., 1988). However, a remarkably large number of reports describe first the risk of confusional and delirious states occurring in old patients treated with antidepressants having anticholinergic properties (Branconnier and Cole, 1981; Breyer-Pfaff and Gaertner, 1987; Kalinowsky and Hippius, 1969). The same applies to elderly patients suffering from Parkinson's disease who received anticholinergics like biperiden or trihexyphenidyl (DeSmet et al., 1982). Only occasionally, and neither regularly nor systematically, were strong effects on memory, orientation or concentration, emphasized (Perlick et al., 1986; Sadeh et al., 1982). These drugs are not "pure" anticholinergics but are simultaneously antihistamines or interfere additionally with the adrenergic or serotonergic systems.

Such reflections, which we have entertained for several years, can probably widen and supplement the hitherto existing concepts and eventually present new therapeutic approaches.

For a long time in the various theories about aging and especially about dementia the question has been ignored as to what extent individual basic components of the neuronal system and their interactions must be preserved so as to guarantee the upkeep of cognitive performances. There is still no convincing answer. But in the last decades it has increasingly been deliberated that the upkeep of nearly all centrally regulated functions is guaranteed in many different ways. This principle permits every creature to adapt constantly to its environment. By numerous compensatory mechanisms, as a rule, sufficient stability, plasticity, and capacity of structures and molecular reactions, exist to maintain homeostasis even under stressed and unfavorable conditions. In this way the continuity during ontogenesis and characteristic reactions to actual situations are always available. But with progressive age the ability to adapt diminishes. This decrease is probably the most important deficiency of ageing and impaired brain functions (Coper et al., 1986). With great interindividual variability the central information processing becomes slower and possibly more incorrect. The norm range narrows and is less flexible, the compensation stability and capacity smaller. The response to exogenous stimuli sometimes appears in surplus, sometimes fails to occur at all, when under high demands or stress adaptive properties are mostly insufficient so that the reaction to particular performance is overtaxed. But remarkably, cognitive infirmities can partly be compensated for by practice, that means, the organism still has a utilizable reserve even in old age (Baltes and Willis, 1982).

Within this concept the nootropics have at least a heuristic value. As a class of their own they have been the subject of discussion for nearly 10 years. The name was introduced in 1972 by Giurgea to describe some special effects of piracetam in animal experiments. Later on the term was taken to comprise a group of substances with a similar sphere of action but different mode of action.

The profile of this group consists in a normalization or stabilization of impaired homoeostastic functions of the central nervous system (Coper and Herrmann, 1988). It includes the restoration of reduced activation and the elimination of premature fatiguability and distractability associated with an improvement of performance and emotional stability. The therapeutic benefit of nootropics was evaluated in 1986 by the "Committee for Geriatric Diseases and Asthenias" at the Health Bureau of the Federal Republic of Germany. The experts did not accept a general efficacy for organic brain syndrome or senile dementia. But it was attested that at least a percentage of patients benefit from treatment with hydergine, piracetam and pyritinol. It must be taken into consideration, however, that often improvement did not occur before 6–12 weeks and that for hydergine higher doses than the usual 1 mg were necessary (Yoshikawa et al., 1983). Furthermore, it is true that the presupposition for an efficient medication is not yet clarified. The same is evident for the mode of functioning. At present great interest has centered around the development of more effective nootropics and there is some hope that such drugs will be available soon.

In a short overview, it is impossible to report on all efforts to improve the therapy of Alzheimer's disease (Crook et al., 1986; Hock, 1987). With the rapid development of new knowledge such completeness is not even necessary. Instead, one should remember that it has taken nearly 80 years to be engaged in this field so intensively. Compared to the history of chemotherapy, regarding the current drug treatment of Alzheimer's disease today, I believe, we are now in the stage of prontosil, the first sulfonamide of clinical applicability. But we can hope that we will need much less time to acquire corresponding effective drugs for dementia than it took to acquire the broad-range antibiotics available today.

References

Baltes PB, Willis SL (1982) Plasticity and enhancement of intellectual functioning in old age: Penn State's adult development and enrichment program (ADEPT). In: Craik FIM, Trehub SE (eds) Aging and cognitive processes. Plenum Press, New York, pp 353–389

Bartus RT, Dean RL, Beer B (1984) Cholinergic precursor therapy for geriatric cognition: its past, its present and a question of its future. In: Ordy JM, Harmann D, Alfin-Slater RB (eds) Nutrition in gerontology. Raven Press, New York, pp 191–225

Bartus RT, Dean RL, Fisher SK (1986) Cholinergic treatment for age related memory disturbances: dead or barely coming of age. In: Crook T, Bartus R, Ferris St, Gershon S (eds) Treatment development strategies for Alzheimer's disease. Mark Powley, Connecticut, pp 421–450

Bartus RT, Dean RL, Pontocorvo M, Flicker C (1985) The cholinergic hypothesis – a historical, current perspective and future direction. Ann NY Acad Sci 444:332–358

Bartus RT, Dean RL, Sherman KA, Friedman E, Beer B (1981) Profound effects of combining choline and piracetam on memory enhancement and cholinergic function in aged rats. Neurobiol Aging 2:105–111

Bowen DM (1981) Alzheimer's disease. In: Davison AN, Thompson RMS (eds) The molecular basis of neuropathology. Edward Arnold, London, pp 64–69

Bowen DM, Davison AN (1981) The neurochemistry of ageing and senile dementia: In: Matthews WB, Glaser GH (eds) Recent advances in clinical neurology, vol 3. Churchill Livingstone, Edinburgh

Bowen D, Smith C, White P, Davison AN (1976) Neurotransmitter-related enzymes and indices of hypoxia in senile dementia and other abiotrophics. Brain 99:459–496

Branconnier RJ, Cole JO (1981) Effects of acute administration of trazodon and amitriptylin on cognition, cardiovascular function, and salivation in the normal geriatric subject. J Clin Psychopharmacol 1:82S–88S

Breyer-Pfaff U, Gaertner HJ (1987) Antidepressiva – Pharmakologie, therapeutischer Einsatz und Klinik der Depression. Wissenschaftliche Verlagsanstalt, Stuttgart

Broks P, Preston GC, Traub M, Poppleton P, Ward C, Stahl SM (1988) Modelling dementia: effects of scopolamine on memory and attention. Neuropsychologia 26(5):685–700

CGDA (1986) The Committee for "Geriatric diseases and asthenias" at BGA: impaired brain functions in old age. AMI-Hefte, 1. Bundesgesundheitsamt, Berlin

Coper H, Schulze G (1987) Charakterisierung und Wirkungsmechanismen von Nootropika. In: Coper H, Heimann H, Kanowski S, Künkel H (Hrsg) Hirnorganische Psychosyndrome im Alter III. Springer, Berlin Heidelberg New York Tokyo, S 3–10

Coper H, Herrmann WM (1988) Psychostimulants, analeptics, nootropics. An attempt to differentiate and assess drugs designated for the treatment of impaired brain functions. Pharmacopsychiatry 21:211–217

Coper H, Jänicke B, Schulze G (1986) Biopsychological research on adaptivity across the life span of animals. In: Baltes PB, Featherman DL, Lerner RM (eds) Life-span development and behavior, vol 7. LEA, New Jersey, pp 207–232

Crook T, Bartus RT, Ferris ST, Gershon S (1986) Treatment development strategies for Alzheimer's disease. Mark Powley, Connecticut

Davies P (1985) A critical review of the role of the cholinergic system in human memory and cognition. Ann NY Acad Sci 444:212–217

De Smet Y, Ruberg M, Serdan M, Dubois B, L'Hermitte F, Agid Y (1982) Confusion, dementia and anticholinergics in Parkinson's disease. J Neurol Neurosurg Psychiatry 45:1161–1164

Drachman DA, Leavitt J (1974) Human memory and cholinergic system. A relationship to aging? Arch Neurol 30:113–121

Drachmann DA (1977) Memory and cognitive function in man: does the cholinergic system have a specific role? Neurology 27:783–790

Duara R, Grady C, Haxby J, Ingar D, Sokoloff L, Margolin RH, Mauring RG, Cutler NR, Rapoport SH (1984) Human brain glucose utilization and cognitive function in relation to age. Ann Neurol 16:702–713

Ferris SH, Reisberg B, Crook T (1982) Pharmacologic treatment of senile dementia: choline, L-dopa, piracetam, and choline plus piracetam. In: Corkin S (ed) Alzheimer's disease: a report of progress. Raven Press, New York, pp 475–481

Flood JF, Cherken A (1988) Effect of acute arecoline, tacrine and arecoline + tacrine post-training administration on retention in old mice. Neurobiol Aging 9:5–8

Ghoneim MM, Mewaldt SP (1975) Effects of diazepam and scopolamine on storage, retrieval and organizational processes in memory. Psychopharmacologia 44:257–262

Giurgea CE (1972) Vers une pharmacologie de l'activité intégrative du cerveau. Tentative du concept nootrope en psychopharmacologie. Actual Pharmacol 25:115–157

Hock FJ (1987) Drug influences on learning and memory in aged animals and humans. In: Mendlewicz J, Pull, Janke W, Künkel H (eds) Biological psychology/pharmacopsychology. Karger, Basel, pp 145–160

Hollander E, Mohs RC, Davis KL (1986) Cholinergic approaches to the treatment of Alzheimer's disease. Br Med Bull 42:97–100

Hollister LE (1986) Summary and conclusions. In: Crook Th, Bartus RT, Ferris St, Gershon S (eds) Treatment development strategies for Alzheimer's disease. Mark Powley, Connecticut, pp 671–677

Hollister LE (1986) Drug therapy of Alzheimer's disease: realistic or not? Prog Neuropsychopharmacol Biol Psychiatry 10:439–446

Hoyer S (1986) Senile dementia and Alzheimer's disease. Brain blood flow and metabolism. Prog Neuropsychopharmacol Biol Psychiatry 10:447–478

Jacobsen A (1958) The use of Ritalin in psychotherapy of depression of the aged. Psychiatr Q 32:475–483

Kalinowsky LB, Hippius H (1969) Pharmacological, convulsive and other somatic treatments in psychiatry. Grune & Stratton, New York

Kopelman MD, Lishman WA (1986) Pharmacological treatments of dementia (non cholinergic). Br Med Bull 42:101–105

Kraepelin E (1910) Das senile und präsenile Irresein; in Psychiatrie. Ein Lehrbuch für Studierende und Ärzte. Verlag Johann Ambrosius Barth, Leipzig

Lassen NA (1959) Cerebral blood flow and oxygen consumption in man. Physiol Rev 39:183–238

Loew DM, Singer JM (1983) Stimulants and senility. In Cresse I (ed) Neurochemical behavioral and clinical perspectives. Raven Press, New York, pp 237–268

Mash DC, Flynn DD, Potter LT (1985) Loss of M_2 muscarine receptors in the cerebral cortex in Alzheimer's disease and experimental cholinergic denervation. Science 228:1115–1117

Mohs RC, Davis BM, Greenwald BS, Mathé AA, Johns CA, Horvath ThB, Davis KL (1985) Clinical studies of the cholinergic deficit in Alzheimer's disease. II. Psychopharmacologic studies. Am Geriatr Soc 33:749–757

Narang PK, Cutler NR (1986) Pharmacotherapy in Alzheimer's disease: basis and rationale. Prog Neuropsychopharmacol Biol Psychiatry 10:519–531

Obrist WD (1979) Cerebral circulatory changes in normal aging and dementia. In: Hoffmeister F, Müller D (eds) Brain function in old age. Springer, Berlin Heidelberg New York, pp 278–287

Parnetti L, Ciuffetti G, Mercuri M, Senin U (1985) Haemorheological pattern in initial mental deterioration: results of a long-term study using piracetam and pentoxifylline. Arch Gerontol Geriatr 4:141–155

Perlick D, Stastny P, Katz I, Mayer M, Mattis St (1986) Memory deficits and anticholinergic levels in chronic schizophrenia. Am J Psychiatry 143:230–232

Pilch H, Müller WE (1988) Piracetam elevates muscarinic cholinergic receptor density in the frontal cortex of aged but not of young mice. Psychopharmacology 94:74–78

Plotkin DA, Jarvik LF (1985) Cholinergic dysfunction in Alzheimer disease: cause or effect? In: van Ree JM, Matthysse S (eds) Prog Brain Res 65:91–103

Rogers RL, Meyer JS, Mortel KF, Mahurin RK, Judd BW (1986) Decreased cerebral blood flow precedes multi-infarct dementia, but follows senile dementia of Alzheimer type. Neurology 36:1–6

Sadeh M, Braham J, Modau M (1982) Effects of anticholinergic drugs on memory in Parkinson's disease. Arch Neurol 39:666–667

Summers WK, Majovski LV, Marsh GM, Tachiki K, Kling A (1986) Oral tetrahydroaminoacridine in long-term treatment of senile dementia, Alzheimer-type. N Engl J Med 315:1241–1245

Szelies B, Herholz K, Pawlik G (1986) Zerebraler Glukosestoffwechsel bei präseniler Demenz vom Alzheimer Typ – Verlaufskontrolle unter Therapie mit muskarinergem Cholinagonisten. Fortschr Neurol Psychiatr 54:364–373

Tachibana H, Meyer JS, Kitagawa Y, Rogers RL, Okayasu H, Mortel KF (1984) Effects of aging on cerebral blood flow in dementia. J Am Geriatr Soc 32:114–120

Tomlinson BE, Blessed G, Roth M (1970) Observation on the brains of demented old people. J Neurol Sci 11:205–242

Yesavage JA, Tinkelenberg JR, Berger PA, Hollister LE (1979) Vasodilators in senile dementia: a review of the literature. Arch Gen Psychiatry 36:220–223

Yoshikawa M, Hirai S, Aizawa T, Kuroiwa Y, Goto F, Sotue I, Idia M, Toyokura Y, Yamamura H, Iwasaki Y (1983) A dose response study with dihydroergotoxine mesylate in cerebrovascular disease. J Am Geriatr Soc 31:1–7

Zornetzer StF (1986) The noradrenergic locus coeruleus and senescent memory dysfunction. In: Crook T, Bartus R, Ferris St, Gershon S (eds) Treatment development strategies for Alzheimer's disease. Mark Powley, Connecticut, pp 337–359

Correspondence: Dr. H. Coper, Institute of Neuropsychopharmacology, Free University Berlin, Ulmenallee 30, D-1000 Berlin 19.

Cognitive enhancing properties of antagonist β-carbolines: new insights into clinical research on the treatment of dementias?

T. Duka, D. N. Stephens, and **R. Dorow**

Research Laboratories of Schering AG, Berlin

Summary

In animal studies, ZK 93 426 antagonised scopolamine-induced decrements in certain tests measuring memory or vigilance, but did not show any effect when given on its own. On the other hand, studies in humans indicated that ZK 93 426 selectively improves delayed recall of pictures shown after its administration. Additionally, evidence has been adduced for an improvement of performance in attentional tasks induced by ZK 93 426.

In a memory test sensitive to scopolamine effects (word list), ZK 93 426 had neither an effect on its own nor did it antagonise scopolamine induced decrements. Further studies in combination with scopolamine are required to clarify this apparent contradiction between the animal and the human data.

Introduction

Although, or perhaps because there is as yet no treatment for dementing diseases, a major research effort is being made to develop therapies, especially for senile dementia of the Alzheimer type. Because of the evidence that one of the earliest changes in SDAT is a loss of cholinergic neurones, the most popular strategy is to develop and test pharmacological probes aimed at improving cholinergic neurotransmission, either, in analogy with the use of L-DOPA as a precursor to stimulate the synthesis of dopamine in Parkinson's disease, by supplying choline or other precursors of acetylcholine synthesis, or by developing muscarinic agonists to stimulate directly muscarinic receptors. To date, neither of these approaches has proved successful, the latter possibly because currently available muscarinic drugs may penetrate poorly into the brain, or because those that do enter the brain in an adequate way may only poorly activate second messenger systems. We have adopted a third approach, to attempt to increase cholinergic neurotransmission by reducing inhibitory influences on surviving cholinergic systems (Sarter et al., 1988).

GABA is the major inhibitory neurotransmitter in the brain, and appears to regulate cholineric neurotransmission at two levels. Firstly, cholinergic cell bodies in the basal forebrain area of Meynert receive a rich innervation of GABA neurones, and express a high density of GABA/benzodiazepine receptors (Sarter and Schneider, 1988; Turner et al., 1990). Secondly, cholinergic nerve terminals in the cortex are subjected to inhibition from cortical GABA interneurones. It might therefore be speculated that cholinergic neurotransmission could be enhanced by reducing the inhibitory effects of GABA at either location. Since inverse agonist benzodiazepine receptor offer a powerful tool for reducing the effects of GABA on chloride channel activity, and because benzodiazepines themselves impair cognitive function by potentiating GABA effects, we have investigated the effects of certain inverse agonists from the β-carboline substance class on cognitive performance. Since certain substances from this class are known to possess convulsant properties in animals and to induce anxiety in both animals and humans, we have concentrated on a single substance, ZK 93426, which has been characterised in biochemical and pharmacological studies in animals as a benzodiazepine receptor antagonist with some weak partial inverse agonist components (Jensen et al., 1987). In early studies in volunteers, we found that ZK 93426 does not induce abnormal EEG characteristics, nor induce an anxiety state (Duka et al., 1987, 1988).

Material, methods and results

Evidence from animals

Effects of ZK 93426 on cognitive function of animals has been found in a number of tests (see Stephens et al., 1990) including antagonism of scopolamine effects on vigilance (Stephens and Sarter, 1988), passive avoidance performance (Jensen et al., 1987), and spontaneous alternation (Sarter et al., 1988), and amelioration of spatial learning deficits induced by lesions of the basal forebrain (Sarter and Steckler, 1989; Hodges et al., 1989). Only three examples will be given here as illustration of these effects.

NMRI mice introduced for the first time into a Y-maze explore the maze systematically, tending to enter each of the three arms in turn. This pattern of exploration is known as spontaneous alteration, and has been hypothesised to depend upon an intact spatial working memory (Sarter et al., 1988). Spontaneous

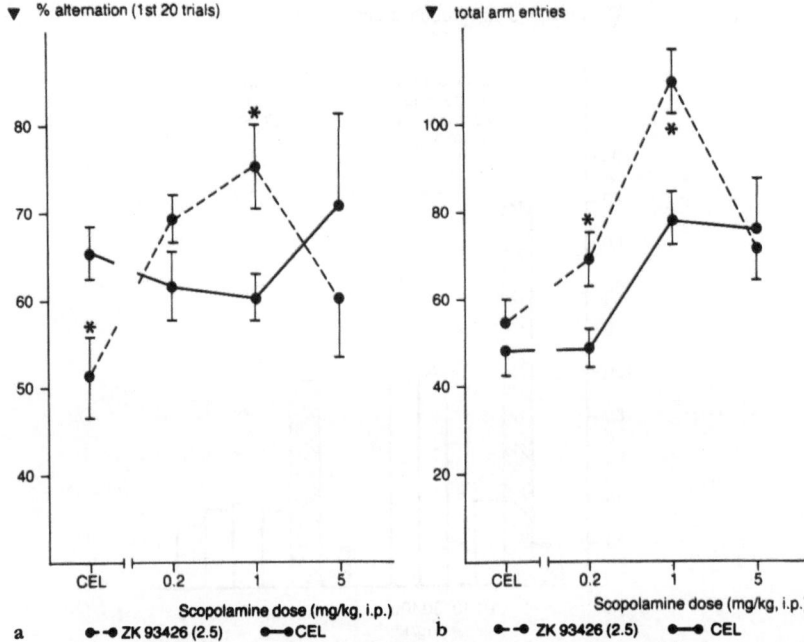

Fig. 1. The effects of the antagonist β-carboline ZK 93 426 (2.5 mg/kg i.p.) in interaction with vehicle (cremophor EL) or scopolamine in different doses on spontaneous alteration performance (**a**) and total arm entries (**b**); * p < 0.05 (Dunnet test)

alteration peformance is disrupted by the muscarinic antagonist scopolamine at low doses, but at higher doses the animals adopt a forced stereotyped pattern of behaviour resulting in an apparent systematic visiting of the three arms (Anisman, 1975, and Fig. 1). ZK 93 426 at a dose of 2.5 mg/kg antagonised the effect of scopolamine on alternation behaviour while increasing the effects of scopolamine on activity (Fig. 1).

When mice are exposed in a two compartment box to foot shock on entering one compartment (the "punished compartment"), on retesting up to several weeks later they avoid entering this compartment. With this paradigm we have tested the effects of ZK 93 426 in mice made amnesic either by treatment with scopolamine (3 mg/kg) or by corneal electroshock (7.5 mA for 2 sec) given following the acquisition. As can be seen in Fig. 2, ZK 93 426 in the doses of 10, 30 and 100 mg/kg antagonised the amnesic effects of scopolamine but not of the corneal electroshock.

In order to test possible vigilance enhancing properties of ZK 93 426 we used a method available for measuring vigilance in animals according to signal detection analysis. In this paradigm the animal "reports" whether a signal has been detected by pressing a bar. Operation of the bar during the signal (correct responses, 'hits') leads to food reward. Operation of the bar in the absence of signal (false alarm; 'misses') does not deliver food and delays signal onset by a

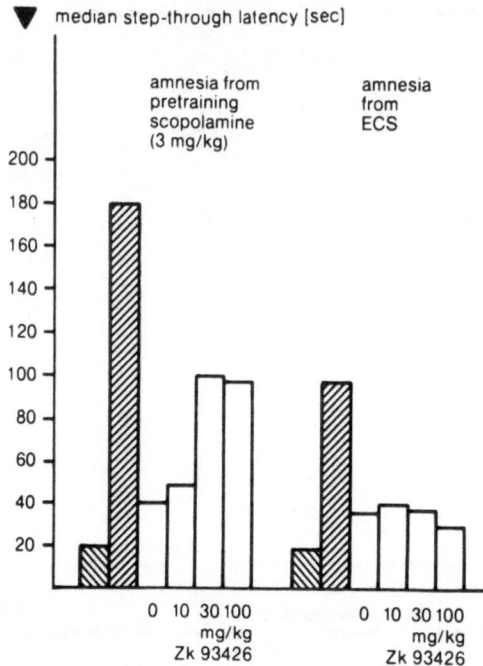

Fig. 2. The effects of different doses of ZK 93 426 injected i.p. on scopolamine-(3 mg/kg i.p.) or corneal electroshock-induced amnesia (decrease in step through latency on retesting day); 30 and 100 mg/kg i.p. of ZK 93 426 were found to antagonise the effect of scopolamine (p < 0.05, Dunnet test)

variable time. From this data, estimates can be made both of the signal detectability and of the tendency of a particular animal to operate the lever at a low or high detection criterion. The response probabilities are used to estimate parameters of sensitivity to the signal (vigilance) and of the tendency to respond, according to the formulae of Grier (1971).

In the experiment demonstrated here signal detection performance was impaired through treatment with scopolamine (0.6 mg/kg). Figure 3 shows that ZK 93 426 in the dose of 10 mg/kg partially antagonised the ability of scopolamine to impair indices of signal detectability (A') calculated on the basis of a formula of Grier (1971).

Evidence from humans

The effects of ZK 93 426 on memory have been tested in a study with healthy males applying tests which have been used to demonstrate effects of benzodiazepine agonists and antagonists (Ott et al., 1988; Dorow et al., 1988). Another

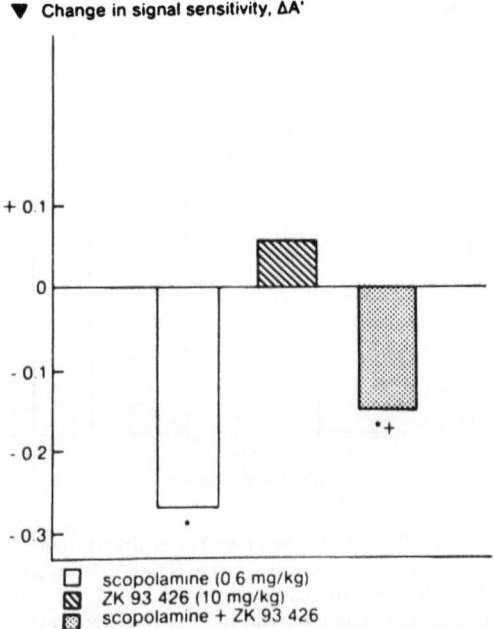

Fig. 3. The effect of ZK 93426 on scopolamine-induced decrease in signal sensitivity. Symbols: * p < 0.05 versus saline, + p < 0.05 versus scopolamine (Student' t-test)

objective of this study was to investigate whether ZK 93426 antagonised the effects of scopolamine on memory. Lists of 6 words were presented according to Buschke restrictive reminding technique (Buschke and Altman Fuld, 1974) and acquisition and learning was evaluated. Visual memory was tested by projecting pairs of simple pictures onto a screen for 5 sec; acquisition was ensured by immediate recall and rehearsal prevented by requiring the subjects to perform simple arithmetic tasks. 2 hours later recall and recognition was tested. Four pairs of pictures and a 6-word list, representing one session, were presented every 15 min for 1 h (five sessions). Scopolamine or saline was administered before any presentation; 60 minutes later the first 6-word list and 4 pairs of pictures were shown; after the second set of presentation (second session) either ZK 93426 (0.04 mg/kg) or the same volume of vehicle (Intralipid®) was administered and 3 more sets were presented to the subjects.

As is shown in Table 1, scopolamine delayed acquisition; ZK 93426 neither had an effect on its own nor antagonised the scopolamine effect. In the test of delayed recall which took place 2 hours later, ZK 93426, as is shown in Fig. 4, improved performance for the pictures shown after drug application but impaired performance for the pictures shown before. Both these effects are opposite to the effects on benzodiazepines on similar memory tests. Scopolamine showed no clear effect on this test.

T. Duka et al.

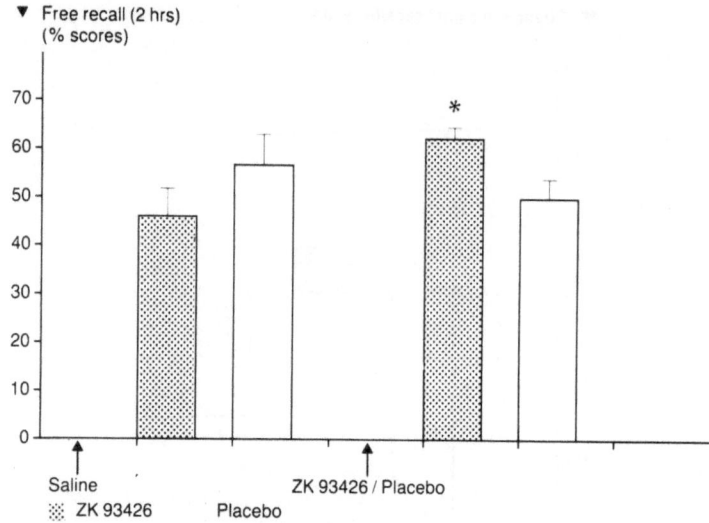

Fig. 4. The effects of ZK 93 426 (0.04 mg/kg i.v.) or vehicle (Intralipid ®) combined with saline sc injection on memory performance during which delayed recall of pictures presented before or after ZK 93 426 or vehicle injection is tested. 100% performance before ZK 93 426 or vehicle administration represents recall of 16 pictures whereas after ZK 93 426 or vehicle injection 24 pictures. * p < 0.07 compared to placebo

Table 1. Acquisition score (mean ± SD). The effects of scopolamine or saline alone and in combination with either ZK 93 426 or Placebo (Intralipid ®) on acquisition score during word-list learning according to Buschke restrictive reminding technique. Acquisition score: Times of word-reading during restrictive reminding. Sessions 1 and 2 take place after either scopolamine or saline injection, sessions 3, 4 and 5 follow the injection of either ZK 93 426 or placebo

Treatments	Sessions				
	1	2	3	4	5
Saline + Placebo	2 ± 1	3 ± 2	3 ± 2	1 ± 1	4 ± 2
Saline + ZK 93 426	2 ± 1	2 ± 1	2 ± 1	3 ± 2	3 ± 1
Scopolamine + Saline	3 ± 2	5 ± 5	5 ± 4	5 ± 4	7 ± 7
Scopolamine + ZK 93 426	3 ± 2	4 ± 2	5 ± 3	4 ± 2	6 ± 4

In another study the effects of ZK 93 426 were tested employing continuous attention tasks. Data will be presented here from an auditory vigilance task (Wilkinson, 1968; Tiplady, 1985) and a visual vigilance task (Tiplady, 1985). Subjects were trained the day before the study to reach equal basal response in the attentional tasks. On the day of the study subjects received 0.04 mg/kg ZK 93 426 or placebo intravenously and 10 min later performed on the auditory

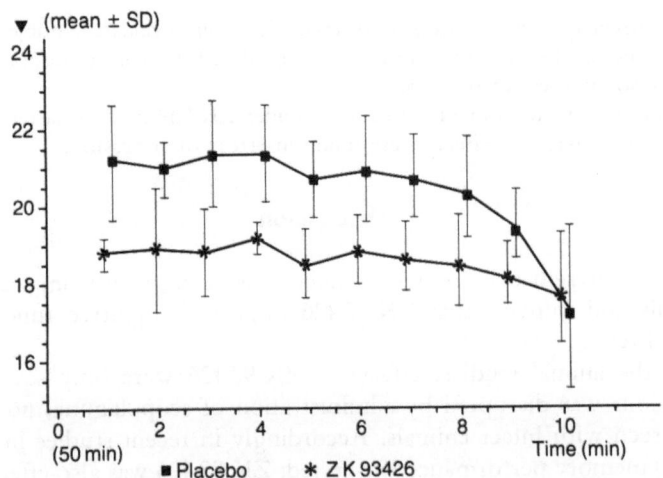

Fig. 5. The effects of ZK 93426 (0.04 mg/kg i.v.) or vehicle (Intralipid®) on the time required to a correct response in a visual vigilance task. Performance was found to be significantly different between the two treatments (p < 0.07 MANOVA)

Table 2. The effects of ZK 93426 (0.04 mg/kg i.v.) or vehicle (Intralipid®) on the errors of commission [median, (min−max)] during performance in a continuous attention task, the auditory vigilance task

Treatment	Time (min) performing				
	0−10 min	10−20 min	20−30 min	30−40 min	Total
ZK 93426 (0.04 mg/kg/i.v.)	0 (0−3)	1 (0−3)	0 (0−1)	0 (0−1)	1 (0−5)*
Placebo	0 (0−2)	0 (0−3)	0 (0−3)	0.5 (0−3)	2.5 (1−5)

* p < 0.05 compared to placebo (Mann-Whitney U-test)

vigilance task for 40 min; following this task subjects were tested on the visual vigilance task which lasted 10 min. During the auditory vigilance task, subjects had to detect a signal (400 ms tone) out of a background (tones of 500 ms). Number of correct responses and errors of comission were evaluated. Subjects were trained the day before until in a 20 min text they made at least 25 correct responses with no more than 1 error of comission.

As can be seen from Table 2 subjects who received ZK 93426 performed with fewer errors of comission. No effects were found concerning number of correct responses.

In the visual vigilance task subjects had to detect a signal (gradual appearance of a shape) out of a background noise (the 'snow' on a TV monitor). The time for a correct response and errors of comission were evaluated. Subjects were trained the day before to a criterion with the time for a correct response more than 16 sec and less than 22 sec.

As can be seen from Fig. 5 subjects under ZK 93426 were faster in their correct responses. No effects were found in errors of comission.

Discussion

In the present brief review evidence has been given from studies in animals and humans that ZK 93426 improves cognitive function in several tests.

In the animal studies, effects of ZK 93426 were only seen when behaviour was disrupted by administration of scopolamine; no effects were seen with intact animals. Accordingly in recent studies in which spatial memory performance was tested, ZK 93426 was also effective in improving performance in animals subjected to basal forebrain lesions but not in the sham operated animals (Sarter and Steckler, 1989).

The data from the human studies were consistent with ZK 93426 exerting effects on memory and vigilance opposite to those of benzodiazepines. In the study investigating memory performance in which scopolamine was given, ZK 93426 did not antagonise the scopolamine induced disruption of performance (Duka et al., in preparation). Lack of such an effect of ZK 93426 could be for several reasons: firstly, the dose chosen either of scopolamine or of ZK 93426 may not have been the right one; for instance in the data from animals presented here (Fig. 1) the dose response curve of scopolamine-induced disruptive behaviour was of U-shape, scopolamine not being active in higher doses. In another study in which different doses of ZK 93426 were used to antagonise the disruptive effects of scopolamine (1 mg/kg ip) (Sarter et al., 1988), the dose response curve of ZK 93426 was biphasic, ZK 93426 not being effective in higher doses; secondly, it could be that the cognitive tests used were not the appropriate ones to detect such an interaction. For instance, scopolamine-disruptive effects were seen in a words list procedure, during which challenges of memory are more related to perceiving and processing the information rather than encoding and storing it. In accordance the effects of ZK 93426 on its own were found in those memory tests (delayed recall) in which encoding and storage were the memory functions challenged.

Unfortunately ZK 93426 itself cannot be developed as a cognitive enhancer to be used in the clinic, due to its poor bioavailability in people. Nevertheless it promises to be the forerunner of class of substances which could be beneficial for the treatment of memory and other cognitive disorders.

References

Anisman H (1975) Dissociation of disinhibitory effects of scopolamine: strain and task factors. Pharmacol Biochem Behav 3:613–618

Buschke H, Altman Fuld P (1974) Evaluating storage, retention and retrieval in disordered memory and learning. J Verb Learn Verb Behav 12:543–550

Dorow R, Berenberg D, Duka T, Sauerbrey N (1987) Amnestic effects of lormetazepam and their reversal by the benzodiazepine antagonist Ro 15-1788. Psychopharmacology 93:507–514

Duka T, Goerke D, Dorow R, Höller L, Fichte K (1988) Human studies on the benzodiazepine receptor antagonist β-carboline, ZK 93426: antagonism of lormetazepam's psychotropic effects. Psychopharmacology 95:463–471

Duka T, Stephens DN, Krause W, Dorow R (1987) Psychotropic activity of the benzodiazepine receptor antagonist β-carboline, ZK 93426, in human volunteers. ·Psychopharmacology 93:421–427

Grier JB (1971) Non parametric indexes for sensitivity and bias: computing formulas. Psychol Bull 75:424–429

Jensen L, Stephens DN, Sarter M, Petersen EN (1987) Bidirectional effects of β-carbolines and benzodiazepines on memory processes. Brain Res Bull 19:359–364

Jensen LH, Petersen EN, Braestrup C, Honore T, Kehr W, Stephens DN, Schneider HH, Seidelmann D, Schmiechen R (1984) Evaluation of the β-carboline ZK 93426 as a benzodiazepine receptor antagonist. Psychopharmacology 83:249–256

Ott H, Rohloff A, Aufdembrincke B, Fichte K (1988) Anterograde and retrograde amnesia after lormetazepam and flunitrazepam. In: Hindmarch I, Ott H (eds) Benzodiazepine receptor ligands, memory and information processing. Springer, Berlin Heidelberg New York Tokyo, pp 180–193

Sarter M, Schneider HH (1988) High density of benzodiazepine binding sites in the substantia innominata of the rat. Pharmacol Biochem Behav 30:193–202

Sarter M, Steckler T (1989) Spontaneous exploration of a 6-arm radial tunnel maze by basal forebrain lesioned rats: effects of the benzodiazepine receptor antagonist β-carboline ZK 93426. Psychopharmacology 98:193–202

Sarter M, Bodewitz G, Stephens DN (1988) Alteration of scopolamine-induced impairment of spontaneous alternation behaviour by antagonist but not inverse agonist and agonist β-carbolines. Psychopharmacology 94:491–495

Sarter M, Schneider HH, Stephens DN (1988) Treatment strategies for senile dementia: antagonist β-carbolines. Trends Neurosci 11:13–17

Stephens DN, Sarter M (1988) Bidirectional nature of benzodiazepine receptor ligands extends to effects on vigilance. In: Hindmarch I, Ott H (eds) Benzodiazepine receptor ligands, memory and information processing. Springer, Berlin Heidelberg New York Tokyo, pp 205–217

Stephens DN, Duka T, Andrews JS (1990) Benzodiazepines, β-carbolines and memory. In: Weinmann J, Hunter J (eds) Memory: neurochemical and abnormal perspectives. Harwood Academic Publishers, London (in press)

Tiplady B (1985) An automated test battery for the detection of changes in mental function and psychomotor performance. Br J Clin Pharmacol 20:305P

Turner JD, Stephens DN, Shivers BD, Hodges H, Mertens NB (1990) Effects of β-carbolines on cognitive processes: mediation by modulation of cholinergic transmission? In: Kewitz H (ed) Pharmacological interventions on central cholinergic mechanisms in senile dementia. Zuckschwert, Munich (in press)

Wilkinson RT (1968) Sleep deprivation: performance tests for partial and selective sleep deprivation. Prog Clin Psychol 8:28–43

Correspondence: Dr. T. Duka, Human Pharmacology, Schering AG, Müllerstrasse 172, D-1000 Berlin 65.

Long term treatment of SDAT patients with pyritinol

S. Knezevic[1], Z. Mubrin[2], G. Spilich[3], G. Vucinic[4], and W. Wannenmacher[5]

[1] Department of Neurology, University of Zagreb Medical School
[2] Memory Center of the University Hospital Centre, Zagreb, Yugoslavia
[3] Washington College, Chestertown, Maryland, U.S.A.
[4] Department of Psychiatry, University Hospital Centre, Zagreb, Yugoslavia
[5] Institute of Psychology, University of Technology, Darmstadt,
Federal Republic of Germany

Summary

A group of 26 patients with the clinical diagnosis of Senile Dementia of Alzheimer's type (SDAT) was randomly assigned in a double blind cross-over trial of pyritinol versus placebo. Psychiatric and neurological examination, psychometric testing and regional cerebral blood flow measurements were used to assess the effects of medication. Pyritinol had a beneficial effect on cognitive performance which was sustained during the long term follow up of one year.

Introduction

The prevalence of Alzheimer's disease is steadily raising with aging of the population and our understanding of the disease is improving, but still the search for treatment has not yielded very promising results. Since there is no specific treatment for SDAT even symptomatic therapy presents a challenge. The aim of this trial was to test the effects of pyritinol in patients with senile dementia of Alzheimer's type (SDAT).

Pyritinol is a pyridoxine derivative with a different pharmacological profile. Pre-clinical research by Hoyer et al. (1977), Martin et al. (1987), Pavlik et al. (1987), Greiner et al. (1988), and clinical research by Hamouz (1977), Cooper et al. (1980), Hermann et al. (1986), and Oswald et al. (1986) suggest the possible beneficial effect of pyritinol on cognitive performance in SDAT patients.

Patients and methods

The effect of pyritinol was evaluated against placebo in a randomized double-blind cross-over trial with a treatment period of 10 weeks. Pyritinol was administered in a dose of 200 mg three times daily. Placebo was administered in the same way and it contained 10 mg of pyritinol which was known to have no pharmacological effects in this low dosage. This was considered necessary because of the unusual taste of the drug. During a two week period before the trial all patients received one coated tablet of placebo 3 times daily (washout period). In the following 10 weeks patients received either the active drug at a daily dose of 600 mg, or a daily dose of placebo. During the second 10 weeks period the drug and placebo administration was reversed without a washout period as no carryover effects were expected. Active drug and placebo were identical in taste and appearance, and they were administered at mealtimes under the supervision of a nurse to maintain compliance.

After the 20 weeks of the cross-over trial the study was continued and patients were left on the second drug of the first part of the study because we wanted to see whether the effects found in the acute trial would last in the follow up. The investigators did not know the individual results of the first part of the trial and were blind to the drug that the patients were using. The follow up was planned to last one year after the first part of the trial. The average duration of the whole investigation was 75 weeks.

Patients

A total of 31 patients were included at the beginning of the trial. Informed consent was obtained from all. Five patients were excluded from the trial (1 died of myocardial infarction, and 4 refused to take part in the CBF measurements). Of the 26 patients who finished the trial 6 were men and 20 women, mean age was 76.2 years (SD = 8.0). They were all residents of an Old Peoples' Nursing Home in Zagreb, Yugoslavia, where all examinations (except rCBF and CT) were performed by the same neurologist and psychologist.

Patients were first screened for dementia using DSM III criteria and then differentiated in those who have the Alzheimer's or MID type of disease. In order to exclude patients suffering from dementing disorders other than AD a detailed history, ECG, extensive laboratory tests and a CT were performed. All patients in whom some other cause of dementia was suspected and all who were continuously on medication that to our knowledge might have caused their symptoms were not included in the study. Beyond that, only patients with mild to moderate dementia with a total score between 9 and 24 on the Short Cognitive Performance Test – SKT (Erzigkeit, 1986) and a total score between 9 and 24 in the factor "Cognitive Disturbances" of the Sandoz Clinical Assessment Geriatric-SCAG (Venn, 1983) were included.

Methods

All patients were examined on 4 occasions: after 2 week wash-out phase, at the end of each of the two consecutive 10 week treatment periods and at the end of the follow up. At each session, psychiatric and neurological examination, and psychometric testing were performed. RCBF measurements at rest and after the mental activation were made at the first three examinations.

SCAG and Alzheimer's Diseases Assessment Scale – ADAS (Rosen et al., 1984) were used for psychiatric evaluation. Neurological examination was conducted with special attention to soft neurological signs such as graphesthesia, stereognosia, two point discrimination, coordination, finger and foot tapping and its synchronization, stance and gait. Each sign was scored and the overall neurological score was the sum of all the scores.

Psychometric testing was performed with three instruments. SKT, a Self Assessment Scale – SASS and an experimental psychological test – Contextual Effects upon Text Memory – CETM (Spilich et al., 1983).

The SKT consists of 9 subtests and is designed to assess cognitive deficits in demented patients, particularly disturbances of memory and concentration.

The SASS consists of 9 items related to general condition, concentration, mood, short term memory, motivation, speed of performing everyday tasks, irritability and sociability. For each item a 10 point scale was presented to the patients and a total score was analysed. The testing was done twice, after each 10 weeks period, and our patients who were mildly to moderately demented understood the task very well.

The CETM was developed by Spilich and Voss (1983) and it is based on recent research in the area of memory and dementia (Cohen, 1979; Cohen et al., 1981; Spilich, 1983, 1985). It was developed to identify major changes in the way in which the brain processes information. It is thought that the disturbances of the "carry-forward" capacity of short term memory change the way in which the individual builds contextual relations among events. Each unit of the CETM is composed of a 3 sentence context followed by a to-be-recalled target sentence, which is clearly marked. The semantic relationship between target and the context varies in three stages from clear and strong through weak but coherent to no context at all. Subjects read both context and target materials aloud. The retrieval part of the test involves only the target sentences and patients are made aware of this through instruction and practice trials. The retrieval is recorded as free recall as well as recognition of sentences among distractors. Additional information about the construction of CETM, and counter balancing procedures between contexts and targets is available in Spilich et al. (1983).

The regional cerebral blood flow measurements (rCBF) were performed always at the same time of day with a standard equipment (Novo Diagnostic Systems) using 32 bilateral detectors and the inhalation method (Obrist et al., 1975; Risberg, 1980, 1985). The first measurement was made during rest with eyes closed. Fifteen minutes later the activation measurement was performed as described by Mubrin et al. (1985 a, b). A slightly modified version of the Word Pair Learning and Recall (WPLR) subtest of the Wechsler Memory Scale (Wechsler, 1945) was used. Twelve word pairs, half highly related and half unrelated, were presented 3 times by tape recorder during the second rCBF measurement. After each presentation, the word pairs were recalled aloud by the patient, while rCBF was measured. The initial slope index (ISI), originally defined by Risberg et al. (1975) was used in the analysis of rCBF data.

Results

Detailed description of the results of the first part of the trial was published by Knezevic et al. (1989).

Results from neurological examination (NEUR), SCAG, ADAS, SKT and SASS are shown in the Table 1. They all show a tendency to improvement on pyritinol, except the neurological examination which is virtually unchanged. After the follow up all the results are in the same range and none show significant differences between treatments.

CETM results, for both free recall and recognition as well as the summary of statistical analysis are shown in Table 2. It is quite clear that

Table 1. Results of the neurological examination and descriptive psychiatric and psychometric measures

		NEUR	SCAG	ADAS	SKT	SASS
Pretreatment	mean	41.39	12.46	23.58	12.42	–
	sd	9.00	2.39	7.31	3.80	–
Placebo	mean	41.19	10.85	19.59	9.19	48.56
	sd	11.98	2.65	8.72	4.61	18.38
Pyritinol	mean	39.50	10.92	18.34	8.23	53.90
	sd	9.85	2.67	6.50	3.94	14.81

Table 2. Descriptive results of CETM

CETM		Recall	Recognition
Pretreatment	mean	3.60	6.52
	sd	2.65	3.87
Placebo	mean	4.24	7.40
	sd	2.80	3.04
Pyritinol	mean	5.36	8.80
	sd	3.29	3.66
Statistics (ANOVA)			
Pretreatment/placebo		$p < 0.05$	$p < 0.04$
Pretreatment/pyritinol		$p < 0.002$	$p < 0.002$
Placebo/pyritinol		$p < 0.002$	$p < 0.001$

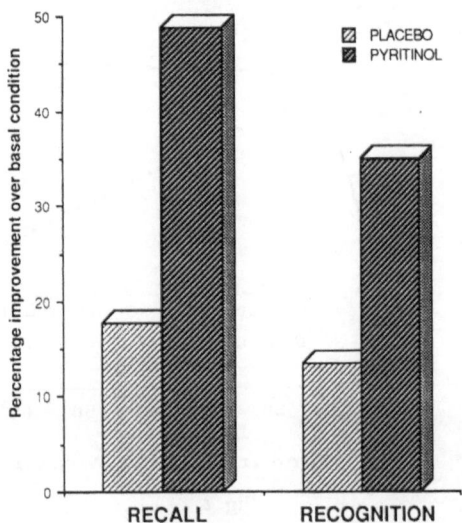

Fig. 1. Improvement as a function of trial and task

after treatment with pyritinol patients recalled and recognized significantly more information than they did following placebo or at the pretreatment examination. The improvement following pyritinol and placebo expressed as per cent of pretreatment scores is illustrated in Fig. 1.

The improvement in recall memory performance for placebo was 17.8%, and for pyritinol 48.8%. The active drug gave 2.7 times the improvement that followed placebo. The corresponding improvement of recognition memory was 13.5% and 35% respectively, thus the active drug was about 2.5 times better than placebo.

Figure 2 shows the results of patients on follow up. It is evident that patients who were taking pyritinol in the second phase of the trial did better, and that the improvement was prolonged in the follow up in which the patients even continued to improve slightly. Patients who continued to take placebo deteriorated quite markedly while on placebo, although in the second phase of the trial they showed some increase in performance.

Investigation of the regional cerebral blood flow measurements was done before the trial, and after both 10 weeks periods. They were reported in details in Knezevic et al. (1989). The results are summarized in Fig. 3.

Control subjects (elderly non demented) show a significant increase in rCBF after the activation in 13 regions of the brain bilaterally mainly in the central and temporal parts of the brain. Demented patients before

S. Knezevic et al.

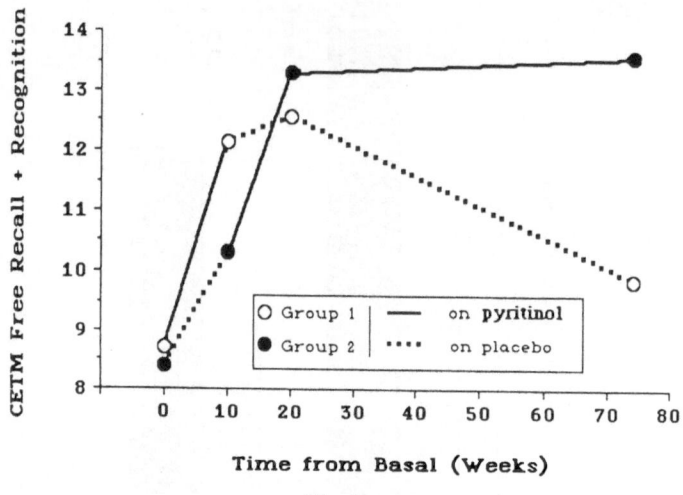

Fig. 2.

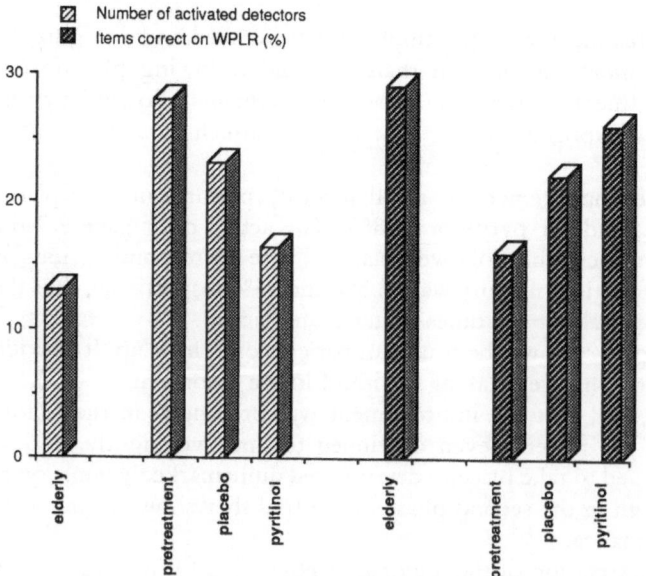

Fig. 3. RCBF and WPLR changes

the treatment show a very increased area of significant activation after the same memory activation procedure. They increase the flow in 28 (out of 32) regions of the brain, so that practically all measured regions of both hemispheres have a very significant increase in rCBF. Placebo treatment results in a decrease of the number of regions in which the rCBF is increased and pyritinol treatment results in a further shrinkage of the activation area so that on placebo 23 regions are activated and on pyritinol only 17. It is clear that pyritinol leads to "normalization" of the pattern of increase in cerebral blood flow following the treatment with pyritinol and the result becomes similar to that found in elderly control subjects. WPLR result was monitored during the same time and Fig. 3 shows that as the rCBF pattern was "normalized" the performance of patients on WPLR increased, approaching that of the controls. Statistical analysis showed all differences in Fig. 3 to be significant at a level of at least $p < 0.05$.

Discussion

The aim of this study was to evaluate the therapeutic effects of pyritinol in patients with SDAT. The results were first assessed with the battery of usual tests: SCAG, SKT, ADAS, neurological examination (Table 1). Although they show a beneficial effect of pyritinol we did not find these tests sensitive enough for measuring the effects of symptomatic treatment. SASS is rarely used in demented patients but we were glad to find that our patients understood the procedure very well and they favourably rated their treatment with pyritinol.

We believe that better tests should be developed for clinical trials in demented patients and our experience in other groups of patients with disturbances of memory showed that CETM might be a more sensitive test. CETM was primarily designed and used to assess changes in working memory operations. Such changes are widely considered to be one of the characteristics of dementing disorders. Our results (Table 2, Fig. 1) clearly indicate that the test is practical and sensitive enough as it shows very consistently the difference between placebo and pyritinol treatment. The effect of pyritinol treatment is about 2.5 times larger than the effect of placebo therapy, and we believe that we can safely conclude that pyritinol exerts a beneficial effect upon cognitive operations in demented individuals.

We were interested to know whether this effect is sustained during a prolonged use of pyritinol and we examined our patients after one year of follow up (Fig. 2). We have demonstrated that patients on pyritinol reach a certain level of cognitive improvement which is sustained for the whole year, while patients on placebo after reaching a similar level on pyritinol and placebo start to deteriorate in memory operations.

RCBF and WPLR results show that the improvement, shown also in WPLR results, is accompanied by a measurable change in CBF after the memory activation. Both measures clearly tend to "normalize" after treatment with pyritinol (Fig. 3). The relation of performance to blood flow changes has been discussed recently by Maximilian and Brawanski (1988). We regret that we were not able to repeat the CBF studies after the follow up period, but our patients found the CBF measurements cumbersome (transportation to the hospital, 90 minutes of lying still on a bed, etc) and even in the first part of the study 4 patients refused to take part in it.

Our conclusion from this study is that pyritinol clearly improves the cognitive performance in patients with SDAT. The improvement is especially striking on CETM which we believe to be a much more sensitive test than the conventional ones. The beneficial effect is also clinically relevant as seen in SASS, and it might have a neurophysiological basis in some metabolic improvements as suggested by rCBF and WPLR measurements. The analysis of individual results which is not presented here shows that a number of patients do particularly well on pyritinol so that in the next phase of this study we believe that a search for pyritinol "responders" should be undertaken.

Acknowledgements

We are grateful to Prof. J. Risberg for his help in the analysis of CBF results, and to the staff of the Old Peoples' Nursing Home in Zagreb.

References

Cohen G (1979) Language and comprehension in old age. Cognitive Psychology 11:412–429

Cohen G, Faulkner D (1981) Memory for discourse in old age. Discourse Processes 4:253–265

Cooper AJ, Magnus RV (1980) A placebo-controlled study of pyritinol (Encephabol) in dementia. Pharmatherapeutica 2:317–323

Erzigkeit H (1986) Der Syndrom-Kurztest zur Erfassung von Aufmerksamkeits- und Gedächtnisstörungen. Vless, Vaterstetten

Greiner HE, Haase AF, Seyfried CA (1988) Biochemical models for the study of the mechanism of action of an encephalotropic drug. In: Kanowski S, Ladurner G (eds) Dementielle Erkrankungen im Alter. Pathogenetische Modelle und Therapeutische Wirklichkeit. Thieme, Stuttgart, S 45–51

Hamouz W (1977) The use of pyritinol in patients with moderate to severe organic psychosyndrome. Pharmatherapeutica 1:398–403

Herrmann WM, Kern U, Rohmel J (1986) The effects of pyritinol on functional deficits of patients with organic mental disorders. Pharmacopsychiatry 19:378–385

Hoyer S, Oesterreich K, Stoll KD (1977) Effects of pyritinol-HCl on blood flow and oxidative metabolism of the brain in patients with dementia. Drug Res 27:671–674

Knezevic S, Mubrin Z, Risberg J, Vucinic G, Spilich G, Gubarev N, Wannenmacher W (1989) Pyritinol treatment of SDAT patients: evaluation by psychiatric and neurological examination, psychometric testing and rCBF measurements. Clin Psychopharmacol 4:25–38

Martin KJ, Vyas S (1987) Increase in acetylcholine concentration in the brain of "old" rats following treatment with pyrithioxine (Encephabol). Br J Pharmacol 90:561–569

Maximilian VA, Brawanski A (1988) Functional and vascular challenge procedures during noninvasive rCBF measurements. In: Knezevic S, Maximilian VA, Mubrin V, Prohovnik I, Wade J (eds) Handbook of regional cerebral blood flow. Lawrence Erlbaum Associates, Hove and London, pp 79–121

Mubrin Z, Knezevic S, Barac B, Poljakovic Z, Pudar-Klein M, Katic Z (1985a) Effect of mental activation on regional cerebral blood flow (rCBF) changes in patients with cerebrovascular disease. Neurologija (Zagreb) 31:107–124

Mubrin Z, Knezevic S, Barac B, Gubarev N, Lazic M, Liscic R, Vidosic S (1985b) Distinct rCBF pattern during different types of short-term memory activation. In: Hartmann A, Hoyer S (eds) Cerebral blood flow and metabolism measurement. Springer, Berlin Heidelberg New York Tokyo, pp 81–87

Obrist WD, Thompson HK, Wang HS, Wilkinson WE (1975) Regional cerebral blood flow estimated by 133Xenon inhalation. Stroke 6:245–256

Oswald WD, Oswald B, Grobe D, Lukaschek K, Sappa J, Fleischmann UM (1986) Recent approaches in the treatment and assessment of psychoorganic brain-syndromes. A double-blind study with pyritinol. In: Ramos GG, Herrmann WM, Otero E, Toledano A (eds) Advances in the field of cerebrovascular disease. Egrat, Madrid, pp 88–96

Pavlik A, Benesova O, Dlohozkova N (1987) Effects of nootropic drugs on brain cholinergic and dopaminergic transmission. Act Nerv Sup 29:62–65

Risberg J (1980) Regional cerebral blood flow measurements by 133Xe-inhalation: methodology and applications in neuropsychology and psychiatry. Brain Lang 9:9–34

Risberg J (1985) Application of the nontraumatic xenon 133 method in neuropsychiatry. In: Hartmann A, Hoyer S (eds) Cerebral blood flow and metabolism measurement. Springer, Berlin Heidelberg New York Tokyo, pp 72–80

Risberg J, Ali Z, Wilson EM, Wills EL, Halsey JH Jr (1975) Regional cerebral blood flow by 133Xenon inhalation. Preliminary evaluation of an initial slope index in patients with unstable flow compartments. Stroke 6:142–148

Rosen WG, Mohs RC, Davis KL (1984) A new rating scale for Alzheimer's disease. Am J Psychiatry 141:1356–1365

Rosen WG, Terry RD, Fuld PA, et al (1980) Pathological verification of ischemic score in differentiation of dementias. Ann Neurol 7:486–488

Spilich GJ (1983) Life-span components of text processing: structural and procedural differences. Journal of Verbal Learning and Verbal Behavior 22:231–244

Spilich GJ (1985) Discourse comprehension across the span of life. In: Charness N (ed) Aging and human performance. Wiley, New York, pp 143–190

Spilich GJ, Voss JF (1983) Contextual effects upon text memory for young, aged-normal, and aged memory-impaired individuals. Exp Aging Res 9:45–49

Venn RD (1983) The Sandoz clinical assessment geriatric (SCAG) scale. Gerontology 29:185–198

Wechsler D (1945) A standardized memory scale for clinical use. J Psychol 27:91–94

Correspondence: Dr. S. Knezevic, Department of Neurology, University Hospital Rebro, Kispaticeva 12, YU-41000 Zagreb, Yugoslavia.

Subject Index

Horst Przuntek, Peter Riederer (eds.)

Early Diagnosis and Preventive Therapy in Parkinson's Disease

(Key Topics in Brain Research)

1989. 59 figures (1 in color). XIV, 442 pages.
Soft cover DM 135,–, öS 950,–
ISBN 3-211-82080-9

Preisänderungen vorbehalten

At the time when "Parkinson's Disease" is diagnosed in a patient roughly two thirds of dopaminergic neurons of substantia nigra are already degenerated. Therefore, the onset of the disease must be much earlier. This book deals with early diagnosis and early preventive treatment which may sustain the process underlying the disease.

By use of psychometric, kinesiologic, physiological, histological, biochemical, endocrinological, pharmacological and imaging techniques including positronemission tomography and brain mapping specialists tried to focus new diagnostic criteria. New methods including psychometric evaluation, apparative measurement of movement, analysis of peripheral blood and urinary constituents have supplemented this approach. Early preventive therapy has been agreed to consist of low dosis L-DOPA plus benserazide, L-deprenyl and dopaminergic agonists.

Springer-Verlag Wien New York

C. G. Gottfries, S. Nakamura (eds.)

Neurotransmitter and Dementia

(Journal of Neural Transmission, Supplementum 30)

1990. 44 figures. VII, 81 pages.
Soft cover DM 70,-, öS 490,-
Reduced price for subscribers to
"Journal of Neural Transmission":
Soft cover DM 63,-, öS 441,-
ISBN 3-211-82190-2

Preisänderungen vorbehalten

Neurotransmitter changes in brains from patients with dementia disorders, mainly Alzheimer type dementia, are reported. Their role in the pathogenesis of Alzheimer's disease is discussed and the neurochemical changes are also considered a base for formulating treatment strategies. By studying markers in the cerebrospinal fluid diagnostic methods may be achieved of value in the diagnosis of subgroups of dementia.

H. Glossmann (ed.)

New Therapeutic Uses of Calcium Channel Blockers

(Journal of Neural Transmission, Supplementum 31)

1990. 21 figures. Approx. 65 pages.
Soft cover DM 65,-, öS 450,-
Reduced price for subscribers to
"Journal of Neural Transmission":
Soft cover DM 58,50, öS 405,-
ISBN 3-211-82200-3

Preisänderungen vorbehalten

This publication presents an overview on new therapeutic uses of calcium channel blockers and future trends for calcium antagonists. The areas covered are tissue-selectivity, myocardial protection, renal effects of calcium antagonists and actions on the peripheral and central nervous system, including cognitive function as well as neuronal repair. Authors, which are competent researchers in the respective fields present own novel data and cover the most recent literature.

Springer-Verlag Wien New York